AF520499

HERBAL TECHNOLOGY
Concepts and Approaches

Herbal Technology
Concepts and Approaches

M. DANIEL
Professor and Head, Department of Botany
The Maharaja Sayajirao University of Baroda
Vadodara - 390002 (INDIA)

2008

SATISH SERIAL PUBLISHING HOUSE
403, Express Tower, Commercial Complex, Azadpur, Delhi-110033 (India)
Phone : 011-27672469, 27672852, Fax : 91-11-27672046
E-mail : info@satishserial.com, hkjain1975@yahoo.com
Website : www.satishserial.com

Published by :

SATISH SERIAL PUBLISHING HOUSE
403, Express Tower, Commercial Complex, Azadpur,
Delhi - 110033 (INDIA)
Phone : 011-27672852 Fax : 91-11-27672046
E-mail : info@satishserial.com, hkjain1975@yahoo.com

ISBN : 81-89304-39-9

Printed at
SALASAR IMAGING SYSTEMS, DELHI-35

Preface

Herbal Technology encompasses all the myriads of ways of utilizing the multifarious potentialities of plants (other than the genetic manipulations) for human welfare. I had the privilege of coining this term and during the past couple of years I was successful in elaborating the concept and practices by conducting seminars and workshops and through publications. This book is the first attempt to codify the principles behind this branch of science and to reach the people who can practice the same and thus better the life of mankind without in any way changing the existing balance of nature and in a eco-friendly and eco-healthy way.

I have included only five aspects in this discipline, though more and more may be added by the brilliant workers who tread this path at a later stage. Plant chemistry, rather Phytochemistry, which is the prime requisition for all type of plant utilisation processes, form the first and foremost division. The Medicinal Plants and their utility are placed second. The other subjects included here are the Natural dyes, Biopesticides, Biofertilizers and Biofuel. A large number of subdivisions, each important on its own merit and might, exist which, if cared and nurtured, will develop to full fledged subjects worthy of teaching and research.

Since all the applications and the ways of utilizing plants and plant products are based on the chemical diversity or chemical virtuosity of the plant, a comprehensive treatise on **Phytochemistry** is deemed fit and thus incorporated. It deals with the different chemical entities which are responsible for the specific properties of the plant and these compounds are narrated in terms of their basic structure, diversity and distribution in the plant world. With due apology to the biochemists, the biosynthesis and related aspects are avoided in this book. Care is taken to name maximum (within the limitations of this book) number of compounds and their source plants. This is true in the case of primary metabolites (sugars, oligosaccharides, polysaccharides, amino acids, peptides and proteins, fatty acids, triacyl lipids, waxes, cutins and suberins) and secondary metabolites. More emphasis is given to secondary metabolites, the reason being the very few treatises available on these compounds. In this section, alkaloids are treated first followed by terpenoids and phenolics. A few minor components which are not comfortable in those groups are treated separate. The new and varied type of compounds which are being added daily in standard publications make it virtually impossible to circumscribe all phytochemicals in one treatise. But I have tried to cover a large share of available literature.

Since there appeared a huge number of books on **Medicinal plants** recently, the botany, history, chemical diversity or properties are not dealt here. (The chemistry of these plants is dealt in one of my recent books "Medicinal Plants: Chemistry and Properties"). On the contrary, emphasis is on the multitude of problems existing in the field of alternative medicine wherever plants are exploited. Thus, the correct botanical identity, name changes, biomarkers (both chemical and pharmacognostic), active principles, cultivation, extraction and formulations are dealt in detail. The need for studies in Pharmacology, Pharmacodynamics and Pharmacokinetics is stressed. The scope of these plants in Aromatherapy, Herbal Cosmetics, Herbal Veterinary Medicine etc. also is dealt with. The emerging fields of antioxidant therapy and Metabolomics are summarized. The need of documentation of the knowledge accrued from the traditional knowledge library to prevent biopiracy also is explained. The chapter ends with a note on the distribution of medicinal plants in India and the available substitutes of the acclaimed medicinal plants.

Natural dyes, which form the third chapter, covers the botany, chemistry and uses of about 150 plants which yield them. A general account of the coloring principles of plants is given in the beginning. The role of proanthocyanins and iridoids, the potentials of which have not yet been exploited, is introduced here. The scope of utilizing natural dyes as colors for food and pharmaceutical preparations, in cosmetics like face powder and lipsticks, in paints and for painting, is also discussed. The various dyeing practices prevalent in different parts of country are given. The analytical methods of common components of dyes are appended.

Biopesticides, one of the largest chapters in this book, gives a bird's eye view on the very many insect/pest repellant plants as well as pesticidal/insecticidal compounds and methods. Though microbial pesticides are explained in the beginning, emphasis is given to the botanical pesticides. More than 300 plants have been explained here in terms of their active principles and utilities and the preparation of extracts of as many as 40 plants also is given. The other ways of driving/destroying the pests such as parasitoids, predators, pheromones, ecdysteroid agonists, nematodes and other substances also are discussed at the end.

The fifth chapter is on **Biofertilisers** wherein emphasis is on soil microbes including cyanobacteria. The utility, efficacy and mode of action of many microbes including nitrogen fixing bacteria, cyanobacteria, phosphate solubilising microorganisms, vascular arbuscular mycorrhizae and mixed cropping are discussed.

Biofuel, the last chapter, delves on the various alternate sources of fuel such as petrocrops, biocrude, liquid resins, vegetable oils and plant oil methyl esters, biogas, alcohols and biological hydrogen. The various researches conducted in many advanced laboratories on biomass conversion and biological hydrogen are narrated and the methods followed for producing various fuels are discussed.

A mere glance on the book indicates that one and the same plant is mentioned in more than one sections. For example neem, *Aloe, Cassia* spp. etc. are cited in medicinal plants, biopesticides and in natural dyes. This brings forth the multifarious activities of the living plants. Now *Jatropha curcas* is known only for its oil (biofuel). But the same plant yields phorbol (an anticancerous drug), dye and possess pesticidal properties. This aspect of using the same plant for many purposes will increase the acceptability of the same in rural areas and will favour increased agricultural practices on these plants and bring more revenues for the common man who manages them. This **value addition** will also encourage large scale agrofarming of these plants and industries based on the products.

This book is just a beginning in this direction. Each of the topics is designed in such a way to rouse the curiosity of the student/reader and a conscientious worker can add many different dimensions in every topic. It is hoped that the subject of Herbal Technology will find a place in the curricula of graduate and post-graduate classes; will initiate a great deal of interest in the scientific community and will lead to many research programmes in each of the topics, be that Biopesticides or Biofuel. The full potential of the plants, when realized and manipulated in an ecofriendly manner, will make the life in this planet (we have only one planet to live!) much more soothing, comfortable and in tune with the nature.

M. Daniel
Department of Botany,
The Maharaja Sayajirao University of Baroda,
Vadodara, Gujarat

Contents

1

Phytochemistry

The phytochemicals elucidated from plants now run to more than 50,000 (De Luca and St. Pierre, 2000) and it is predicted that the final figure for the plant kingdom will approach or even exceed 200,000 (Fiehn, 2002). All these molecules are conventionally classified into two groups, **primary** and **secondary metabolites**. The primary metabolites are all those compounds which are continuously turned over, *i.e.* synthesized regularly depending on the need and also catabolised regularly. They provide a metabolic pool from which molecules are being picked up for the manifold daily functions of the plant, like growth, respiration, flowering etc. They are almost universal in plant kingdom being present in each and every plant. The secondary metabolites, on the other hand, are the compounds produced by specialized metabolic pathways for specific functions. It is believed that once produced, they never get metabolized. They may be different in different plant groups. Though this classification is widely accepted the boundaries of these two groups are still fluid. Some of the primary metabolites are found to assist the plant in survival, a function attributed to secondary metabolites, while some of the latter show diurnal or seasonal variations indicating a regular turn over of these compounds. But for the present, this classification is accepted and followed in this treatise.

1.1. Primary Metabolites

Carbohydrates, proteins and saponifiable lipids are the three main classes of primary metabolites. Among the various carbohydrates, some of the polysaccharides like gums and mucilages do not take part in the metabolism and are of adaptational in nature. Same is the case with certain structural proteins and some enzymes specific to secondary metabolism. Waxes, cutins and suberins are the saponifiable lipids which once formed do not break down and contribute to the survival strategies of the plants.

1.1.1. *Carbohydrates*

Carbohydrates are polyhydroxyaldehydes or ketones. Some of the sugar alcohols, acids, amines and similar compounds do not fit exactly into this definition, but have been included here owing to the similarity in biosynthetic pathways or because of their derivation from simple carbohydrates. The majority of carbohydrates contain hydrogen and oxygen in the ratio 2:1 like that of water (and so the designation-hydrate of carbon). In addition to C, H, and O, which all these compounds invariably possess, a few of them contain nitrogen (aminosugars), sulphur (sulphoquinovose) or phosphorous (as the phosphate ester).

Being the primary product of photosynthesis, carbohydrates are the key intermediates in the metabolism of green plants. Practically all the biochemical compounds are directly or indirectly derived from them. Apart from this, they function as storage food, mechanical framework and sometimes modify many physical and chemical characters of various other groups of compounds by combining with them.

Carbohydrates are classified based on the basic unhydrolysable units-monosaccharides-they possess. Thus they are mono, di, oligo or polysaccharides depending on whether they contain one, two, few or many basic units.

Monosaccharides

These are simple sugars which cannot be converted to simpler units by mild acidic/alkaline/enzymic hydrolysis. They may be **aldoses** or **ketoses** depending on the aldehyde or ketogroup they contain. All monosaccharides are colourless, crystalline solids soluble in water. They exhibit optical isomerism and impart a sweet taste. The suffix-**ose** denotes that the compound is a sugar. Monosaccharides are further classified on the basis of the number of carbon atoms contained in them. Thus we have monoses, bioses, trioses, tetroses, pentoses, hexoses, heptoses, octoses and nonoses, subjected to one, two, three, four, five, six, seven, eight and nine carbon atoms present in them.

The optical activity exhibited by the sugars is attributed to the asymmetric carbon atoms present and the value '2^n' represents the number of possible optical isomers, 'n' denoting the number of asymmetric carbon atoms. (Monoses and bioses do not show optical activity because of the lack of asymmetric carbon atoms in them.) Thus glyceraldehyde exists in two possible configurations, in one the hydroxyl group attached to the single asymmetric carbon atom is directed towards the right side and in the other towards left. The former one is referred to as D-glyceraldehyde and the latter, L-glyceraldehyde. This system of conformational method is proved to be so useful that the whole monosaccharide group can be arranged in either **D-series,** wherein the penultimate carbon atom (the highest numbered asymmetric carbon atom - note that in sugars the numbering starts from the top carbon atom *i.e.* the aldehydic carbon atom or the single carbon on one side of ketonic carbon) is similar to D-glyceraldehyde (*i.e.* towards right), or **L-series** if the compound is similar to L-glyceraldehyde. It is to be noted that the direction of plane of polarized light is denoted by (+) or (-) signs as dextro- or laevorotatory respectively and is independent of D- or L-series. D- and L-denote conformation and (+) or (-), the direction of the rotated plane polarized light. Thus the name D- (+) glyceraldehyde indicates that it belongs to D-series and is dextrorotatory. L-forms are mirror images (*enantiomers*) of D-forms. D-glucose and L-Glucose rotate polarized light dextro and levo-as D-and L-glyceraldehyde. But it should be noted that the direction of rotation may be opposite to the prefix. This is true in the case of D- and L-arabinose and D- and L-fructose. (D-Fructose is known as levulose due to the strong laevorotation it exhibits.)

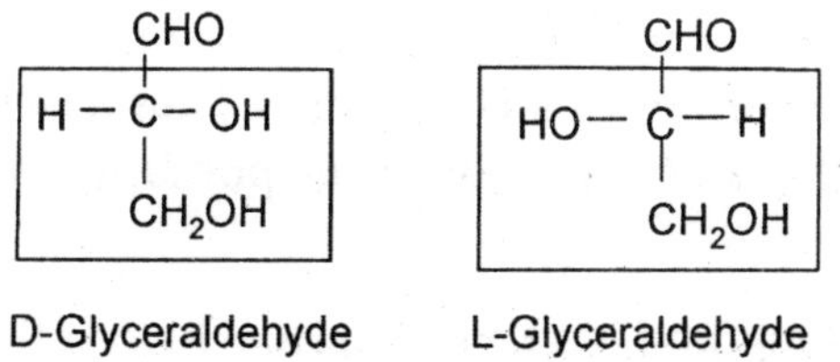

D-Glyceraldehyde L-Glyceraldehyde

The fact that glucose and related sugars fail to give Schiff's test for aldehydes and the stability noted in the case of solid sugar (no auto-oxidation expected of compounds with a free aldehydic or ketonic group is observed here) prompted the formulation of closed ring systems instead of open chain structures. This cyclic form is possible by the formation of a hemiacetal linkage between the aldehydic/ketonic group and the hydroxyl group of the 4th or 5th (in the case of hexoses and pentoses) or the last carbon atom (in pentoses). Thus an aldehexose may exist as a pyran-ose (a sugar having pyran- 6-membered oxygen containing heterocylic ring) if the linkage formed is between the hydroxyl groups of the first and the fifth carbons or a furanose (sugar containing a furan-5-membered oxygen heterocyclic ring) if the linkage is between the carbons of first and the fourth position. The sugars are designated by adding the suffix-pyranose or furanose – to the trivial names *e.g.*, glucopyranose, fructofuranose etc. In the Fischer formula, which is two-dimensional, the carbon chain is written in a vertical straight line in a plane parallel to the plane of the page on which it is written and the oxygen bridge formed is drawn towards the right in the sugars of D-series and towards left in L-sugars. The newly formed hydroxyl group at carbon 1, is known as the *glycosidic hydroxyl* group, because of the readiness with which it reacts with another alcoholic group to form an acetal linkage – a *glycosidic bond.* This hydroxyl group and the oxygen bridge formed provide the carbon atom, to which it is attached (C_1 or C_2), a center of asymmetry; and resultant isomers formed are known as α- and *β-anomers.* These two stereoisomers are formed because the glycosidic hydroxyl group can exist either on the side of the oxygen bridge or in the opposite side. In α-anomers of D-sugars this hydroxyl group and the oxygen bridge are on the right side (*cis*-form) while in β-anomers of D-sugars they are on opposite sides (*trans- i.e.* the oxygen bridge is towards the right side and hydroxyl group to the left). But in α-anomers of L-sugars both the groups are on the opposite sides (the oxygen bridge on the left side and the hydroxyl group on the right side) whereas in β-anomer they are on the same side. In simple terms, the OH group is on the right for α-D- and β-L-isomers. The two anomers of a D-sugar are *enantiomers* (mirror images for the configuration of C-1) of the corresponding two anomers of the L-sugar.

In Haworth's method of presentation, which is three-dimensional, the carbon chain in the ring coils and the plane of the ring is kept perpendicular to the page and those functional groups written towards the right in Fischer formula are represented below the plane of the ring while the groups on the left (in F. formula) are written above. In this form the -OH group in the hemi-acetal group (glycosidic hydroxyl or the -OH group of first asymmetric carbon - chiral center of aldohexoses) of the D-sugars points down for the α-isomer and up for β-isomer. In the L-series, the -OH group of α-isomer is up and of β-isomer is down.

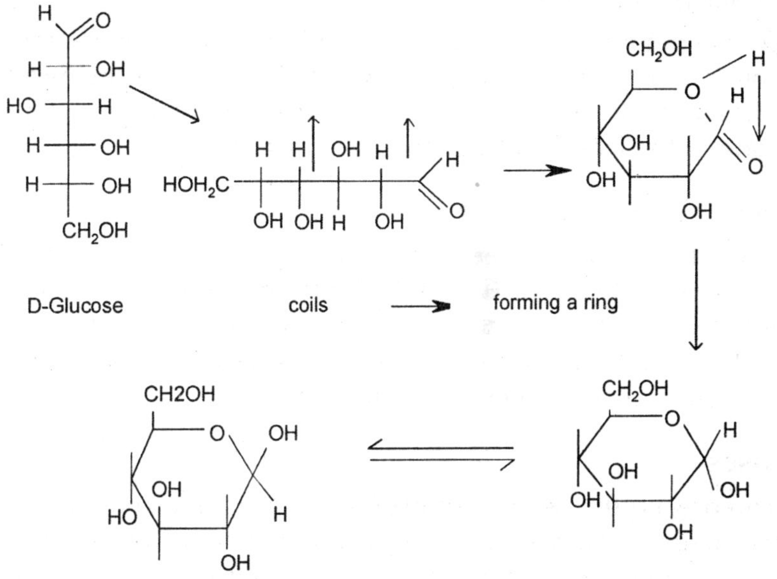

D-Glucose coils → forming a ring

beta-D-Glucopyranose ⇌ *alpha*-D-Glucopyranose

In solution, the open chain form exists in equilibrium with the ring structures which occur in both β- and α-forms. When α-D-(+)-glucopyranose is dissolved in water it shows a specific rotation of +113° which slowly falls to +53°. But a solution of β-D-(+) glucopyranose shows an initial rotation of +18.7° which gradually raises to 53°. This phenomenon of **mutarotation** is due to the equilibrium established between α- and β-forms in the solution. At this stage the α-form amounts to 36% and β-form , 64%.

Trioses and tetroses do not form stable hemiacetals and are best considered as hydroxy aldehydes or hydroxy ketones. Most of the monosaccharides exist in nature as the D-enantiomers. L-Arabinose is the only common sugar existing in L-form.

Certain sugars contain non-hydroxylated (rather non-oxygenated) carbons-termed deoxy sugars (*e.g.*, Deoxyribose). But if such groups are present in the terminal carbons, then they are often named as methyl sugars (*e.g.*, methyl pentoses). A few monosaccharides possess a branched carbon chain instead of a linear chain *e.g.* Hammamelose and apiose.

Altogether 60 monosaccharides are reported to be present in nature.

Monoses and Bioses

Formaldehyde, which can be considered as a monose (and the only one), is reported to occur free in volatile oils of many plants. Glycolic aldehyde (glycolose), the only biose possible, is present in the free state in the leaves of Caprifoliaceae members.

Trioses

These three carbon sugars are invariably present in all living cells as their phosphate esters. Glyceraldehyde and dihydroxy-acetone are the only two members in this group; the former is an aldotriose and the latter a ketotriose. These compounds are very important intermediates in carbohydrate metabolism.

Tetroses

Threose, erythrose and erythrulose are the three known members in this category. They occur free as well as bound forms. Erythrose and erythrulose play an important role in photosynthesis.

CHO	CHO	CH_2OH
—OH	HO—	=O
—OH	—OH	—OH
CH_2OH	CH_2OH	CH_2OH
D-Erythrose	D-Threose	D-Erythrulose

Pentoses

This important group of five carbon sugars plays very varied roles in the structure and function of plants and is widely distributed. They rarely occur in the free form. Ribose, deoxyribose, arabinose, xylose and lyxose are the known aldopentoses while ribulose and xylulose are the ketoses. Of these, the former two are of universal occurrence as components of nucleic acids. Ribose occurs in many coenzymes and vitamins. Ribulose is an intermediate in photosynthesis (as RuDP - ribulose diphosphate) and lyxose is a component of yeast nucleic acid. Arabinose and xylose are better known by their polymers, the arabans and xylans, the wood components. Arabinose occurs in free form (?) in many exudates and is an ingredient of gums and mucilages. Xylose is a constituent of primeverose, certain cardiac glycosides and a few saponins. Xylulose-5-phosphate is an intermediate in carbohydrate metabolism.

D-Ribose	D-Deoxyribose	D-Arabinose	D-Xylose
CHO	CHO	CHO	CHO
—OH	—	HO—	—OH
—OH	—OH	—OH	HO—
—OH	—OH	—OH	—OH
CH_2OH	CH_2OH	CH_2OH	CH_2OH

D-Ribose D-Deoxyribose D-Arabinose D-Xylose

A number of methyl pentoses (also known as 6-deoxyhexoses) are reported free or in bound state from various plants. L-Rhamnose (6-Deoxy-L-mannose) is the most widespread among them occurring as the principal component of many glycosides and pectin. L-Fucose is the monomer of a cell wall component of Brown algae and is present in the gum of *Melia azaderach* L. (D-Fucose occurs in many cardiac glycosides and glycolipids such as convolvulin, jalapin etc). D-Quinovose (6-deoxy-D-glucose) is the sugar component of many glycosides in *Cinchona* bark. A group of methyl pentoses of special interest occurs in cardiac glycosides. They are unique in that they are reported from nowhere else in the plant kingdom. The important ones in this group are antiarose (6-Deoxy-gulose), allomethylose (6-Deoxy-D-allose), epifucose (6-Deoxy-D-galactose) etc. The methyl esters of pentoses which also are specific to cardiac glycosides are described in "ethers and glycosides". Dideoxy sugars also are having a strange distribution. 2, 6-Dideoxy sugars are mostly confined to cardiac glycosides, whereas 3, 6-dideoxysugars are seen in bacteria. Abequose, tyvelose etc. are the 3, 6-deoxy sugars seen in bacteria while the corresponding 2, 6-deoxysugars from higher plants are boivinose (2, 6-dideoxy-D-xylohexose), digitoxose (2, 6-dideoxy-D-ribo-hexose), canarose (2, 6-dideoxy-D-arabinohexose) and 2, 6-dideoxy-L-lyxohexose.

Apiose (3-C (hydroxymethyl)-D-glycero-aldotetrose), the sugar component of apiin, a flavone glycoside, is a branched chain monosaccharide, widely distributed in higher plants. It is considered as a cell wall component of many aquatic monocotyledons (Gibbs,1973). Hamamelose (2-C-(hydroxymethyl)-D-ribose), first reported from *Hamamelis virginiana* as an ester with hamamelitannins, occurs free in many angiosperms like *Primula, Buddleja, Tilia* etc.

Pentoses exist in both pyranose and furanose forms. In free state the pyran ring is preferred while in bound form (glycosides), furanose form is prevalent.

Hexoses

Hexoses form the largest group of sugars, both in number and distribution. Most of the polysaccharides abundant in plant kingdom are polymers of these compounds. Of the eight D-aldosugars possible; only three, *i.e.*, glucose, galactose and mannose, occur naturally. All these eight sugars have been synthesized in the laboratory.

D-Glucose (grape sugar, dextrose) is universally present in plant saps and in fruits and is considered as one of the most abundant organic compounds in nature. It is the sugar frequently encountered in glycosides. The two most important polysaccharides, starch and cellulose, are the polymers of this sugar. D-Mannose is known only in its polymeric form – the mannans (a component of hemicelluloses). While D-galactose is present in lactose (a disaccharide) and galactans (pectins), L-galactose occurs as a component of agar. Of the four ketohexoses known, D-fructose (fruit sugar, levulose) is the most important one, widespread both free and in oligo- and polysaccharides like inulin and related fructans. L-Sorbose is reported from fruits of *Sorbus* (Rosaceae) and Tagatose from *Sterculia setigera* gums. D-Psicose is believed to occur in molasses and in Saxifragaceae members.

Similar to pentoses all these sugars favour pyranose forms in free state. The less stable furanose form is met with in glycosides only.

Glucose	Fructose	Mannose	Galactose
CHO	CH_2OH	CHO	CHO
–OH	=O	HO–	–OH
HO–	HO–	HO–	HO–
–OH	–OH	–OH	HO–
–OH	–OH	–OH	–OH
CH_2OH	CH_2OH	CH_2OH	CH_2OH

Glucose Fructose Mannose Galactose

Higher sugars

The higher sugars (*i.e.*, heptoses upwards) are named by adding two or more prefixes derived from the lower sugars. Each sugar molecule is arbitrarily divided to an upper part containing four asymmetric carbons (5-carbons in total) designated by a term denoting its similarity to a hexose *i.e.*, allo-, altro-, gluco-, manno-, gulo-, ido-, galacto- and talo-; and a lower part containing remaining asymmetric carbons and CH_2OH groups. Heptoses having one asymmetric carbon in the lower half will be denoted by adding glycero-(D/L) to the first half; octoses-erythro- or threo- and nonoses, ribo-, arabino-, xylo- and lyxo-. The prefix named first is that of second part (lesser carbons).

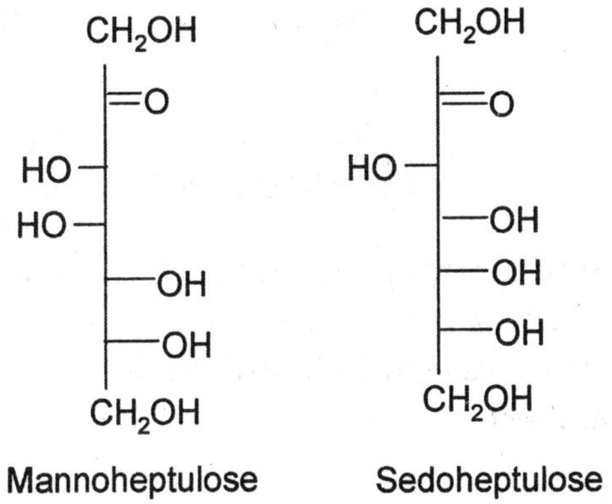

Mannoheptulose	Sedoheptulose
CH_2OH	CH_2OH
=O	=O
HO–	HO–
HO–	–OH
–OH	–OH
–OH	–OH
CH_2OH	CH_2OH

Mannoheptulose Sedoheptulose

Somehow, the sugars containing more than 6-carbons are infrequent in plants although sedoheptulose (D-altro-D-fructo-heptose) and D-mano-heptulose are common metabolic intermediates. L-Galo-D-mannoheptose and D-taloheptulose are the two other heptoses known, the former from bacteria and latter from the leaves and fruits of *Persea gratissima* L. (Lauraceae), D-Glycero-D-mannooctulose, the only octose known and D-erythro-L-gluco-nonulose, the single nonose known also are reported from *Persea.*

All the higher sugars discovered in nature are ketoses.

Properties of Monosaccharides

1. Monosaccharides are reducing agents owing to their free or potential aldehydic or ketonic group. This property is observed when they are treated with metal ions like copper or silver in alkaline solutions. When sugars are treated with Fehling's solution, the Cu^{++} ions (present as its tartarate complex) get reduced to Cu^{+} causing the precipitation of brick red Cu_2O.

2. Mild oxidizing agents oxidize the aldehydic group of D-monosaccharides producing monocarboxylic – *'onic'* acids, while strong oxidizing agents oxidize both the terminal carbons to produce a dicarboxylic *'aldaric'* acid. Thus glucose on mild oxidation gives gluconic acid and on strong oxidation yields saccharic acid.

$$\underset{\substack{\text{-ose}\\ \textit{e.g.}\ \text{Glucose}}}{\begin{array}{c}CHO\\ |\\ (CHOH)_4\\ |\\ CH_2OH\end{array}} \xrightarrow{+(O)} \underset{\substack{\text{-onic acid}\\ \textit{e.g.}\ \text{Gluconic acid}}}{\begin{array}{c}COOH\\ |\\ (CHOH)_4\\ |\\ CH_2OH\end{array}} \xrightarrow{+(O)} \underset{\substack{\text{-aric acid}\\ \textit{e.g.}\ \text{Saccharic acid}}}{\begin{array}{c}COOH\\ |\\ (CHOH)4\\ |\\ COOH\end{array}}$$

3. When heated with strong mineral acids, hexoses yield a furfural derivative which reacts with amines or α-naphthol producing coloured derivatives (**Molisch test**). In these conditions, pentoses yield furfural itself

$$\begin{array}{c}CHO\\ |\\ (CHOH)_4\\ |\\ CH_2OH\end{array} \xrightarrow{+H_2SO_4} HOH_2C\text{-(furan ring)-}CHO$$

Hydroxy methyl furfural

4. The hemiacetal group of monosaccharides yields acetals on condensation of the hydroxyl group with another alcoholic group of a sugar or an aliphatic alcohol. *e.g.*, when glucose is treated with methanol in presence of HCl, methyl glycoside is formed. These **acetals** are termed glycosides and the linkage, glycosidic bond. Such a reaction is of paramount importance due to the innumerable glycosides found in the plant kingdom. Complete methylation and acetylation of all the hydroxyl groups are possible by heating with dimethyl sulphate in alkaline conditions or acetic anhydride in presence of a catalyst. This methylation/ acetylation technique is extensively used in the structure determination of oligo- and polysaccharides.

5. Periodic acid is another reagent used for structure determination of higher carbohydrates. This reagent cleaves carbon bonds of adjacent carbon atoms bearing hydroxyl groups or a hydroxyl and an amino group, and converts the hydroxyl groups into aldehydes. When three adjacent carbon atoms bear hydroxyl groups, the central carbon is oxidized and released as a molecule of formic acid which could be estimated. This method is used in estimating the end groups of the sugar chains.

6. The **osazones** formed, when the monosaccharides are treated with phenyl hydrazine, exhibit characteristic crystal shapes and colors used effectively for the identification of sugars. In this reaction first a molecule of phenyl hydrazine reacts with the aldehydic or ketonic group to produce a hydrazone which is oxidized by a second molecule of phenyl hydrazine yielding ketohydrazone, aniline and ammonia. This ketohydrazone reacts with a third molecule of phenyl hydrazine to produce the osazone. Glucose and fructose produce the same osazone.

Derivatives of Monosaccharides

These compounds, biosynthetically derived from sugars, possess other reactive groups such as carboxylic and amino groups or have ester or ether linked groups. Their roles are varied, though some are components of polysaccharides alongwith typical sugars.

Sugar alcohols (Glycitols)

These compounds are the reduction products of common sugars in which the aldehyde or keto-groups are reduced to alcohols. These polyhydric alcohols resemble the sugars in general properties but for the reducing nature. Thus we have glycerol from glyceraldehyde or dihydroxyacetone, erythritol from erythrose, ribitol from ribose and ribulose, sorbitol from glucose, fructose and gulose, mannitol from mannose and dulcitol from galactose. Glycerol is universally present as the component of fats and phospholipids. Erythritol is reported from algae and grasses and ribitol is the sugar component of riboflavin. Ribitol amounts to 4% of the plant *Adonis,* giving the name adonitol to it. Sorbitol (glucitol) is comparatively widespread but abundant in fruits of Rosaceae and in algae and fungi. Mannitol also is widely distributed in higher plants, especially in plant exudates and amounts to 6% of stem bark of *Randia.* Mannitol is one of the storage foods of Brown algae. Dulcitol is another sugar alcohol found in algae (Rhodophyaceae) and fungi and has a restricted distribution in higher plants in which they are frequent only in members of the Celastraceae. Threitol (from threose), arabitol (arabinose) and allitol (from allose) are the lesser known sugar alcohols reported from algae, lichens and the genus *Itea* (Saxifragaceae) respectively.

Two anhydro-hexitols also are reported from higher plants. They are polygalitol (1, 5-anhydro-D-glucitol) reported from *Protea* and

Polygala and styracitol (1, 5-anhydromannitol) from *Styrax*. Of the two known heptitols, perseitol (D-α-mannoheptitol) is known from *Persea* and sedoheptitol (D-volemitol) from algae, fungi and *Primula.*

Sugar alcohols function in the plant as storage food, in osmoregulation, in translocation of food in phloem and protect the plants from frost damage and dessication.

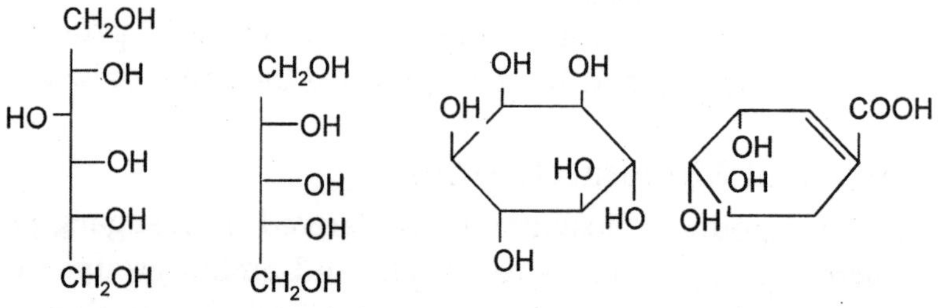

Sorbitol   Ribitol   *myo*-inositol   Shikimic acid

The polyhydric cyclohexanes - **Cyclitols** - resembling the sugar alcohols in many respects, are widely distributed in higher plants. About a couple of dozens of cyclitols are known from the plant kingdom. They may be tetrols having 4 hydroxy groups *e.g.* 1-leucanthemitol; pentols with five hydroxy groups *e.g.* d-quercitol or hexols, having six hydroxy groups *e.g. myo*-inositol & scillitol. There are monomethyl ethers (l-bornestiol and d-pinitol) and dimethyl ethers (liriodendritol) of hexols. Myo-inositol (*meso*-inositol), by far the most common cyclitol, is a lipid constituent occurring in phytic acid, a member of vit. B complex. It is a known growth regulator. Phytic acid is myo-inositol esterified with upto six phosphate groups. Pinitol (3-O-methyl-d-inositol) is universal in pines while quebrachitol (2-D-methyl-l-inositol) is abundant in dicots. Both these compounds are not recorded from monocotyledons. 1-Bornestiol (l-O-methyl myo-inositol) is having a restricted distribution among dicot families. Galactinol (l-O-D-galactosyl-D-myo-inositol), the only glycoside known, is found to be an intermediate in the synthesis of oligosaccharides like verbascose and stachyose in many labiatae members, beet, *Buddleja* etc.

Shikimic acid (3, 4, 5-trihydroxy-D-1, 6-cyclohexene-1-carboxylic acid) and quinic acid (1, 3, 4, 5-tetrahydroxy-cyclohexane-1-carboxylic acid), though not strictly cyclitols, are included here due to their similarity to common cyclitols and the biogenetic relations with carbohydrates. Shikimic acid and quinic acids form important intermediates in the biosynthesis of aromatic compounds from carbohydrates especially glucose. This route, often termed as **shikimic acid pathway**, leads to the formation of aromatic amino acids, quinones, flavonoids and related compounds. It is proved that glucose gives rise to D-erythrose-4-phosphate which combines with phosphoenol pyruvate (also from Glucose) to produce 2-keto-3-deoxy-7-phospho-D-glucoheptonic acid. This seven carbon sugar is quantitatively converted to 5-dehydroshikimic acid and 5-dehydroquinic acid which give rise to shikimic acid and quinic acid respectively. These two acids normally occur as depsides (esters derived from two phenolic acids).

Sugar acids

These polyhydroxy carboxylic acids are formed by the oxidation of the first or the last carbon atom or both. This oxidation of C-1 *i.e.* the aldehydic group yields an aldonic acid *e.g.* gluconic acid, whereas that of the last carbon forms - uronic acid *e.g.* glucuronic acid. The dicarboxylic acids formed by the oxidation of both the terminal carbons are called 'saccharic' acids. Somehow, in the plant kingdom uronic acids are found to be abundant when compared to aldonic or saccharic acids.

Glyoxylic and glycolic acids-the simplest of sugar acids - form important intermediates in the organic acid metabolism. The importance of glyceric acid and its phosphate in sugar metabolism is rather too well known to be discussed here. The 4-carbon tartaric acid occurring in fruits (tarmarind, grape etc) corresponds to erythrose.

Glucuronic acid is reported from many glycosides and mannuronic, galacturonic and guluronic are the components of many polysaccharides like hemicelluloses, pectins and gums. These sugar acids, at times, function as intermediates in hexose-pentose conversion by decarboxylation. Saccharic acid (from glucose) is reported from the latex of *Ficus elastica* (Moraceae) and mucic acid (from galactose) from the fruits of *Prunus* and *Elaeocarpus*.

The most distinguished sugar acid, of course, is L-ascorbic acid, a 6-carbon compound which is structurally related to xylose. This acid, occurring in the form of a lactone, does not possess a free carboxyl group and the acidity is attributed to the two enolic hydrogens. Ascorbic acid is the antiscorbutic factor (vit. C) in man, but in plants it has varied roles such as, a coenzyme in thioglucosidases and control of mitosis. An oxidized form of ascorbic acid, dehydroascorbic acid, is neutral in its properties. Fruits like *Citrus*, paprikas and leafy vegetables are rich sources of ascorbic acid. It is biosynthetically related to glucose.

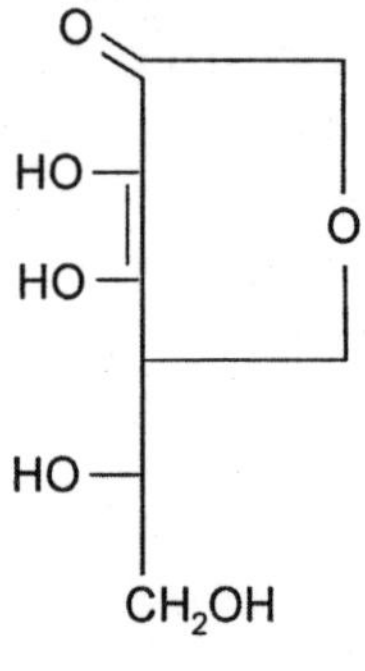

L-Ascorbic acid

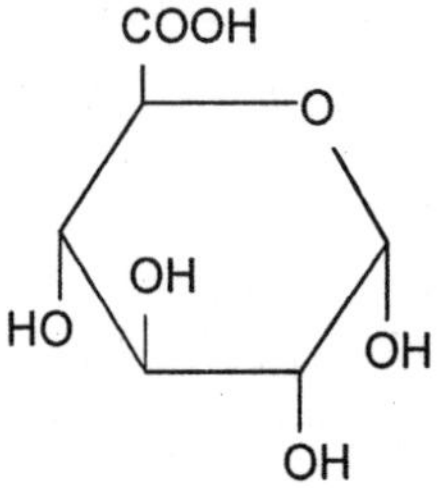

Glucuronic acid

Aminosugars

These compounds are characterized by an amino group replacing a hydroxyl group of 2-position (in higher plants) or 3-position (in fungi and bacteria). Of the 20 or more compounds known, glucosamine (2-amino-2-deoxy-D-glucose) and galactosamine (2-amino-2-deoxy-D-galactose), by far, are the commonest among them. Glucosamine is a constituent of several polysaccharides (e.g. Chitin) and amounts to 90% of sugar components of glycoproteins in chloroplast membranes. It is also reported from vicilin and legumin, two storage proteins of pulses and its N-methyl derivative (N-methyl-L-glucosamine) is a constituent of the antibiotic streptomycin. Galactosamine amounts to 10% of sugars in the chloroplast membrane glycoproteins, and is reported from certain gums and as UDP derivative from *Dahlia* tubers. As mentioned earlier, 3-amino-sugars are components of many antibiotics; *e.g.* 3-deoxy, 3-amino-D-glucose, is present in kanamycin and tobramycin while 3-amino, 3-deoxyribose in puromycin.

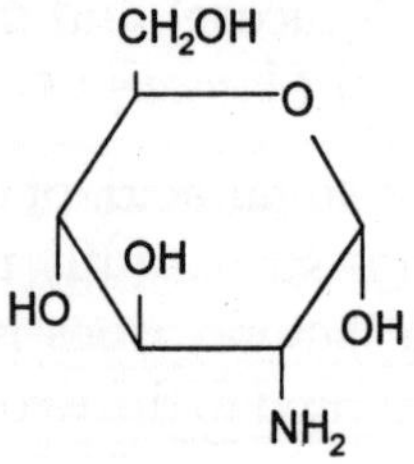

Glucosamine

Ethers and Glycosides

Sugars form ethers with a wide variety of compounds with great ease. In this process, the hydroxyl group of the hemiacetal group at C-1 gets condensed with the hydroxyl group of a second compound to form an acetal derivative. Such a condensation product is termed as a glycoside - more specifically O-glycoside- and the linkage is known as a glycosidic bond. When such as linkage is formed between two monosaccharide units, disaccharide results and in cases where many sugar units are involved a polysaccharide results. The glycoside in which a sugar and a non-sugar component are participating is known as a **heteroside** and the non-sugar moiety is designated as an **aglycone** (Glycone denotes the glycosides). The sugar component exists in furanose or pyranose form, the former being more common. As the sugar moiety exhibits α- and β-forms either α- or β-glycosides result. Those glycosides are easily hydrolysed by acids or enzymes. The enzymic cleavage of glycosides is more specific *i.e.* the α-glycoside is hydrolysed by α-glycosidase only and β-linkage by β-glycosidase.

The number of aglycones occurring as O-glycosides is legion. The important groups recognized are phenols, anthocyanins, flavonoids, coumarins, saponins, alkaloids, cardiac glycosides, anthraquinones, organic cyanides, lipids, etc. In many cases the glycosides are better known than the aglycones.

In sulphur containing glycosides the reducing group of a sugar is linked to a thiol and such a glycoside is known as *S-glycoside* (and not O-glycoside). These thioglycosides are having a very restricted distribution. Similar to S-glycosides, these are *N-glycosides* which involve linkage between the hydroxy group of a sugar and an amino group of an aglycone *e.g.* nucleosides. In addition to all these, there are *C-glycosides* in which the sugar is inseparably linked to the aglycone by C-C-linkage. Such glycosides are very stable towards normal acid hydrolysis.

The sugars in a heteroside need not always be a true sugar. Sugar derivatives such as uronic acids and sugar alcohols are reported to take part in this linkage. *E.g.* Glycyrrhizin contains 2 molecules of glucuronic acid and riboflavin, one mol. of ribitol. Glucose is the most common hexose found in glycosides closely followed by the pentose, rhamnose. Di- and oligosaccharides also take part in glycosidic linkage with many aglycones (e.g. cardiac glycosides).

The simplest of the glycosides which are encountered are O-methyl sugars occurring in the free state, in polysaccharides or in combination with other aglycones. 3-O-Methyl-D-galactose is reported from the mucilage of slippery elm (*Ulmus*) and 3-O-Methyl-D-glucose from *Streblus asper*. Many of the unique sugars in cardiac glycosides are 3-methyl esters of pentoses. Acofriose (3-O-methyl-L-rhamnose), cymarose (3-O-methyl digitoxose), thevetose (6-deoxy-3-O-methyl-D-glucose) are some among them. Dimethoxy sugars like 2, 3-Di O-methyl-D-fucose and 2, 3-di-O-methyl-D-glucose are reported from cardiac glycosides of *Streblus* while acetyl dimethyl hexoses from C-glycosides of *Mansonia*.

Esters

The alcoholic groups of a sugar may form esters with carboxylic groups of any organic acid, many of which are very important for the plants. The sugar phosphates - the intermediates in carbohydrate metabolism - are too well known to be described here. Many of the co-enzymes and nucleotides contain phosphorylated sugars. The esters of aromatic acids like gallic acid and ellagic acid constitute the hydrolysable tannins. Galactose sulphate is a major component of agar, a polysaccharide from algae.

Disaccharides

These dimers contain two monosaccharide units joined by a glycoside linkage. Of the 3 dozen or more compounds known, the common ones are dihexoses in which the linkage occurs between, (1) two reducing groups *e.g.* sucrose; (2) C-1 of one molecule to C-4 of another *e.g.*, Maltose or (3) C-1 of one mol. to C-6 of second one *e.g.*, Melibiose. The uncommon combinations are C-1-C-2, C-1-C-3 and C-1-C-5. The reducing property is lost when the linkage is between two reducing groups, but it is retained, if only one of the reducing groups is involved in the glycosidic linkage. The former group is known as non-reducing sugars and the latter, reducing sugars.

The glycosidic linkage is designated as α- or β-, depending on the position of the hydroxyl group of hemiacetal carbon in the first sugar component. If this hydroxyl group is in α-position the linkage is α-linkage and if in β-position, β-linkage. This aspect of glycosidic link, at times, influences the physical and chemical properties of the sugar.

e.g. Maltose in which two glucose units are linked by α-(1-4) link (*i.e.* the first glucose molecule is an α-anomer) is sweet in taste and soluble in water whereas cellobiose-a dimer of glucose-in which the linkage is β-(1-4) (*i.e.* the first glucose molecule is in β-form), is the component of insoluble cellulose. Even enzymes are specific for α- and β links *e.g.*, α-glycosidases split only α-linkage and β-glycosidases; β-linkage. The reducing disaccharides, in turn, exist in α- or β-forms based on the nature of the 2nd molecule in the dimer. Thus maltose occur in β-form and gentiobiose in α-form.

The non-reducing disaccharides are known as **glycosyl glycosides** and reducing ones are **O-glycosyl glycoses.** The sites of linkages may be indicated by an arrow pointing from the first sugar to the second. Thus maltose, 4 O-α-D-glucopyranosyl-D-glucopyranose is written as O-β-D-glucopyranosyl - (1→4)-D-glucopyranose.

About 3 dozen disaccharides are known occurring free or as **heterosides** (in combination with aglycones). All are colorless crystalline solids with a sweet taste. Those occurring in α- and β-forms exhibit mutarotation. Some of the important disaccharides are described below.

1. Sucrose (Saccharose, cane and beet sugar)

Sucrose, α-D-glucopyranosyl-β-D-fructofuranoside, is a non-reducing sugar universally present in the free form or combined in raffinose group of sugars. Dilute acids or the enzyme invertase hydrolyses sucrose to equimolecular mixture of D (+)-glucose and D (-) fructose-commercially known as invert sugar. Sugarcane, sugar beet, palms, nectar etc are the commercial sources of sucrose.

2. Trehalose (Mycose)

Another non-reducing sugar, trehalose yields two molecules of glucose on hydrolysis. The non-reducing nature denotes that the two reducing groups are engaged in glycosidic linkage and so this sugar is α-D-glucopyranosyl-α-D-glucopyranoside. Trehalose, a sugar abundant in yeast, fungi and algae, is seldom reported from higher plants.

3. Maltose

Maltose, 4-O-α-D-glucopyranosyl-D-glucopyranose, is a dimer of glucose in which two molecules of glucose are joined by α-(1→4) linkage. Being a reducing sugar, it occurs as the β-anomer. Maltose is rarely present in the free from but is obtained by the partial hydrolysis of starch. An enzyme maltase cleaves the maltose molecule to two glucose units.

4. Isomaltose

It differs from maltose in the linkage which is (1→6) and so isomaltose is 6-O-α-D-glucopyranosyl-D-glucopyranose. This sugar also does not occur naturally, but obtained by the partial hydrolysis of amylopectin fraction of starch.

5. Gentiobiose

Gentiobiose resembles isomaltose in consisting of two glucose units linked by 1→6 fashion. But here the linkage is β-as against the α-linkage of isomaltose. It occurs free in gentian root and as a constituent of a glycoside amygdalin, some cardiac glycosides and the trisaccharide, gentianose.

6. Cellobiose

This dimer differs from maltose in having the β (1→4) linkage and so is 4-O-β-D-glucopyranosyl-D-glucopyranose. On acid hydrolysis or by the action of an enzyme cellobiase it yields two molecules of glucose. This reducing α-sugar is obtained by the partial hydrolysis of cellulose.

7. Lactose (Milk Sugar)

Lactose-4-β-D-galactopyranosyl-D-glucopyranose, once believed to be absent from the plant kingdom, is now reported from the fruits of many plants such as *Ceratonia, Achras, Mimusops* and in the saponins of *Sapindus*. Lactose occurs in the β-form. On hydrolysis it yields galactose and glucose.

8. Melibiose

Also is a disccharide containing galactose and glucose residues. The linkage here is α (1→6) in contrast to β (1→4) of lactose, and so is designated as 6-O-α-D-galactopyranosyl-D-glucopyranose. Melibiose occurs free in leaves of scots pine, beet molasses, and cotton seed hulls and as a constituent of the trisaccharide raffinose.

The less common disaccharides containing hexoses are:

1. Nigerose

A dimer of glucose with 1-3 linkage, found in beer and in the pollen of *Typha*. Technically the sugar is 3-O-α-D-glucopyranosyl-D-glucopyranose.

2. Planteobiose

Obtained by the mild acid hydrolysis of planteose, this sugar occurs free in *Fraxinus*. The constituents are galactose and fructose joined by α (1→6) linkage.

3. Sophorose

Sophorose is 2-O-β-D-glucopyranosyl-D-glucopyranose-occurring in glycosides like kaempferol-3-sophoroside, stevioside and in many anthocyanins.

Apart from these, there are many disaccharides reported from various glycosides especially cardiac glycosides. An enumeration of those sugars is considered beyond the scope of this book. But a few disaccharides containing pentoses are listed below.

1. Apiosyl glucose

Linkage is 1→2 and is reported from apiin. On hydrolysis this sugar yields apiose and glucose.

2. Primeverose

Contains xylose and glucose (both in pyranose form) linked by β(1→6) glycosidic bond. Occurs in many glycosides esp. that of anthraquinones. (*e.g.*, alizarin from *Rubia*).

3. Rutinose

6-O-α-L-Rhamnopyranosyl-D-glucopyranose occurring in glycosides like rutin, datiscin and many anthocyanins.

4. Sambubiose

β (1→2) - Xylosyl glucose occurring in many anthocyanins.

5. Vicianose

(6-O-α-L-Arabinopyranosyl-D-glucose) occurs in many glycosides like vicianin (a cyanogenic glycoside).

Aldobiouronic acids include disaccharides containing uronic acids in glycosidic linkage among themselves or with one or more units of sugars. They are the common constituents of gums and mucilages and are obtained by the hydrolysis of these polymers. They rarely occur free in plants. 2-α-Galacturonosyl-L-rhamnose (obtained from gums of *Khaya, Sterculia* and *Cochlospermum* and mucilages of *Plantago, Hibiscus* etc.), 6-β-glucuronosyl-galactose (obtained from gums of *Prunus, Acacia, Ferula* etc.), 4-α-Galacturonosyl-galacturonic acid (from various pectins), 2-β-glucuronosyl-mannose (from gums of *Hakea, Anogeissus* and *Albizzia*) are some of the common aldobiouronic acids reported in nature.

Oligosaccharides

Oligosaccharides are polymers containing 3-10 simple sugars linked together. Similar to disaccharides most of the Oligosaccharides contain hexoses as the constituents. Reducing or non-reducing, these are crystalline water soluble solids exhibiting optical activity. These compounds occur alone or combined in glycosides.

About two dozen **trisaccharides** are known to us, of which half a dozen occur in various glycosides. The common trisaccharides are:

1. Raffinose

Raffinose is the most abundant oligosaccharide reported from lower as well as higher plants. On hydrolysis it yields one molecule each of D-fructose, D-glucose and D-galactose. On partial hydrolysis D-fructose and melibiose are obtained from raffinose but the enzyme α-glycosidase splits this sugar to galactose and sucrose while the enzyme invertase produces fructose and melibiose. With the help of these facts the sequence of sugars is interpreted as galactose-glucose-fructose and so the structure of raffinose is O-α-D-galactopyranosyl (1→6)-O-α-D-glucopyranosyl-(1→2) β-D-fructofuranoside. Raffinose is abundant in beet root and cotton seed.

2. Gentianose

Controlled acidic or enzymic hydrolysis by invertase produces D-fructose and gentiobiose from Gentianose; α-glycosidase converts it to sucrose and D-glucose. On complete hydrolysis, 2 molecules of D-glucose and one molecule of D-fructose are obtained showing the arrangement of sugars as glucose-glucose-fructose. The structure of gentianose is O-β-glucopyranosyl (1→6)-O-α-D-glucopyranosyl-(1→2)-β-D-fructofuranoside. Gentianose occurs free in *Gentiana* roots.

3. Melezitose

This trisaccharide on mild hydrolysis yields D-glucose and the disaccharide turanose. On vigorous hydrolysis, two molecules of D-glucose and one molecule of D-fructose are produced. The structure of melezitose is thus O-α-D-glucopyranosyl-(1→3)-O-β-D-fructofuranosyl (2→1)-α-D-glucopyranoside. Melezitose occurs in nectar of various trees like poplars, lime etc. and manna of *Alhagi* and *Fraxinus*.

Some of the remaining trisaccharides are: (1) Planteose: composed of glucose, fructose and galactose obtained from the seeds of *Plantago, Sesamum* and *Theobroma*; (2) Manninotriose (galactose-galactose-glucose) from *Fraxinus* (manna) and *Betula*; (3) Mannotriose (3-molecules of mannose) from *Fraxinus* & *Borassus* and (4) Rhamninose (galactose-rhamnose-rhamnose) from *Centella* as the sugar component of asiaticoside. Some of the trisaccharides present in cardiac glycosides are: (1) Chacotriose (Rhamnose-glucose-Rhamnose) from chacoin (a cardiac glycoside of *Solanum*); (2) Odorotriose (Gentiobiose-diginose) from odoroside-K; (3) Scillatriose (Rhamnose + 2-glucose) in glucoscillaren-A and (4) Strophanthotriose (Gentiotriose + glucose) in strophanthin-K.

Tetrasaccharides

About a dozen tetrasaccharides are known to occur in higher plants. Except for a few, all are having a very restricted distribution. Possibly the lack of systematic surveys is the reason of our limited knowledge on the types and distribution of the oligosaccharides. A common tetrasaccharide encountered in higher plants is Stachyose (manneotetrose), first reported from *Stachys japonica* (Labiatae) and manna of *Fraxinus ornus*. This compound on hydrolysis yields two molecules of galactose and one molecule each of glucose and fructose. The structure is found to be O-α-D-galactopyranosyl-(1→6)-O-α-D-galactopyranosyl-(1→6)-α-D-glucopyranosyl-(1→2)-β-D-Fructofuranoside.

Lychnose (Galactose-(1→4)-glucose-(1→2)-fructose-(1→1)-galactose) obtained from many caryophyllaceous genera; Maltotetrose (Glucose-(1→4)-glucose-(1→4)-glucose-(1→4)-glucose) present in malt extracts and Scordose (4 mols of fructose) reported from Liliaceae plants are a few tetrasaccharides worth mentioning. The primary (tetra) glycosides of *Digitalis* also contain tetrasaccharides in combination with cardiac genins.

The higher oligosaccharides-about 10 are known-do not occur in heterosides but remain free. They are believed to be intermediates in polysaccharide synthesis or breakdown. A pentasaccharide very widely distributed in higher plants is Verbascose first reported from *Verbascum* (Scrophulariaceae). This oligosaccharide is O-D-galactopyranosyl-(1→6)-O-α-D-galactopyranosyl (1→6)-O-α-D-galactopyranosyl-(1→6)-O-α-D-glucopyranosyl-(1→2)-β-D-fructofuranoside.

A novel system of classifying oligosaccharides is to arrange them in their respective "homologous series". Our incomplete knowledge of the types and distribution of these oligosaccharides does not permit us to arrange all the known compounds in this manner. Only one series is completely recognized and a mere glance at this series gives the scope and utility of system. This is "**Raffinose series**" (Raffinose family), the simplest sugar is sucrose and all the higher members are derived by adding one galactose molecule each in α (1→6) linkage to the preceding one. All the known members of this series are presented below.

1. Sucrose
2. Raffinose = sucrose + 1 galactose
3. Stachyose = sucrose + 2 galactose
4. Verbascose = sucrose + 3 galactose
5. Ajugose = sucrose + 4 galactose
6. Penta-D-galactosyl sucrose
7. Hexa-D-galactosyl sucrose
8. Hepta-D-galactosyl sucrose
9. Lactosin = sucrose + 8-10 galactose

It is peculiar that all these sugars are present in Scrophulariaceae and related families.

Polysaccharides

Polysaccharides are polymers containing ten to several thousand monosaccharide units always arranged in orderly sequences. Judging by the very many different combinations of sugars observed in the case of disaccharides, one may expect a wide array of polysaccharides in nature. But only a few such polymers are actually met with in living kingdom. This discrepancy is because (1) relatively fewer hexoses and pentoses are taking part in this polymer formation; (2) the glycosidic linkage is restricted to a specific number of hydroxyl groups; (3) a number of repeating units of oligosaccharides are formed within the molecule of a polysaccharide.

The polysaccharides may be branched or unbranched. The branched ones tend to form viscous colloidal systems in water by getting easily dispersed in water. The linear polysaccharides are insoluble or slightly soluble in water. This property is effectively made use of in plants in that the structural polysaccharides which give the rigidity and strength to the plants are always linear polysaccharides while the nutrient polysaccharides are invariably branched. A polysaccharide may be homopolysaccharide when it contains only one type of sugar residue or heteropolysaccharide when it consists of more than one type of sugar residues. Starches and celluloses are homopolysaccharides while gums and galactomannans are heteropolysaccharides. A polysaccharide is generally designated by adding a suffix-*an* to the monomer. Thus all polysaccharides are glycans while starch and cellulose are glucans and inulin, a fructan.

a. Nutrient polysaccharides

These glycans act as metabolic reserves of monosaccharides, being met with in all living cells. Despite their omnipresence, certain storage organs like tubers, seeds and stems are particularly abundant in them. Starch is the most bounteous nutrient polysaccharide in higher plants; however plants such as algae and fungi contain different glycans like laminarin, glycogen etc.

Starch

Starch is a glucan which can be represented by a general molecular formula $(C_6H_{10}O_5)_n$. On complete acidic hydrolysis starch gives rise to glucose molecules and on slow hydrolysis the first products are dextrins (smaller polymers of 25-30 glucose units) which later break down to maltose units. Maltose finally is hydrolysed to glucose. The enzyme diastase (β-amylase) splits the starch msolecule to maltose which in turn is acted upon by maltase to produce glucose.

Starch is essentially composed of two polysaccharide fractions, α-amylose and β-amylose (amylopectin). These two components can be separated by adding n-butanol to a hot colloidal aqueous solution of starch. On cooling the mixture, α-amylose gets precipitated. Amylopectin, remaining in solution can be precipitated by the addition of methanol.

α-Amylose is found to be a linear chain of about 300 glucose molecules linked by α (1→4) glycosidic bonds. Methylation by dimethyl sulphate (which attack all the free hydroxyl groups to produce acid stable methoxyl groups) followed by acid hydrolysis, amylose yielded 99.7% 2, 3, 6-trimethyl glucose and 0.3% 2, 3, 4, 6-tetramethyl glucose. This indicates that in the case of 2, 3, 6 - trimethyl glucose, the hydroxyl groups at C-1 and C-4 are engaged in the chain formation while that of C-5 is used up in ring formation. The presence of 2, 3, 4, 6 - tetramethyl glucose denotes that they are the non-reducing end units of glucose chains. (The tetramethyl glucose formed by the glucose residues of the reducing end possesses an acid labile methoxyl group at C-1, and this methoxyl group gets cleaved by the acid acting on amylose chain resulting in the formation of 2, 3, 6 trimethyl glucose). This means that there are about 0.3 end groups for every chain of 100 glucose molecules or each amylose chain is about 300 glucose units in length and the units are linked by 1→4 glycosidic bonds. The hydrolysis of amylose by α-amylase to maltose units proves that the maltose units are held by α-(1→4) linkages (Maltose itself being α-(1→4) glucosyl glucose).

Amylose is inhomogeneous in composition. Some of the amylose preparations contain upto 6000 glucose residues. The amylose chains in potato amylose vary in length containing 2000-6000 residues and possess a few (one per thousand glucose residues) α- (1→6) branching points. The stereochemistry of α-(1→4) linkage does not permit amylose to remain as a straight chain; instead it coils to form an α-helix having six residues per turn and a diameter of 13.7 A°. Helices having seven or eight residues per turn with diameters of 15 A° or 16.2A° respectively also are reported. The lumen inside the helix can accommodate fatty acids, low molecular weight alcohols or iodine and thus starch acts as an inclusion complex (clathrate) when these compounds are present. The characteristic blue colour produced when starch solution is treated with iodine is due to the trapping of the iodine molecules within amylose helix. The disappearance of the blue colour on heating the starch-iodine solution is due to the uncoiling of the helix and subsequent liberation of iodine. On cooling the solution, clathrates are formed again and the blue color reappears. When both fatty acids and iodine are present, fatty acid-starch complex is preferred over iodine-starch complex and therefore the blue colour is not formed. Amylose takes up to 19% of its weight of iodine and this absorption of iodine is sometimes used for estimating starch. In solution, amylose chains tend to clump together due to the development of intermolecular hydrogen bonds and precipitate gradually. This crystallization reduces the shelf-life of many starchy foods and causes the staling of bread. Introduction of phosphate groups on the side hydroxyl groups prevents such hydrogen bonding and prolongs the shelf-life of foods and starchy preparations.

The amount of amylose in starch varies with the source plant; potato starch contains about 20% amylose and it is found to be the case in many other starches. The starch of waxy corn is almost free of amylose. Potato amylose is inhomogeneous in composition containing amylose chains of variable length (2000-6000 units) and having a few 1, 6-glucosidic interchain linkages.

β-Amylose (Amylopectin). In-contrast to the linear nature of amylose, amylopectin gives a highly ramified structure. The end group assay by methylation yields 91% 2, 3, 6-trimethyl glucose, 4% 2, 3, 4, 6-tetramethyl glucose and 5% 2, 3-dimethyl glucose. The larger yield of 2, 3, 6-trimethyl glucose as in the case of amylose, clearly indicates that the major linkage in this polymer is 1-4. The 4% tetramethyl glucose denotes that one end group occurs for every 25 glucose residues or in other words there are many short chains of 25-30 glucose molecules. Furthermore, the formation of 2, 3-dimethyl glucose gives clear indication that in some of the glucose molecules the 6th position also are involved in glycosidic linkage. Thus the amylopectin is visualized as a highly branched structure in which many shorter linear chains of about thirty α- (1→4) linked glucose residues are cross linked by α-(1→6) glycosidic bonds. (It is also possible that amylopectin is a long chain of α- (1→4) linked glucose units with regular α-(1→6) linked branches on every 24th or 25th glucose residue.) The amylopectin of rice starch contains 80-90 such cross-linked chains. An average amylopectin molecule is estimated to contain about 3000 glucose units and the nature of ramification varies with the source. The formation of isomaltose during certain enzymatic hydrolysis of amylopectin conclusively proves the presence of α- (1→6) linkage in the molecule. The isolation of minor quantities (about 0.1%) of nigerose (O-α-D-glucopyranosyl (1→3) glucopyranose) from the hydrolysates of certain samples of amylopectin suggests the usage of α-(1→3) bonds for ramification in those starches. The presence of phosphate groups in starches is conclusively proved. But exact location of these groups is still very much unknown. Amylopectin is soluble in water and gives a violet color with iodine.

Starch occurs in the form of grains in storage organs. The central point of a starch grain is proteinaceous-hilum-around which starch is deposited in layers. Starch grains may be concentric (regular discs) as in wheat, eccentric (oval in shape with hilum on one side as in potato) or platelets (in rice).

Dextrins

Dextrins are large oligosaccharides composed of D-glucose units. They are derived from starches by partial hydrolysis. There are many types of dextrins like α-, β-, γ- and ϕ-dextrins. The structure of these compounds depends on the nature of preparations, but on an average, dextrins contain about 30 glucose residues. These compounds are used as adhesives.

Dextrans

Dextrans are polyglucans elaborated by certain bacteria such as *Leuconostoc mesenteroids* growing on a sucrose substrate. These polymers have a mol. wt. of 40000-75000 and the glucose units are linked predominantly by α- (1→6) glycosidic bonds with occasional α- (1→3) and α- (1→4) cross links. Dextrans are antigenic substances and used as plasma expanders.

Glycogen

Glycogen is a collective term given to a group of nutrient polysaccharides widely distributed in fungi and animals. They closely resemble amylopectin in having branched chains of glucose and with its reaction to iodine in giving a brown color. But glycogens are highly ramified in which shorter linear chains of 11-18, α- (1→4) linked glucose residues are interconnected by α- (1→6) glycosidic bonds and the molecule is much larger than amylopectin.

Inulin

Inulin is a fructosan (fructan) consisting of 25-30 D-fructofuranose units linked by β-(2→1) glycosidic bonds often terminated by a glucose unit which forms a non-reducing (sucrose) end group. The tubers of Dahlias, chicory, artichoke etc. are particularly rich in this polymer. It occurs in the cell sap and on immersion of the fresh tubers in alcohol for some time, aggregates forming characteristic spherical crystals. Inulin and related fructans replace starch as the principal food reserve in most of the Compositae members. Fructans of shorter chain lengths such as **levans** [β-(6→2) linked 20-25 units] occur in grasses. Some of these fructans are branched while others are linear. All these polymers are soluble in water (Inulin is soluble in hot water only). Some amount of glucose also is reported from these fructans which probably function as end groups.

Mannans

Mannans serve as reserve food in the seeds of a few palms such as ivory nut palm (*Phytelephas*), but act as structural polysaccharides in lower plants (yeast), ferns and conifers. Mannans of ivory nut consists of a linear chain of 75-80 β- (1→4) linked D-mannopyranose units with some β-D-glucose end groups. These polymers amount to 16% dry wt. of the seeds. The mannans of yeast cell wall contain mannose units in α- (1→2) and α- (1→6) linkages.

Galactomannans

These heteropolysaccharides, yielding galactose and mannose on hydrolysis, consist of chains of β-(1→4) linked mannopyranose units with α-(1→6) linked galactopyranose units as side appendages. Often considered as mucilages, these polymers of varying length occur as a vitreous layer on the inner side of seed coats of many legumes such as carob (*Ceratonia siliqua*) and guar (*Cyamopsis tetragonolobus*).

Lichen starch is considered to be built of cellobiose units linked 1→3.

b. Structural polysaccharides

The structural diversity is larger in this group of polymers than the nutrient polysaccharides. The heterogeneity exhibited in the cell wall composition is necessary to withstand the magnitude of stress and strain the cell-wall is subjected to. Xylose and arabinose are more frequent in structural polysaccharides.

Cellulose

Cellulose is the most important cell wall constituent being present in all plants except fungi. It is the most abundant organic compound in nature and represents a large share of combined carbon in plants. The ratio of cellulose with other cell wall constituents varies with plant groups. Cotton fibres form almost pure cellulose (93%) while the wood of higher plants contains 40-45% of this polymer.

Cellulose is a glucan, consisting of a linear chain of β- (1→4) linked D-glucopyranose units. A cellulose chain is zigzag in nature in that each successive glucose residue is rotated 180° in relation to the plane of the preceding one. This facilitates the formation of hydrogen bond between 3-OH of one residue with ring (pyranose) oxygen of the next one. This impedes the free rotation of the glucose residues around the glycosidic bond and produces a stiff band like molecule with the puckered (chair) faces of the pyranose, all in the same plane. This arrangement and the close packing of the cellulose chains coupled with the intermolecular hydrogen bonds developed, are responsible for the great tensile strength and rigidity of the cellulose fibres. Cotton cellulose molecule contains about 3000 glucose units and the native cellulose about 10000. In the cell wall, cellulose molecules are arranged parallel to form micelles of about 60A°. Parallely arranged micelles form a micro fibril. The core of the micro fibril, consisting of about 50 cellulose molecules, has a 3-dimensional array and crystalline lattice. Surrounding the core is the paracrystalline phase in which the molecules exhibit no 3-dimensional array. This region, cortex, contains about 100 polysaccharides including some hemicelluloses.

A unit cell of the crystal lattice (core) consists of 5 cellobiose units from 5 cellulose molecules. Of these, four are arranged at 4 angles of a square and one in the center. Hydrogen bonds between the -OH in 6 position of a glucose residue of a molecule and the glycosidic oxygen of the adjacent molecule stabilize the lattice further.

On hydrolysis cellulose yields cellodextrins (about 30 glucose) → cellobiose→glucose. The enzyme cellulase cleaves cellulose to cellobiose units.

Cellulose forms the basis of many industries like pulp and paper. In addition, innumerable cellulose derivatives are prepared which are commercially exploited.

Matrix polysaccharides (Hemicelluloses)

Matrix polysaccharides is a collective term given to a number of polysaccharides occurring in close association with cellulose especially in lignified tissue, which can be extracted by 17.5% sodium hydroxide solution. The term hemicellulose was attributed to them since they were believed to be the precursors of cellulose. The fact that they have no role at all in the cellulose biosynthesis necessitated a rethinking on this terminology and therefore, now they are designated as matrix polysaccharides. They include a large number of polysaccharides like xyloglucans and glucuronoarabinoxylans, xylans, mannans, glucomannans and arabinogalactans. The main function of these compounds is to cross link cellulose microfibrils. They may coat microfibrils, but are also long enough to span the distance between microfibrils and link them together to form a network.

1. Xyloglucans and glucuronoarabinoxylans

These are two cross-linking polymers present in all primary cell walls. The **Xyloglucans** are seen in most of the dicots and about 50% of the monocots. But the **glucuronoarabinoxylans** are the specific to primary cell walls of the Commelinoid monocots such as Commelinaceae, Bromeliaceae, Arecaceae, Zingiberales, Cyperaceae and Poaceae. The Xyloglucans consists of blocks of 6-11 sugars having a linear chain of β(1- 4) linked glucose units bearing α-D xylose units at regular intervals to the C_6 of the glucose units. In some species, the xylose units are replaced by α-L-arabinose or β-D-galactose units. There exists three types of Xyloglucans: 1.*Fucogalactoxyloglucans* This group is widely common, and consists of four $\beta(1\rightarrow4)$ linked glucose backbone wherein the subsequent glucose units are turned 180° with respect to the preceding one and β-D-xylosyl units are added to three contiguous glucosyl residues of the backbone to produce a heptasaccharide unit structure. On about one half of these unit structures a disaccharide α-L-Fucose-$(1\rightarrow2)$ β-D-Galactose is added to the C_2 of Xylose side group forming a nonasaccharide unit. 2. *Arabinoxyloglucans* seen in the Solanaceae and Lamiales have a hexasacchaRide unit. Here only two of every four glucose units bear a xylose unit each and the xylose residues may contain one or two arabinose side chains. There is always an acetyl group replacing the xylose in the third glucose unit. 3. The third group, present in small amounts in Commelinoid monocots, contain random additions of xylose units to the xyloglucans.

The **glucuronoarabinoxylans** contain a β-$(1\rightarrow4)$ linked xylose backbone having a large number of arabinose side chains and a few glucose side chains. Some of the arabinose side chains contain ferulic or another hydroxycinnamic acid group (Carpita and McCann, 2000)

There is another glucan found in grasses termed **mixed linkage β(1→3) and β(1→4) glucans (β-glucans).** These are unbranched polymers consisting of cellotriose and cellotetrose (β-1→4 linked tri- and tetrasaccharides of glucose) units in the ratio 2:1 and connected by β(1→3) linkages. These cellotriosyl and cellotetrosyl units together form "corkscrew"-shaped polymers of 50 residues long which are spaced by oligomers of four or more cellotetrose units.

2. Xylans

Xylans are abundant in lignified cell walls and on hydrolysis yield D-xylose as the principal component and (in many cases) a few molecules of L-arabinose and D-glucuronic acid. Xylans differ in proportion and types of side chains and in their mode of attachment. Wheat straw xylan is a branched polysaccharide in which straight chains of β- (1→4) linked D-xylopyranose units are cross-linked by β-(1→3) linkages. Each straight chain contains about 40 D-xylose, 5-L-arabinose and 3-D-glucuronic acid units. Corn cob xylan consists of unbranched β- (1→4) linked chains of about 200 D-xylose units. The acidic xylans contain larger number of uronic acids attached as side groups on the xylose backbone. In European beach wood every 10th xylose unit carries a single 4-O-methyl glucuronic acid linked to C_2. Generally softwood xylan contains more 4-O-methyl glucuronic acid (15-20%) than those from hardwood (8-15%). Arabinose is found linked at C_3. Xylans are found to occur in cereal gums, exudates, gums, and some mucilages too.

3. Mannans and glucomannans

These polymers are all linear chains of 1, 4-β-D-mannose with 1, 4-β-D-glucose as a main structural feature. In conifers, mannans form half of the hemicellulosic fraction and here the chains are at times branched. The ratio of mannose: glucose is about 3:1 in softwood while in hardwood it is 2:1. Though at certain places there is a regular arrangement of 4 mannose and 2 glucose units, generally there is a random placing of these two monomers along the chains. Some softwood mannans contain a few units of galactose also, linked 1, 6- or 1, 3- to glucose.

4. Galactoglucomannans and galactomannans.

Galactoglucomannans consist of a backbone of blocks of di- or tetrasaccharides of glucose alternating with similar blocks of mannose with galactose units as side chains to O-6 of some mannose residues. Galactomannans consist of β(1→>4) linked mannose units with α-D-galactose as side chains of O-6 of mannose residues.

5. Galactans, arabinogalactans and arabinoxylans

These polysaccharides are widespread in conifers. In contrast to glucomannans, arabinogalactans present a highly ramified picture with α 1, 3-linked galactopyranose backbone having a variety of side groups like L-arabinofuranose and D-galactopyranose. In some of the larches 1,6-linked galactose backbone also is reported. Similar arabinogalactans are met with in exudate gums such as gum Arabic in which the inner chains show 1, 3-linkages and outer 1, 6- linkages. Arabinogalactans are also found to occur in algae like *Cladophora, Chaetomorpha* and *Codium.* Arabinoxylan of wheat consists of a number of oligosaccharides (4 - 10 sugar units) consisting of xylose backbone β (1→4) on which one or two arabinose units are attached as side chains at random.

Pectic polysaccharides

Pectic polysaccharides comprise a family of acidic and neutral polymers found associated with the walls of living cells. Middle lamella, the first wall formed during cytokinesis, is composed of the calcium salt of pectic compounds. These polysaccharides are abundant in soft tissues of fruits, flowers and seeds where they are concentrated in primary walls and in intercellular regions.

Pectic polysaccharides are the most complex class of wall polysaccharides in which several neutral, oligo and polysaccharides are covalently attached to the rhamnogalacturonan or homogalacturonan backbone primarily through the rhamnopyranosyl residues. Although the neutral polysaccharides such as galactans, arabinans and arabinogalactans are isolated in the free form, they are considered extraction artifacts formed from a highly complex molecule. The *rhamnogalacturonan backbone* consists of blocks of 8 α- (1→4) linked galacturonic acid residues joined to one another by a trisaccharide, L-rhamnopyranosyl-α-(1→4)-D-galactopyranosyl-α-(1→2)-L-rhamnopyranose. The terminal galacturonic acids are linked at rhamnose (of the trisaccharide) by α- (1→2) linkages. In some cases both galacturonosyl and rhamnopyranosyl residues occur in a repeated sequence. For example, a rhamnogalacturonan isolated from *Acer* suspension cultures has upto 300 alternating residues of each sugar of which half of the rhamnopyranosyl residues have side chains attached through C_4. Homogalacturonans which form another important structural feature in many types of pectin consists of three types. 1) Uninterrupted homogalacturonans as seen in carrot pectin, (one of the isolated homogalacturonan consists of about 200 methylated D-galacturonosyl units joined by α- (1→4) glycosidic links); 2) xylogalacturonan having a galacturonan backbone with xylose units attached to O-3 of galacturonic acid residues and 3) rhamnogalacturonan II (RG II). RG II has a wide variety of sugars such as apiose, aceric acid (3, C′-carboxy-5-deoxy-L-xylose), 2-O-methyl fucose, 2-O-methyl xylose, Kdo (3-deoxy-D-mann-2-octulosonic acid) and Dha (3-deoxy-D-lyxo-heptulosaric acid). Dimers of RG II have been found to be cross-linked by two diester bonds per Boron between apiose units in complex side groups.

A number of neutral sugars such as D-galactose, L-arabinose, D-xylose and less frequently L-fucose and D-glucuronic acid occur as mono/oligo/polysaccharide side chains. In most of the pectins the galacturonic acid residues are methylated or acylated (sugar beet pectin). In gel-forming conditions both rhamnogalacturonan and homogalacturonan backbones fold into helix conformation. This allows calcium to bind cooperatively with carboxyl groups of unbranched, unesterified segments of the polygalacturonosyl chains.

The neutral polysaccharides occur either as side chains attached to O-4 of many rhamnose units of the rhamnogalacturonan back bone. Pure *arabinans* isolated from mustard seeds are branched polysaccharides having an α-(1→5) linked arabinofuranose backbone with α-(3→1) linked arabinofuranose side chains on blocks of 3 arabinose units separated from one another by one arabinose *i.e.* the side chains occur in 1, 2, 3, 5, 6, 7; 9, 10, 11 numbered arabinose units. Even α- (1→2) linked side-chains occur in some pectins. *Galactans* are seen in the compression wood of a number of gymnosperms. These polymers are unbranched chains in which about 120 galactopyranose units are joined by β(1→4) linkages. Galactans containing 5-10% uronic (galacturonic or glucuronic) acids also are located in some plants. *Arabinogalactans* are more frequent than pure galactans. They contain β-(1→4) linked D-galactopyranose chains with upto 25% of L-arabinofuranose side-chains. In pectins isolated from *Centrosema* seeds and soyabean cotyledons arabinogalactans form the major components of pectin. Another type of arabinogalactans, having short galactan chains of ß-(1→3) and ß-(1→6) linkages and interconnected to each other, are found associated with specific proteins called Arabinogalactan proteins

De Vries *et al.*, (1982) proposed a model in which pectin molecules are considered to contain homogalacturonan regions comprising 90% galacturonan back bone and the so called "hairy" (ramified) regions with a galacturonan backbone rich in rhamnose and carrying the major parts of the neutral sugars of the pectins. The hairy regions are studied by Voragen *et al.*, (1993) who found that there are at least three repeating units,

(1) Xylogalacturonan units containing an (1-4) α-D-galacturonan backbone with some rhamnose inserted and carrying single unit side chains of xylosyl residues ß-linked to O-3 of galacturonosyl residue,

(2) Residual stubs of pectic backbone containing arabinan side chains,

(3) Rhamnogalacturonan chains with alternating (1, 2) linked α - L - rhamnose and (1→4) linked α- D - galacturonic acid residues with part of the rhamnosyl residues carrying single O - 4 linked galactose side chains.

Ferulic acid residues also have been found to occur esterified to neutral polysaccharides isolated from spinach and sugar beet. In these polymers approximately 1 in 250 sugar residues are feruloylated.

Callose

Callose is another structural polysaccharide found to occur in granular form deposited around sieve plates and on the sides of sieve tube pores. It is an unbranched ß- (1→3) glucan with some uronic acids (~2%, mainly glucuronic acid) occurring as end groups.

In addition to these polymers some other polysaccharides which occur in fungal cell wall are: (1) Chitin, ß(1→4) linked N-acetyl glucosamine units (an amino cellulose), (2) Chitosan, ß-(1→4) linked glucosamines, (3) ß-Glucans, ß-(1→3) and ß-(1→6) linked glucose and (4) α-Glucans, α-(1→3) and α-(1→4) linked glucose.

A number of structural polysaccharides are obtained from algae. Laminarin, algin, agar and carrageenin are some of the important ones in this group. Laminarin is a linear polymer of ß- (1→3) linked glucose units with occasional branch points at C-6 and terminal groups of mannitol.

Algin

Algin or sodium alginate is the sodium salt of alginic acid, a polysaccharide obtained from brown seaweeds, mainly *Macrocystis pyrifera* Ag., native of coasts of U.S.A. and Canada. Alginic acid, which forms the principal cell wall component, is extracted by macerating plant parts with dilute sodium carbonate solution, diluting with water, filtering and then precipitating with dil. sulphuric acid or with $CaCl_2$. The free alginic acid is then converted to its sodium salt by neutralizing with Na_2CO_3.

Sodium alginate is a tasteless and odorless, yellowish powder readily soluble in water, whereas alginic acid is relatively insoluble in water. Alginic acid is a linear copolymer of D-mannuronic acid and L-guluronic acid. These two types of monomers occur in three different block structures- homopolymeric blocks of D-mannuronic acids residues, homopolymeric blocks of L-guluronic acid and blocks with an alternating structure (1-4 linked). The ratio of mannuronic acid to guluronic acid varies from 3:1 to 1:3 depending upon the source plants. This variation may be due to the relative amounts of varying block structures.

The alginates, especially sodium alginate, are extensively used in the food industry (mainly for ice creams) as a smoothening agent, sizing of paper, cosmetics and as a tablet binder and thickening agent. Calcium alginate (alginate fibres) is used as a haemostatic dressing or as a swab for pathological work.

Agar

Agar is the hydrocolloid obtained from the various red algae like *Gelidium cartilagineum* Gaillon and other species of *Gelidium* (in Japan) or *Gracilaria confervoides* Grevil or *Petrocladia* sp. (in U. States and Australia). Agar is extracted by boiling the algae with acidulated water, filtering and cooling when the agar separates out as a jelly. The traces of water are removed by freeze-drying. Agar is manufactured only in winter.

Agar occurs in the form of translucent yellowish-white odorless flakes, stripes or powder. It is a heterogeneous polysaccharide containing two components, agarose and agaropectin. Agarose is a neutral galactose polymer consisting of 1-3 linked β-D-galactopyranosyl (some methylated at 6) and 1→4 linked 3, 6-anhydro-α-L-galactopyranosyl units. The structure of agaropectin is poorly understood, but is considered to be a polymer similar to agarose, in which the anhydrogalactose units are partly esterified with sulphuric acid. Agaropectin contains a few molecules of glucuronic acid and 1% pyruvic acid in acetal linkage to D- galactose residues.

The most important use of agar is as gel in culture media. It is also used as an emulsifier, in cosmetics and paper industries and in medicine as a laxative.

Carrageenin

Carrageenin is a group of hydrocolloids obtained from various red algae like *Chondrus crispus* Stackhouse, *Gigartina mamillosa* J. Agardh or related species, native to coasts of Ireland and United States. The algae growing on rocks are collected, bleached with the help of sun & dew (sometimes with the help of SO_2), dried and stored.

The dichotomously branched thalli contain at least four polysaccharide complexes. They are galactans like agar, but contain D-galactose instead of L-sugar and higher content of sulphate esters. Carrageenins on treatment with 0.25% KCl solution yield an insoluble fraction and a soluble fraction. The insoluble fraction is designated as k-carrageenin having an alternate structure of 3, 4-linked β-D galactose and 3, 6-anhydro α-D-galactopyranosyl (1→4)-3, 6, anhydro-D-galactose) and contains more than one sulphate ester group per disaccharide unit-probably at 6-position of D-galactose. The polymer chain, at times, is branched at every tenth D-galactose residue and the terminal non-reducing units are believed to be 3, 4- of 3, 6-disulphated galactose molecules. The KCl soluble fraction yields soluble and insoluble components on alkali treatment. The alkali soluble polysaccharide, λ-carrageenin is found to be a highly sulphated galactan usually with a low content of 3, 6-anhydro D-galactose. This polymer contains equal amounts of β-1, 3-and α-1, 4-linked galactose units with the 4-linked units 6-sulphated or 2, 6-disulphated. The alkali insoluble fraction is found to be identical with k-carrageenin, considered to be a precursor of the latter and is named μ-carrageenin. The precursor of λ-carrageenin is not yet isolated but designated l-carrageenin.

It is interesting to note that the sporophytes of *Chondrus crispus* contain mainly λ-carrageenin and the gametophytes *k*, and μ-carrageenins.

Carrageenins are used extensively in the manufacture of cosmetics and as an emulsifying agent in medicine.

Gums and mucilages

Gums and mucilages include all the hydrocolloids obtained from plants and are polysaccharides consisting of more than one type of monosaccharide residues. The gums are considered as pathologic products, produced in response to injury by a process known as "gummosis", whereby the cell walls and their ingredients are dissolved to from a colloid, which serves as a protective layer over the wounded tissue and later occur as exudates from the various plant parts especially the trunk. Mucilages, on the other hand, are classified as natural plant products produced by the plant for the imbibition and retention of water. But from a chemical point of view gums and mucilages are almost identical and it is nearly impossible to draw a line demarcating one from the other.

The solutions of gums and mucilages are laevorotatory. On hydrolysis they yield sugars like arabinose, galactose, glucose, mannose and xylose along with various uronic acids and methyl sugars. The sugar acids, when present in appreciable amounts, tend to lower the pH of the solution enabling the gums to occur frequently as salts of sodium, potassium, calcium or magnesium. In some cases, the sugar components are methylated (gum Tragacanth) or acetylated (Karaya gum). The trace amounts of nitrogen (0.08-5.6%), at times encountered in certain samples of gum, are considered to be due to be presence of proteins or sugar amines like glucosamine.

Gums containing linear polysaccharides are found to be less soluble in water, producing very viscid solutions. They tend to precipitate in course of time because of the inter-molecular hydrogen bonding facilitated by the parallel alignment of molecules, they form colloidal gels -**sols**- possessing low surface tension and therefore act as important protective colloids and stabilizing agents. Even though the monomer components of most of the gums are known, the data on detailed structure are wanting in many cases. The great difficulty in isolating the polysaccharide components is the factor responsible for this situation. The final structure, when attained, may prove to have much relevance to the taxonomy of the group concerned.

1.1.2. *Amino acids and Proteins*

Proteins are high molecular weight polymers of amino acids. Amino acids are carboxylic acids containing at least one amino/imino group. There are two types of amino acids met with in nature. The first group consists of protein amino acids which occurs free as well as bound in a polymeric form in the proteins. The second group includes all the other amino acids which occur predominantly in the free form or in a combined form as low molecular weight peptides or related compounds.

Protein aminoacids

Despite the fact that the number of aminoacids reported from proteins is 25, only 16-20 are present frequently and only 22 are coded by DNA. All these aminoacids can be represented by a general formula R CH (NH_2) COOH in which the R-group is the distinguishing feature of every amino acid and many of the properties of aminoacids and many proteins are decided by it. Except glycine in which the R-group is a hydrogen atom, all amino acids contain an asymmetric carbon atom thereby exhibiting optical activity. They exist either in D- or L-form depending upon whether they have a configuration same as that of D- or L- glyceraldehyde. But it is interesting to note that all protein aminoacids belong to L-series.

Aminoacids normally contain C, H, O and N, but sometimes sulphur (Cysteine) or iodine (thyroxine). They contain more than one amino group (lysine) or more than one acidic group (aspartic acid). A few aromatic (phenylalanine) and heterocyclic (histidine) amino acids are also present in proteins.

Aminoacids are colourless crystalline solids soluble in water and possess high melting points with decomposition on heating. (The solubility is reduced with longer aliphatic or aromatic side chains). These properties evidently being very different from those of any organic acid or amine, it is proposed that these compounds exist in the form of an internal salt. This salt, variously known as *Zwitterion, ampholyte* or *dipolar,* is a double charged ion which accounts for the lack of acidic or basic properties of aminoacids. A neutral solution contains all the three forms of an amino acid in equilibrium. In an alkaline solution the conjugate base predominates and in acidic solution the conjugate acid. But at a particular pH the net charge of an amino acid becomes zero, it becomes electrically neutral and the amino acid does not migrate when placed in an electric field. This specific pH is known as the isoelectric point of that amino acid. For monoamino monocarboxylic amino acids (neutral amino acids) this is between 5-6, but for basic amino acids containing more than one amino group like lysine and arginine it is 9-11 whereas for acidic aminoacids containing more carboxyl groups such as glutamic and aspartic acids the isoelectric point is about 3.

Aminoacids are amphoteric forming salts with acids and bases. As mentioned earlier they migrate in an electric field under certain conditions, a property used effectively for their isolation by electrophoresis. All protein aminoacids are α-aminoacids i.e., both the amino-and the carboxyl groups are attached to the first (α) carbon atom of the molecule.

Plants are able to synthesise all of the protein amino acids, a feature not shared by the animals. About 10 aminoacids are there which are produced in insufficient amounts in many animals especially mammals necessitating their supply from an external source. These compounds are named **'essential aminoacids'**. They are:-

M - Methionine
I - Isoleucine
L - Leucine
L - Lysine
P - Phenyl alanine
A - Arginine
T - Tryptophan
H - Histidine

T - Threonine
V - Valine

The protein aminoacids are classified based on the nature of R-groups. The following groups are recognized:

A) Aliphatic amino acids

These compounds containing an aliphatic side chain are neutral in nature. Those aminoacids having a branched R-group are present in lesser concentrations than those of straight chains and therefore are the 'essential amino acids'.

(1) *Glycine* (Amino acetic acid) $H_2N\ CH_2COOH$

The simplest of amino acids, glycine is a major component (upto 25%) of scleroproteins such as collagen. N-Methyl glycine is a major component of arachin and conarachin (the two peanut proteins) and the antibiotic actinomycin. Betaine $(CH_3)_3\ N^+CH_2\text{-}COO^-$ a trimethyl derivative of glycine is a major constituent of *Atriplex* and *Beta* leaves (upto 5% dry weight).

(2) *Alanine* (α-Amino propionic acid)

Also known as methyl glycine, alanine is an important constituent of protein. Here the R-group is a methyl group.

(3) *Valine* (α-Amino isovaleric acid)

Although widely distributed among proteins, its lesser amount in plant proteins makes it an essential amino acid

(4) *Leucine* (α-Amino isocaproic acid)

One of the more abundant acids of plant proteins, leucine is presumed to have some role in hemiterpene synthesis. It is an essential amino acid. It differs from valine in having an additional methylene group.

(5) *Isoleucine* (α-Amino-β-Methyl Valeric acid)

A structural isomer of leucine, isoleucine, though widely distributed, is present only in small amounts in proteins. Thus is an essential amino acid.

(6) *Serine* (α-Amino-β-hydroxy propionic acid, β-amino glyceric acid)

In serine, the methyl group of alanine is converted to an aliphatic alcoholic group. It is present in large amounts in seaweed proteins.

(7) *Threonine* (α-Amino-β-hydroxy butyric acid)

Threonine resembles serine but for a methyl group and so is considered as higher homologue of serine. Another essential amino acid.

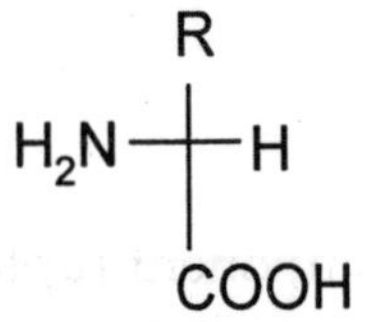

L-amino acid (general formula) 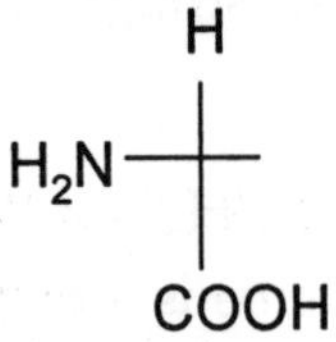 | Glycine 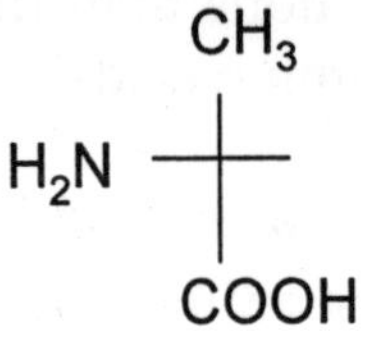 | Alanine 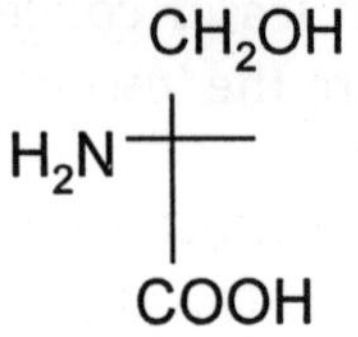 | Serine

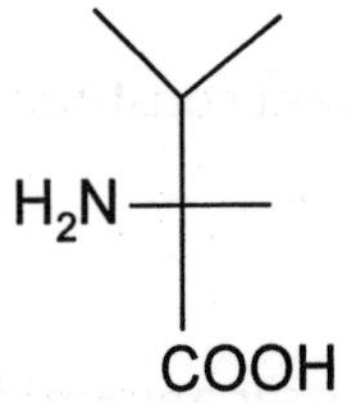

Valine 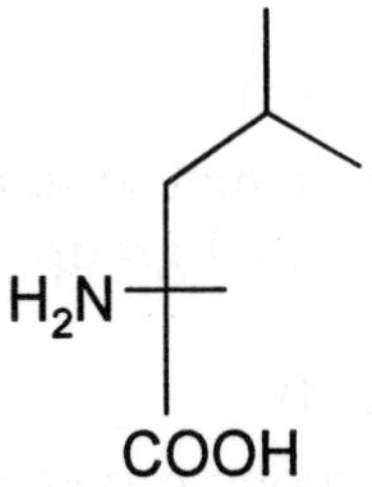 | Leucine 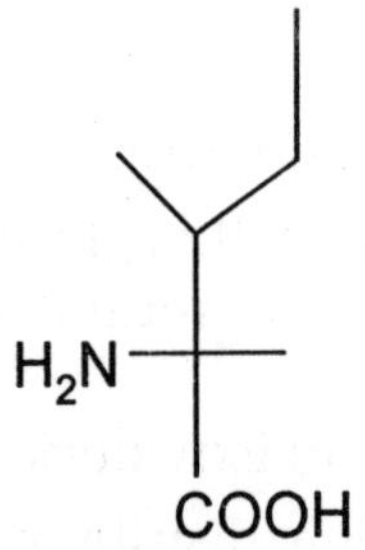 | Isoleucine 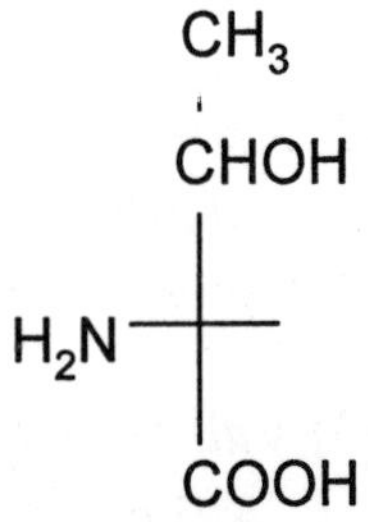 | Threonine

(B) Sulphur containing amino acids

Only three sulphur containing amino acids are known.

1. *Cysteine* (α-Amino-β-mercapto propionic acid)

Cysteine differs from serine in the replacement of the – OH group with an – SH group. Cysteine may be absent from protein hydrolysates because it gets easily oxidized to cystine. Cysteine is related to penicillamine (an antibiotic) and sulphur containing compounds of *Allium*.

2. *Cystine*

Cystine is considered a dimer of cysteine and is formed by the disulphide linkage of two cysteine molecules. This disulphide linkage is important is stabilizing protein structure.

3. *Methionine* (α-Amino -γ-methyl thiobutyric acid)

Another essential amino acid, methionine is a constituent of many proteins and enzymes.

S——S

CH_2SH CH_2 CH_2 S

H_2N H_2N H_2N H_2N

COOH COOH COOH COOH

Cysteine Cystine Methionine

(C) Acidic amino acids

As the name implies they contain more carboxyl groups normally two- and are distinctly acidic, having an isoelectric point about 3. The acidic proteins contain large amounts of these acids. Only two are known in the group and both are important intermediates of carbohydrate and protein metabolism.

1. *Aspartic acid* (α-Amino succinic acid)

Abundant in storage proteins, aspartic acid along with its amide asparagine form the chief soluble nitrogenous reserve in many plants.

2. *Glutamic acid* (α-Amino glutaric acid)

Considered as a higher homologue of aspartic acid, glutamic acid occurs in large amounts (upto 45%) in seed proteins. This amino acid is one of the key intermediates in metabolism. Its amide, glutamine, also is present in many proteins.

COOH $CONH_2$

COOH $CONH_2$

H_2N H_2N H_2N H_2N

COOH COOH COOH COOH

Aspartic acid Asparagine Glutamic acid Glutamine

(D) Basic amino acids

These aminoacids, containing more than one amino group, are distinctly basic in nature and are the major components of basic proteins like histones. Lysine, hydroxylysine, arginine and histidine are the members of this group.

1. *Lysine* (α-ε-Diaminocaproic acid)

This amino acid may reach upto 6% in plant proteins, but sometimes entirely absent from them (*e.g.* Zein). N-Trimethyl lysine is present in cytochrome c of wheat germ.

2. *Hydroxy lysine* (α-ε-diamino-δ-hydroxy caproic acid)

This aminoacid is having a very restricted occurrence having been reported from only animal proteins such as collagen and gelatin.

3. *Arginine* (α-Amino-δ- guanidino-valeric acid)

Arginine is the principal aminoacid of the basic nuclear proteins histones and protamines. It is the strongest base with an isoelectric point of 10.76 and serves as a nitrogen reserve in underground organs and stems. On treatment with alkali arginine gives rise to ornithine (α-δ-diamino valeric acid), an important aminoacid in alkaloid biosynthesis. Citrulline (α-amino-δ-carbamido-valeric acid) is another aminoacid derived from arginine under basic conditions. All these three, together known as "ornithine family", seem to be interconvertable in the plant. Arginine is responsible for the Sakaguchi reaction (On treatment with α-naphthol and sodium hypochlorite, protein gives a red color.) exhibited by proteins.

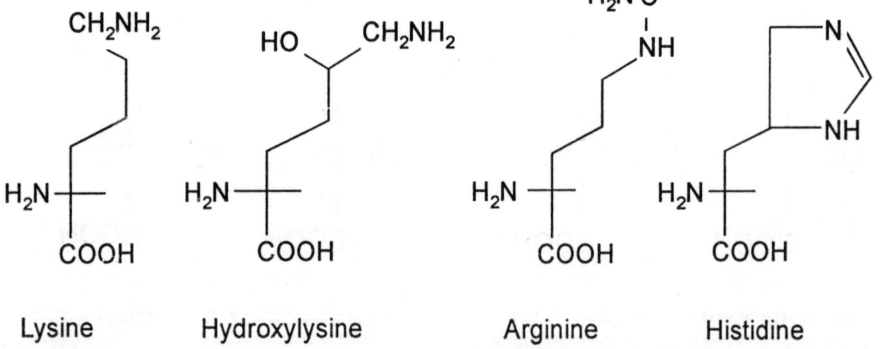

Lysine | Hydroxylysine | Arginine | Histidine

4. *Histidine* (α-Amino-β-imidazolyl-propionic acid)

This amino acid contains an imidazole ring and therefore considered as a heterocylic aminoacid. Histidine is responsible for the Pauly reaction (On treatment with α-naphthol and sodium hypochlorite, protein gives a red color.) Histamine is the decarboxylation product of this aminoacid.

(E) Aromatic amino acids

These acids containing aromatic rings are responsible for the UV absorption exhibited by proteins. Phenylalanine, tyrosine and tryptophan are the members of this group.

1. *Phenyl alanine* (α-Amino-β-phenyl propionic acid)

This aminoacid contains a phenyl group attached to the methyl group (R group) of alanine. This essential aminoacid sometimes amounts to 5% in certain plant proteins.

2. *Tyrosine* (α-Amino-β [*p*-Hydroxyphenyl]-propionic acid; *p*-Hydroxy phenyl alanine)

The phenol ring present in this aminoacid is responsible for the Xanthoprotein (When protein is treated with conc.HNO_3 a white precipitate occurs which on heating and subsequent addition of alkali becomes dark yellow.) or Million reaction (When protein is treated with a mixture of mercurous nitrate, mercuric nitrate and nitric acid, a white precipitate is formed which on heating become red.) characteristic of proteins. Zein of maize contains more than 10% of tyrosine. DOPA (3, 4-dihydroxy phenyl alanine), a widely distributed non-protein amino acid is an oxidation product of tyrosine. Thyroxine (an animal hormone) is an iodinated derivative of tyrosine.

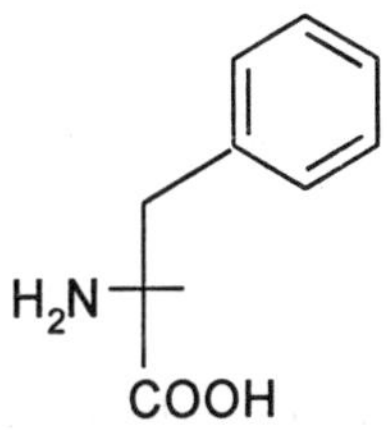

Phenyl alanine 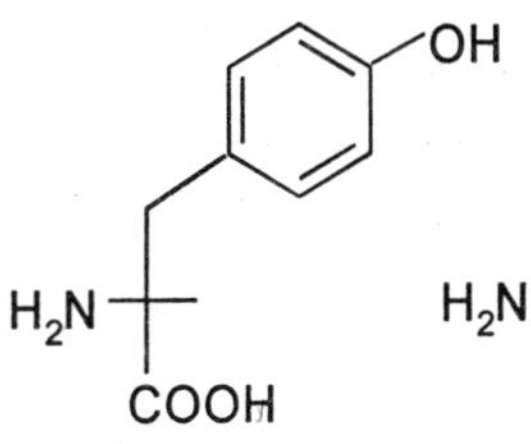 | Tyrosine 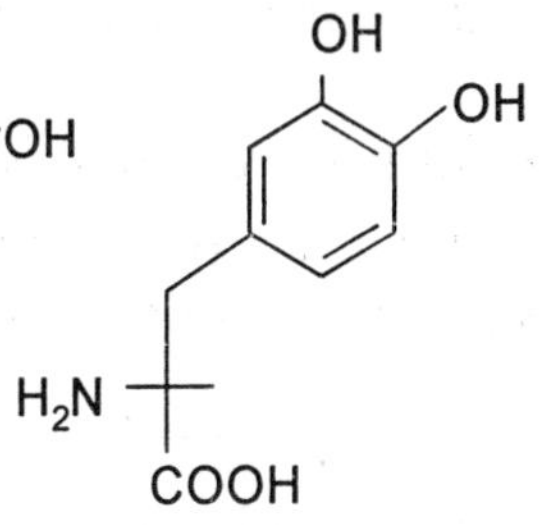 | 3,4-Dihydroxy phenyl alanine (DOPA)

3. *Tryptophan* (α-amino-β-indolyl propionic acid)

It contains an indole ring as a side chain. Despite its presence in many proteins the amount reported is very low. Tryptophan is a precursor for many primary metabolites and indole alkaloids.

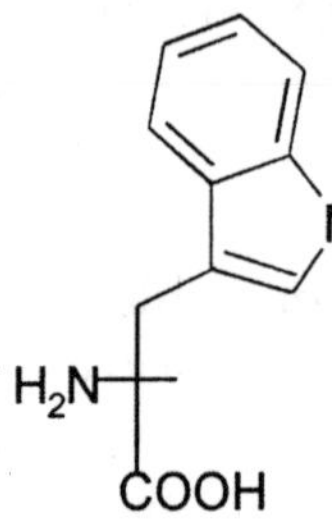

NH
H_2N
COOH
Tryptophan

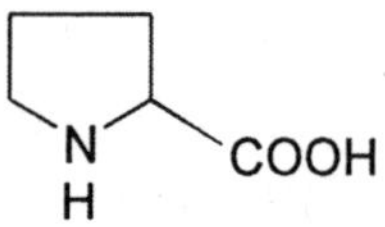

N
H
COOH
Proline

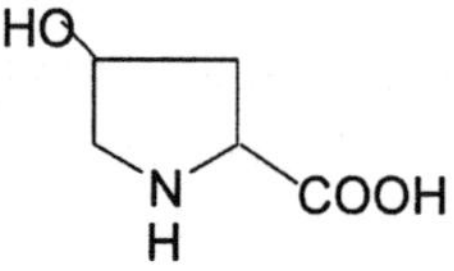

HO
N
H
COOH
Hydroxyproline

(F) Heterocyclic amino acids

These aminoacids are characterized by an imino group rather than an amino group. Proline and hydroxy proline are included in this group.

1. *Proline* (Pyrrolidine-2-carboxylic acid).

A widely distributed aminoacid, proline occurs in large proportions in proteins like Zein, gliadin etc. It is soluble in alcohol.

2. *Hydroxyproline*

This 4-hydroxylated derivative of proline is of limited distribution.

Properties of amino acids

Aminoacids act as electrolytes and the reactions which they exhibit may be due to either the basic or acidic group.

Reactions due to the carboxyl group

1. Aminoacids form salts with cations (Na^+, Ca^{++}) which are sparingly soluble in water.

   $$R\ COOH + NaOH \longrightarrow R\ COONa + H_2O$$

2. On dry distillation or on heating with barium oxide aminoacids get decarboxylated and form amines.

   $$RCH\ (NH_2)\ COOH \longrightarrow RCH_2NH_2 + CO_2$$

3. The carboxyl group is esterified on treatment with an alcohol in presence of acid catalysts. The methyl or ethyl esters, formed by reacting with methyl or ethyl alcohols, being volatile, are extensively used in the gas chromatographic separation of amino acids.

$$\text{H-}\underset{\text{R}}{\overset{\text{NH}_2}{\text{C}}}\text{-COOH} + CH_3OH \longrightarrow \text{H-}\underset{\text{R}}{\overset{\text{NH}_2}{\text{C}}}\text{-COOCH}_3 + H_2O$$

4. When a carboxyl group of one aminoacid reacts with the amino group of another aminoacid a "peptide linkage" is formed with the elimination of a water molecule and the resulting product is a peptide. This peptide bond is the basis of protein structure.

$$H_2\text{N-}\underset{\text{R}}{\overset{\text{H}}{\text{C}}}\text{-COOH} + H_2\text{N-}\underset{\text{H}}{\overset{\text{R}}{\text{C}}}\text{-COOH} \longrightarrow H_2\text{N-}\underset{\text{R}}{\overset{\text{H}}{\text{C}}}\text{-}\overset{\text{O}}{\overset{\|}{\text{C}}}\text{-}\underset{\text{H}}{\text{N}}\text{-}\underset{\text{H}}{\overset{\text{R}}{\text{C}}}\text{-COOH} + H_2O$$

Peptide bond

Dipeptide

Reactions due to amino group

1. The amino group can react with inorganic acids to form salts which are sparingly soluble in water.
2. Amino group can be acetylated with acetyl chloride or acetic anhydride in an alkaline solution. These acetyl derivatives are found to be acidic in nature.
3. When heated with HI at 200°C the amino group of an aminoacid is eliminated with the formation of a fatty acid.

$$RCH(NH_2)COOH \xrightarrow[200^\circ C]{HI} RCH_2COOH + NH_3$$

4. Nitrous acid reacts with the amino group and liberates gaseous nitrogen quantitatively. This N_2 may be measured manometrically and can be used in the estimation of α-amino groups in aminoacids as well as in proteins.

5. 2, 4- Dinitrofluorobenzene reacts with the amino groups and produces intensely colored derivatives. This reaction is commonly used for isolating the N-terminal amino acid of a polypeptide or a protein. Since 2, 4-DNF reacts only with the free amino groups (of the terminal aminoacids) the protein is first treated with 2,4-DNF followed by hydrolysis and chromatography, the terminal amino acid will form a colored derivative which could be identified. Thus by careful controlled hydrolysis and repeated treatment with this reagent all the amino acids of a polypeptide chain may be identified.

$$H_2N\text{-}CH(R)\text{-}COOH + F\text{-}C_6H_3(NO_2)_2 \longrightarrow O_2N\text{-}C_6H_3(NO_2)\text{-}NHCH(R)COOH + HF$$

Colored derivative

6. The reaction with ninhydrin is very characteristic of aminoacids. All the aminoacids except proline and hydroxyproline produce a purple / violet coloured derivative with ninhydrin (Proline and hydroxyproline give yellow products). Firstly one molecule of ninhydrin (tri-keto hydrindene hydrate) reacts with the amino group to produce a reduced ninhydrin molecule and α-imino acid which in turn is hydrolysed to form an α-keto acid with the liberation of ammonia. (This α-keto acid is then decarboxylated to produce an aldehyde and CO_2). The ammonia molecule reacts with equimolar quantities of oxidized and reduced ninhydrin yielding a violet product which is quantitatively related to the aminoacid present.

Non-Protein aminoacids

The number of non-protein aminoacids detected reaches well above 700. Some of these compounds are widely distributed, while others are of restricted distribution. The structural variations found in this group include ethylenic and acetylenic linkages, cyclopropane, azetidine, and pyrimidine and pyrazole ring systems. Many homologous members of protein aminoacids are also seen in this group. The relatively low concentration of these compounds in plants delayed their discovery. But a few like lathyrine are present in such high concentrations to become toxic to animals. Though they are often present in the leaves, which are the principal metabolic centers, they tend to accumulate in seeds and fleshy plant parts where they act as nitrogen reserves. In seeds they may reach a concentration of 8% dry wt. and up to 50% of nitrogen present. Non-protein aminoacids occur free or as low molecular weight peptides. The lack of uniformity and the great variations they exhibit makes it a difficult proposition to classify them. The grouping done by Robinson (1975) is followed here, in which seven broad categories are recognized.

1. *D-Aminoacids*

The common D-aminoacids are enantiomers of the protein amino acids. D-Alanine occurs as a peptide component of *Staphylococcus* and as N-malonyl derivative in higher plants. D-Tryptophan is isolated as N-γ-L-glutamyl derivative. D-Lysine is reported from *Nicotiana*. The D-aminoacids may act as inhibitors of plant metabolism due to their antimetabolite (analogue) actions in place of protein aminoacids. D-Leucine, D-phenylalanine and D-glutamic acids are found to be the components of antibiotics while D-proline is present in ergot alkaloids.

2. *Non-α-Aminoacids*

These compounds have the amino group in any other carbon other than the α-carbon. γ-Aminobutyric acid ($H_2N\text{-}CH_2CH_2CH_2\text{-}COOH$) is the most widely distributed aminoacid in this group. Its homologue β-alanine also occurs free in plants but is more frequent in bound form in animal peptides. α-Methylene-γ-amino butyric acid, β-amino isobutyric acid and γ-amino α-hydroxy butyric acid are some members of this group widely distributed in plants.

3. *Protein aminoacid derivatives*

DOPA (dihydroxy phenyl alanine) is the most common member in this group, often present in high concentrations to be toxic to animals. Canavanine (α-amino-γ-hydroxy guanidino butyric acid) - similar to arginine - known only from legume seeds, is cleaved by arginase to yield another aminoacid canaline (α-amino-γ-O hydroxyl-amino-butyric acid) and urea. Canavanine, α, γ-diaminobutyric acid and homoarginine (α-amino-ε-guanido-caproic acid) form the major constituents of dry legume seeds wherein they may function as a form of storage nitrogen. γ-Methylene glutamic acid and γ-methylene glutamine – two common aminoacids of the Liliaceae- appear only after the seed germination in peanut. The later compound forms 95% total nitrogen in the exuding saps, clearly indicating its function in nitrogen transport. N-N-Dimethyl tryptophan is a known plant inhibitor. Some of the remaining members of this group are 5-hydroxy tryptophan, *m*-carboxy glycine and *m*-carboxyphenyl alanine.

4. *New heterocylic and alicyclic aminoacids*

The homologues of proline, azetidine-2-carboxylic acid and pipecolic acid are reported from many plants. The former, an aminoacid widespread in the Liliaceae, amounts to 50% of total N_2 present in the rhizomes of *Polygonatum* sp. while the latter is common in plants belonging to the Leguminoseae and Liliaceae. Azetidine-2-carboxylic acid is a growth inhibitor to non-liliaceous plants.

The two known cyclopropane aminoacids are α- (2 carboxcycloprop-l-yl) glycine (isomeric with γ-methylene glutamic acid) and hypoglycemic β-(methylene cyclopropyl) alanine, (hypoglycin A) both reported from Sapindaceous members like *Blighia* and *Litchi.*

Mimosine (β-(1, 4-dihydro-3-hydroxy-4-oxopyrid-l-yl)-L-alanine, another heterocyclic aminoacid causes hair loss in cattle and acts as a growth inhibitor for other plants. Lathyrine [β-(2, aminopyrimidin-4-yl) alanine]-a pyrimidine derivative, has toxic properties and causes 'lathyrism' in man. Both these compounds are thus protective in function in the plant.

5. *Dicarboxylic acids and amides*

α-Amino adipic acid (HOOC $(CH_2)_3$ CH (NH_2) COOH) and α-amino pimelic acid are widespread in higher plants. The former one is believed to be present in the seed protein of Maize.

6. *Sulphur containing aminoacids*

These compounds are mostly S-substituted Cysteines commonly met with in members of the Liliaceae, Cruciferae and Leguminoseae, S-Methyl-cysteine is reported from *Astragalus* and *Phaseolus* and S-methyl-cysteine sulphoxide (CH_3C SO CH_2 $CH(NH_2)$ COOH) from cabbage and turnip. Alliin – the pungent principle of garlic – is S-allyl cysteine sulphoxide and cycloallin reported from onions is its cyclized form. Djenkolic acid present in Djenkol beans and *Pithecellobium dulce* is an S-alkyl derivative of cysteine. All these compounds elicit insecticidal and fungicidal properties.

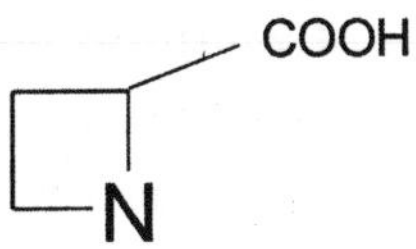

Azetidine-2-carboxylic acid

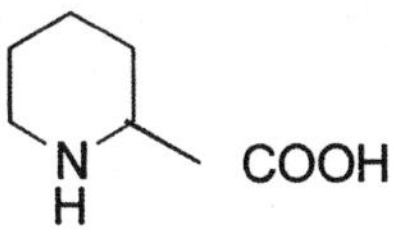

Pipecolic acid

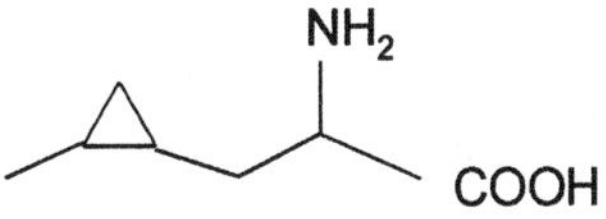

Hypoglycin A

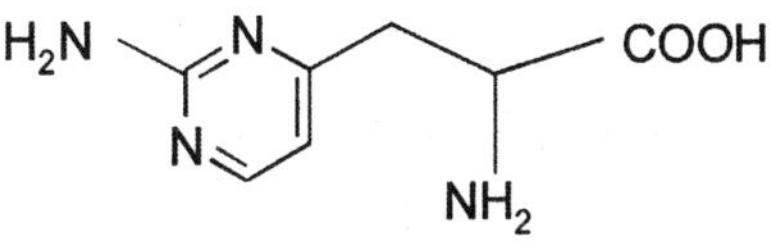

Lathyrine

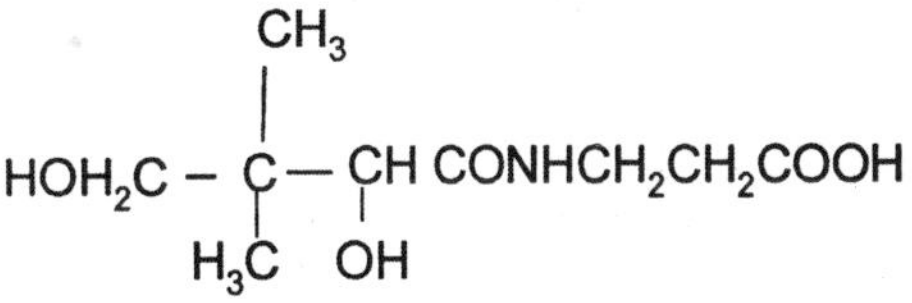

$HOH_2C-C(CH_3)(H_3C)-CH(OH)\ CONHCH_2CH_2COOH$

Pantothenic acid

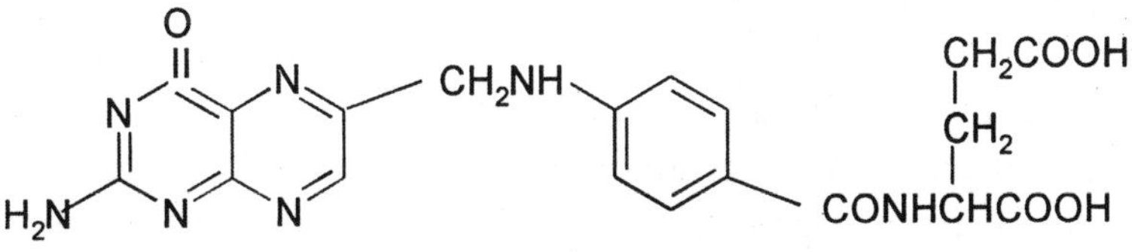

Folic acid (Pteroyl-L-glutamic acid)

7. Hydroxy aminoacids

These compounds are characterized by additional hydroxyl groups to the otherwise common compounds. γ-Hydroxy glutamic acid discovered from *Phlox* and many members of the Liliaceae and β-γ-dihydroxy glutamic acid from seeds of lettuce and leaves of rhubarb are the common examples of this group. β-Hydroxy aspartic acid is a microbial metabolite. Homoserine (γ-hydroxy-α-amino butyric acid) is believed to act as an intermediate in formation of methionine and threonine from aspartic acid. γ-Hydroxy ornithine ($H_2N-CH_2-CHOH-CH_2-CH (NH_2) COOH$) and γ-hydroxy arginine are reported from *Vicia* while 4-and 5-hydroxy pipecolic acids are reported from *Acacia* and *Albizzia*.

β-Cyanoalanine, reported from *Vicia*, is the only aminoacid containing a cyanogroup and is presumably toxic to higher organisms.

Peptides

The term peptide denotes all condensation polymers of aminoacids with a molecular weight less than 6,000. The number of peptides reported from plants is much less. These compounds were normally overlooked in a systematic survey because of the minute quantities in which they were located or due to the lack of understanding about the function of these compounds in nature. There are peptides which contain only amino acids while some others consist of a non-protein component in addition to aminoacids. The former group may be called simple peptides or homomeric peptides and the second one as conjugated peptides or heteromeric peptides. Non-protein amino acids are frequent in these compounds.

Simple peptides (homomeric) may be di-, tri-, tetra or polypeptides. The simplest group (of course the largest also) is of dipeptides. The dipeptide may be a homodipeptide in which the same aminoacid is repeated (glycyl glycine) or a hetero dipeptide in which two different aminoacids are involved (γ-L-Glutamyl-β-alanine). γ-L-Glutamic acid is a frequent partner in this linkage; *e.g.*, all the 18 dipeptides listed by Gibbs (1973) from higher plants invariably contain this amino acid. Some of the known dipeptides are γ-L-glutamyl γ-aminobutyric acid (*Lunaria* seed), γ-L-glutamyl-β-amino isobutyric acid (*Iris* bulbs) and γ-L-glutamyl β-cyano-L-alanine (a toxin from *Vicia sativa*). In addition, a series of dipeptides are found in nature formed between γ-L-glutamic acid on one hand and L-isoleucine, L-leucine, L-methionine (*Phaseolus*) and L-tyrosine (*Glycine max*) on the other hand. γ-L-Glutamyl-5-methyl-L-cysteine is a common constituent of beans like *Phaseolus vulgaris* and *Vigna sinensis*. Many sulphur containing dipeptides are characteristic of the genus *Allium* (e.g. γ-L-Glutamyl-S-allyl-L-cysteine and γ-L-glutamyl-3-(prop-l-enyl) L-cysteine).

Glutathione (γ-L-glutamyl-cysteinyl-glycine), the first tripeptide isolated from a lower organism (Yeast), is found to be widely distributed among higher plants. This peptide seems to act as a prosthetic group of some enzymes. Homoglutathione (γ-L-glutamyl-L-cysteinyl-ß-alanine) is reported from the seeds of *Phaseolus aureus,* while γ-L-glutamyl-S (2-carboxy-l-propyl)-cysteinyl glycine is found to occur in onion bulbs.

Some of the tripeptides elaborated by lower plants function as antibiotics (e.g. Polymyxins) or as poisons to higher organisms. Phalloidine is one such toxic peptide, occurring in *Amanita phalloides* which on hydrolysis yields cysteine, alanine and allo hydroxyl-L-proline.

Our knowledge of higher peptides is very scanty. Actinomycins and gramicidins are cyclic decapeptides synthesized by lower plants which exhibit antibacterial properties. The toxins of mistletoes (*Viscum* and *Phoradendron*) are found to be of peptide nature. The alkaloidal peptides reported from some members of the Rhamnaceae, Rubiaceae and Urticaceae are found to be true peptides. Canthiumine is one such tetrapeptide isolated from *Canthium* (Rubiaceae) consisting of N, N-dimethyl phenyl alanine, *p*-hydroxy styryl amine, L-proline and L-threo ß-phenyl serine, while scutianine, a hexapeptide obtained from *Scutia* (Rhamnaceae) contains N-N-dimethyl phenyl alanine, phenyl alanine, ß-hydroxy leucine, proline, leucine and glycine.

Heteromeric or conjugated peptides contain diverse non-aminoacid components. Pantothenic acid of coenzyme A molecule consists of pantoic acid and ß-alanine. In folic acid (a member of vit. B. complex), pteroic acid occurs in peptide linkage with L-glutamic acid. Both these compounds are peptide linkage with higher plants. γ–L-Glutamyl-ß-aminopropionitrile, otherwise known as "Lathyrus factor", occurring in certain legumes causes 'Lathyrism', due to the neurotoxic effects it exhibits.

The peptide alkaloids of ergot contain D-lysergic acid combined with simple peptides. Zizyphine and related alkaloids reported from *Zizyphus* represent another group of conjugated peptides in which the non-aminoacid component is pyrocoll.

Proteins

Proteins are linear polymer chains of aminoacids having a mol. wt of 6,000 to millions. These polypeptides may remain as linear chains or coiled to form spherical to oval structures. In linear polymers the stability is achieved by the development of intermolecular linkages developed between different chains. The globular nature seen in many proteins is the result of innumerable folding/coiling effected by the formation of various intra-molecular linkages. As these compounds contribute to the structure and functions of the living cell, they occur as independent units as well as in combination with lipids, nucleic acids, carbohydrates and many other compounds. The first products of hydrolysis of a protein are simple peptides which on further hydrolysis yield the amino acids. The elementary composition of protein definitely varies with the source but most proteins contain 46-55 % carbon, 6-9 % hydrogen, 12-13 % oxygen. 10-32% nitrogen and 0.2-0.3% sulphur. A number of proteins yield about 16% of nitrogen gas on oxidation and this is conveniently made use of for the estimation of proteins in Kjeldahl's method.

Similar to aminoacids, proteins are amphoteric substances with specific isoelectric points. This amphoteric nature arises due to the free acidic and basic groups present in the R-groups of aminoacid units present. The isoelectric point, in which both the positive and negative charges are equally balanced, is decided largely by the acidic or basic aminoacids present in a particular protein. At this pH, the protein is least soluble and is easily precipitated. Any agent like heat, acid, salts or the presence of other immiscible solvents also causes this precipitation. Such a precipitated protein is said to be denatured. Denaturation is an irreversible process and is believed to be due to the changes in the conformation (unfolding of the coiled molecule) of the protein. The denatured protein loses its biological activity completely. Many of its physical properties also are altered at this state. A few proteins refold or renature under specific circumstances by a process known as annealing.

Proteins exhibit optical activity too.

Structure of proteins

Although proteins are simple polypeptides formed by the condensation of a number of repeating α-aminoacid units, the large number of these monomers and the innumerable cross-links formed between separate chains and between different parts of the same chain make it a highly complicated macromolecule. For a better understanding of this molecule, 4 structural levels are proposed.

(a) Primary structure

The primary structure is the proper sequence of aminoacid residues in a long linear chain of a polypeptide. One end bearing the free amino group is named N-terminal (and N-terminal aminoacid) and the other end with a free carboxyl group is C-terminal. In general methods of representation the N-terminal aminoacid is written to the left. R-groups forming the side chains alternate on each side of the chain. The aminoacids in a chain may be represented by symbols consisting of the first three letters of the name with arrows indicating the direction of linkage from carboxyl to amino group (Asparagine and glutamine are represented by Asn and Gln).

(b) Secondary Structure

Secondary structure denotes the nature of conformation the polypeptide backbone assumes. The linear chain of aminoacid residues coils and forms an α-helix due to the intramolecular hydrogen bonds developed between the carbonyl (CO) group of an amino acid and the imino (-NH) group of the fourth aminoacid residue. Although the hydrogen bond is weak in nature, a multitude of these bonds reinforce one another and produce a stable and rigid structure. (Proline is incapable of hydrogen bond formation and helix becomes interrupted in places where this aminoacid is present). The α-helix normally is right handed and it is calculated that 3.6 aminoacid residues are present for every turn of the helix.

In skeletal proteins, where the coiling is definitely disadvantageous, a pleated sheet arrangement or β-conformation is favored. In this the polypeptide chain is in the extended (straight) form and the stability is achieved by intermolecular hydrogen bonds formed between two or more chains. Two types of pleated sheets are envisaged (1) Parallel, in which the two chains run parallely in the same direction (e.g. Keratin) and (2) antiparallel, in which the chains run alternately in opposite directions (e.g. fibroin).

(c) Tertiary structure

Tertiary structure denotes the folding and extensive coiling of the entire molecule to produce a more stable arrangement. This arrangement imparts the rigidity, complexity and the characteristic shape to the proteins. The folding is done by the formation of a wide variety of bonds formed among the R-groups present in the aminoacid residues. The various linkages formed are salt bridges, hydrogen bonding, hydrophobic interactions, dipole-dipole interactions and disulphide linkages.

Salt bridges (ionic bonds or electrostatic forces) are developed between the polar groups. The amino groups of basic aminoacids like lysine and arginine are attracted towards the carboxylic groups of the acidic aminoacids, like aspartic and glutamic acids. *Hydrogen bonds* may develop between the hydroxyl groups of tyrosine residues and the carboxyl groups of acidic aminoacids. *The hydrophobic interaction* is seen between the alkyl side chains and aromatic side chains. These groups, being hydrophobic, tend to remain together away from the hydrogen bonding water molecules in the medium (cell). Thus, two benzyl groups of phenylalanine residues or the alkyl side chains of neutral (aliphatic) aminoacids meet and remain together due to the mutual repulsion of the solvent and create "hydrophobic bonds". The hydrophobic groups tend to remain in the inner side of the coiled protein molecule while the polar groups are directed towards outside. Due to *Dipole-dipole* interactions two permanent dipoles like - CH_2OH group of serine are attracted towards each other. In *Disulphide* linkages - CH_2SH side chains of two cysteine residues interact to form one cystine residue (Disulphide linkage in primary structure is the reason for cyclic nature of some peptides).

.The tertiary structure is sometimes known as the *"native conformation"* of the protein. Of the two common molecular shapes i.e. fibrous and globular proteins, the former contains many helices having a little or no folding resulting in a rod like structure while the letter contains lesser helical sections facilitating more folding-mostly at random and assumes a spherical shape.

(d) Quaternary structure of proteins

Quaternary structure depends on the degree of polymerization a polypeptide chain undergoes to achieve more stability. Each chain of polypeptide in a protein is known as a protein subunit or protomer. The enzyme phosphorylase *a* contains four identical subunits (each one individually inactive as to the enzymic action concerned) and such a protein is designated as a homogenous quaternary structure. But in the case of haemoglobin there are two identical α-chains and two identical β-chains, a case of heterogeneous quaternary structure. The nature of many enzymes is ably explained by this arrangement, *i.e.* the quaternary structure is the active enzyme and at the time of inactivity the protomers remain separate thereby losing the power of enzyme activity.

Classification of proteins

Proteins are omnipresent in the living organism, each cell containing several hundred to several thousands of enzymic proteins and millions/billions of structural proteins (Half of the dry matter of protoplasm is protein). Many of the proteins are of universal occurrence while some like the enzyme myrosinase (which hydrolyses glucosinolates to mustard oils) are of limited distribution being specific to a few plant groups. The same type of protein may have different aminoacid sequences in different plant groups *e.g.*, the aminoacid sequence of Cytochrome *c* of monocots is different from that of dicots in at least 10 amino acid residues.

Proteins are classified differently. They may be grouped on the basis of function as enzymic or non-enzymic, the latter including storage as well as structural proteins. In a second method of classification, the criterion used is the molecular weight of individual proteins, in which arbitrary groups of varying mol. wt. are formulated. Thus cytochrome *c* has a mol. wt. of 12,000, alcohol dehydrogenase-60,000 and urease 500,000. In another system of classification the molecular shapes are considered and two groups -fibrous and globular proteins- are recognized. In fibrous proteins elongated α-helical forms are predominant which are joined laterally by various cross links to form a stable but insoluble structure. The globular proteins, on the other hand, have a much-folded polypeptide chain producing an elliptical or a spherical shape. Most of the structural proteins are of the former type and the enzymes and storage proteins, the latter type.

The most widely accepted system of classification is based on the structural components of proteins. In this, the proteins are divided into simple and conjugated proteins. The simple proteins are composed of aminoacids only while the conjugated proteins contain a non-protein component also in addition to the aminoacids. A third group also is sometimes recognized as derived proteins.

(A) Simple proteins

These proteins yield the protein aminoacids or their derivatives on hydrolysis. Further classification of these macromolecules is based on their solubility characteristics. They are:

1. *Albumins*

Albumins are the simplest of proteins. They are soluble in water, dil. acid, dil. alkali or alcohol and contain a little or no glycine *e.g.* egg albumin.

2. *Globulins*

These proteins are insoluble in water but soluble in dilute acid/ alkali/ salt solutions *e.g.* vegetable globulins.

3. *Prolamines*

Insoluble in water or salt solutions, prolamines are dissolved in dil. acid/alkali/70-90% alcohol *e.g.* Cereal proteins such as zein (Maize), gliadin (Wheat) and hordein (Barley).

4. *Glutelins*

Glutelins are soluble in dil. acid or dil. alkali and are rich in arginine, proline and glutamic acid *e.g.*, glutenin (Wheat) and oryzenine (Rice).

5. *Basic proteins*

The proteins included here are strongly basic due to the abundance of lysine and arginine. They are not coagulated by heat. Two types of basic proteins are recognized. (1) *Histones* are soluble in water and dil. acids but not in dil. ammonia and are rich in histidine and arginine.

Histones contain sulphur containing amino acids cysteine and methionine in traces only and tryptophan is entirely lacking. They are hydrolysed by the enzyme pepsin; (2) *Protamines* are more basic than histones, soluble in water, dil. acid and ammonia. Sulphur is absent in protamines. They are stable towards pepsin.

(B) Conjugated proteins

Conjugated proteins consist of simple proteins in combination with a non-protein counterpart known as 'prosthetic' group. The various members in this category are differentiated by the nature of the prosthetic group.

1. *Nucleoproteins*

Here the prosthetic group is a nucleic acid *e.g.*, DNA, RNA.

2. *Chromoproteins*

A colored chromophore forms the prosthetic group of these proteins. Most of the chromophores contain metals like iron and magnesium *e.g.*, Phytochrome, cytochrome, haemoglobin etc.

3. *Glycoproteins* (Mucoproteins)

Here the proteins occur in combination with carbohydrates *e.g.*, Barley albumin, lectins etc. A number of glycoproteins are found to form networks with the carbohydrates of the plant cell wall. These structural proteins are hydroxy proline- rich glycoproteins, proline-rich glycoproteins and glycine-rich glycoproteins. They are rod-shaped or β-pleated sheets and are placed at the interface between the cell wall and the plasma membrane.

4. *Lipoproteins*

They have the lipids as the prosthetic groups *e.g.*, proteins in the chloroplast.

5. *Matalloproteins*

They contain metals like copper and zinc. *e.g.*, Nitrate reductase.

Derived proteins are the degradation products of proteins formed by the action of acid/alkali/enzymes. They are denatured proteins, proteoses, polypeptides or simple peptides. They seldom occur in nature.

Plant storage proteins

Plant proteins generally contain lesser sulphur aminoacids and lysine. The storage proteins are essentially simple proteins. Cereals contain all the types of simple proteins except the basic ones. The gluten which forms 70-80% of storage proteins in barley, corn and wheat is a combined aggregate of prolamines and glutelins in the ratio of 1:1. The remainder of seed proteins in these grains consists of 50% albumins and globulins. In rice, the major group of protein is glutelins (50%) and the rest consists of equal quantities of albumins, globulins and prolamines. But in oats the major protein is globulin. All these storage proteins are normally composed of many similar or dissimilar component fractions. *E.g.*, Gliadin- the prolamine - of wheat is composed of at least eight components while in rye gliadin 5 fractions are present.

Legume proteins are large molecules containing uncommon aminoacids at times. Peanut protein arachin contains sarcosine; mung bean cytochrome *c* contains ε-trimethyl lysine and hydroxy proline in bulk protein of various pulses. Most of the proteins in pulses are globular proteins and vary from 12,000 to several million in molecular weight. Jack bean urease, one of the bigger proteins, has a molecular weight of several millions and is a polymer of many monomeric polypeptides. A single polypeptide may contain upto 700 aminoacid residues.

The storage proteins of pulses occur as aggregates, aleurone grains or protein bodies. All these proteins are characterised by large amounts of aspartic acid, glutamic acid, leucine, basic aminoacids and amides. The abundance of basic aminoacids and amides clearly indicates the function of these proteins as nitrogen reserves. Vicilin and legumin are two common storage proteins of *Pisum sativum* and *Vicia faba* with a mol. weight of 186,000 and 331,000 respectively. The majority of storage proteins (excepting vicilin and legumin) contain two distinct major components with sedimentation coefficients 7s and 11s resp. These compounds show reversible dissociation/association. Some proteins have a smaller component of 2s along with a larger component of 15-18s. The major storage proteins of *Arachis* and *Glycine* are named as arachin and glycinin respectively.

Albumins and globulins always predominate over prolamines and glutelins in legume seeds. Although the ratio of globulin to albumin varies in different seeds, globulins are the major proteins. In *Vicia faba*,

80% of the total protein consists of globulins and albumins while glutelins form 15% and prolamines less than 5%. This ratio is closely followed in majority of pulses.

Lectins

Lectins/Phytohemagglutinins - a group of glycoproteins which agglutinates erythrocytes, are present in some seeds of economic importance. Abrin (in *Abrus*), crotin (in *Croton tiglium*), ricin (in *Ricinus*) and soyin (in *Glycine max*) are some of the important members of this group. Structurally many of these proteins are globulins. Similar toxic proteins of low toxicity are reported from species of *Phaseolus, Ervum, Pisum* and *Vicia*.

1.1.3. Fats, oils and waxes

Fats, oils and waxes, collectively known as triacyl glycerols or triacyl lipids, form the saponifiable fraction of the total lipids of the cell. **Lipids** is a general term to include all the compounds which are insoluble in water and soluble in organic solvents like petroleum ether, benzene or chloroform. Lipids contain two types of compounds; 1) **saponifiable lipids** and 2) **non-saponifiable lipids**. Saponifiable lipids are compounds which produce soaps on treatment with sodium/ potassium hydroxide (see saponification which is explained later) and are esters of fatty acids with monohydric fatty alcohols or the trihydric alcohol, glycerol. The esters of monohydric fatty alcohols are waxes, cutins and suberins while the esters of glycerol are the fats and oils. The latter group was earlier known as triglycerides but now by a technically correct term triacyl glycerols. The non-saponifiable lipids are the terpenoids which are explained somewhere else.

There exists very little difference between fats and oils but for their physical state. Both are triglycerides of fatty acids, but fats are solids at ordinary temperature while oils are liquids. These esters act as reserve food materials which are commonly stored either in the fruit walls or in the seeds. The preference for fats and oils as reserve foods of plants is better explained by the fact that, being hydrophobic, no osmotic problems are involved in their storage and that they are readily converted to acetyl CoA and incorporated directly into glyoxalate and citric acid cycles. The waxes resemble fats in physical properties but are esters of monohydric aliphatic long chain alcohols and fatty acids. Waxes along with the related cutins and suberins have a protective function in the plant, occurring on the surface layers of leaves, fruits, stems etc. The properties of all these esters are generally decided by the fatty acid components they contain.

Fatty acids

Much of the physical and chemical properties of the saponifiable lipids are decided by their fatty acid components. These fatty acids are aliphatic/alicyclic, mono/poly-carboxylic acids and normally insoluble in water (The lower members like butyric and caproic acids are slightly soluble in water).

Table 1. Saturated straight chain fatty acids

No. of C. atoms	Common name	IUPAC name	Formula	M.P. -°C
4	Butyric	n-Tetranoic (Butanoic)	$CH_3(CH_2)_2$ COOH	-4.7
6	*n*-Caproic	Hexanoic	$CH_3(CH_2)_4$ COOH	-1.5
8	*n*-Caprylic	Octanoic	$CH_3(CH_2)_6$ COOH	+16.5
10	*n*-Capric	Decanoic	$CH_3(CH_2)_8$ COOH	+31.3
12	Lauric	Dodecanoic	$CH_3(CH_2)_{10}$ COOH	+43.6
14	Myristic	Tetradecanoic	$CH_3(CH_2)_{12}$ COOH	+58.0
16	Palmitic	Hexadecanoic	$CH_3(CH2)_{14}$ COOH	+62.9
18	Stearic	Octadecanoic	$CH_3(CH_2)_{16}$ COOH	+69.9
20	Arachidic	Eicosanoic	$CH_3(CH_2)_{18}$ COOH	+75.2
22	Behenic	Docosanoic	$CH_3(CH_2)_{20}$ COOH	+80.2
24	Lignoceric	Tetrascosanoic	$CH_3(CH_2)_{22}$ COOH	+84.2

TABLE 2. Unsaturated (olefenic) fatty acids

No. of C. atoms	Common name	IUPAC name	Formula
4	Crotonic	But-2-enoic	$CH_3CH=CH\ COOH$
16	Palmitoleic	*cis*-9-Hexadecenoic	$CH_3\ (CH_2)_5\ CH = CH\ (CH_2)_7\ COOH$
18	Petroselinic	*cis*-6-Octadecanoic	$CH_3\ (CH_2)_{10}\ CH = CH\ (CH_2)_4\ COOH$
18	Oleic	*cis*-9- Octadecanoic	$CH_3\ (CH_2)_7\ CH = CH\ (CH_2)_7\ COOH$
18	Vaccenic	*cis*-11- Octadecanoic	$CH_3\ (CH_2)_5\ CH = CH\ (CH_2)_9\ COOH$
18	Linoleic	*cis-cis*-9, 12- Octadecanoic	$CH_3\ (CH_2)_4\ CH = CH\ CH_2\ CH=CH\ (CH_2)_7\ COOH$
18	Linolenic	*cis, cis, cis*-9, 12, 15-Octadecatrienoic	$CH_3\ CH_2\ CH=CH\ CH_2CH = CHCH_2CH=CH(CH_2)_7\ COOH$
18	Elaeostearic	Octadeca-*cis*-9, *trans*-11, *trans*-13-trienoic acid	$CH_3\ (CH_2)_3\ (CH=CH)_3\ (CH_2)_7\ COOH$
18	γ-Linolenic	*Cis, cis, cis*-6, 9, 12-Octadecatrienoic	$CH_3\ (CH_2)_3\ (CH_2CH=CH)_3\ (CH_2)_4\ COOH$
20	Arachidonic	*Cis, cis, cis, cis*-5, 8, 11, 14-Eicosatetra-enoic acid	$CH_3\ (CH_2)_3\ (CH_2CH=CH)_4\ (CH_2)_3\ COOH$
22	Erucic	*Cis*-13-Docosenoic	$CH_3\ (CH_2)_7\ CH=CH\ (CH_2)_{11}\ COOH$

The fatty acids, which are more abundant in nature, almost always possess a straight chain of even number carbon atoms. The odd carbon counterparts (mostly C_7-C_{15}) have also been located in plant lipids, free or as esters. The amounts of such compounds are often too small to be located in a traditional survey.

The fatty acids present in various ester combinations can be classified as follows:

(1) *Saturated straight chain fatty acids*

All the important saturated fatty acids present in fats are presented in Table I. The lower members are liquids at ordinary temperature while the higher members are solids. The melting point also increases with the addition of more carbon atoms and these saturated members are always denoted by the suffix-*anoic*. The fats contain a larger quantity of these acids than oils. The acids having a chain length of C_{26}-C_{34} are normally met with in waxes only (Table III).

(2) *Unsaturated straight chain fatty acids*

The introduction of olefinic bonds (double bonds) lowers the melting points of fatty acids considerably. Thus oleic acid, having one double bond (octa dec-9-enoic acid) has a low melting point *i.e.* 14°C, when compared to its saturated counterpart stearic acid which melts at 69.9°C. Linoleic acid having two double bonds melts at 5°C and linolenic acid with three double bonds melts at –11°C. All the acids with olefinic bonds are designated as-*enoic* (ethenoid) acids and they can be conveniently grouped as dienoic (diethenoid), trienoic (triethenoid) or tetraenoic (tetraethenoid) depending upon the two, three or four double bonds they possess. The presence of double bonds facilitates *cis-trans* isomerism also. Thus *cis*-9-octadecenoic acid is oleic acid and *trans*-9-octadecenoic is elaidic acid. The *trans*-forms normally have higher melting points than their *cis* counterparts. Thus when oleic acid melts at 14°C, elaidic acid melts at 44°C. The common unsaturated straight chain fatty acids are listed in Table II.

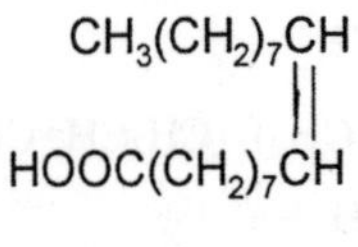

$$\begin{array}{c} CH_3(CH_2)_7CH \\ \| \\ HOOC(CH_2)_7CH \end{array} \qquad \qquad \begin{array}{l} CH_3(CH_2)_7CH \\ \quad \| \\ \quad CH(CH_2)_7COOH \end{array}$$

Oleic acid 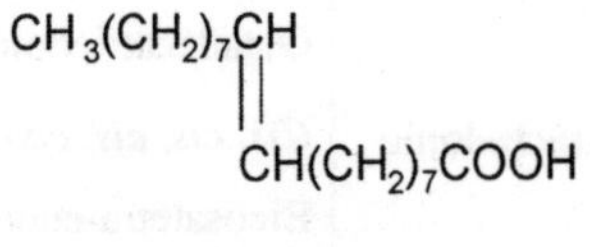 | Elaidic acid

The unsaturated fatty acids containing more than one double bond are important in animal metabolism because they are not synthesized in the body in sufficient amounts; are to be supplied externally and so are termed "essential fatty acids". Linoleic, linolenic, γ-linolenic and arachidonic acids are examples of such acids.

(3) *Acetylenic fatty acids*

These fatty acids, characterized by the presence of acetylenic (triple) bonds, are denoted by adding a suffix-*ynoic* acid. They have a very restricted occurrence, reported only from the seed fats of Simaroubaceae, Oleaceae and Santalaceae. Tariric (octa-dec-9-ynoic) acid is located in the oils of *Picrasma* (Simaroubaceae) and is believed to be confined to this family. Ximenynic acid (Octa dec-*trans*-11-ene-9-ynoic acid-having an olefinic as well as acetylenic bond), also known as santalbic acid, first reported from *Ximenia* forms the major component of the seed oils of Sandalwood tree and is found to be confined to the Santalaceae. Mycomycin (trideca-3(*trans*), (*cis*-5)-7, 8, tetraene-10, 12-diynoic acid), an antibiotic produced from an actinomycete and crepenynic acid (*cis*-9-octadecen-12-ynoic) from the seed fats of *Crepis alpina* (Asteraceae) are other examples of acetylenic fatty acids.

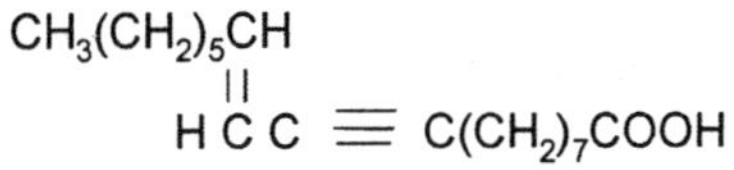

$$CH_3(CH_2)_5CH{=}CH\,C \equiv C(CH_2)_7COOH$$

Xymenynic acid

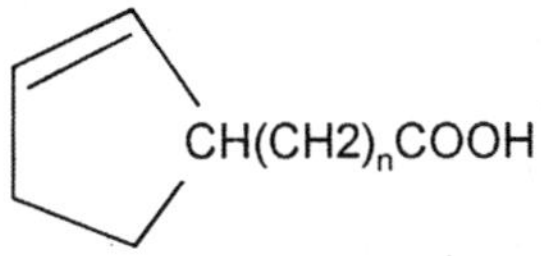

CH(CH2)$_n$COOH

n= 10, Hydnocarpic acid
n= 12, Chaulmoogroc acid

(4) *Branched chain fatty acids*

Branched chain acids occur as minor components in a few fats and oils especially those derived from lower plants and animals. Isovaleric (3-methyl butanoic), 11-methyl lauric, 13-methyl myristic and 10-methyl stearic are some of the low molecular weight representatives of this group. Mycolic ($C_{88}H_{176}O_4$), phthienoic and mycoserosic acids are examples of high molecular weight branched chain fatty acids.

(5) *Epoxy acids*

Epoxy acids form a small group, probably confined to the seed/ fruit fats of some of the Asteraceae and Brassicaceae members. Vernolic acid (*cis*-12, 13-epoxy-*cis*-octadec-9-enoic) is the major acid present in the seed oil of *Vernonia anthelmintica* Willd. Coronaric (*cis*-9, 10-epoxy octadec-*cis*-9-enoic), *cis*-9-10-epoxy octadecanoic and *cis*-12, 13-epoxy octadec-*cis*-9-*cis*-12-deienoic are the other known members of this group.

(6) *Ketoacids*

Licanic acid, the major component of *Licania rigida* Benth seed fat, is the lone representative in this group.

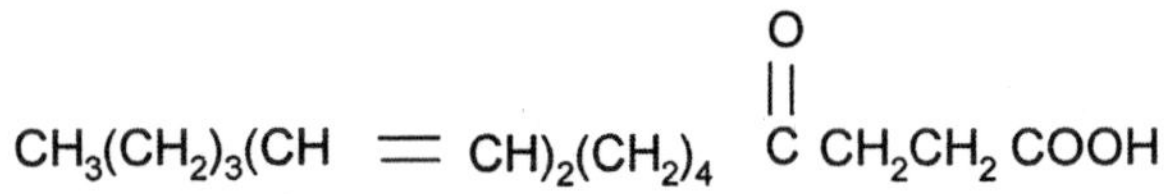

$$CH_3(CH_2)_3(CH = CH)_2(CH_2)_4\overset{O}{\overset{||}{C}}\,CH_2CH_2\,COOH$$

4-Keto $\Delta^{9,11,13}$ octadeca trienoic acid (Licanic acid)

(7) *Hydroxy acids*

Fatty acids having hydroxyl groups are abundant in waxes. Among seed fats, ricinoleic, D-(+)-12-hydroxy octadec-*cis*-9-enoic acid, is the only one occurring in appreciable amounts (upto 85%) in castor oil. Other hydroxy acids reported in traces from seed fats are 14-hydroxy eicos-*cis*-11-enoic, 9-hydroxy octadeca-*trans*-10-*trans*-12-dienoic and 9, 10-dihydroxy octadecanoic (phloionic) acids. Hydroxy acids are abundant in cutins and suberins.

(8) *Cyclic acids*

Fatty acids with cyclic skeletons in the side chain are reported from the seed fats of a few members belonging to families Flacourtiaceae, Malvaceae, Sterculiaceae and from an aerobic bacterium. A terminal cyclopentenyl ring is characteristic of the oils from Flacourtiaceae members *e.g.* hydnocarpic and chaulmoogric acids, the two principal components of Hydnocarpus oil. Lactobacillic (*cis*-11, 12-methylene octa decanoic) and *cis*-9, 10-methylene hexadecanoic acid are two cyclopropane acids isolated from the anaerobic bacteria – *Lactobacillus*. Cyclopropene acids, sterculic and malvalic acids – are characteristic to members of the Malvaceae and Sterculiaceae.

Triacyl glycerols

Triacyl glycerols (fats and oils) are esters of fatty acids and the trihydric alcohol glycerol. One molecule of glycerol reacts (esterifies) with 3 molecules of any fatty acid resulting in the formation of a triglyceride.

The fat containing only one type of fatty acid (*i.e.* the three molecules of one fatty acid esterifying with all the 3 hydroxy groups of glycerol) is a *symmetrical triacyl glycerol*. e.g., tristearin yields 3 mols. of stearic acid; triolein, 3 mols. of oleic acid. Such fats are less common in nature and a majority of fats and oils consists of *mixed triacyl glycerols* in which two or three different fatty acids are found esterified with the different hydroxyl groups of glycerol. In such cases the carbon atoms of the glycerol are designated as α-, β- and α′- to denote the position of individual fatty acid in a mixed triglyceride.

$$
\begin{array}{lll}
\alpha & - & CH_2O\ OC\ C_{15}H_{31} \\
 & & \quad| \\
\beta & - & CHO\ \ OC\ C_{17}H_{33} \\
 & & \quad| \\
\alpha' & - & CH_2O\ OC\ C_{17}H_{35}
\end{array}
$$

α- Palmito-β- oleo- α′- stearin

The unsaturated fatty acids exhibit a strong affinity towards the β-position. The mixed triglyceride having different fatty acid residues in α- and α'-positions exhibit optical activity due to the asymmetric β-carbon atom and remain in D- or L- conformation.

The natural fats always contain many different triglycerides, although one may be predominant. As noted earlier, fats contain a higher percentage of saturated fatty acids while oils contain more unsaturated acid units.

Fats and oils, being the reserve food materials, are present in all living cells, but in abundance only in fruits and seeds. They occur in the form of oil drops or as crystals. (In the endosperm of *Elaeis* palm, short needle shaped crystals of fat are reported).

Oils are classified into three groups based on their drying nature. They are (1) *drying oils,* rich in linoleic and linolenic triglycerides (2) *semidrying oils,* rich in oleic and linoleic acids and (3) *non-drying oils* containing oleic acid in large amounts. The drying property is dependent on the extent of polymerization the fatty acid undergoes in producing a tough protective film which is resistant to water, fungus etc. The degree of polymerization can directly be measured by the degree of unsaturation in the oil, as evidenced by tung oil which contains more than 70% of α-elaeostearic acid- an unsaturated acid having a conjugated triene system - which gets readily polymerized and impart the oil its characteristic drying property.

Rancidity

Rancidity is a broad term denoting the various chemical changes taking place within the fats after its removal from the parent tissue. This deterioration is of two different types; hydrolytic and oxidative.

Hydrolytic rancidity is caused by the traces of acids, bases or enzymes present in the oil. In presence of moisture, any of these agents cleaves the triglycerides with the liberation of free fatty acids of unpleasant taste and odour. The oils such as coconut oil, containing many short chain acids, are very susceptible to this type of rancidity. The enzyme lipase often accompanies the oil during extraction process, and a prior heating of the oil with steam will help in destroying such enzymes. The growth of moulds like *Penicillium, Aspergillus,* etc. which contain lipolytic enzymes in abundance, favour rancidity. Increased temperature also cleaves the triglyceride and prolonged heating causes the production of acrolein from glycerol. Acrolein is having a foul and irritating odour.

The hydrolytic rancidity can be prevented by storing the oil at lower temperatures, pasteurizing the fat or adding salt or other antimicrobial agents and keeping them in a sterile, moisture free environment.

Oxidative rancidity begins with the attack of oxygen on an allylic (the methylene group associated with double bonds) carbon atom resulting in the formation of a hydroperoxide, which in turn undergoes secondary reactions to produce epoxides. This epoxide may either favour polymerization or undergoes disruption with the formation of aldehydes, glycols or other secondary products having a 'rancid' flavor.

Rapid oxidation by enzymes like lipoxidase cleaves the unsaturated fatty acids at double bonds resulting in the formation of shorter chain fatty acids, causing rancidity.

β-Oxidation or ketone formation is another form of oxidative rancidity occurring mostly in saturated fatty acids. This is essentially an enzymic process wherein a peroxidase enzyme present in certain moulds like *Penicillium* attacks the fatty acids and converts them into methyl ketones. Warmer temperatures, moisture and darkness are prerequisites for this reaction. n-Undecyl, methyl-n-nonyl and methyl-n-heptyl ketones are the oxidation products of lauric, capric and caprylic acids respectively. The strong perfume they emit tempted some oil technologists to name this type of rancidity as 'perfume rancidity'. Nut fats like coconut or palm kernel fats are more susceptible to β-oxidation.

Oxidative rancidity can be prevented by the addition of 'anti-oxidants' such as *o*-, *p*-, di- or polyphenolic derivatives, hydroxy acids including phosphoric acid, tocopherols, flavones, ascorbic acid etc. These compounds being easily oxidisable, get oxidized themselves and thus protect the unsaturated glycerides.

Hydrogenation of the double bonds of unsaturated fatty acids in presence of suitable catalysts like palladium or platinum at high temperature and pressure, results in the formation of corresponding saturated acid. Thus oleic, linoleic and linolenic acids yield stearic acid on hydrogenation and the oil containing them is converted to a solid fat. This process - hardening of oils - is often utilized for the manufacture of butter substitutes such as margarine.

Soaps

Soaps are the salts of fatty acids. When oil is treated with sodium or potassium alkali, hydrolysis of the triglycerides takes place with the liberation of free glycerol and the salts of the fatty acid. The process is termed *saponification*. Potassium soaps are softer than sodium soaps. The presence of a polar end (hydrophilic) and a non-polar (hydrophobic) end makes a soap molecule a good emulsifying agent. It is this property that helps in soap action. Once the oily substance is removed from clothes by mechanical action the soap molecules attach themselves to the oil particles by the hydrophobic end and disperse the lipid molecules in water. Such an alignment prevents the oil globules from getting redeposited to the clothes or any other substrate and they get washed off in a stream of water.

The metallic ions like Ca^{++} and Mg^{++} present in hard water replace the Na or K producing insoluble soaps and thus prevent the foam formation, leaving an insoluble scum on the surface of water. In such cases more soap molecules are needed to produce the desired foaming.

Phospholipids

Phospholipids are triacyl glycerols in which one hydroxyl group of glycerol, normally that of α'- position is esterified with a molecule of phosphoric acid. These compounds, occurring in L-configuration, are considered to be the derivatives of L-α-glycerophosphoric acid. The simplest compound in this series is phosphatidic acid. Phospholipids are more abundant in seeds where they reach upto 2% of dry wt. They contain more unsaturated acids and so are oily liquids. Linoleic acid, by far, is the most characteristic acid of the seed phosphatides.

Lecithins and cephalins contain a nitrogenous compound linked as an ester with the phosphate group. Choline is the N-containing group in lecithins while in cephalins, it is ethanolamine or serine. Lecithins are peculiar in having a dipolar group at one end of the molecule - *i.e.* the nitrogen being quaternary, possesses a positive charge while the phosphate is negatively charged - and thus function as one of the best emulsifiers. Both lecithins and cephalins are waxy solids very liable to get oxidized in the air.

The other noteworthy phospholipids present in the plants are phosphatidyl glycerol, diphosphatidyl glycerol and phosphatidyl inositol. The former two are abundant in chloroplasts and mitochondria and latter amounts to 20% of the total lipid fraction of the photosynthetic tissue. Phospholipids are also found in combination with glucosamine, uronic acid and phytosphingosine.

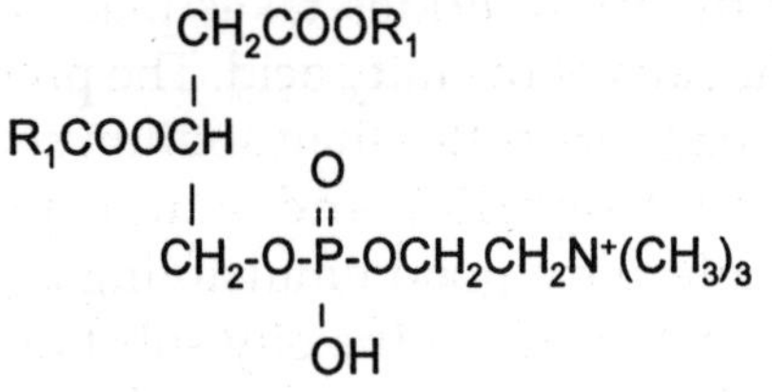

$$\begin{array}{l} CH_2COOR_1 \\ R_1COOCH \quad O \\ CH_2\text{-}O\text{-}\overset{\|}{P}\text{-}OCH_2CH_2N^+(CH_3)_3 \\ \quad\quad\quad OH \end{array}$$

Phosphatidyl choline

(Lecithin)

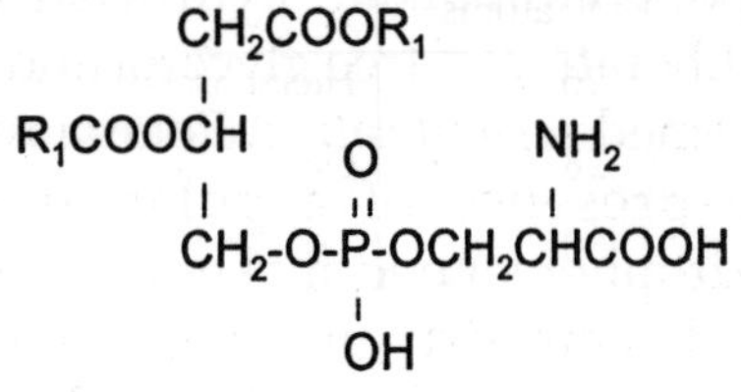

$$\begin{array}{l} CH_2COOR_1 \\ R_1COOCH \quad O \quad\quad NH_2 \\ CH_2\text{-}O\text{-}\overset{\|}{P}\text{-}OCH_2CHCOOH \\ \quad\quad\quad OH \end{array}$$

Phosphatidyl serine

(Cephalin)

Phospholipids play a very important role as components of protoplasmic membranes, wherein they act as bridges or binding agents between polar and nonpolar ingredients.

Glycolipids

These glycerides contain sugar molecules in place of phosphate groups. The number of sugar molecules ranges from 1-5 and normally they occur linked to one another forming an oligosaccharide in the α-position of the glyceride. They also are soluble in organic solvents. Glycolipids are more concentrated in chloroplasts and in one instance *i.e.* in alfalfa chloroplasts, 50% of the membrane lipids are composed of glycolipids. Monogalactosyl triglceride is a typical example of alfalfa glycolipid. Similar to phospholipids the acid present is normally unsaturated and linoleic acid, by far, is the most preferred one. Glycolipids serve as active ion transport agents.

L-D-Galactopyranosyl-2-glycerol of *Irideae laminarioides* (a marine red alga) and β-galactopyranosyl-l-glycerol from wheat flour represent glycolipids without a fatty acid residue.

Waxes

Waxes are esters of monohydric long chain aliphatic alcohols with fatty acids. The acids participating in this ester formation may be the same as that of fats and oils, but more often are saturated members having a chain length of C_{24}-C_{36}. The higher fatty acids and their corresponding alcohols are presented in Table III.

TABLE 3. The higher fatty acids and their corresponding alcohols present in waxes

No. of C atoms	Acids	Alcohol
26	Hexacosanoic (ceric/cerotic)	Hexacosanol (ceryl)
28	Octacosanoic (montonic)	Octacosanol
30	Triacontanoic (melissic)	Triacontanol (myricyl)
32	Dotriacontanoic (lacceroic)	Dotriacontanol (lacceryl)
34	Tetratriacontanoic	Tetratriacontanol

Waxes exhibit great variations in their chemical composition. As a rule, they are mixtures of complex esters. Apart from the long chain saturated aliphatic ester components, there exist varying amounts of alkanes, free acids, alcohols (primary and secondary), aldehydes, ketones and acetals. The various other components reported from different wax sample are:

1. Disubstitued derivatives like hydroxy fatty acids, β-diketones, ketols, α, ω-diols and dicarboxylic acids.
2. Trisubstituted derivatives like hydroxy β-diketones.
3. Triterpenoids like ursolic acid, oleanolic acid, α- and β-amyrins and β-sitosterol; especially in fruit waxes *e.g.* sultana grape wax contains upto 70% oleanolic acid.
4. Phenolics like flavonoids and cinnamic acids, free, esterified or as glycosides and glycerides.

Although the alkanes present in plant waxes range in chain length from C_7 to C_{62}, only those having a carbon number between C_{25} and C_{35} are present in appreciable amounts. In contrast to fatty acids, the content of odd carbon number alkanes is usually greater and the major constituent is often C_{27}, C_{29}, C_{31} or C_{33} n-alkane. The iso-(2-methyl) and anteiso-(3-methyl) alkanes are relatively rare.

The presence of large amounts of hydroxy fatty acids in certain waxes favours the formation of lactones (internal esters) or polymeric esters linking several hydroxy acids to each other. The lactones are more frequent in the waxes of higher plants. Typical examples of such lactones are aparajitin from the leaf of *Clitoria* and exaltolide from *Angelica* roots. Ambrettolide – a components of vegetable musk oils, which confers the oils their distinctive odour – is a lactone of 16-hydroxy-Δ^7-hexadecenoic acid.

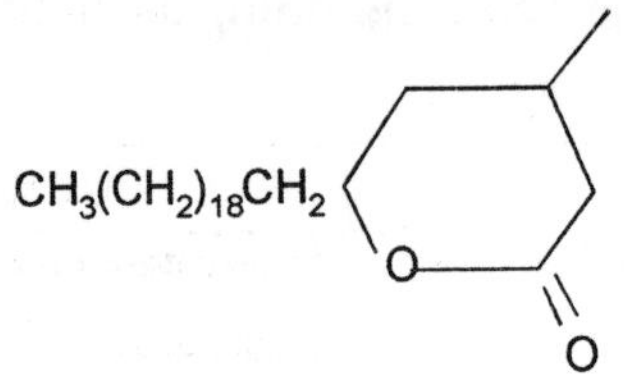

Aparajitin

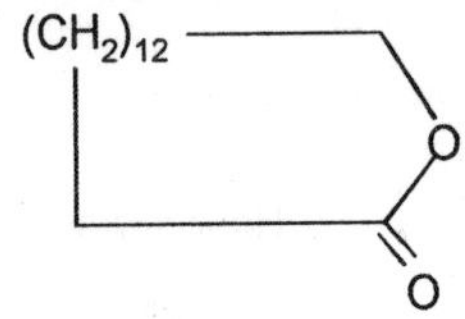

Exaltolide (a macro-cyclic lactone)

The polymeric esters formed by cross linking these hydroxy acids; known as estolides or etholides, may be linear or cyclic, and as such are located only in conifers. C_{12}, C_{14} and C_{16} ω-hydroxy acids are the most common representatives of this group, *e.g.*, Juniperic and sabinic acids.

$HOCH_2(CH_2)_{10}COOH$ — Juniperic acid

$HOCH_2(CH_2)_{14}COOH$ — Sabinic acid

In *Picea pungens*, a cyclic estolide is isolated; composed of 4, ω-hydroxy acids inter-esterified to form a ring structure

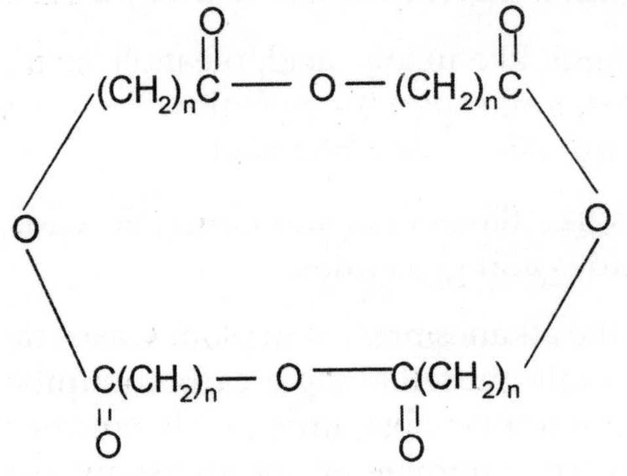

Estolide of Picea (n=11/13/15)

Wax occurs as an easily separable layer on cuticle and is extruded from the epidermal cells in liquid form through small pores (ectodesmata) in the cuticle. In the epicuticular region this liquid solidifies and is deposited as: (1) fine needles to form a smooth layer; (2) granulated layer with large interstitial spaces; (3) rodlets or (4) as smooth membranaceous layers. Waxes, being hydrophobic, reduce the water losses of the plants and act as physical barriers to pathogens like fungi and bacteria.

Cutins and suberins

Cutins and suberins, polyesters of high molecular weight, are differentiated chiefly by their positions in the plants; the former occurs in the cuticles of leaves, fruits and other fleshy organs and the latter in the cork cells of bark. Both remain tightly bound to the cell wall constituents namely cellulose, pectin, lignin, tannins etc. and are not easily isolated as wax, and that is the reason why the knowledge of their chemistry is still obscure.

Cutin accounts for 50% wt. of the cuticle and yields little glycerol on hydrolysis. On saponification, cutin yields many C_{14}-C_{22} acids with 1-4 hydroxy groups, possibly occurring as cross-linked polymers (etholides) in intact cuticle.

The principal hydroxy fatty acids obtained from cutin are as follows:

C_{16} acids	1. 16-Hydroxy hexadecanoic
	2. 10, 16-Dihydroxy hexadecanoic
	3. 9, 10, 16-Trihydroxy hexadecanoic
C_{17} acids	4. 10, 17-Dihydroxy heptadecanoic
C_{18} acids	5. 18-Hydroxy octadecanoic
	6. 18-Hydroxy octadec-9-enoic
	7. 18-Hydroxy octadeca-9-12-dienoic
	8. 10, 16-Dihydroxy octadecanoic
	9. 9, 18-Dihydroxy octadecanoic
	10. 9, 18-Trihydroxyl octadecanoic
	11. 9, 10, 16-Trihydroxy octadecanoic
	12. 9, 10, 12, 18-Tetrahydroxy octadecanoic
C_{20} acid	13. Dihydroxy eicosanoic

Depending on the increasing hydroxylation of the constituent acids, three types of cutins are recognized.

The first type is present in delicate cuticular membranes and contains esters of simple monobasic acids as seen in the leaves of *Bryonia dioica* L. and *Lactuca sativa* L. where the cutin contains predominantly hexadecanoic acid.

Second type, which forms the largest group, is characterized by high contents of 10, 16-dihydroxy hexadecanoic acid (upto 70%) as in the case of cutins in *Citrus aurantifolia* Swingle.

The third type is seen in thick, ridged cuticular layers like that of *Agave americana* leaf, wherein 9, 10, 18-trihydroxy octadecanoic acid predominates.

The cutins of different organs of the same plant are found to vary in chemical compositicn.

Cutin is not directly secreted by the cell layers. The precursor of cutin - procutin - is secreted from the outer cellulosic wall in fine droplet form and this gets polymerized by the establishment of ester and peroxy cross-linkages in ultraviolet light. The cutin thus formed - a high molecular weight polymer - is found invariably bound to the other components of the cell wall *i.e.* carbohydrate or pectin. Cuticle is more resistant than suberin to chemical reagents or enzymes.

Suberin comprises about 50% dry weight of the cork. Along with most of the fatty acids present in cutins, suberin of angiosperms on saponification yields acids like 9, 10-dihydroxy octadecane-1, 18-dioic (phloionic acid), 22-hydroxy octadecane-1, 18-dioic and octadec-9-ene-1, 18-dioic acid. Suberins of conifer bark yield mostly C_{18} acids like octadecane-1, 18-dioic, 18-hydroxy octadecanoic and octadec-9-ene-1, 18-dioic acid.

The presence of epoxy acids like 9, 10-epoxy-18-hydroxy octadecanoic acid in suberin is conclusively proved. Glycerol and ferulic acid also are reported from a number of suberins. Suberin is now defined as polyester of hydroxy fatty acids and phenolic acids. These phenolic acids presumably bind suberin to lignin.

As noted earlier, the conifer suberins are characterized by C_{16} and C_{18} mono-and di-basic acids. The angiosperm suberin contains mostly C_{18} and C_{22} acids having two or three hydroxy groups. Suberins of different species are found to vary in detailed chemical composition.

1.2. Secondary Metabolites

Secondary metabolites all those compounds which do not get involved in the primary metabolism of plants, and are often stored in cell vacuoles or in metabolically inactive or dead plant tissues. The role of these compounds, which exist in wide variety, in plants has always been underestimated. The poor understanding on the function of these compounds lead to naming them by erroneous terms like "waste products", "end products", "excretory products" etc. But now there is a genuine realization of the varied roles of these compounds, most of which are being adaptational and survival mechanisms. Thus alkaloids are defensive agents, deterring the herbivores due to their bitter taste; volatile oils act as pheromones (for attracting insects for pollination) and protect the plant from microbes (and sometimes suppress neighbouring plants as in case of *Eucalyptus*) and diterpenes or triterpenes, present in resin, perform wound healing and antimicrobial functions. The insect repellants, limonoids and quassinoids, protect the plant (especially wood) from insect attack. Phenolics are the best antioxidants protecting cellular membranes and tissues containing lipids from oxidation. Among them anthocyanins and flavonoids are flower pigments, which make the flowers visible to the "color blind" insects, and thus act as pollinator guides. All the sulphur-containing compounds are antimicrobial in nature.

The manifold functions of these compounds in plants reiterate that they are not aberrations in biosynthesis in plants. They are synthesized by genetically controlled mechanisms. They are never formed in any of the primary metabolic pathways but are prepared by **specially designed pathways, the secondary metabolic pathways.** All the molecules formed by these specialized secondary metabolic pathways are designated secondary metabolites. The great adaptive significance of each one of these compounds is being understood now. It is now confidently proclaimed that the plants survive and evolve due to their secondary metabolites. Each group of plant specialized in one or the other group of compounds which enabled them to survive better, be it the pollination of flowers, dispersal of seeds, warding of the pathogens and herbivores or removing competition by suppressing the neighbouring plants. It is these compounds that made the plants face the adversities, enemies or competition. The plants specialized in such compounds, performed better, evolved further and became the large taxa.

Though it is generally believed that secondary metabolites are never metabolized, studies on the dynamics of secondary product metabolism indicate that there is a definite **turnover** of these compounds that is evidenced by diurnal variation, seasonal variation and disappearance at various stages of development (as in the case of bitter tannins which are lost when certain fruits ripen and disappearance of anthocyanins when the leaves mature). Pulse labeling experiments proved that the concentrations of isoflavones (*Phaseolus aureus*), alkaloids (*Atropa* and *Papaver*) and flavonols and cinnamoyl derivatives of many plants remain at constant levels as a result of simultaneous and equally rapid synthesis and subsequent consumption.

Alkaloids, terpenoids and **phenolics** form the three major classes of secondary metabolites. All the other compounds which occur in minor amounts are included in a fourth section on "Minor Components".

1.2.1 Alkaloids

Alkaloids are basic plant products possessing a nitrogen-containing heterocyclic ring system and showing marked pharmacological activity. The term 'alkaloid' means 'alkali like substance', and they form the largest class of secondary plant substances; at present numbering more than 12,000. They form a very heterogeneous group exhibiting a wide range of chemical structures and properties.

The basic nature of alkaloids is due to nitrogen. The nitrogen atoms present may be single as in piperine; or as many as 5, as in ergotamine. The N may occur as a primary amine, i.e., RNH_2, as in mescaline; a secondary amine, i.e., R_2NH, as in pelletierine; as a tertiary amine, i.e. R_3N, as in codeine; or as a quaternary ammonium ion, i.e., $R_4N^+X^-$, as in tubocurarine chloride. It is the pair of unshared electrons in N (except for quaternary ammonium compounds), which makes the alkaloids alkaline. The degree of basicity of alkaloids varies from slight to moderate to strong, depending on the position and number of nitrogen atoms in the skeleton, and on that of other functional groups. In the cell sap, alkaloids exist as cations of the salts of various organic acids such as oxalic, succinic, malic or any other acid.

Most alkaloids contain O also in addition to C, H and N and are, crystalline solids. The alkaloids like coniine and nicotine, which lack O, are liquids at room temperature. Almost all are colourless; exceptions being berberine and serpentine, both being yellow and sanguinarine, brownish red. Alkaloids are usually insoluble in water, or only sparingly so, but are freely soluble in ether, $CHCl_3$ and other relatively non-polar solvents. Their salts, formed by their treatment with acids, are soluble in water and insoluble in non-polar solvents. This property is made use of in the extraction, purification and estimation of alkaloids. Alkaloidal salts are crystalline, often with characteristic crystal forms and habit. Alkaloids can be precipitated out of their aqueous or acid solutions by a number of substances like picric acid and tannic acid.

Optical activity is another property seen in many alkaloids and usually they are *l*-rotatory. Generally, only one of the isomers occurs naturally in a plant, and only occasionally do racemic mixtures occur (for e.g., atropine is a racemic form (*dl*) of hyoscyamine). In most cases, the isomers differ in their physiological activities. Almost all alkaloids absorb UV light, and possess characteristic absorption spectra. This property is useful in characterizing the group.

All alkaloids, as a rule, occur in plants only, though a few of them are seen in animals too. Almost all of them can be synthesized chemically. Present studies indicate a rather restricted distribution of alkaloids amongst plants. Among Cryptogams they occur in some fungi like *Claviceps* and *Amanita* and some ferns. *Cephalotaxus, Ephedra* and *Cycas* are the Gymnosperms having alkaloids. Among Angiosperms, monocots generally have a lesser incidence of alkaloids, seen in the Amaryllidaceae, Liliaceae, Stemonaceae, Arecaceae, Poaceae and Orchidaceae. The Ranunculaceae, Menispermaceae, Berberidaceae, Papaveraceae, Rutaceae, Rhamnaceae, Fabaceae, Rubiaceae, Asteraceae, Apocynaceae, Asclepiadaceae, Loganiaceae and Solanaceae, include most of the alkaloid-yielding plants. Usually alkaloids with complex structures are characteristic of specific plant families, *e.g.*, benzyl isoquinolines in Ranales, lupinanes in Fabaceae, tropanes in Solanaceae, ergot alkaloids in Convolvulaceae and colchicine in the Liliaceae. Certain interrelationships appear exist between alkaloids and volatile terpenes in that they tend to be mutually exclusive.

Alkaloids generally occur in groups which may differ only slightly in physical and chemical characters. In the plants, alkaloids may be systemic, *i.e.*, distributed throughout (*Ephedra*), or restricted to specific organs like roots (Aconite, Belladonna), rhizomes and roots (Ipecac, *Withania*), stem barks (*Cinchona*, Pomegranate), leaves (*Adhatoda, Tylophora*), fruits (Pepper, *Conium*) or seeds (Nux-vomica, *Areca*).

Three classes of alkaloids are being recognized from a chemotaxomical view point, (Hagnauer, 1966). They are the following.

True Alkaloids

These have nitrogen containing heterocyclic ring skeleton derived from a biogenetic amine formed by decarboxylation of an amino acid.

Protoalkaloids

These are derived from amino acids or biogenetic amines or their methylated derivatives, but lack heterocyclic nitrogen. The nitrogen may be present in an aliphatic side chain, as in colchicine.

Pseudo alkaloids

These basic compounds do contain nitrogen, but the carbon skeleton is derived from mono-, di or triterpenes, sterols, or acetate derived polymers.

Alkaloids are classified generally on their chemical structures. The structural complexity ranges from monocyclic (*e.g.*, arecaidine) to heptacyclic (e.g., kopsine). The most common skeletons found in alkaloids are indole, isoquinoline, quinoline, pyridine, piperidine, tropane, purine, pyrrole, pyrrolidine and steroid.

Amino acids are found to be the precursors of many alkaloids. They are:

1. Phenylalanine and tyrosine, in the biosynthesis of isoquinoline alkaloids and protoalkaloids.
2. Tryptophan, in the biosynthesis of indole and quinoline alkaloids.
3. Ornithine, in the biosynthesis of tropane and pyrrolidine alkaloids,
4. Lysine and cadavarine, in the biosynthesis of piperidine and lupine alkaloids.

The alkaloids were once considered as non-toxic excretory products of toxic compounds or nitrogen reserve food materials. They may serve as nitrogen source, in that the seed alkaloids partly get degraded during germination and seedling stage. But recently they were found to have great adaptational significance as the defensive chemicals of plants which deter herbivores, insects, parasites etc. Some of these compounds serve as allelochemics also. A few of them have certain physiological functions such as growth regulators (lupine alkaloids inhibit germination), removing the inhibitory effects of tannins, and even as growth stimulators. Being basic, they help the plant to maintain its ionic balance by replacing mineral bases.

A number of **physiological actions** of the alkaloids are now known. The alkaloids derived from phenylalanine are found to mimic and superimpose on neurotransmitters like serotonin, nor-adrenaline, dopamine, histamine etc. Pyrrolizidine alkaloids and aristolochic acid alkylate DNA and alter metabolic activation in liver. Berberine, emetine and sanguinarine etc. being planar and lipophilic, intercalate DNA. Colchicine and vinblastine inhibit the assembly or disassembly of microtubules and induce apoptosis. Indolizidine alkaloids inhibit carbohydrate processing enzymes in EPR.

The **pharmacological actions** of alkaloids are very diverse. Some of them are:

1. CNS depressants : Gelsemicine, Piperine
2. Expectorants : Vasicine, Tylophorine
3. Local anaesthetics: Cocaine,
4. Myotics : Physostigmine, Atropine,
5. Antihaemorrhagic : Hydrastine,
6. Antispasmodics : Atropine, Hyoscyamine,
7. Skeletal Muscle relaxant : Tubocurarine
8. Vermifuges, e.g., Pelletierine,
9. Aphrodisiacs, e.g., Yohimbine,
10. Anti-hypotensives, e.g., Ephedrine,
11. Tranquilisers : Celapanin
12. Anti-hypertensives, e.g., Reserpine,
13. Cardiac repressants, e.g., Quinine,
14. Diaphoretics, e.g., Pilocarpine,
15. Muscle paralyzers, e.g., Tubocurarine,
16. Oxytocics : Sparteine
17. Antitumor agents, e.g., Camptothecine., Harringtonine
18. Immunostimulants: Uncarine
19. Antileukemic: Vincristine, Maytensine and Vinblastine
20. Anticancerous: Cissampeline
21. Cardiac stimulants: Ephedrine and Pseudoephedrine
22. Analgesics and narcotics, e.g., Morphine and Codeine,
23. Antipyretic: Berberine
24. Emetics, e.g., Emetine,
25. Central stimulants, e.g., Strychnine, Brucine,

However, most alkaloids are toxic at higher concentrations, and the physiological response they elicit depends chiefly on the dosage.

The main groups of alkaloids are the following :

Acridine alkaloids

Acridines are characterized by a pyridine ring flanked by two aromatic rings on either side. About 10 alkaloids like melicopicine, xanthevodine or arborinine are seen in members of the Rutaceae such as *Glycosmis, Acronychia,* and *Evodia* etc. They are derived from anthranilic acid.

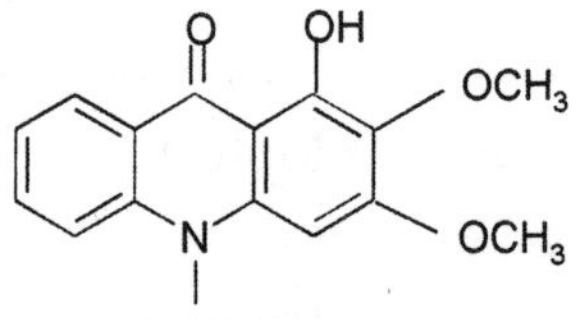

O OH OCH$_3$ N OCH$_3$

Arborinine

Alkaloidal amines

These compounds possess a primary amino group (aliphatic N-atom) numbering about fifty or more, do not possess heterocyclic nitrogen and therefore named 'protoalkaloids' They normally possess a simple structure and are similar to phenyl alanine. Though the Chenopodiaceae and Cactaceae (*Lophophora*-mescal buttons) are particularly rich in these compounds they are present in Fabaceae (*Alhagi*), Malvaceae (*Sida*) and a few like ephedrine (*Ephedra*) are widely distributed in plant kingdom. Mescaline, Phenyl ethylamine etc are other important members in this group.

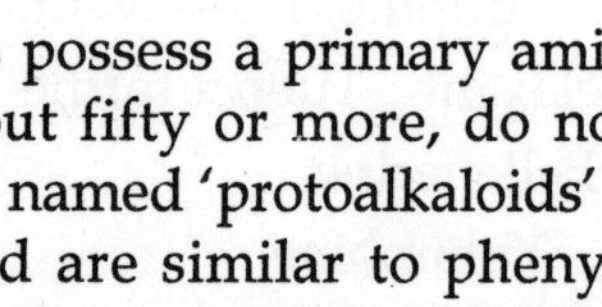

OH N

Ephedrine

Alkaloidal peptides

This group of alkaloids, having a number of peptide linkages, is once thought to be confined to the Rhamnaceae, but they have now been reported from widely separate families like Euphorbiaceae, Pandaceae, Rubiaceae and Urticaceae. Canthiumine from *Canthium* (Rubiaceae) yields N, N-dimethyl-L-phenyl alanine, *p*-hydroxy styrylamine, L-proline and L-*threo*-α-phenyl serine on hydrolysis. Zizyphine (*Ziziphus*) and scutianine (*Scutia*) are some other examples of this group.

Amaryllidaceous alkaloids

These alkaloids, forming a large group, are confined to the members of the subfamily Amaryllidoideae of Amaryllidaceae and absent from other subfamilies like Agavoideae, Hypoxidoideae and Compynematoideae. They form a diverse group including pyrrolo-(de)phenanthridines (Lycorine from *Amaryllis, Narcissus, Pancratium* etc), 2-benzopyranoindoles (Hippeastrine from *Amaryllis, Lycoris* etc.), dibenzofurans (Galanthamine from *Crinum, Haemanthus, Zephyranthes* etc.), spirostructures (Tazettine from *Crinum, Hymenocallis* etc.) ethanophenanthridines (Crinine from *Crinum*), N-benzyl-N-β-phenethylamines (Belladine from *Amaryllis*) and a number of alkaloids of unknown structure. Most of these alkaloids are seen generally in the bulbs or other underground parts indicating the fungicidal nature of these compounds. (Lycorine exhibits marked fungicidal activity).

Apart from the Amaryllidoideae, similar alkaloids having a lycorine like structure (benzopyranoindoles) is seen in Menispermaceous members such as *Cocculus* and *Anamirta* (these alkaloids possess an indolizidine ring also).

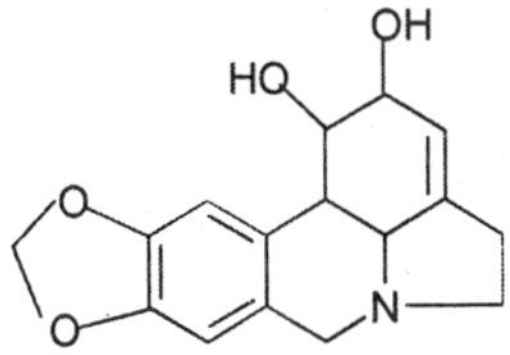

OH
HO
O
O
N

Lycorine

Dihydroquinolines

This small group, isolated from *Melodinus* of Apocynaceae has a keto group at *ortho*- position to the quinoline nitrogen.

Diterpenoid alkaloids

Theses alkaloids, of a different biosynthetic pathway, i.e. mevalonic acid pathway, are having very complex structures and are restricted to Ranunculaceous genera like *Aconitum* and *Delphinium*, Leguminous genus *Erythrophleum*, Asteraceae genus *Inula* and Garryaceous *Garrya*. Most of these alkaloids possess a perhydrophenanthrene skeleton (similar to steroids) and in fact the alkaloids of *Erythrophleum* have the same type of pharmacological activity as the *Digitalis* group of steroid glycosides. Many of these alkaloids are highly toxic.

Aconitum and *Delphinium* are found to contain a majority (more than 70) of the total 90 diterpenoid alkaloids known. All of them are found to be derived from complex polyhydric alcohols probably based on $C_{19}H_{28}NH$, which is having a diterpenoid skeleton. The pharmacologically important alkaloids are only three, aconitine, atisine and ajacine.

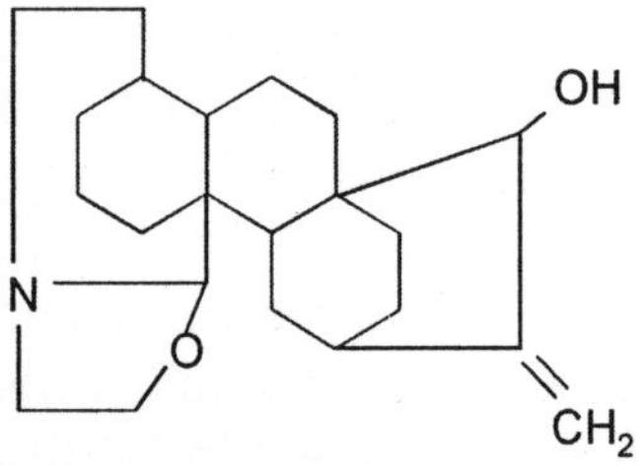

Atisine

Imidazole alkaloids

Imidazole alkaloids are a small group of 13 alkaloids reported from the families Rubiaceae (*Pilocarpus*), Euphorbiaceae (*Glochidon*) and a few members of Fabales like *Cassia* and *Acacia*. Pilocarpine (*Pilocarpus*) is the only pharmacologically important alkaloid. All these compounds possess an imidazole-glyoxaline nucleus, very similar to that of histidine. Purines also possess such a ring, but they are treated separate because of the differences existing in biosynthetic pathways.

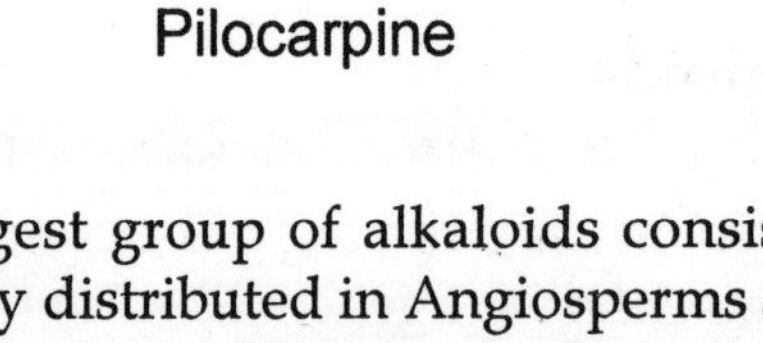

Pilocarpine

Indole alkaloids

This is the largest group of alkaloids consisting of about 1000 members, are widely distributed in Angiosperms and *Claviceps*. Many of these alkaloids are formed by the condensation of indole nucleus (from tryptophan) and a monoterpenoid moiety. All the indole alkaloids usually contain two nitrogen atoms; one indolic and the other generally 2 carbons removed from the β-position of the indole ring. Gibbs (1974) recognized as many as 23 groups of indole alkaloids. They are the following.

a) Simple Indole Bases

They are simple derivatives of indoles closely related to the routes of biosynthesis of indole acetic acid (IAA) or tryptophan. They are widely distributed, but most of these occur in Fabales. E.g. abrine (*Abrus*), serotonin (*Mucuna*) and many methyl derivatives of tryptamine.

b) Carboline alkaloids

They have a pyridine skeleton attached to the indole ring, and thus are tricyclic. Many of them occur in Rubiaceae and a few in related Loganiaceae and Apocynaceae. E.g. adifoline (*Adina*), harmaline (*Peganum*), harmine (*Banisteriopsis*), harmol (*Passiflora*) etc.

c) Quinazolinocarbolines

These alkaloids, probably restricted to the family Rutaceae, possess a quinazoline group and a carboline group. They are generally pentacyclic. E.g. evodiamine (*Evodia*)

d) Hexahydro-pyrrolo indole alkaloids

They contain an additional hexahydropyrrole ring (making the molecule tricyclic). E.g. physostigmine (*Physostigma*) and echitamine (*Alstonia*). This group contains dimeric or trimeric compounds also. (Calycanthine is dimeric and hodgekinsine is a trimer)

e) Canthin-6-ones

This is a small group having an additional ketonic ring attached to pyrrole and pyridine rings. They are seen in Rutaceae, Simaroubaceae and Amaranthaceae.

f) Eburnamine-Vincamine alkaloids

These are related to canthin-6-ones, but mostly with another ring making them pentacyclic. This group is formed from an indole and a monoterpenoid component. E.g. Vincamine (*Vinca minor*). All are confined to the Apocynaceae.

g) Oxindole group

This group is characterized by a keto group at *ortho*-position to indolic nitrogen. Restricted to the Gentianales. E.g. gelsemine (*Gelsemium*), routindifoline (*Mitragyna*) etc.

h) Aspidosperma group

This is a large group of 100 and more alkaloids further subdivided into 6 groups. This is a terpenoid alkaloid in that indolic moiety is combined with *aspidosperma* type monoterpene. A majority of these alkaloids are from the Apocynaceae and the rest from *Strychnos.* E.g. Aspidospermine from *Aspidosperma,* vindoline from *Vinca* and strychnine from *Strychnos.* Some of these alkaloids are characterized by 7 rings (kopsine, strychnine)

i) Ajmaline–Sarpagine alkaloids

These alkaloids, also known as yohimbinoid alkaloids because of the yohimbinoid type monoterpenoid precursor, are seen in Apocynaceae and Loganiaceae only. They can be easily understood as products of carboline and yohimbinoid monoterpene. E.g. Ajmaline and sarpagine (both from *Rauwolfia*). Most of the members are from the Apocynaceae.

j) Yohimbane alkaloids

These are pentacyclic alkaloids in which a carboline ring is fused with a yohimbinoid monoterpene. Seen in Apocynaceae and Rubiaceae E.g. reserpine (*Rauwolfia*), yohimbine (*Corynanthe*).

k) Heteroyohimbane group

This also is a group of pentacyclics as yohimbane alkaloids, but here the 5th ring (E) is oxygen heterocyclic. They are restricted to the Gentianales. The important members are ajmalicine (*Catharanthus*), alstonine (*Alstonia*) and serpentine (*Rauwolfia*)

l) Corynane (16, 18-seco yohimbane) group

These are derived form yohimbane alkaloids by the opening of the E ring. They also are restricted to the Gentianales. E.g. corynanthine (*Corynanthe*), gambirine (*Uncaria*) etc.

m) 2, 2′-Indolyl quinuclidine alkaloids

This small group of alkaloids contains a quinuclidine group and is restricted to Rubiaceous members like *Cinchona, Isertia* and *Ramijia.* E.g. cinchophyllamine (*Cinchona*).

n) Ergoline alkaloids

All these alkaloids possess a lysergic acid skeleton to which a peptide or any other amino compound is esterified. They are formed from tryptophan and mevalonic acid. They are first reported from *Claviceps* (ergot) and now are wide spread in the member of the Convolvulaceae. Ergotoxine, ergometrine etc.

o) Iboga alkaloids.

This also is a terpenoid alkaloid group in which tryptophan or it's derivative is linked to a monoterpene named Iboga type (Which resembles geraniol). They are seen in the Apocynaceae only. Catharanthine (*Catharanthus*), ibogaine (*Tabernanthe*) and voacamine (*Voacanga*)

p) Aspidosperma- iboga dimers

These are dimeric alkaloids probably confined to *Catharanthus*, such as vinblastine, neoleurocristine etc. Both the monomers may be similar or dissimilar in structure.

q) Alstonia alkaloids

Some of the alkaloids of *Alstonia* like macralstonidine and macrophylline are included here.

r) Erythrinane alkaloids

These alkaloids, common in all the species of *Erythrina* (Fabaceae), are tetracyclic. Both indolic and isoquinoline nuclei can be visible here. E.g. erysopine (*Erythrina*) and cocculolidine (*Cocculus*).

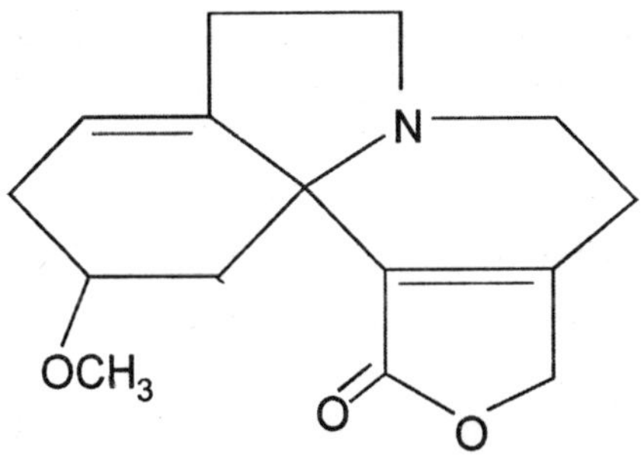

Cocculolidine

s) Orchidaceae alkaloids

They also are tetracyclic with a lactone ring forming one of the rings. These alkaloids are unique to the Orchidaceae, that also to *Dendrobium.*

t) Mesembrine group

This small group of alkaloids, confined to the genus *Sceletium* (a segregate from *Mesembrayanthemum,* Aizoaceae), have an aromatic ring attached to the 4-position of the indole ring. E.g. mesembrine.

u) Stemona group

Another small group of alkaloids, characterized by a 7-membered ring attached to the 8th carbon and the indolic nitrogen and a lactone ring to 5 & C. Restricted to *Stemona* (Stemonaceae) E.g. Tuberostemonine.

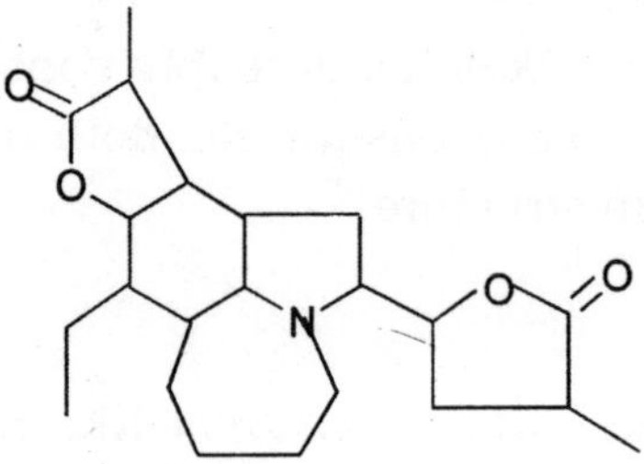

Tuberostemonine

v) Indolo (2,3 α) quinolizine group

An extremely small group characterized by both indole and quinolizidine rings. Reported from *Dracontomelum* (Anacardiaceae).

w) Carbazole group

This small group contains a carbazole (dibenzo -pyrrole) skeleton and confined to the Rutaceae. E. g. girinimbin, murrayanine (*Murraya*) etc.

Isoquinoline alkaloids

Second largest group of alkaloids, this class of compounds possess an isoquinoline nucleus. The three common ring structures, i.e., the isoquinolines, benzyl isoquinolines and phenanthrenes and their various modifications and derivatives make this a very complex assemblage necessitating a further classification. Out of the many classification systems, the one proposed by Gibbs (1974) appears to be comparatively simple and convenient wherein 12 groups of isoquinolines are recognized.

a) Simple isoquinolines

These alkaloids possess a tetrahedroisoquinoline skeleton. Only about 20 alkaloids are known at present. (E.g. anhalamine from *Lophophora williamsii* (Cactaceae).

b) Benzyl isoquinolines

These have a benzyl group attached to the position 1 of isoquinoline nucleus through a carbon. There are about 60 members here. (E.g., papaverine from *Papaver somniferm* L.)

c) bis-Benzyl isoquinolines

These are dimers of benzyl isoquinolines wherein two benzyl isoquinoline nucleii are joined by one, two or three ether linkages. They are believed to have originated through enzymatic dehydrogenations of benzyl isoquinoline units. Known structures are about 80. (E.g., tiliacorine from *Tiliacora racemosa*-Menispermaceae).

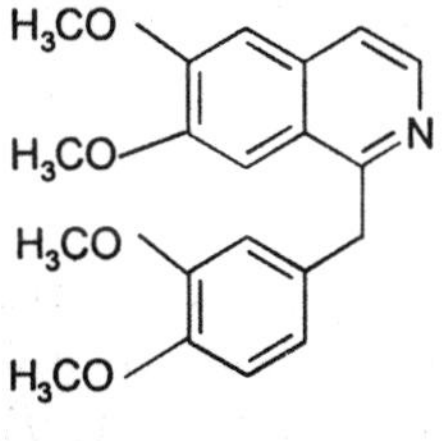

Papaverine

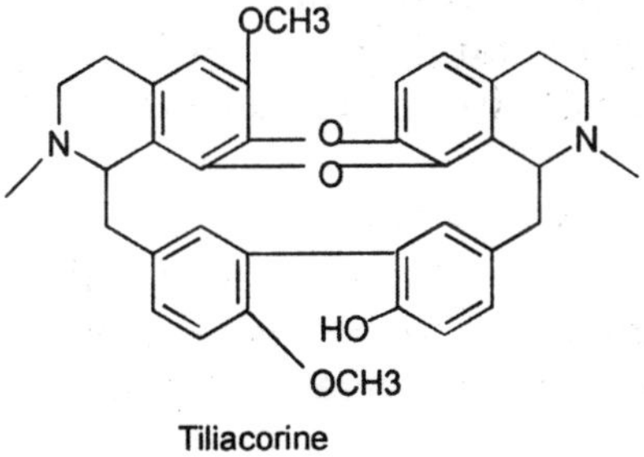

Tiliacorine

d) Aporphines

Mostly restricted to Magnoliales and Ranunculales, these alkaloids arise naturally by oxidation of phenols of the benzyl tetrahedro-isoquinolines (E.g., Glaucine in *Thalictrum minus* L. - Ranunculaceae).

e) Cularines

These also are derivatives of benzyl isoquinolines in which a phenolic oxygen is incorporated into a seven-membered ring. A very small group of 4 compounds reported from *Dicentra* and *Corydalis* (*E.g.*, Cularine, *Dicentra cucullaria* Bernh.-Papaveraceae).

f) Morphines

These possess a phenanthrene nucleus along with isoquinoline ring. These are reported mostly from Papaveraceae (E.g., codeine from *Papaver somniferum* L).

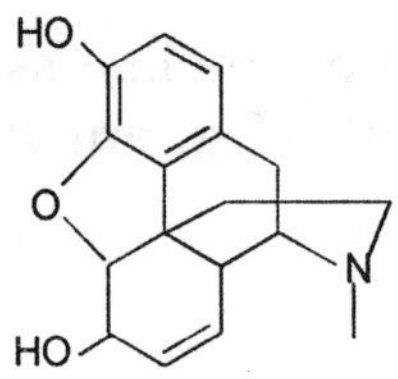

Morphine 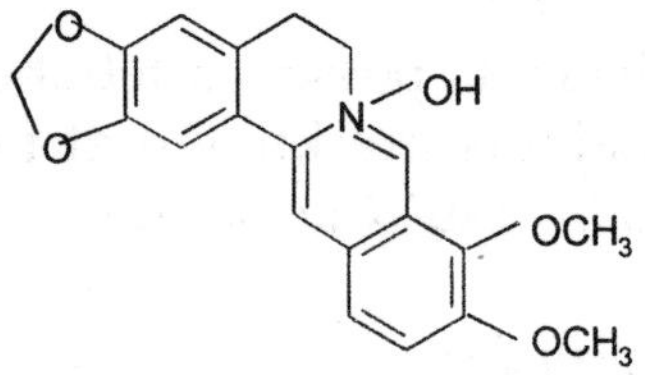 Berberine

g) Protoberberines

There are about 50 members, derived by condensation of benzyl isoquinolines and formaldehyde molecules. They are frequent in Papaveraceae and reported from many families of Magnoliales and Ranunculales. (Berberine from *Berberis.*)

h) Protopine group

These alkaloids are devoid of an isoquinoline nucleus, but are included here because they are derived from isoquinolines and are reconvertible. Protopines are characterized by a 10-membered heteroring containing one carbonyl group. Only 14 compounds, isolated mostly from Papaveraceae, are known so far. (Protopine)

i) Phthalide isoquinolines

These are benzyl isoquinolines derived by attaching a furanolactone ring to the benzyl group. Only 11 alkaloids are reported so far, all from Papaveraceae. (e.g., narcotine from *Papaver*).

j) Emetine group

This group is characterized by the presence of both isoquinoline and quinolizidine nucleii. All the 17 compounds known are isolated from Rubiaceae. (e.g., emetine from *Cephaelis ipecacunanha* Rich.)

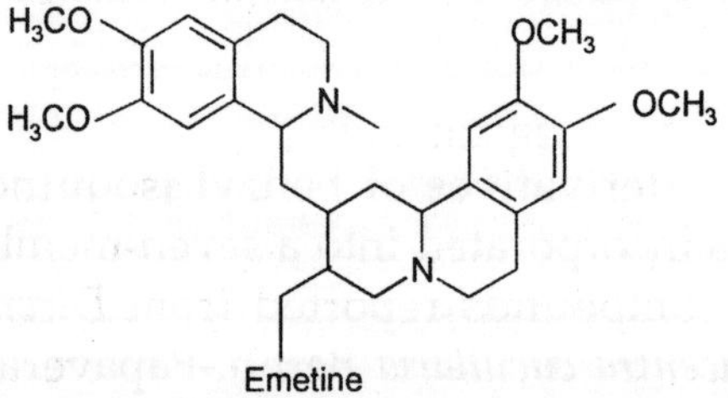

Emetine

k) α-Naphthophenanthridines

These possess tetracyclic ring systems probably derived from tetrahydroberberine systems. About 18 have been isolated mostly from Papaveraceae (e.g., chelidonine from *Chelidonium majus* L.)

l) Miscellaneous isoquinolines

The rest of the isoquinolines, which cannot be grouped in the above classes, are placed in this group, e.g., Thalicarpine - a benzyl isoquinoline-aporphine alkaloid from *Thalictrum dasycarpum* Fisch. (root).

Maytensinoids

This is a small group of alkaloids characterised by an ansamacrolide (19-membered amide structure) isolated from *Maytenus ovatus* and *M. buchananii* (Celastraceae). Of these Maytensine and related esters exhibit great antileukemic activity.

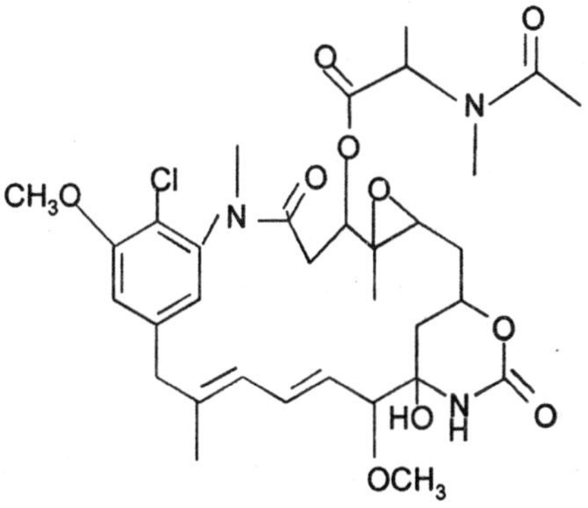

Maytensine

Monoterpenoid alkaloids

These alkaloids are closely related to the monoterpenoid glycoside aucubin. They are found in families such as Valerianaceae, Gentianaceae and Bignoniaceae. *Valeriana officinalis* (Valerine - tranquilizer), *Enicostemma hyssopifolium* (erythrocentaurine - antidiabetic) and *Tecoma stans* (tecomanine) are the important medicinal plants containing them.

Oxazole alkaloids

This also is a small group isolated from the members of the Rutaceae.

Papaverubines

These are similar to isoquinolines, but differ in having a 7 – membered ring containing nitrogen. They give intense color with acids and thus is the name papaverubines. *Papaver* is the only genus containing these alkaloids. E.g. Papaverubine, rhoeadine etc.

Phenanthro – indolizidines

These alkaloids possessing both phenanthrene and indolizidine nuclei are restricted to a few asclepiadaceous members like *Tylophora, Cynanchum* and *Vincentoxicum,* and one member of Moraceae, i.e., *Ficus septica* Forst. Tylophorine (*Tylophora indica*) and related alkaloids are powerful vescicants.

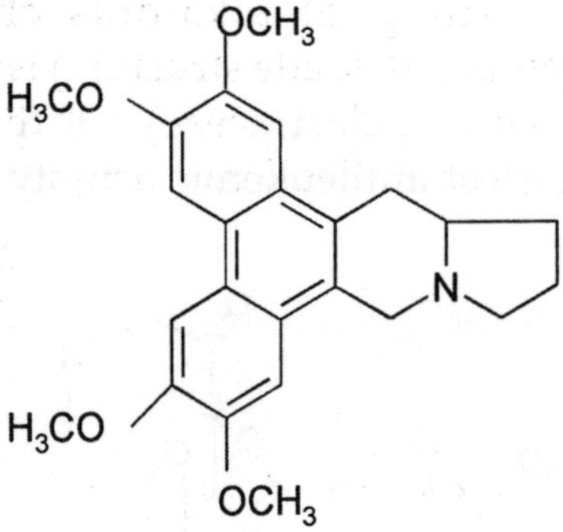

OCH_3 H_3CO N H_3CO OCH_3

Tylophorine

Purine alkaloids

Purine alkaloids form a small group of alkaloids similar to the nuclear bases such as adenine and guanine, having a purine skeleton *i .e,* a heterocyclic nucleus consisting of a six membered pyrimidine ring fused with a five membered imidazole ring. The important alkaloids of this group are the methylated derivatives of xanthine (2, 6-dioypurine) such as caffeine (Coffee, Tea), theobromine (Cocao) and theophylline (Tea). Free adenine is reported from plants like *Thea, Beta, Saccharum* and coffee. Crotonoside from *Croton tiglium* is another interesting purine derivative.

Pyrazoles

They contain a 5- membered heterocyclic ring having two nitrogens. Withasomnine (*Withania*) is one of the few alkaloids in this group.

Pyridine and piperidine alkaloids

These alkaloids, possessing an unsaturated or saturated ring, form a fairly large group widely distributed in plant kingdom. They are subdivided based on the position of substitution such as:

Alkaloids with substitution at position 1- e.g. alkaloids of pepper.

Alkaloids having substitutions at positions 1 & 3- e g. alkaloids of areca nut

Alkaloids substituted at 1& 2 and 1,2 and 6- e.g. pomegranate alkaloids

Lobelia alkaloids having substitutions at 1, 2 and some times 6.

Alkaloids with substitution at 1,3 & 4; 1,2,3 & 4 and at 1,2 &5- e.g. mimosine (*Mimosa*)

Alkaloids with substitution at positions 1,2,3 & 6- e.g. yulocrotin (*Julocroton*)

Alkaloids with substitution at position 2 - e.g. conium alkaloids

Alkaloids having substitution at 2 & 6- e.g. prosopine (*Prosopis*)

Alkaloids having substitutions at 2 & 6- e.g. 2,6-dimethyl piperidine (*Nanophyton*)

Alkaloids substituted at 3 and sometimes else where- e.g. tobacco alkaloids.

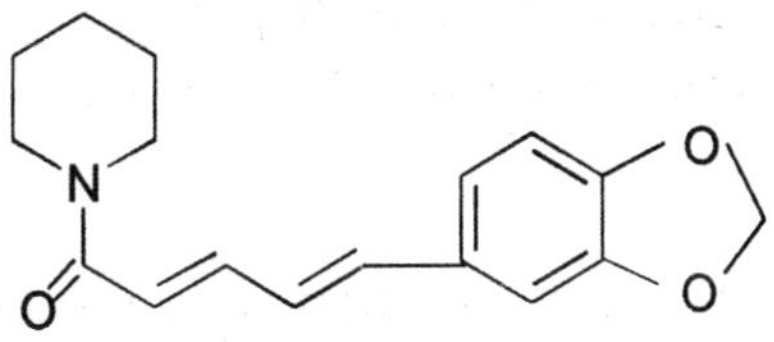

N O O O

Piperine

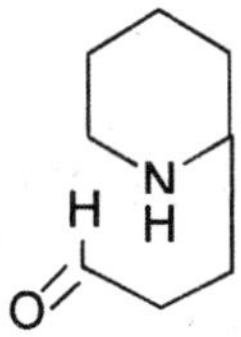

N H H H O

Pelletierine

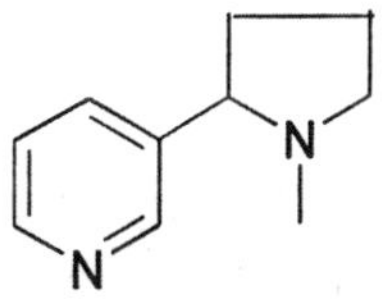

N N

Nicotine

N H

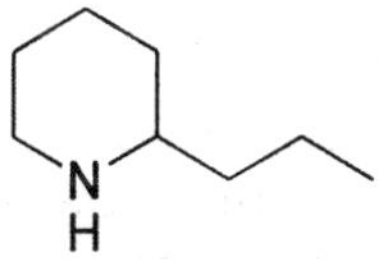

Coniine

Pyridoquinolines

This group represented by alkaloids of *Festuca* and *Lolium,* has a 3-ring system with two of the rings containing nitrogen e.g. perlolidine (*Lolium*).

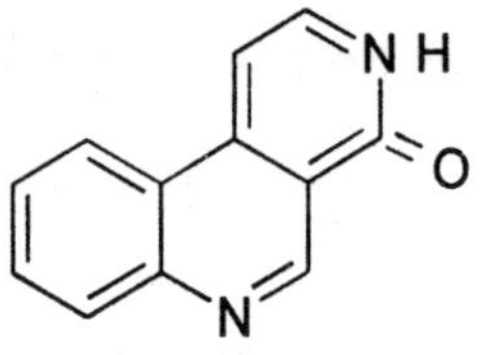

Perlolidine 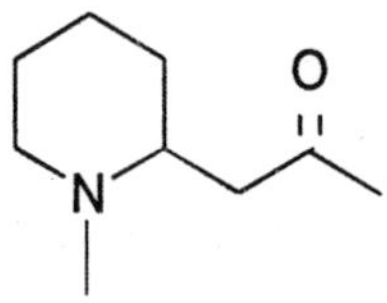 Hygrine

Pyrrolidines

This group is characterized by one or two pyrrolidine rings. Some of the interesting members here are (a) Ficine giving raise to the flavone chrysin on treatment with alkali and (b) Gerrardine containing sulphur, found in Rhizophoraceae. Hygrine (*Erythroxylum*) and stachydrine (*Stachys*) are some other common members of this group.

Pyrrolizidines

This is a fairly large group reported from the Boraginaceae (*Heliotropium, Cyanoglossum* etc), the Asteraceae (*Senecio*), Fabaceae (*Crotalaria*) and Rhizophoraceae (*Cassipourea*). All of them contain a recine (having a pyrrolizidine nucleus with a-CH_2OH group at position 1, esterified with one, two or three necic acids (mono/ dicarboxylic acids). A wide variety of necic acids such as senecic, trachelanthic, angelic, latifolic, sarracinic acids etc. are seen in plants. E.g. Retronecine (*Crotalaria*), lolin (*Lolium*), laburnine (*Laburnum*) etc.

Quinazolines

This small group of alkaloids is fairly widely distributed in Acanthaceae, Rutaceae, and Fabaceae etc. The important members are vasicine (peganine), glycosmine, aniflorine etc.

Quinoline Alkaloids

These alkaloids, characterized by a quinoline nucleus, are by and large confined to the Rutaceae except for *Cinchona* (Rubiaceae). Quinoline alkaloids are subdivided to 1) Simple quinolines (Casimiroine- *Casimiroa*), 2) Furoquinolines (Dictamine in *Dictamnus,*

Skimmianine in *Aegle* and Kokusaginine in *Acronychia*), 3) Quinolyl-quinuclides (alkaloids of *Cinchona*) and 4) Quinolyl-imidazoles (Macrorungine in *Macrorungia*)

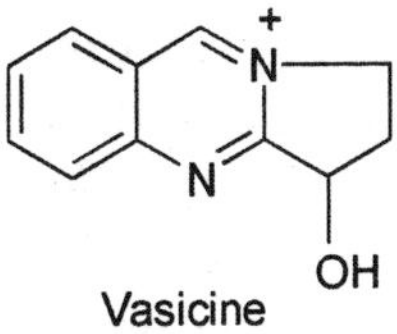

Vasicine

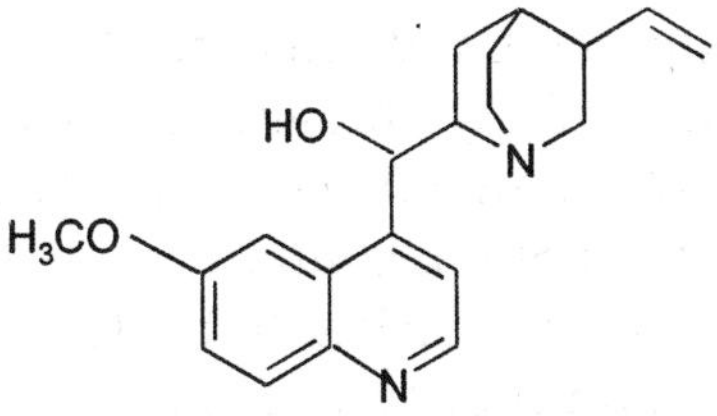

Quinine

Quinolizidines / Lupinane alkaloids

Having a quinolizidine nucleus, these alkaloids are known as nor-lupinane or octahydropyridocoline also. They are very common in the Fabales, but also are seen in Chenopodiaceae, Ranunculaceae, and Nymphaeaceae etc. Since many of them occur in the genus *Lupinus*, they are also known as Lupinane alkaloids. These alkaloids are seen mostly in the aerial parts of the plant and are toxic to herbivores. E.g. cytisine (*Cytisus*), lupinine (*Lupinus*), sparteine (*Spartium*) etc.

Sesquiterpene alkaloids

This is a small group of which celapanin and related alkaloids from *Celastrus paniculatus* as well as muscicapines from *Croton muscicapa* are the pharmacologically important compounds. Sesquiterpene nicotinoyl alkaloids such as putterine A, mayteine etc, reported from *Maytenus* sp., are characterised by a dihydroagarofuran sesquiterpene bonded as a diester to a substituted nicotinic acid. These compounds possess marked insecticidal activity (Schaneberg *et al.*, 2001).

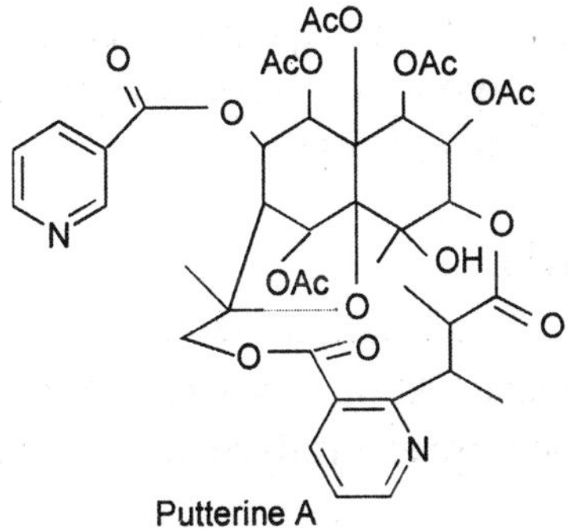

Putterine A

Steroidal alkaloids

This is a fairly large group characterized by a steroidal skeleton (derived by terpenoid pathways) and a nitrogen containing ring (pyridine, lupinane, indolizidine etc.). Many of these alkaloids occur as esters of acids like acetic, angelic, methyl butyric, vanillic and veratric acids. Some of the members from Solanaceae occur as glycosides also. Tomatine, solanine etc contain an aglycone of steroidal alkaloid and an oligosaccharide attached to the C_3, (similar to saponins in structure and function and they form emulsions with water). *Solanum* alkaloids are now used as starting materials for the commercial preparation of steroidal hormones. Some of the important members in steroidal alkaloids are conessine (*Holarrhena*), germine, jervine (*Veratrum*) solanine (Tomato) and zygacine (*Zigadenus*)

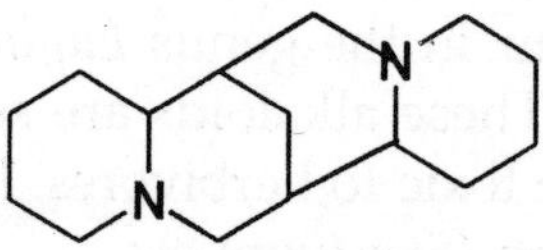

Sparteine

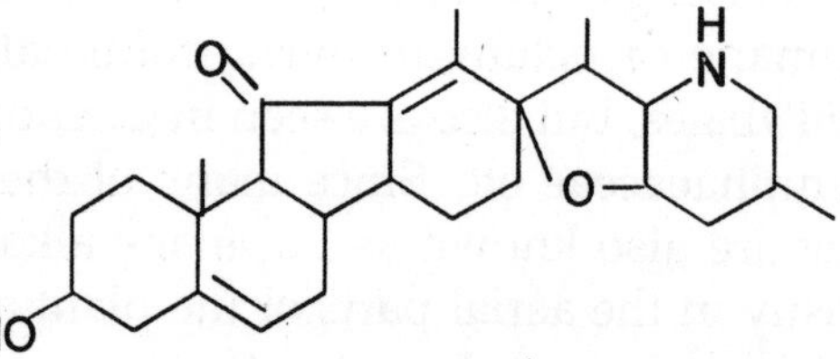

Jervine

Tropane alkaloids

Tropane alkaloids form a small group, of which more than half occur in plants belonging to Solanaceae, a dozen in Erythroxylaceae and one each in Euphorbiaceae, Convolvulaceae, Lentibulariaceae, Brassicaceae, Rhizophoraceae and Asteraceae. The tropane nucleus is bicyclic and is formed by the condensation of ornithine (a pyrrolidine precursor) and three acetate units. The fact that both pyrrolidine and piperidine ring systems can be visualized in the tropane skeleton prompted some authors to merge these alkaloids in piperidine or pyrrolidine groups. Hyoscine, hyoscyamine and cocaine are the important members of this group.

Tropolone alkaloids

This is a group of protoalkaloids possessing a tropolone ring, confined to the Liliaceae. About 25 structures are known, of which colchicine is the lone representative having significant medicinal importance.

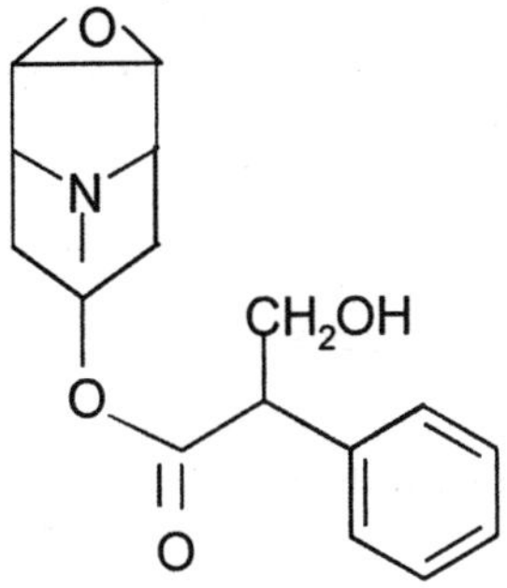

Hyoscine 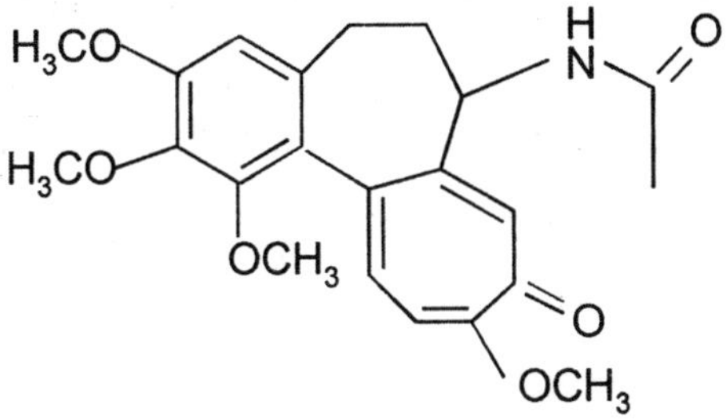 Colchicine

1.2.2. *Terpenoids*

Terpenoids form another large class of compounds with basic skeletons derived from mevalonic acid, or a closely related precursor. The number of known terpenoids exceeds 15,000. They are considered built up of isopentane or isoprene units and thus are called "isoprenoids" or "isopentanoids". The participant isoprene units are linked in various ways, with different modes of ring closure, unsaturation and different functional groups to produce a large variety of compounds accomplishing widely differing duties in the plant. The isoprene residues are usually linked in a 'head to tail' fashion, though head to head, tail-to-tail and irregular linkages do occur.

Not all terpenoids comprise of carbon atoms in multiples of five as can be expected of them, considering that they are composed of isopentane units. Some compounds have additional or missing carbon atoms, the modifications (probably) being managed in the course of the long biosynthetic pathways. Sometimes isoprenoid side chains occur attached to non-terpenoid central skeletons.

The classification of terpenoids is present in Table 4.

Monoterpenoids

This is a heterogeneous class of compounds differing in many of their chemical and physical properties. The volatile monoterpenes, generally comprising of the volatile oils were the only group known earlier. But the discovery of glycosidic iridoids and the seven-membered tropolones made the group very versatile in terms of the nature and properties. The volatile components are treated as normal monoterpenes here.

Table 4. Classification of Terpenoids

Class	*Number of carbon atoms*	*Number of isoprenes*	*Sources*
Hemiterpenoids (C_5H_8)	5	1	Volatile oils, esters
Monoterpenoids ($C_{10}H_{16}$)	10	2	Volatile oils, glycosides, mixed terpenoids
Sesquiterpenoids ($C_{15}H_{24}$)	15	3	Volatile oils, bitter principles
Diterpenoids ($C_{20}H_{32}$)	20	4	Resins, Chlorophyll
Sesterpenoids ($C_{25}H_{40}$)	25	5	Rare (mostly in animals)
Triterpenoids ($C_{30}H_{48}$)	30	6	Resins, waxes, steroids, saponins, cardiac glycosides
Tetraterpenoids ($C_{40}H_{84}$)	40	8	Carotenoids
Polyterpenoids ($(C_5H_8)\eta$)	$\propto$	n	Rubber and gutta

(a) Normal monoterpenes: These are colorless, steam distillable, water insoluble liquids with characteristic aroma, with boiling points ranging from 140°C to 180°C. These C_{10} compounds are formed by various combinations such as the head to tail, head to head or tail-to-tail condensation of two isoprene residues. These compounds with every possible mode of ring closure, various degrees of unsaturation, and substitution of different functional groups produce a large contingent of more than 2300 members which can be classified as derivatives of 15 common types of basic skeletons and 15 less common types. They include all aliphatic and cyclic (having basically a 6-membered ring), steam distillable monoterpenes. These occur usually in their free state in steam distillable oils, and possess distinct aroma. Recently, non-steam distillable monoterpene glycosides, e.g., β-D-glucosides of geraniol, neral, citronellol, thymol and carvacrol, have been found to exist naturally.

Normal monoterpenes are classified as (i) acyclic, (ii) monocyclic, and (iii) bicyclic which may occur as pure hydrocarbons or have various functional groups such as aldehydes, alcohols, ketones or oxides.

i) *Acyclic normal monoterpenes:* Majority of the acyclic compounds are alcohols or aldehydes. The alcohols occur freely, or as esters.

(ii) *Monocyclic normal monoterpenes*: A majority of monocyclic monoterpenes possess a *p*-menthane skeleton.. The various compounds differ in the number and positions of double bonds and functional groups. Geometrical isomerism is common, resulting from the formation of asymmetrical carbon atoms due to attachment of various functional groups.

iii) *Bicyclic normal monoterpenes* : By additional ring closures of the monocyclic skeletons the bicyclic forms are formed. These are of 7 main structural types.

Normal monoterpenes are employed pharmacologically as skin stimulants, sedatives, expectorants, antiseptics etc., and to flavour foods and medicines with disagreeable tastes. A few are insecticidal, and hence are used as pesticides.

(b) Iridoids (Cyclopentanoid monoterpenes): A cyclopentano-pyran ring nucleus is the characteristic feature of these compounds. About 200 iridoids are known today widely distributed in advanced families such as Scrophulariaceae, Verbenaceae, Loganiaceae, Apocynaceae, etc. In contrast to the normal monoterpenes which occur free (in volatile oils), most of these compounds occur as β-D glucosides (loganin, asperuloside etc.) Only those of the nepetalactone type are without a sugar moiety and are volatile, occurring in essential oils. Cleavage of the cyclopentane ring of loganin results in secoiridoids (gentiopicroside, oleuropin and swertiamarin etc.).

Both iridoids and secoiridoids are mostly bitter in taste and contribute much to the bitterness of many medicines. Both these groups of compounds are found to exhibit a wide range of medicinal properties. For example, all of them are found antimicrobial in nature, while allamandin from *Allamanda cathartica* is antileukemic, oleuropin from olive as well as valeopicrates (a group of non-glycosidic compounds) from *Valeriana* spp. are found to be hypotensive and sedative; herpagide from *Herpagophytum procumbens* is analgesic and geniposide from *Gardenia jasminoides* is found to be laxative. Amarogentin and amaroswerin, two secoiridoid glycosides from gentian root, are among the bitterest substances. Some of the iridoids are found to be anti-inflammatory in that they interfere with the production of inflammation promoters like prostaglandins and leukotrienes.

Acyclic monoterpenes

OH

CHO

CH_2OH

Myrcene Linalool Geraniol Neral

Monocyclic monoterpenes

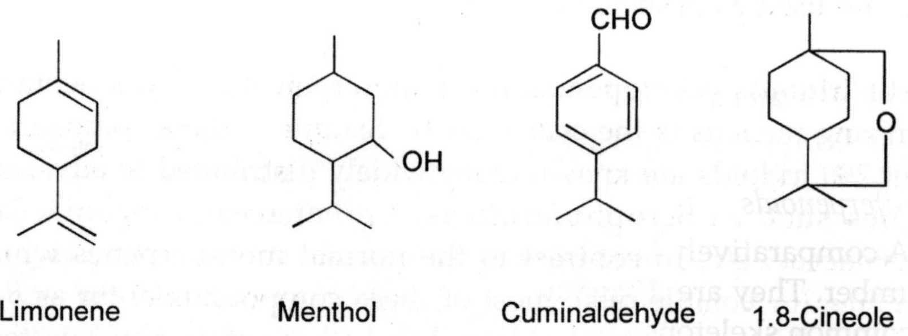

Limonene Menthol Cuminaldehyde 1.8-Cineole

Bicyclic monoterpenes

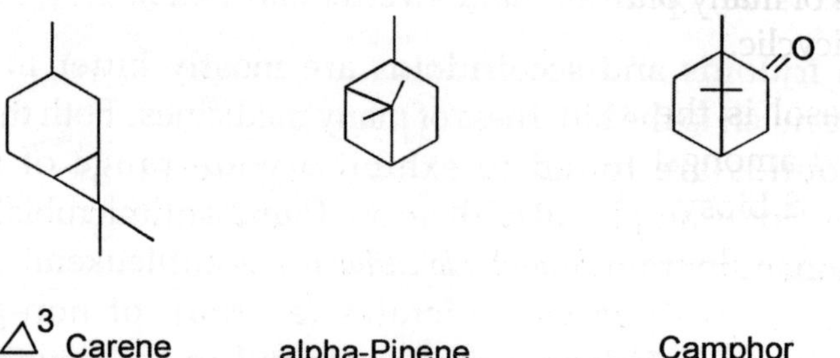

Δ^3 Carene alpha-Pinene Camphor

(c) Tropolones: Tropolones possess a seven membered ring with a double bond system conjugated with a keto group and a hydroxyl group, as their basic structure. They resemble phenolics in many properties. They are of restricted distribution, identified as yet only in certain fungi and conifers, e.g., γ-thujaplicene in *Thuja*. They are strong fungicides. They occur freely or linked with other compounds.

The C_{10} skeleton of monoterpenes are seen in combination with a large number of other compounds also. Such compounds, sometimes named as **mixed terpenoids,** are found among phenols (cannabinoids); flavonoids (rotenone, calophyllolide etc.); alkaloids (many of the indole alkaloids such as iboga, aspidosperma and corynanthe types); ketoalcohols (pyrethrins) or coumarins (marmelosin).

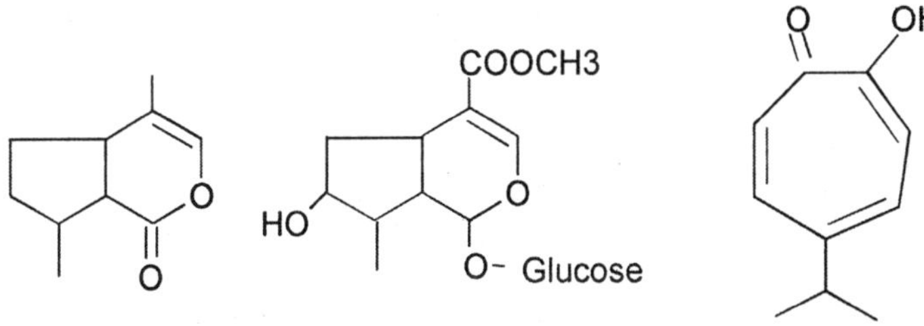

Nepetalactone Loganin Thujaplicene

Sesquiterpenoids

A comparatively larger group, sesquiterpenes are more than 5000 in number. They are distributed in 30 main skeletal structures and 70 less common skeletons.

Sesquiterpenes are composed of three isoprene units, *i.e.,* are C_{15} compounds; and occur in steam distillable volatile oils, and in the bitter principles of many plants. Sesquiterpenes are (a) acyclic, (b) monocyclic and (c) bicyclic.

Farnesol is the most important acyclic sesquiterpene widely distributed amongst plants. Its pyrophosphate is a key intermediate in terpenoid biosynthesis (Robinson, 1975).Abscissic acid, a growth regulator, is a monocyclic sesquiterpene carboxylic acid having a carotenoid skeleton resembling violaxanthin.

Bicyclic sesquiterpenes may be bicyclic naphthalenic (having a perhydronaphthalene system - cadinene, dimenol, erimophilone groups) or bicyclic azulenic (having a cyclopentane and cycloheptane rings), macrocyclic (germacranolide group) or dilactones (vernolepin).

Sesquiterpenes are known to exhibit a number of biological and pharmaceutical properties. Guaiazulene and chamazulenes are anti-inflammatory in nature. Bisabolol is the antiphlogistic and spasmolytic agents as also, petasin and isopetasin (the latter two from *Petasites hybridus*). Illudin M. and S., tumigillin etc. show antibiotic activity.

Being bitter in taste, many are the constituents of a few bitter medicines. The Asteraceae are particularly rich in sesquiterpenes especially sesquiterpene lactones responsible for the allelopathic properties of many a plants there. A large number of sesquiterpenes such as germacranolides, guaianolides and pseudoguaianolides, elemanolides and related compounds are found to exhibit antitumour activities.

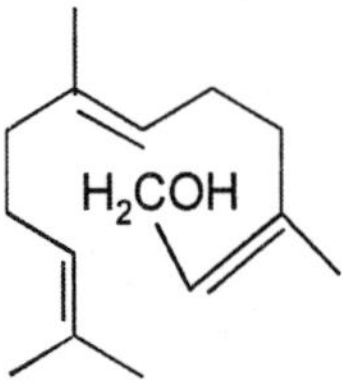

Farnesol 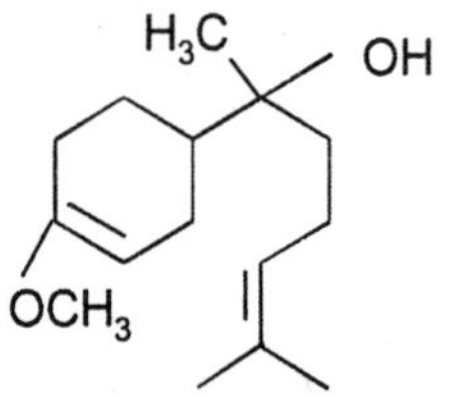 Bisabolol 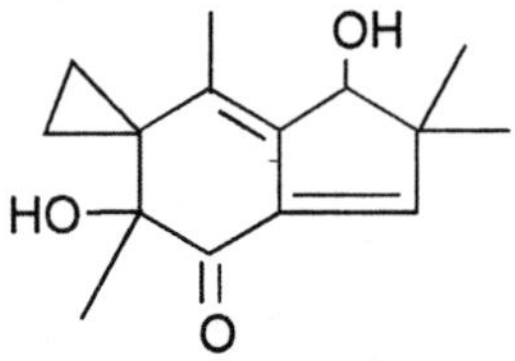 Illudin-M

The **volatile oils** of plants contain both monoterpenes and sesquiterpenes. Monoterpenes form the major components of large number of oils, though sesquiterpenes are present in many oils in sizable amount or in a few, they predominate. Functionally the volatile oils have a number of roles. Being antimicrobial, they protect the plants from microbes. Since a number of monoterpenes in certain volatile oils are pheromones, they attract pollinator insects and the specific monoterpene/s contribute much to the pollinator specificity, highly significant for insect pollination. The particular smell they emit help them being picked up by various birds or animals which in turn act as dispersal agents.

In medicine, the different essential oils and their constituents are generally used against infections particularly those of the bronchial and urinary tracts and in preventing sepsis of burns and wounds. The antiseptic activity of thymol is about 20 times more than phenol. Eugenol, menthol are employed in toothpastes for their antiseptic and rubefacient actions that are beneficial to gums while thymol and carvacrol are used in mouthwashes.

Ascaridole is an important anthelmintic. Due to the irritant properties, some are used as counter irritants and as rubefacients in the form of embrocations and liniments. They cause an initial feeling of warmth and smarting, which is often followed by a mild local anesthesia making them useful in antipruritic preparations. Such preparations are useful to alleviate rheumatic pain and neuralgia and in common colds and bronchitis. Due to their mild irritation of the bronchial glands, they are also used as inhalants with expectorant and cough stimulant properties. Volatile fractions of poppy seeds, sacred basil and cumin seeds are useful as anticancer agents by interaction with glutathione-S-transferase, an enzyme that inhibits cancer development (Aruna and Sivaramakrishnan, 1996).

Certain oils are found to act on central nervous system and thus can be used for their central stimulating, central sedative and narcotic effects. The sedative effect exhibited by oils of valerian, calamus, melissa and lavender are due to citronellal, geraniol, citronellol and citral, the first showing the greatest activity. Central stimulating effect is exhibited by camphor. The carminative property exhibited by certain volatile oils is due to local irritation (by which the enzymes are released) and by the spasmolytic effect.

Diterpenoids

Diterpenes, the dimers of monoterpenes, are C_{20} compounds formed by the condensation of 4 isoprene residues. The known 2500 or more diterpenes are grouped into twenty major common groups and 4 less common groups. Here also various substituents such as alcohols, aldehydes, ketones, and acids are all known in this group, besides the hydrocarbons. Phytol, which constitutes the lipidic tail of the chlorophyll molecule, is the only acyclic diterpene known and most of the diterpenoids are regarded as derived from it by ring closures, oxidation and substitutions. The monocyclic derivatives are known by a few like artemisin, wormwood diterpene and camphorene. The rest of diterpenes are having larger ring structures. The important groups found in this category are: (1) bicyclic (labdane, abietane, pimarane, (2) tricyclic (cassane, kaurene, gibbane), (3) tetracyclic (bayerene) and (4) pentacyclic (trachylobane). Gibberellins (growth regulators), ryanodine (insecticide) and phorbol (purgative) are some of the physiologically active diterpenes.

Diterpenes include a number of compounds having very interesting properties. Stevioside, a diterpene glycoside isolated from leaves of *Stevia rebaudiana* Bert. (Asteraceae), is at least 300 times sweeter than sugar and used as a sweetener for diabetic patients and dieters. Marrubin and columbin are some of the bitterest principles. A number of other diterpenes like phorbol, are highly toxic while another such member isolated from *Taxus brevifolia* Nutt., taxol, is one of most acclaimed antileukemic principles. Other compounds exhibiting anticancerous properties are taxodone and taxodione from *Taxodium distichum,* nagilactones B – E from *Podocarpus,* triptolide and tripdiolide from *Tripterygium wilfordii,* ingenol-3, 20-dibenzoate from *Euphorbia esula* and phorbol, 12-tiglate, 13-decanoate from *Croton tiglium.* Ginkgolides of Ginkgo are platelet-activating factor (PAF) antagonists.

Taxol

Stevioside

Ginkgolide A

Triptolide

Triterpenoids

The triterpenoids are C_{30} compounds or their derivatives made up of two sesquiterpene units linked tail to tail. This is a large and structurally diverse group, biosynthesized from squalene, oxidosqualene or bis-oxidosqualene. More than 5000 triterpenoids are isolated from plants. Nearly hundred types of skeletons are known in this group. Acyclic members are rare, represented by squalene, oxidosqualene or bis-oxidosqualene and lower ring systems are comparatively rare. Most triterpenoids are tetracyclic (6-6-6-5 or 6-6-6-6-5) or pentacylic (6-6-6-6-6). Achilleol A and camelliol C are the two monocyclic triterpenes known, isolated from *Achillea odorata* (Asteraceae) and *Camellia sasanqua* respectively (Akihisa *et al.*, 1999). Bicyclic members are represented by polypoda –8 (26), 13, 17, 21-tetraen-3-ol and polypoda-7,13,17,21- tetraen-3-ol, isolated from *Cratoxylum cochinchinense* [Clusiaceae], as well as polypoda-13,17,21-trien-3,8-diol from *Pistacia* (Boar *et al.* 1984). No tricyclic triterpene is isolated from the nature so far though a tricyclic malabarica-14(27), 17,21-trien-3-diol is synthesized (Jakupovic *et al.*, 1987, Xu, et al., 2004). Various functional groups such as alcohols, aldehydes, ketones or carboxylic acids are seen in them. Many are colourless crystalline compounds with high melting points. Optical isomerism is a common phenomenon exhibited by these natural products.

Generally the **tetracyclic triterpenes** (6-6-6-5) are characterized by a cyclopentanoperhydrophenanthrene nucleus and an 8-carbon side chain at C_{17}. As such they resemble steroids and thus some workers have grouped them as sterols and named them accordingly, e.g., lanosterol. They occur free in laticiferous plants and cutins and as glycosides in some saponins and cardiac glycosides. Cucurbitacins, limonoids and quassinoids are some of the other important members in this group. Withanolides are a group of steroidal lactones of important pharmacological activity.

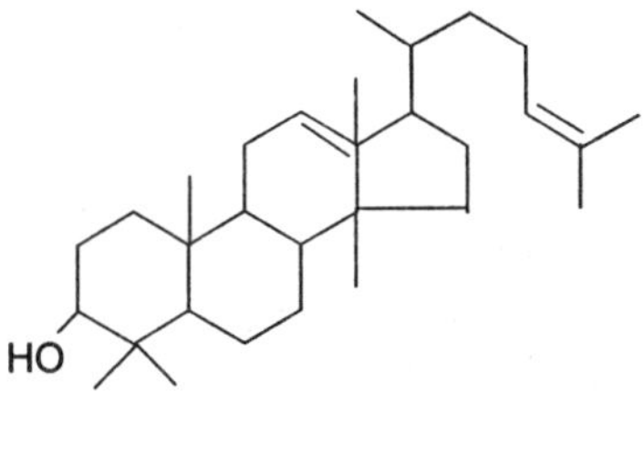

HO

Lanosterol

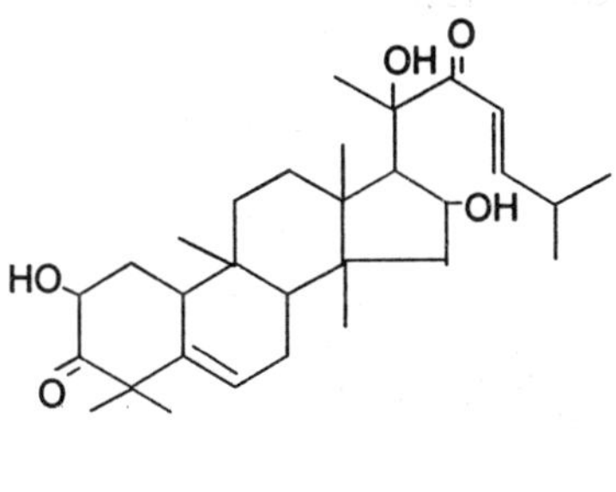

OH O
OH
HO
O

Cucurbitacin E (Elaterin)

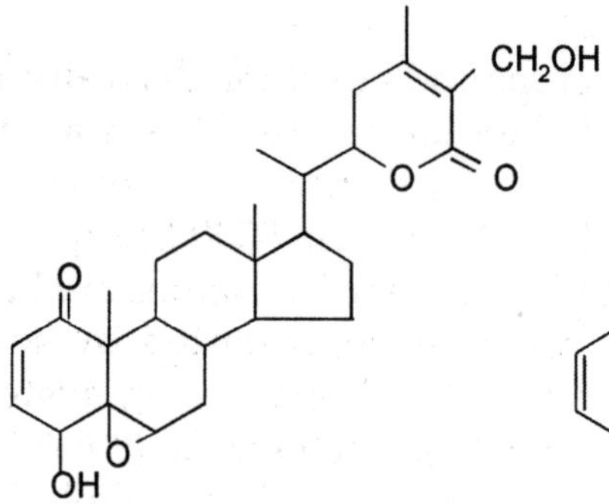

Withferin A 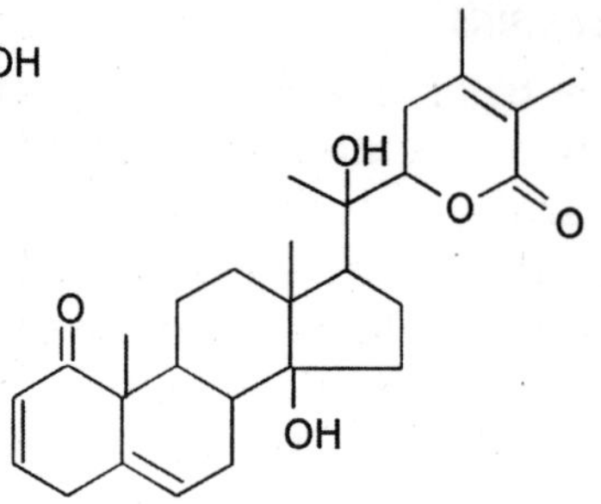   Withanolide G

Pentacyclic triterpenes (6-6-6-6-5 or 6-6-6-6-6) are widely distributed in higher plants, occurring free or as glycosides (saponins). The aglycones, found in the excretions (resins) and cuticle, have a protective as well as water-proofing function e.g., ß-amyrin. Invariably all the members are oxygenated at the C_3 position. Gymnemic acids, which can destroy the ability to taste sweet substance, form another interesting group.

a. Steroids

Steroids also possess the cyclopentanoperhydrophenanthrene nucleus (6-6-6-5) and 8-carbon side chain at C_{17}. But they differ from the tetracyclic triterpenes by having only two methyl groups, at positions 10 and 13. These compounds, widely distributed in animal kingdom and thought to be rare in plant kingdom, are extensively located in higher plants now, where they occur free, as esters or as other derivatives. The typical "animal sterols" like cholesterol and hormones like estrone and ecdysones are reported from many plants, but their role in plants is still very ambiguous. Ergosterol is confined to fungi whereas sitosterol, stigmasterol and campesterol are widely distributed in the plant kingdom. More than 300 plant sterols have been identified.

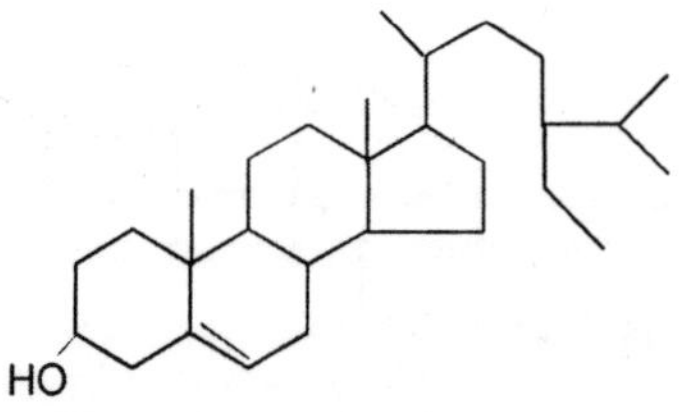

Sitosterol 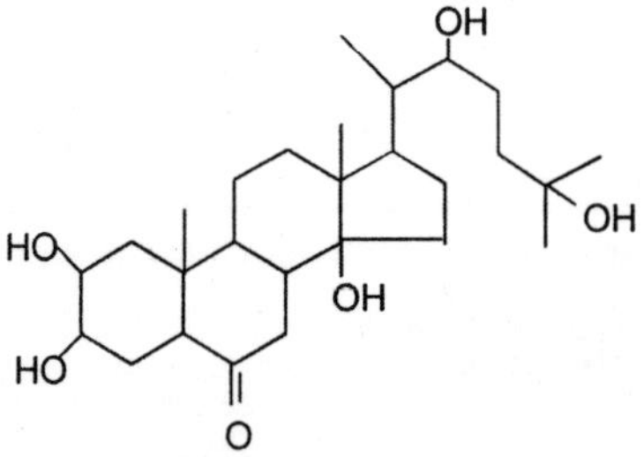   Ecdysone

b. Brassinosteroids

This group includes about 40 steroidal compounds exhibiting various physiological responses like proton pump activation, reorientation of cellulose micro fibrils, xylogenesis etc and morphological characteristics such as increased stem elongation, pollen tube growth, unrolling and bending of grass leaves. These are C_{27}- C_{29} steroids with different substitutions on the A and B rings and side chain. Brassinolide, a steroid with a B ring which is an O-containing 7-membered ring, is the first brassinosteroid isolated from the pollen of *Brassica napus.* Costasterone from the insect galls of Chest nut (*Castanea crenata*) is another important member in this group. Brassinosteroids are found to occur in ferns, Gymnosperms and Angiosperms, but are absent from microorganisms.

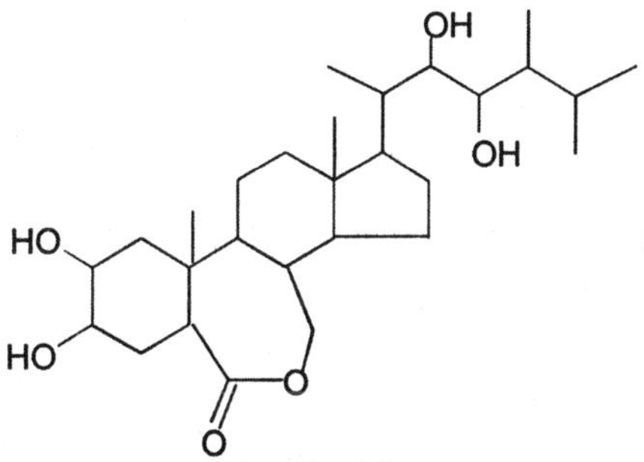

OH OH HO HO O O Brassinolide

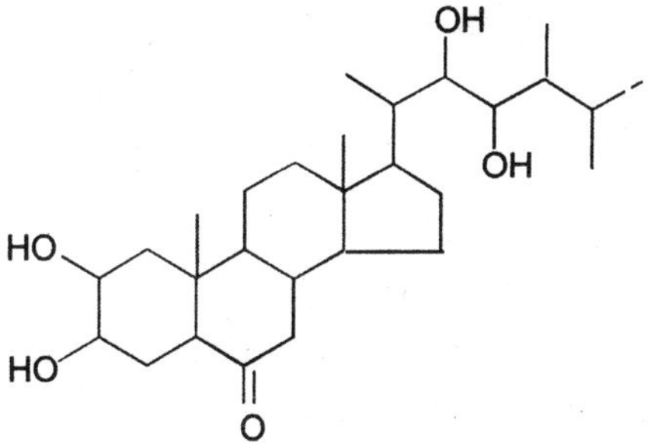

OH OH HO HO O Castasterone

c. Sterolins

Sterolins are sterol mono/biglycosides and occur along with free sterols in waxes and unsaponifiable lipid fractions. Unlike saponins, they neither form emulsions with water nor haemolyse RBC nor are non-toxic to fish. It is due to the lesser number of sugar moieties, they are insoluble in water. They possess high m.p. like sterols but are comparatively less soluble in lipid solvents. β-Sitosterol glycoside is the commonest sterolin.

d. Saponins

Saponins are the glycosides in which an oligosaccharide consisting of 2-5 sugar units and one glucuronic acid molecule is linked to C_3 position of the aglycones, are soluble in water and like soaps and produce stable forth (emulsions) when shaken with water. They are characteristically toxic to fish and haemolyse RBC. Sapogenins, the aglycones, may be either a triterpene like α-amyrin or a steroid possessing a spiroketal side chain like saraspogenin. A series of saponins may have different sugar components but same sapogenin.

Saponins are found widespread in higher plants, having been detected in well over 70 plant families and 3000 species. Steroid saponins are frequent in monocot families, whereas triterpene saponins are predominant in dicots. Gymnosperms are virtually free of saponins.

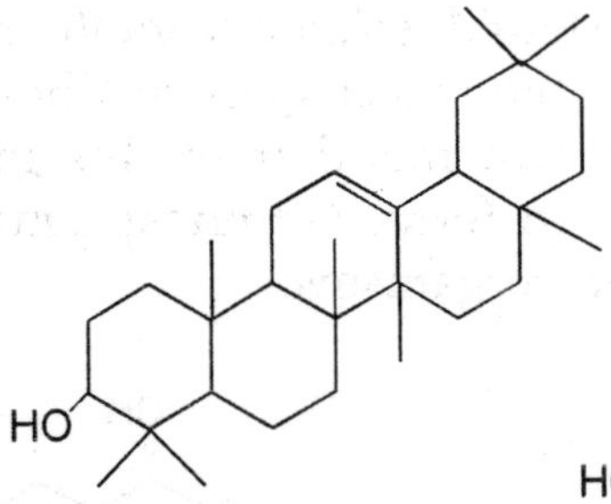

HO

beta-Amyrin

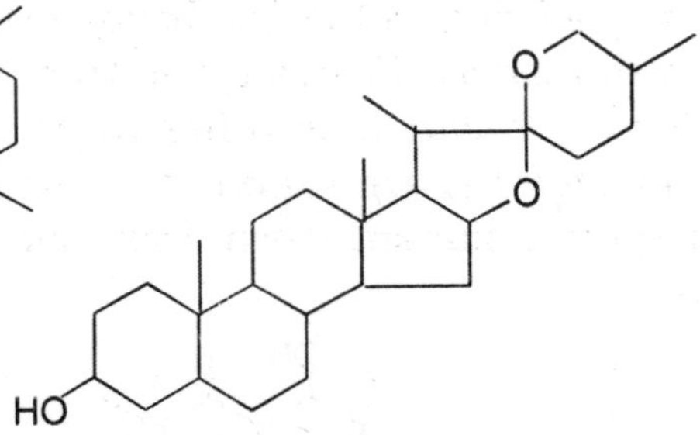

HO

Sarasapogenin

e. Limonoids and Quassinoids

All these compounds are formed by the loss of a few carbon atoms from C_{30} triterpenoids and thus known as **nor-triterpenoids.** Limonoids are characteristic to the families Rutaceae, Meliaceae and Cneoraceae, whereas quassinoids are seen in the Simaroubaceae. Limonoids are tetra-nor-triterpenoids formed by the conversion of 8-carbon side at C_{17} to a furan ring with subsequent loss of four atoms. But the first group, the **Tirucallane group** has an intact triterpenoid skeleton (C_{30}) where there is no reduction in carbon number. Grandifoliolenone (*Khaya*) and glabretal are examples of this group.

Other groups are based on the cleavage of various rings. **Azadirone group,** the second class, has all the 4 rings intact. 12α-hydroxy amoorstatin (cytostatic, from *Aphanomixis*) and trichilins (antifeedant from *Trichilia*) are some members of this group showing biological activity. The third group, **Gedunin group,** wherein ring D is cleaved and has a pyran skeleton there, has gedunin and merogedunol as representatives. Nimbin, azadirachtin and nimbandiol (from *Azadirachta* and *Melia*) represent the fourth **Azadirachtin group** which are well-known antifeedant compounds. **Toonafolin group,** characterized by cleavage of ring B, has andirobin, methyl angolensate and ekebergo lactones (from *Ekebergia*) besides toonafolin (from *Toona*).

Tricoccin group, the sixth group, represented by tricoccin S_{22} has the A ring cleaved. The next group, **surenolactone group** has both A and B rings cleaved; contains prieurianin, dregianin and hispidin (all from *Trichilia* spp.). The **C_{25} group** (penta-nor-triterpenoids) contains tricoccin S_{14}, cneorin etc and is seen in members of Cneoraceae such as *Cneora* and *Neochamaelea*.

All the various types of limonoids (expect for the last group of C_{25}) are seen in members of Meliaceae. The limonoids of Rutaceae belong mostly to the 3rd and 6th group i.e. A and D ring cleaved (**seco-limonoids**). Obacunone, deacetyl nomilin, nomilin, limonin, deoxy limonin, harrisonin, 12-ß acetoxy harrisonin, azadirachtin, salanin, trichilin B, toonacilin and 6-acetoxy toonacilin are some of the limonoids found to have antifeedant activity on many insects.

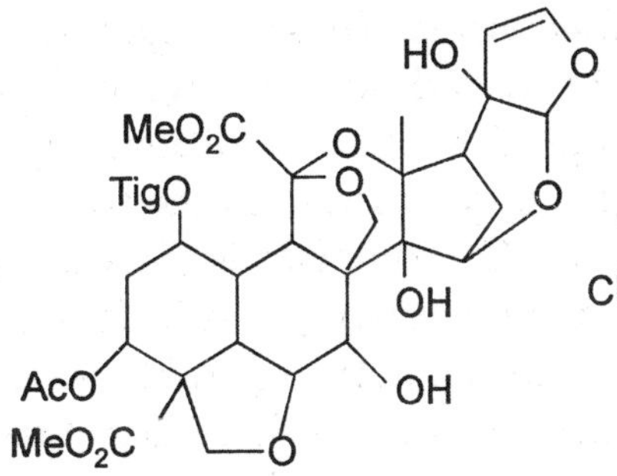

Azadirachtin 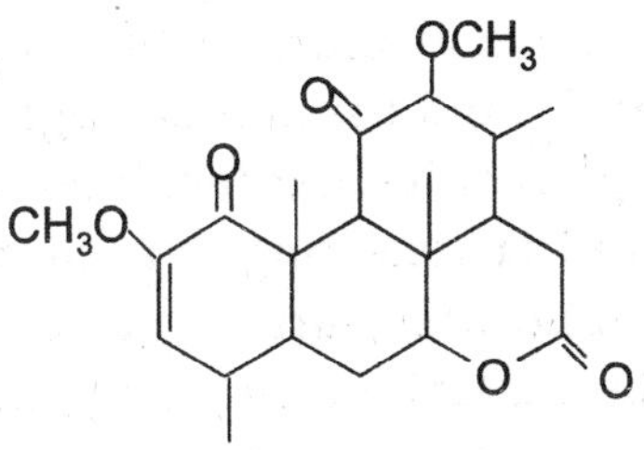 Quassin

A majority of **quassinoids** possess C_{20} or C_{19} skeletons though a few possess C_{25} skeleton. All the quassinoids are heavily oxygenated lactones. Altogether four groups are recognized here. They are:-

C_{25} quassinoids: They are considered intermediates in the biosynthesis of C_{20} quassinoids. Simarinolide and guanepolide are the examples here.

C_{20} quassinoids. They form the largest group. Bruceine, glaucarubolone, sergeolide are the important members here.

C_{19} quassinoids. E.g. Samaderin C

C_{18} quassinoids. E.g. Samaderin A

Many of the quassinoids exhibit antifeedant and insecticidal properties. Some are found active against *Plasmodium* and quite a few of them are antimicrobial in nature.

Of the 2000 or more plants belonging to the Rutaceae Meliaceae, Simaroubaceae and Cneoraceae, where the limonoids / quassinoids are widely distributed, only a couple of dozen plants have been studied for the presence of these compounds. This is an area which holds great promise in bioprospecting and a systematic survey will unearth a large number of plants containing insecticidal / antifeedant / repellant activity.

f. Cardiac glycosides

These also are triterpenoid glycosides similar to saponins in forming emulsions in water and a definite effect on the heart mostly promoting heart beats. Due to this property they were once used as arrow poisons in that, in high concentrations they cause a violent contraction of the heart muscle and resultant cardiac arrest. These compounds differ from saponins in having an unsaturated lactone ring at C_{17} of the nucleus, a *cis* juncture of the rings C & D, an additional hydroxyl group at C_{14} and the unique sugars they possess. Of the two types of cardiac glycosides, cardenolides are C_{23} compounds possessing a 5-membered lactone ring (e.g. sarmentocymarin) and bufadienolides (scilladienolides) are C_{24} compounds with a doubly saturated 6-membered lactone ring (e.g., scilliroside). Both cardenolides and bufadienolides are mutually exclusive in plant kingdom.

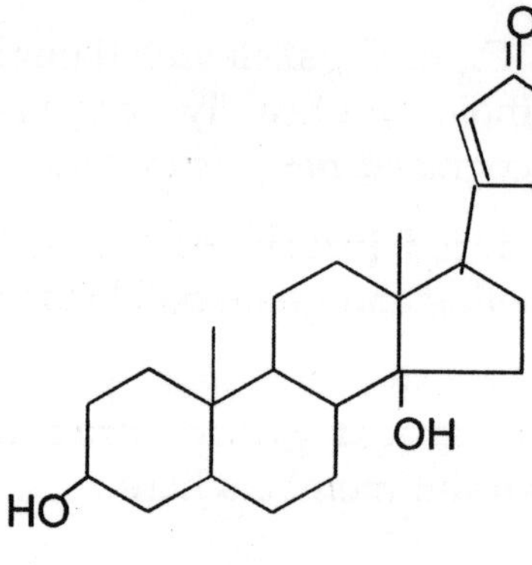

Digitoxigenin

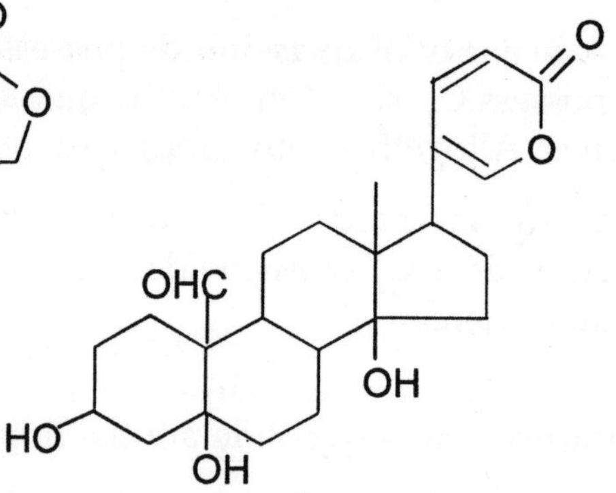

Hellebrigenin

The cardiotonic activity increases with the increase in hydroxyl groups of these glycosides. Their genins also exhibit similar properties.

All the sugars in the oligosaccharide, linked at C_3 position, possess methyl and / or methoxy groups (e.g., digitalose) and the number of sugar units present in a glycoside ranges from 1 to 4.

Carotenoids

Carotenoids are the C_{40} tetraterpenoid polyenes, formed by tail-to-tail condensation of two diterpene moieties. About 500 carotenoids are seen in plant kingdom. They are extremely widely distributed in plants, present in all plants and in a few animals through dietary intake. Carotenes are the hydrocarbons and their derivatives containing oxygen in different forms such as alcohols, aldehydes, ketones, carboxylic acids etc. are the xanthophylls. All the carotenoids are colored yellow or red, though a few like phytoene and phytofluene possessing lesser double bonds are colorless. In most of the carotenoids, double bonds are all *trans*. Higher ring systems are absent in carotenoids and all of them are either acyclic, monocyclic or bicyclic. Lycopene is the acyclic basic molecule from which all the other carotenoids are derived. Cyclisation of lycopene at one end gives γ-carotene while cyclisation at both ends produces bicyclic β-carotene. Xanthophylls may contain substituents like alcohol (monohydroxy - lutein; dihydroxy- zeaxanthin), epoxy (violaxanthin) or, in rare cases, keto (rhodoxanthin) or carboxyl (crocetin) groups. A phenolic carotenoid, 3, 3′-dihydroxy isorenieratene, is reported from *Streptomyces* sp. Generally cyclisation in carotenoids produces a cyclohexane ring, but capsanthin is peculiar in having a cyclopentane ring at one end. Most of the xanthophylls occur as esters of fatty acids (palmitic, oleic or linoleic acids). Glycosides of carotenoids with rhamnose or glucose are reported from various algae (Herzberg and Liaan-Jensen, 1971). In higher plants crocin is the only glycoside known.

β-Carotene is an essential dietary requirement as provitamin A. One molecule of β-carotene is found to give rise to only one molecule of C_{20} retinol (Vit. A). α–, and γ-Carotenes as well as cryptoxanthin containing a β-ionone residue also can give rise to vitamin A. All the carotenoids are excellent antioxidants and can be employed as food colors. Lycopene is known for its role in treating prostrate cancer. Crocin and bixin are other two carotenoids (or derived carotenoids due to the lesser number of carbon atoms) used as dyes and having marked medicinal properties.

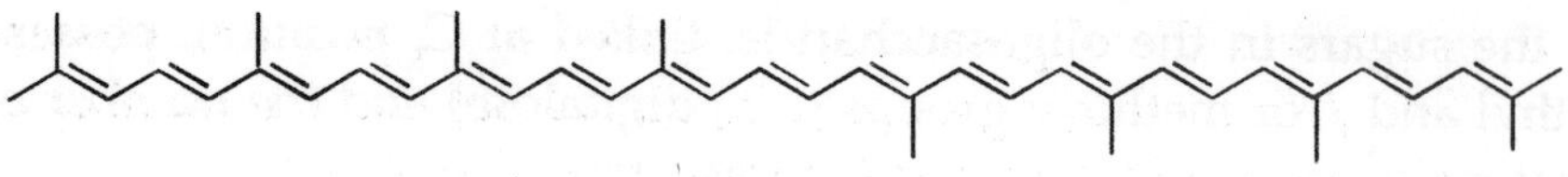

Lycopene

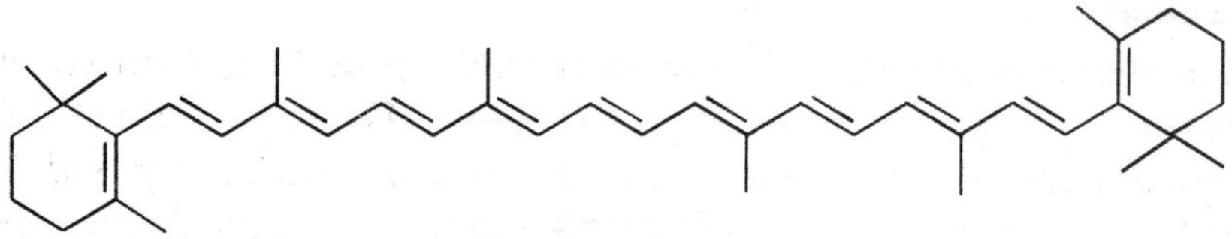

beta-Carotene

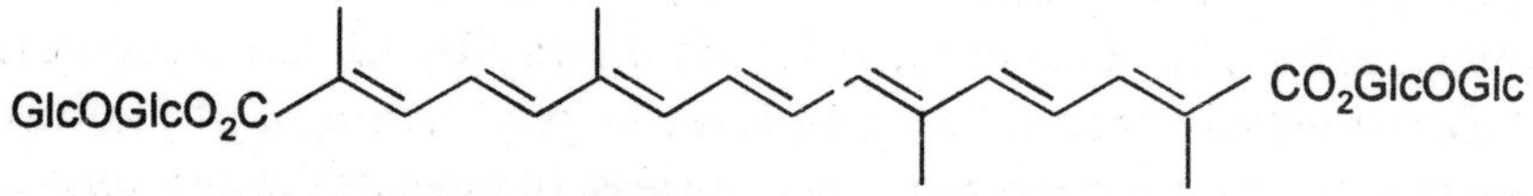

GlcOGlcO$_2$C CO$_2$GlcOGlc

Crocin

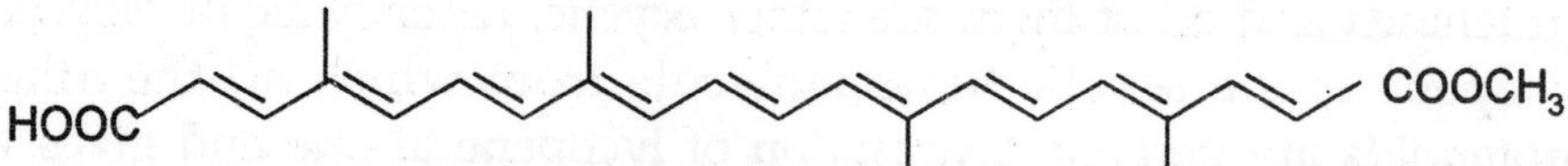

HOOC COOCH$_3$

Bixin

1.2.3. Phenolics

The term 'phenolics' includes all the aromatic (benzenoid) compounds, which possess a hydroxyl group or its substituents. More than 7,000 compounds are known in this group. This group, considered once as waste products or by-products, attracted attention after its roles as antioxidants and antimicrobials and the various physiological and pharmacological responses it stimulated in plants and animals were discovered and focused. Phenols are now found to have a universal occurrence and as such there is no plant or animal without these compounds. The several hundreds of phenols discovered are classified mainly on the number of carbon atoms they contain and the biosynthetic pathways by which each group is derived. Phenols are derived by the shikimic acid pathway from 5-dehydroquinic acid. But at times, highly complex phenols result due to the incorporation of acetate units (anthraquinone), isoprenoid chains (gossypol, ubiquinone etc), or amino groups. We still do not know much about the structure of many polyphenols like lignin, melanins and tannins. As the phenols are considered harmful (?) if brought in direct contact with the protoplasm, the phenolic function is often blocked by glycosylation, methylation or esterification. The glycosylation makes them water-soluble and helps in cellular transport.

1. Simple Phenols – C_6/C_7

These compounds contain a single aromatic ring and include all the phenolic alcohols, aldehydes, ketones and their glycosides. They are colourless solids easily getting oxidised on exposure to air or in slightly alkaline conditions. Free phenols tend to occur in woody tissues while glycosylated phenols are common in metabolically active sites (Hopkinson, 1969).

About a dozen phenolic alcohols are known, occurring free or as glycosides. Phenol itself is reported from the essential oils of pines, tobacco, mango and *Ruta*. Catechol is present in the leaves of *Ephedra*, scales of onion and rind of lemon. Resorcinol is isolated from onion scales and tea leaves, while phloroglucinol (as a glucoside) occurs in *Citrus* fruits, and as a tautomeric triketone in many members of Pteridaceae and in the essential oils of certain Angiosperms. A large number of complex and widely distributed polyphenols contain catechol or phloroglucinol units in their structure. Hydroquinone is the most widely distributed compound in this group, reported from many members of the Rosaceae, Proteaceae and Asteraceae. The phloroglucinol derivatives of *Aspidium* (Male fern) is known for their anthelmintic properties, while that of hops (*Humulus lupulus* L.) with isoprenoid side chains are the flavoring and preservative substances or beer. Phenols with methoxy groups or with short (C_3) or long (C_{15}) aliphatic side chains also are seen in plant kingdom. Thus *p*-cresol (4-methyl phenol) occurs in essential oils of several plants while 4-allyl phenol is present in *Prunus lusitanica* and urushiol and bhilawanol, two catechol derivatives with a C_{15} side chain, are the vesicant principles of *Rhus toxicodendron* and marking nut (*Semecarpus anacardium*).

A number of simple phenols occur as glycosides also. Salicin is the β-D-glucoside of salicyl alcohol and occurs in many plants. Populin and glycosmin – two derivatives of salicin – are reported from *Populus* and *Glycosmis* respectively. The glucoside of phloroglucinol, phlorin occurs in the *Citrus* peel. Arbutin is the glycoside of hydroquinone present in leaves of pear. Glucovanillin, which yields vanillin (vanillic aldehyde) the odoriferous principle of vanilla – is obtained only from monocots. Piperonal is another odoriferous principle occurring in heliotrope. β-D-Glucosides of gentisic alcohol and *p*-hydroxy acetophenone are present in *Salix*.

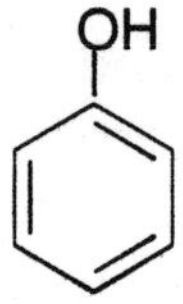

Phenol 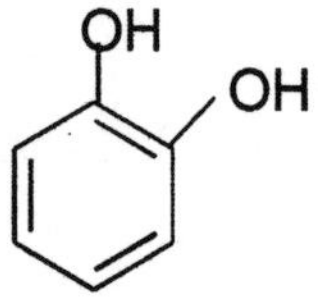 Catechol 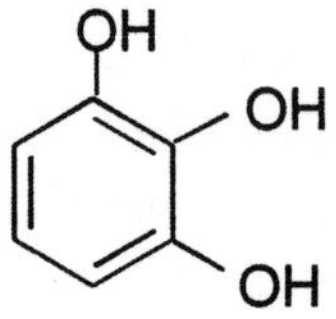 Pyrogallol

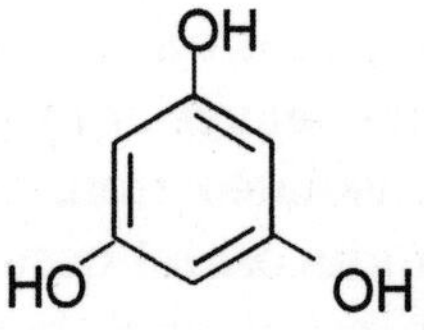

Phloroglucinol  Hydroquinone 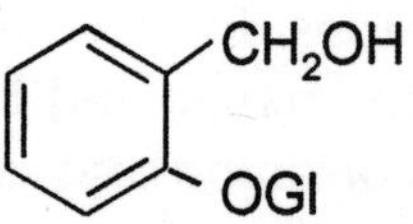 Salicin

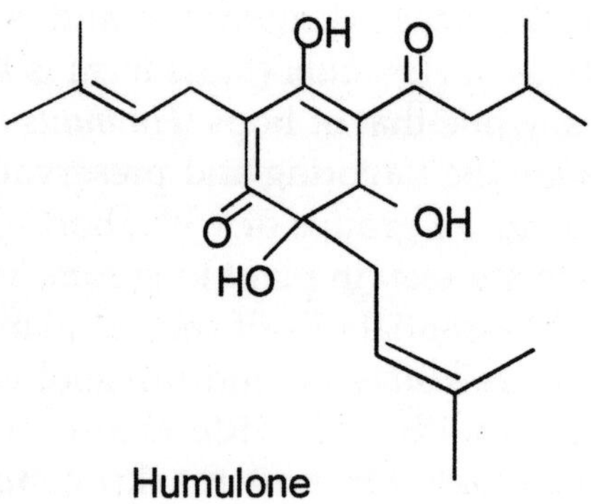

Humulone

2. *Phenolic acids* – C_7, C_8, C_9.

The term phenolic acid includes both the benzoic and cinnamic acids and they are a group widely distributed in plant kingdom. The common benzoic acids, *p*-hydroxy benzoic, vanillic and syringic acids are normally derived from *p*-coumaroyl, coniferyl and sinapyl residues present in lignin and so located in all angiosperms.

Gymnosperm lignin lacks syringyl residues. Gentisic and protocatechuic acids are two other phenolic acids common in higher plants. Salicylic and *o*-pyrocatechuic acids are frequently met with in Ericaceae. Salicylic acid occurs as its methyl ester in the oil of wintergreen (*Gaultheria procumbens*). Salicylic acid and its derivatives inhibit cyclooxygenase, a key enzyme of prostaglandin biosynthesis and therefore used in cardiac problems. Gallic and digallic acids along with the dimeric ellagic acid are the phenolic components of gallo- and ellagitannins. Hexahydroxydiphenic acid is another component homologous with ellagic acid found in tannins. These acids are said to be absent in lower groups of plants. Some phenolic acids with unusual hydroxylation patterns such as 2-hydroxy-4-methoxy, 2-hydroxy-5-methoxy and 3, 5-dimethoxy acids are seen in *Primula officinalis*. Similarly, 2-hydroxy-6-methoxy benzoic acid and 2,4-dihydroxy benzoic acid are reported from *Gloriosa superba* whereas 2-hydroxy-6-methoxy benzoic acid is present in *Colchicum autumnale*. Aldehydes such as *p*-hydroxy benzaldehyde, vanillin and salicaldehyde (corresponding to the common phenolic acids *p*-hydroxy benzoic, vanillic and salicylic acid) also are seen in many plants.

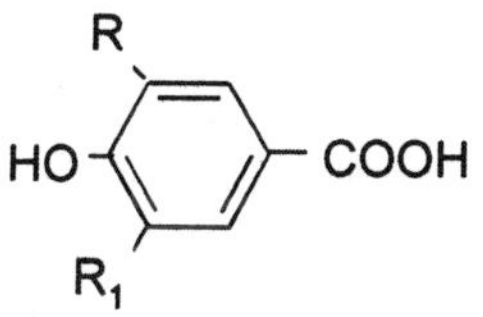

R=R_1=H, para-Hydroxy benzoic,
R=OH,R_1=H, Protocatechuic,
R=OMe, R_1=H, Vanillic,
R=R1=OH, Gallic,
R=R_1=OMe, Syringic acid.

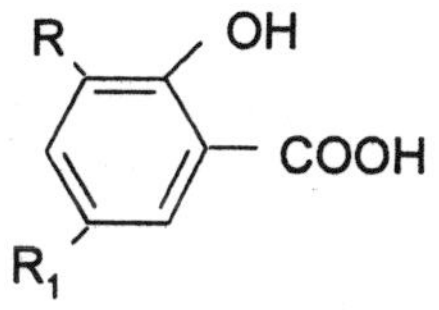

R=R_1= Salicylic,
R=OH,R_1=H,o-Pyrocatechuic,
R=H,R_1=OH, Gentisic acid,

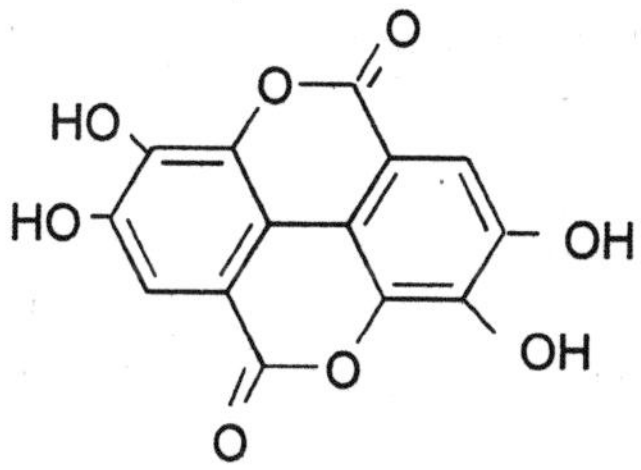

Ellagic acid

Ginkgolic acid, reported from the fruits of *Ginkgo biloba,* is a derivative of salicylic acid with a C_{15} side chain attached to the 6th position. Phthalic acids are seen from a number of fungi.

Cinnamic acids are described in phenyl propanes.

3. Phenyl acetic acids

A few phenyl acetic acids also are reported from plants. 2-Hydroxy and 4-hydroxy phenyl acetic acids are reported from *Astilbe* and *Taraxacum* respectively, while 3, 4-dihydroxy and 2, 5-dihydroxy acids are mould metabolites.

4. Acetophenones

These are C_6-C_2 ketones (aromatic ketones), a few of them are found to be widely distributed. Of the 15 known compounds, acetophenone itself forms the most widely distributed compound. Some of these ketones like acetovanillone, paeonol and 4-hydroxy acetophenone occur as glycosides. Acetophenone is found to be strongly allelopathic.

5. Phenyl propanoids – C_6-C_3

These compounds occur both in open and closed chain forms with or without combining with sugars. Some of the open chain forms are most widespread among the phenolics. *p*-Coumaroyl alcohol, coniferyl alcohol and sinapyl alcohol (resembling *p*-hydroxy benzoic, vanillic and syringic ring structures) are well-known components of lignin and are present, in the free form, in cambial saps of the higher plants. Phenyl propenes, a group of volatile components, occur associated with essential oils of a number of spices. Of the various cinnamic acids known, caffeic, *p*-coumaric, ferulic and sinapic acids are abundant in the leaves of higher plants free or in combination with anthocyanins. Depsides are esters formed of two or more phenolic acids. Depsidones are depsides with one O-linkage making a third ring. These compounds are very frequent in fungi while their abundance in higher plants is yet to be proved. Chlorogenic acid (3-O-caffeoyl-quinic acid), by far, is the most common representative of this group. Chicoric acid of chicory contains two molecules of caffeic acid and one mol. of tartaric acid. Chichoric, chlorogenic and caffeic acids belong to a new class of HIV medications known as **integrase inhibitors**, which block the enzymes those allow viruses to invade cells and "integrate" their DNA with the cell's DNA.

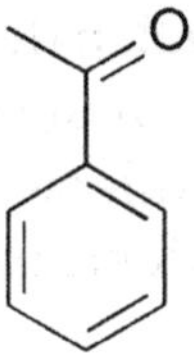

Acetophenone

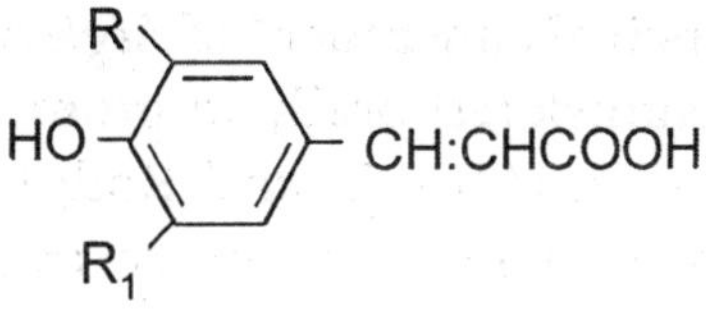

$R=R_1=H$, *p*-Coumaric,
$R=OH, R_1=H$, Caffeic,
$R=OMe, R_1=H$, Ferulic,
$R=R_1=OMe$, Sinapic.

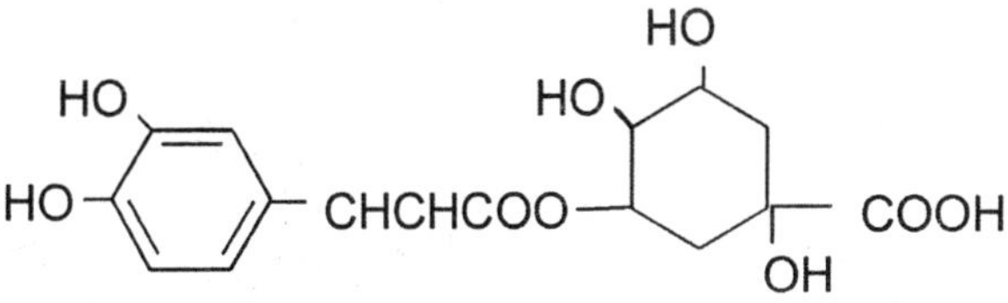

Chlorogenic acid

6. Coumarins

Coumarins are basically phenylpropanes formed by the lactonisation of *o*-hydroxy cinnamic acid. More than 1200 coumarins are known from higher plants but are more common in dicotyledons. Coumarin itself is present as an aromatic constituent in many plants especially Tonka beans and as a glycoside in *Melilotus*. Almost all the naturally occurring compounds are O-substituted at C_7. 7-Hydroxy coumarin – umbelliferone- is widely distributed in the Apiaceae while daphnetin (7, 8-dihydroxy coumarin) is common in the Thymeliaceae and Euphorbiaceae. Mammein, with isoprenoid side chains (from *Mammea americana* L.) is a known insecticide. Scopoletin is the most common coumarin in higher plants.

Furano- and pyranocoumarins have a furan or pyran ring fused with the benzene ring of coumarins. 6, 7-Furanocoumarins (psoralens) are abundant in the Apiaceae and Rutaceae. Some of the furanocoumarins are employed as fish poisons while some others are effectively used in medicine as spasmolytic and vasodilatory agents.

Bergaptene is the bergamot camphor and imperatorin (marmelosin) is a principal constituent of *Aegle marmelos* (Fruit). Psoralen derivatives are administered orally to promote the susceptibility of sun tanning of the skin (Musajo and Rodghiero, 1972). Isopsoralens are 7, 8-furocoumarins. Pyranocoumarins are also designated as chromanocoumarins.

Isocoumarins are a small group of α-pyrones reported from the Saxifragaceae and related groups. Bergenin is the most widespread member in this group. Phyllodulcin, a dihydroisocoumarin, is the sweetening agent in *Hydrangea.*

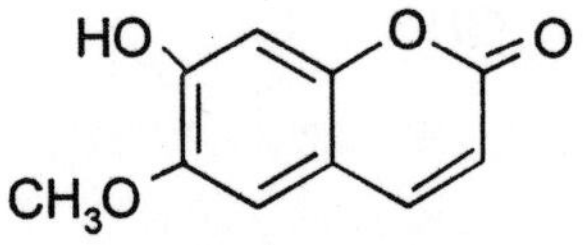

Scopoletin

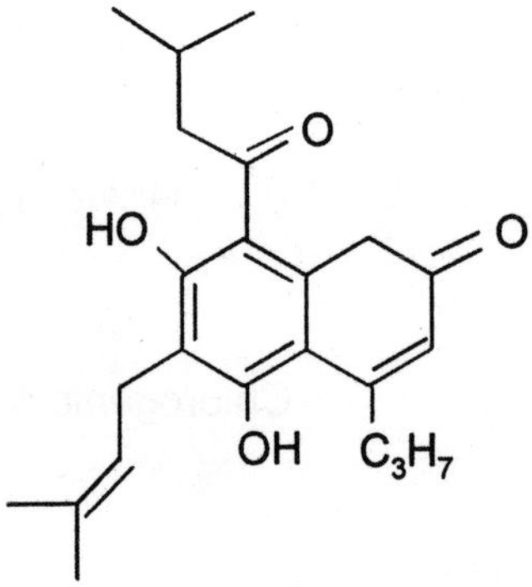

Mammein

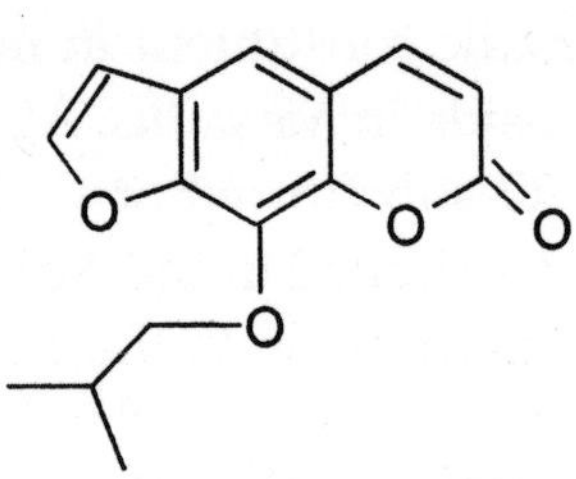

Marmalosin

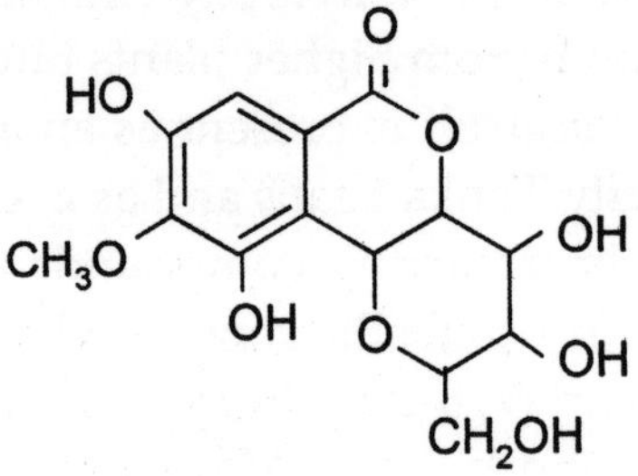

Bergenin

Coumarins reach a concentration of up to 20% in some members of the Apiaceae. Furocoumarins can intercalate DNA and upon illumination with UV light can form cross-links with DNA bases and proteins.

7. Benzophenones

These are C_6-C_1-C_6 compounds having a very limited distribution among higher plants. Only 8 such compounds are known till now. Cotoin (2, 6-dihydroxy-4-methoxy-benzophenone) and hydrocotoin (6-hydroxy-2, 4-dimethoxy benzophenone) found in coto bark (*Aniba*) are used as dyes and astringents in medicines. Fustic and Osage orange are well- known dyes due to the presence of maclurin (2, 4, 6, 3′, 4′-pentahydroxy benzophenone). Some of the hydroxy benzophenone derivatives are condensed tannins.

Cotoin

Maclurin

8. Xanthones

Xanthones resemble benzophenones in having a C_6-C_1-C_6 skeleton but differ in having a heterocylic oxygen ring in between the two benzene rings. More than 80 compounds are known of which almost all are hydroxylated (Carpenter *et al.*, 1969). These compounds have a very restricted distribution and are similar to flavonoids in chromatographic behavior and colour reactions. The Gentianaceae are found to be rich in these compounds. Gentisin is a yellow pigment obtained from *Gentiana lutea* L. roots. A xanthone C-glycoside, mangiferin, is reported from *Mangifera, Hedysarum, Madhuca* and *Iris* (Harborne, 1964).

Mangiferin

Khellin

9. Chromones

Chromones resemble coumarins in overall structure and properties but differ in the biosynthesis and position of the ketogroup. Chromones are formed by acetic acid pathway (by the condensation of five acetate molecules) while coumarins are derived from shikimic acid pathway. About a dozen simple chromones are known to occur naturally. These compounds, generally, have a methyl group at C_2 and are oxygenated at C_5 and C_7. An example of simple chromone is eugenin (in *Eugenia caryophyllata*). Khellin, visangin and khellol are some furochromones obtained from *Ammi visnaga* Lam. and *Eranthis hyemalis* Salisb., having certain marked pharmacological activity. Khellin is a powerful vasodilator and antispasmodic.

Flavones and isoflavones are 2-, and 3-phenyl chromones respectively.

10. Stilbenes

Stilbenes are C_6-C_2-C_6 compounds and about 9 are known, occurring mainly as heartwood constituents. Most of these compounds are toxic to fungi, fish and insects. Some of the stilbene derivatives are considered condensed tannins. Rhapontigenin from rhubarb, pterostilbene from *Santalum* and pinosylvin from *Pinus* are some of the interesting members of this group. Combretastatins are a group of anti-neoplastic compounds reported from a number of Combretaceae members

Lunularic acid is a dihydrostilbene found in lower plants, functioning as a growth inhibitor analogous to abscissic acid.

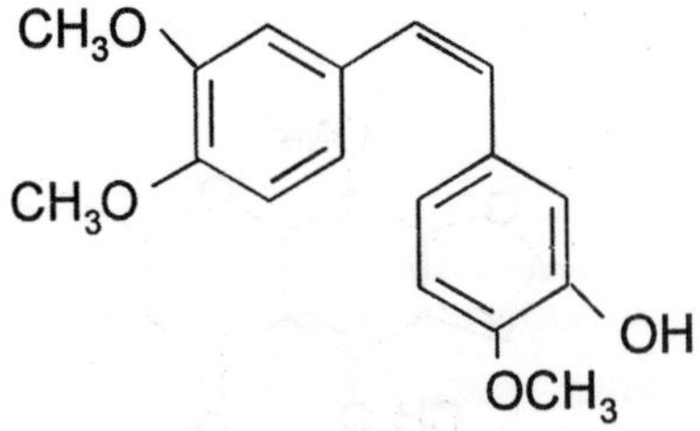

CH_3O, CH_3O, OH, OCH_3

Combretastatin A-4

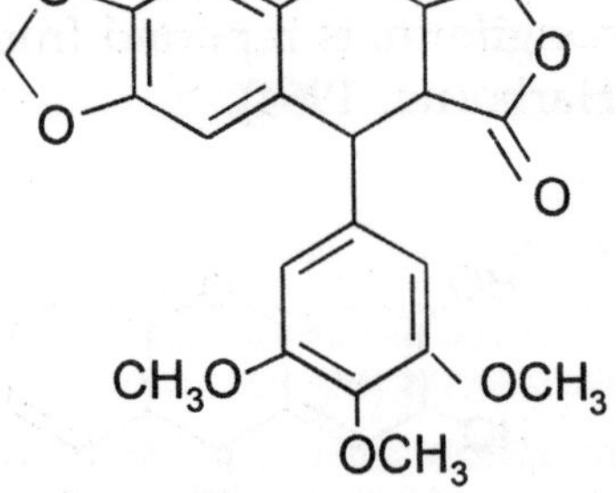

OH, O, O, O, O, CH_3O, OCH_3, OCH_3

Podophyllotoxin

11. Lignans

Lignans are dimers formed by linking two cinnamic acids/ cinnamyl alcohols, through the β-carbons of their aliphatic side chains. The aromatic rings are often oxygenated and additional rings are also formed in the molecule. Optical activity is exhibited by many lignans. Once considered as typical heartwood constituents associated with lignin (so the name lignans), these compounds are located in leaves, fruits and in exudates of many plants. Very few glycosides are reported from this group of which about 85 members are known to exist naturally in higher plants. They are extensively used as antioxidants in food. The stability of sesame oil is attributed to the lignan, sesamin, present in it. The same lignan forms the synergistic ingredient in pyrethrum insecticides. Podophyllotoxin, the principal component of *Podophyllum* resin along with α- and β-peltatins exhibits powerful cathartic action. Podophyllotoxin, being cytotoxic, is found useful in treating human malignancies (neoplasms). A large number of lignans exhibit anticancer as well as antimicrobial properties.

Lyonia - xyloside occurring in the sapwood of *Lyonia* (Ericaceae) and arctiin (arctigenin glucoside) in the fruit of *Arctium* (Asteraceae) are two examples of lignan glycosides. Though lignans are considered intermediates in the formation of lignin, they are also produced as a result of fungal attack and thus serve as phytoalexins.

The lignans derived from propenyl phenols or allyl phenols are designated as *neolignans* of which about 130 members are known. Otobain, present in the Otoba butter (*Myristica otoba* H&B.) is used as an antifungal factor. Schizandrin is one of the active principles of *Schizandra chinensis* C. Koch. used as a stimulant, while nor-dihydroguaiaretic acid and *l*-guaiaretic acid from *Guaiacum* Linn. are extensively used as anti-oxidants in lipids. Trimeric and tetrameric lignans have also been reported in nature (Wagner *et al.*, 1975). These compounds on detailed analysis would help immensely our understanding of the complexity of lignin and related polyphenols.

12. Flavonoids

The term flavonoid, in a wider sense as it is used here, includes not only the compounds based on flavone (2-phenyl chromone) skeleton, but also a number of related or derived compounds like anthocyanins, isoflavones, neoflavones etc. They are formed by the condensation of a phenyl propane with a polyketide moiety. This is one of the most widespread groups of phenolics present from *Chara* to the most advanced Angiosperms. More than 4,000 compounds have been discovered so far. The different classes on these C_6-C_3-C_6 compounds are recognised chiefly by the oxidation pattern of the middle C_3 fragment, the additional oxygen heterocylic rings and the glycosylation. The spectra of properties exhibited by these compounds include colouring of flowers and fruits (anthocyanins, flavone/ols), as accessory pigments which absorb and pass on the solar energy from lower wavelengths to chlorophyll in leaves (flavones), astringency (flavonones) estrogenic (isoflavones) antioxidants, anticancerous (quercetin) and so on. The color of the fruits and flowers are mainly due to anthocyanins, chalcones and aurones which are intensely colored compounds. Flavonols and flavones, being lighter in colour, help in the production of diluted colors of the anthocyanins and other compounds. Proanthocyanidins and related flavans are the astringent compounds present in many fruits. Citrus flavonones are bitter in taste. Some of the flavonoids act as growth regulators and most of the flavonoids contribute much to the disease resistance, and some like pisatin are well-known phytoalexins. The isoflavonoids such as genistein and diadzin are estrogenic. Flavan-3-ols, flavan-3,4,-diols and their polymers such as proanthocyanidins are now viewed as potential inhibitors of angiogenesis and inflammations. The phenolic hydroxyl groups of flavonoids can interact with proteins to form hydrogen and ionic bonds and thus modulate a multitude of molecular targets. Thus they are protein modifying compounds.

Flavonoids often occur as glycosides in plants, but a number of them are present in the free state too. The sugars involved in glycosidic linkages rarely are unusual ones. C-Glycosides also are not uncommon in this group.

The various classes of flavonoids are the following:

a. Chalcones

b. Flavan 3-ols, -3, 4-diols and proanthocyanidins

c. Flavonones, and flavononols

d. Flavones and flavonols

e. Anthocyanins

f. Aurones

g. Biflavones

h. Isoflavones, isoflavonones and homoisoflavones

i. Neoflavonoids

a. Chalcones

Chalcones and aurones are together known as 'anthochlor pigments' because their presence in a colored plant portion is detected by fuming the plant part with ammonia when they turn orange or red (Anthochlor test). The labeling experiments prove them to be the first C_{15} precursors formed in flavonoid synthesis. These are yellow pigments possessing an open chain system for the middle C_3 fragment. About two dozen chalcones are known to occur in plants. They are seen in even primitive vascular plants like ferns. In acid solution chalcones are converted to flavonones while the reverse reaction occurs in alkaline medium. Thus the butein (chalcone)-butin (flavonone) system is inter-convertible.

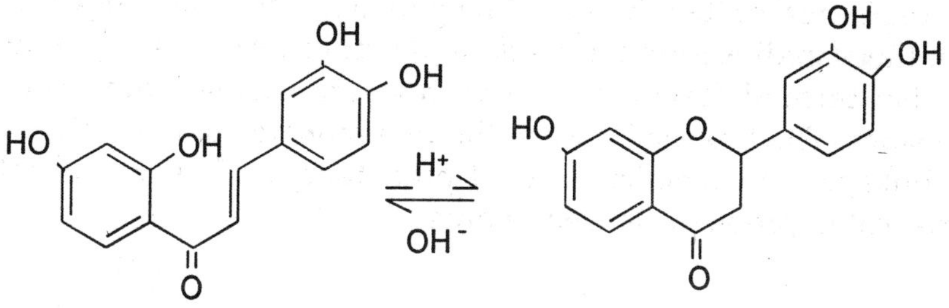

Butein ⇌ (H^+ / OH^-) Butin

A hydroxyl group is always present in position-2, corresponding to the hetero oxygen atom of flavones. Phlorizin, (the 6-glucoside of phloretin - 2, 4, 6, 4'-tetrahydroxy dihydrochalcone) occurring in *Malus*, causes glycosurea in animals. Carthamin, butein and isoliquiritigenin glycosides are the colouring principles of safflower, *Butea* and liquorice respectively. Xanthohumol, a constituent of Hops, has an isoprenoid side-chain and carthamone, the red quinone coloring principle of Safflower is a complicated derivative of chalcone. Isoliquiritigenin is the coloring principle of liquorice root. Angolensin, obtained from the heartwood of *Pterocarpus angolensis* is the only example of an iso-chalcone.

b. Flavan-3-ols, flavan-3, 4-diols and proanthocyanidins

All these compounds are devoid of the keto group at C_4 and possess a saturated C_3 fragment, making them 'flavans' instead of flavones. They are optically active (the activity due to the asymmetric nature of C_2 and C_3), colourless compounds omnipresent in higher plants and concentrated especially in woody plants. Catechins are flavan-3-ols occurring free in certain gums or bound with flavan-3, 4-diols in proanthocyanidins (known earlier as 'leucoanthocyanins'). Depending upon the relative positions of the two H-atoms at C_2 and C_3, the isomer of catechin, epicatechin is recognized. Both (+) and (-) catechins contain 2 and 3 hydrogens *trans* whereas in epicatechin they are in *cis* form.

Flavan-3, 4-diols are components of proanthocyanidins. A proanthocyanidin molecule may be a dimer of flavan 3, 4-diol or contain a catechin and a flavan-3, 4-diol. These compounds, as the name signifies, yield anthocyanidins (25%) on acidic or alkaline hydrolysis along with complex polymeric structures known as 'phlobaphenes' or 'tannin reds'. The polymeric phlobaphenes and the proanthocyanidins form the condensed tannins. All these compounds are also termed flavolans. Proanthocyanidins seldom occur as glycosides. The linkage between the two monomers, most often, is a C-C linkage. The beneficial role of proanthocyanidins in preventing heart-related problems is well known.

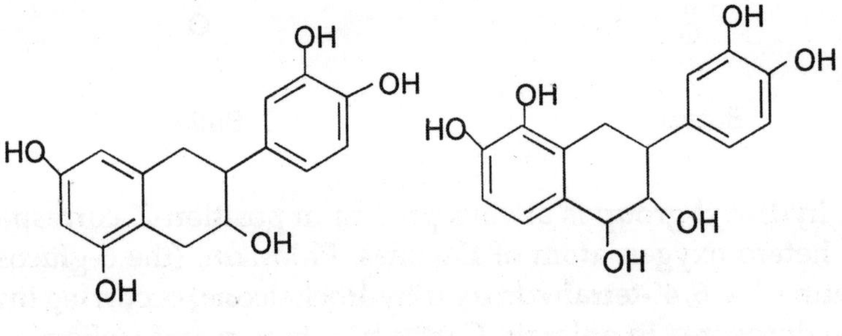

OH OH HO OH OH

Catechin

OH OH OH HO OH OH

Melacacidin(Flavan-3,4-diol)

c. Flavonones and flavononols (Dihydroflavonones and dihydro flavonols)

These compounds, though colourless, occur in appreciable quantities in several economically important plant products. They do not contain the keto group at C_4. Similar to flavans they contain a saturated C_3 fragment, which render them colourless. About 80 flavonones and 40 flavononols are known today.

These compounds are optically active and a majority of them are *laevo* rotatory. Flavonone with only one asymmetric carbon (C_2) can exist in two forms while flavononols (3-hydroxy flavonones) with two asymmetric carbons (C_2 and C_3) are capable of existing in four forms.

Flavonones are stable in acidic medium but are converted to chalcones by warm alkali solutions. The flavonones corresponding to apigenin and luteolin; naringenin and eriodictyol are widely distributed among higher plants, but their low concentration is the reason for their not being detected in a routine survey. Flavononols are a much neglected group, of which very little is known. Taxifolin (dihydroquercetin) is the most common compound in this group.

Hesperidin and naringin, two glycosides of the flavonones hesperetin and naringenin, are responsible for the bitter taste of the juice and peel of *Citrus* fruits. Both flavonones and flavononols are fungitoxic compounds useful in wood preservation. Both these compounds are considered obligate intermediates in flavonoid biosynthesis. Despite the similarity with flavans no condensed tannins are known from this group.

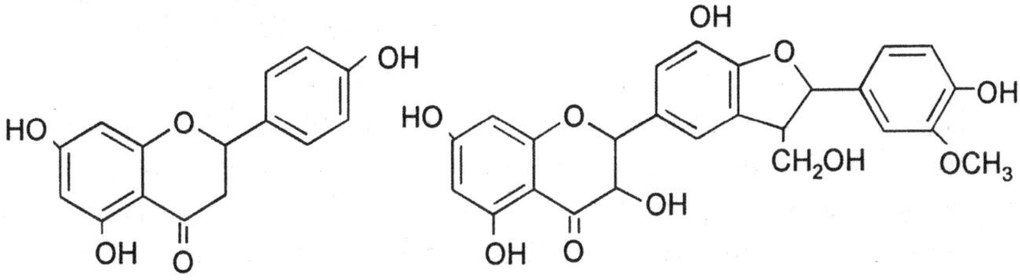

Naringenin Silychristin

Silymarin from *Silybium* is a group of flavonolignans (in which a flavone skeleton is linked with a phenylpropane group) having marked antihepatotoxic activity.

d. Flavones and flavonols

These compounds, together known as "anthoxanthins" (yellow flower pigments), are by far, the most abundant group of flavonoids. Most of them are yellow in colour and are responsible for the colour of many flowers ranging from yellow to white. When they co-occur with anthocyanins they exert a blueing effect on flower colour. Flavone itself occurs in primrose flowers and its various hydroxylated and methoxylated derivatives contribute to the light yellow or white colour of many flowers. Apigenin, luteolin and their methoxy derivatives are widespread in plant kingdom. The glycosylation normally occurs at C_7, and the sugar component may be a single sugar like glucose (most frequent), galactose or rhamnose or a disaccharide (contrary to the anthocyanin diglucosides, two sugars are never attached to different hydroxyl group in a flavone). The glycosides are present dissolved in cell sap while the free aglycones occur in the secretion products.

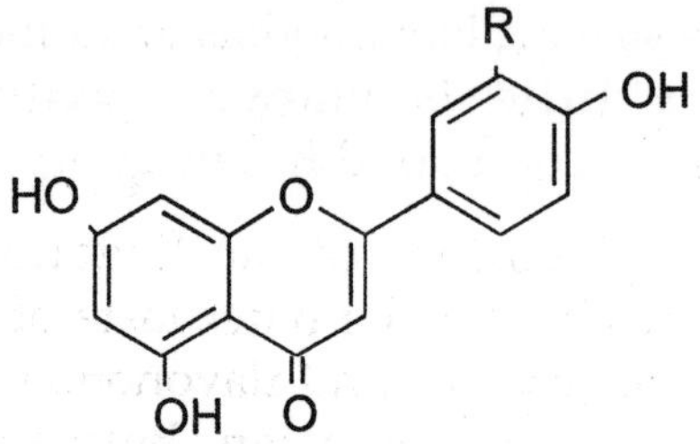

Flavones
R=H, Apigenin
R=OH, Luteolin

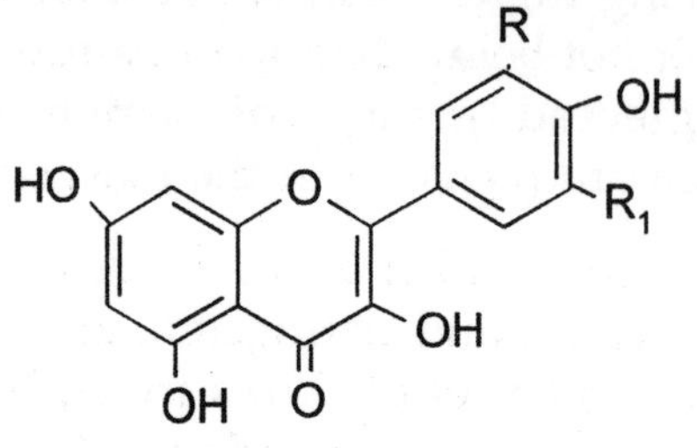

Flavonols
$R=R_1$=H,Kaempferol,
R=OH, R_1=H, Quercetin,
$R=R_1$=OH, Myricetin

Flavonols are 3-hydroxy flavones. They are darker in colour and contribute to the deep yellow colour of many flowers. Excepting that the glycosylation is common in C_3; the properties of these compounds are very similar to flavones. The three common members of this group are kaempferol, quercetin and myricetin. Rutin (vitamin. P) is the 3-O-rutinoside of quercetin. Quercetin is an anticancerous drug which exhibits antihistamine and anti-inflammatory properties also.

e. Anthocyanins

Anthocyanins are glycosides of anthocyanidins-derivatives of 2-phenyl benzopyrilium. These compounds are responsible for the various shades of red, purple, pink and blue colours of flowers and fruits. In general, they are red in acid solutions but change into blue in basic solutions. They occur as cations in the vacuolar sap and are thought to be oxonium compounds with the positive charge residing on the heterocyclic oxygen. The quinonoid form, resulting by the action of alkali, gets rapidly oxidized in air resulting in the formation of a colourless derivative. This explains the unstability of anthocyanins in alkaline solutions.

About 16 anthocyanidins are known to occur in nature. Most of them occur as glycosides while a few are products of hydrolysis of proanthocyanidins. It is found that the increase in the number of methyl groups in a compound is the cause of redness and increase in the –OH groups or 5-glycosylation causes blueness. The association of metals and/or pH markedly alters the colour of an anthocyanin. The 3-position is frequently oxygenated. The preferred position of glycosylation is C_3 for monosides and 3 & 5 for diglucosides (Among flavonoids only anthocyanins are glycosylated simultaneously in two positions). In acylated anthocyanins, cinnamic acids like *p*-coumaric, ferulic and caffeic acids are seen linked to the sugar component, which in turn, is linked at C_3 position of the anthocyanidins.

The common anthocyanidins are cyanidin, delphinidin and pelargonidin. These compounds differ in the hydroxylation pattern of the B ring. Petunidin, malvidin and peonidin are the mono/dimethoxy derivatives of these compounds. The two 3-deoxy anthocyanidins, apigenidin and luteolinidin, are characteristically present in ferns and mosses (Crowden and Jarman, 1974). The anthocyanins of fruits like blueberry are found to delay cataract and related problems of the eye and increase night vision. The red anthocyanins of wine are considered a reason for the low incidence of cardiac problems in French people who are known to consume large quantities of wine.

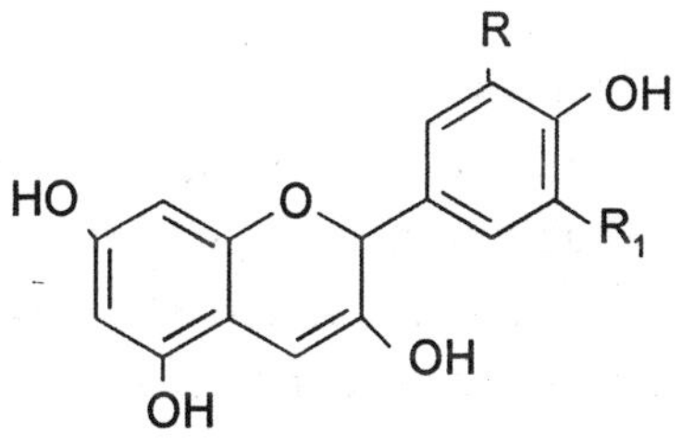

$R=R_1=H$, Pelargonidin,
$R=OH, R_1=H$= Cyanidin,
$R=R_1=OH$, Delphinidin

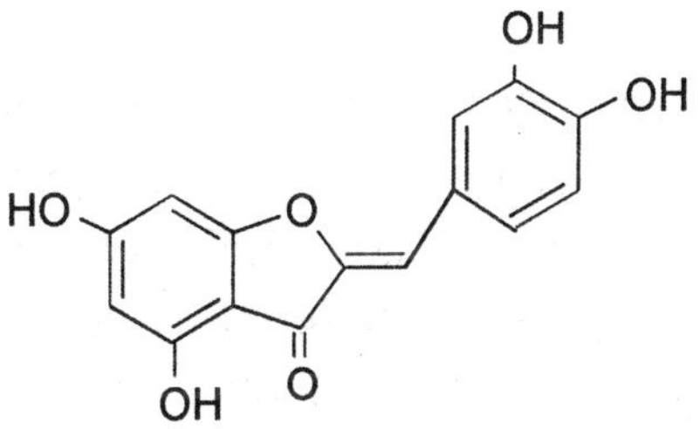

Aureusidin

f. Aurones

Aurones are benzyl coumaranones. Alongwith related chalcones these compounds are known as 'anthochlor pigments' due to the bright yellow/orange colour they exhibit. Aurones posses a furan ring instead of the pyran ring of other flavonoids. They are reported from flowers, bark, wood and leaves. The known aglycones are seven in number and they are mostly confined to the advanced families like the Cyperaceae and Asteraceae. The most widespread compound is aureusidin.

g. Biflavones

These are dimeric compounds in which two flavones are linked through a C-C linkage (amentoflavone) or an oxygen bridge (hinokiflavone). Biflavones are widely distributed in the gymnosperms and are rare in angiosperms. Almost all the known compounds, about sixty, are dimers of apigenin. Morelloflavone contains a flavone and flavonone units.

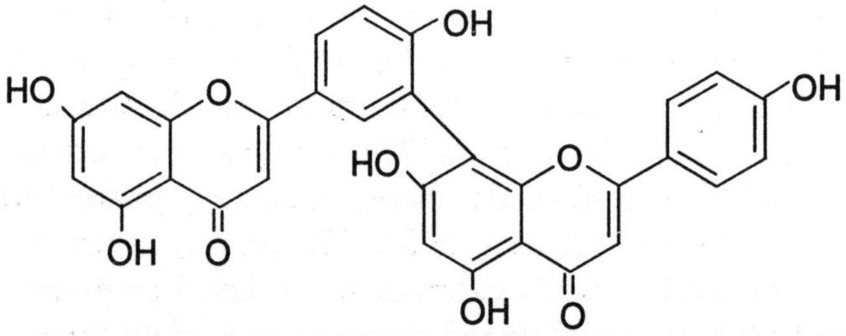

Amentoflavone

h. Isoflavones, isoflavonones and homoisoflavones

Isoflavones are 3-phenyl chromones. They are isomeric with flavones and differ in the placement of B ring to the C_3 position. The hydroxylation pattern also is similar to flavones. Genistein and orobol correspond to apigenin and luteolin. The total number of isoflavones isolated exceeds 100.

Isoflavones are known to be the heartwood constituents of Fabaceae members where they are particularly abundant in the subfamily Lotoideae. These compounds are of great interest, eliciting diverse physiological responses. 5, 7-Dihydroxy isoflavone is a strong inhibitor of oxidative phosphorylation. Pisatin and trifolirhizin are produced in response to injury or fungal attack and are protective in nature (phytoalexins). Genistein and pterocarpans are estrogenic and fungicidal. Genistein and diadzin are two isoflavonoids widely used in treating and preventing breast cancer. Rotenone is a known insecticide.

The different structural variations found in isoflavones include (1) *isoflavonones* (e.g., padmakastein from bark of *Prunus puddum* Roxb.)

(2) *rotenones* – with a 4 ring chromanochromanone system as in *Derris,* (3) pterocarpans having a coumaranochroman ring (4) *reduced isoflavans* e.g. duratin from *Dalbergia,* (5) *coumestans* with highly oxidised coumaranocoumarin structure e.g. coumestrol and (6) 3-aryl, 4-hydroxy coumarins eg. lonchocarpic acid from *Derris.*

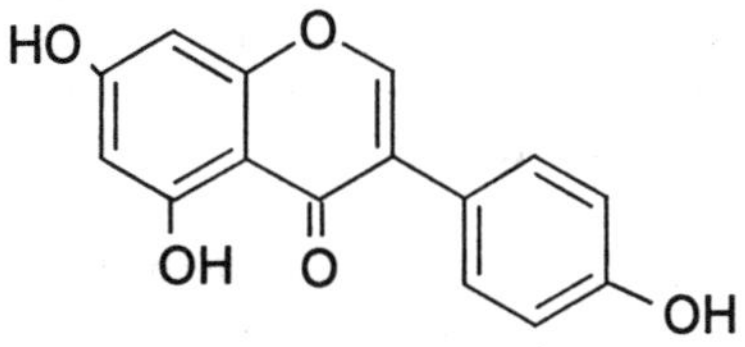

Genistein

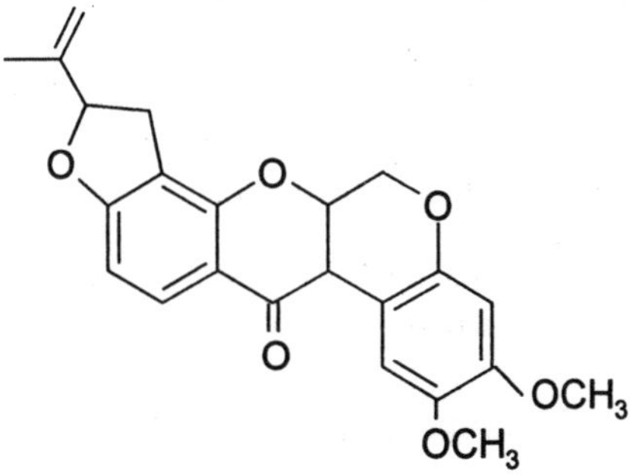

Rotenone

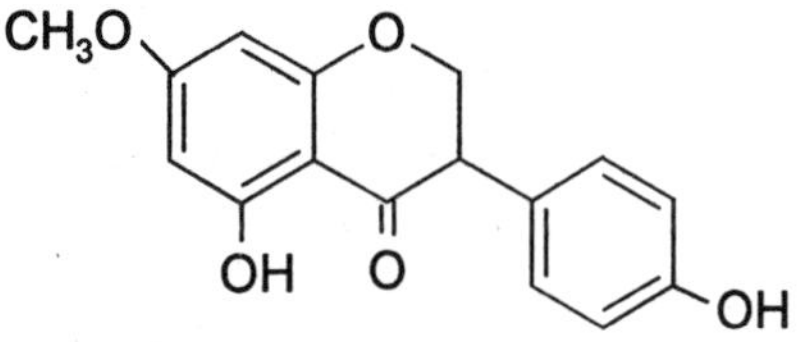

Padmakastein

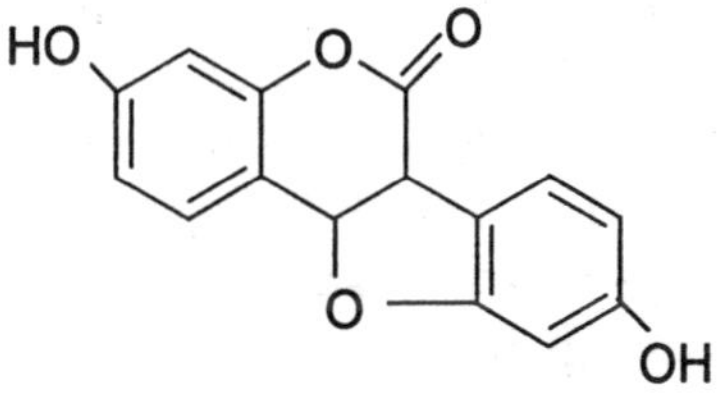

Coumestrol

Homoisoflavones are higher homologues (C_{16}) of isoflavones. Only two compounds, eucomin and eucomol are known, both from *Eucomis bicolor* Baker (Liliaceae).

i. Neoflavanoids

This is a newly recognized group, characterised by a 4-phenyl chroman skeleton. The structural variations in this group include (1) *4-phenyl coumarins* like calophyllolide obtained from Guttiferae members like *Calophyllum, Mesua* and *Mammea* (2) *dalbergiones,* a group of benzoquinones isolated from *Dalbergia, Machaerium* (Fabaceae) and *Exostemma* (Rubiaceae); and (3) *aryl chromans* like brazilin and haematoxylin. Calophyllolide and related compounds are piscicidal while brazilin and haematoxylin are dyes.

13. Quinones

Quinones, the aromatic diketones, form the largest class of natural colouring matters and are formed by the polyketide pathway. Of the total 800 and more compounds known, about 50% occur in higher plants and of the rest, a large percentage are known from fungi. Though quinones are the sole colouring pigments of fungi, in higher plants they play a subsidiary role, being mostly confined to the bark or underground portions and if present in other parts, make very little contribution to the colour, due to the masking by other pigments. Colourless quinones and their derivatives also exist in plants possessing different physiological roles.

Quinones may be benzo, naphtho or anthraquinones depending on the mono-, bi- or tri-cyclic ring systems they possess.

a. Benzoquinones

These compounds possess an aromatic ring with or without an aliphatic side chain. A few of them are medicinally important, e.g., embelin, a dihydroxy quinone isolated from the dried fruits of *Embelia ribes* Burm. F., is an anthelmintic used against tapeworm and for skin diseases. The physiologically active representatives of this group are plastoquinones, ubiquinones and tocopherols. Plastoquinones are believed to act as redox carriers in photosynthesis; ubiquinones as hydrogen carriers in respiratory chain and tocopherols as antioxidants or growth regulators.

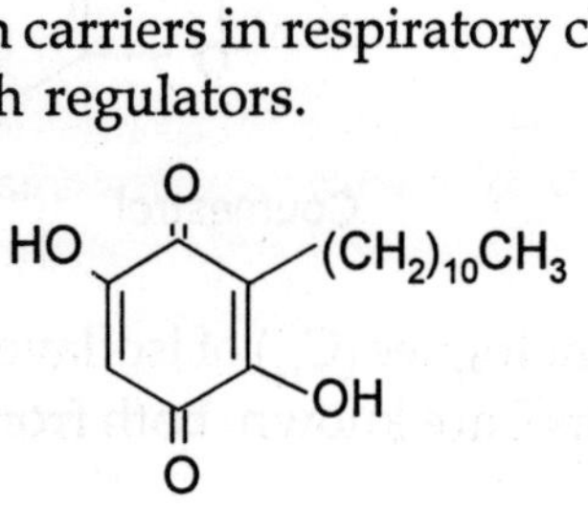

HO, O, $(CH_2)_{10}CH_3$, OH, O

Embelin

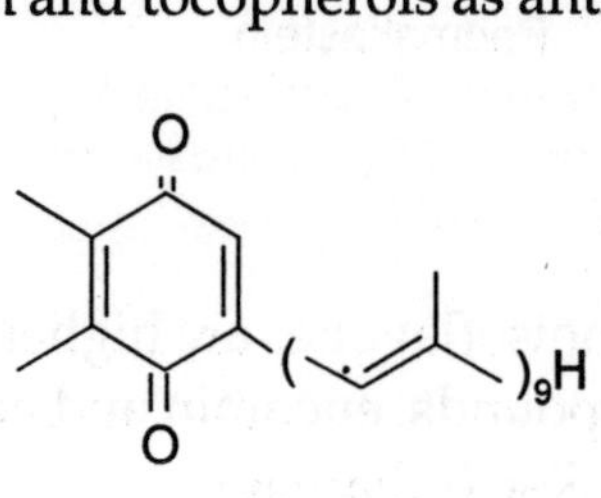

O, $)_9H$, O

Plastoquinone-9

b. Naphthaquinones

Most of the naphthaquinones are yellow or red plant pigments. A few like chimaphilin, plumbagin and eleutherin are toxins or antimicrobial agents; while some others like lapachol are important dye stuffs. Vitamin K and related compounds are having very specific roles in electron transport systems. Arnebin, another naphthaquinone isolated from the roots of *Arnebia nobilis* roots, is a potent anticancer agent.

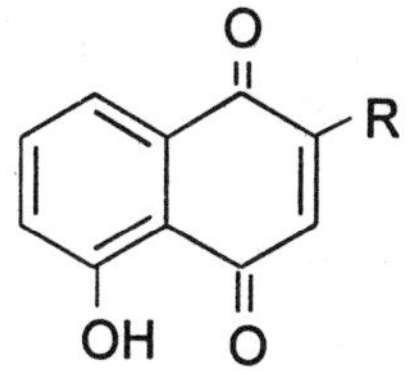

R=H, Juglone,
R=CH_3, Plumbagin

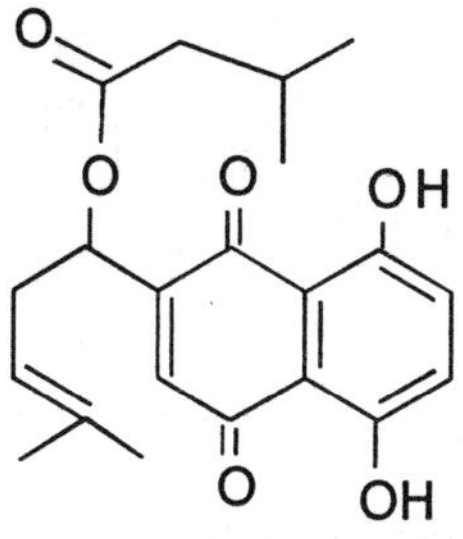

Arnebin

c. Anthraquinones

Of this large group of quinones, about 50% are distributed in higher plants especially Rubiaceae, Rhamnaceae and Polygonaceae and the rest in fungi. This group provides some important dyestuffs and many purgatives.

The hydroxylated anthraquinones occur as glycosides. The common compounds are dihydroxy phenols like chrysophanol and alizarin, trihydroxy derivatives such as emodin and aloe-emodin or methoxylated compounds like physcion. Carboxylated anthraquinones such as rhein also are reported. But the pharmacologically active anthraquinones are found to be anthranols that are formed as a result of slow hydrolysis of the glycosides during storage. Anthranols also, in course of time, get oxidised to anthraquinones. The pharmacological action of anthraquinone is due to its conversion to anthranols by intestinal bacteria (Fairburn, 1964).

In plants, the function of quinones is not properly understood. It is assumed that they play some role in oxidation-reduction reactions. For further details about these compounds the reader is referred to Thomson (1971).

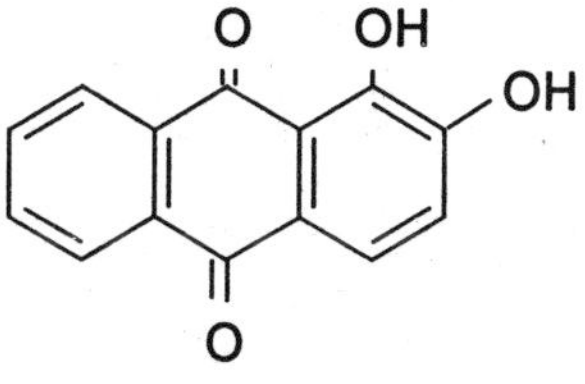

Alizarin

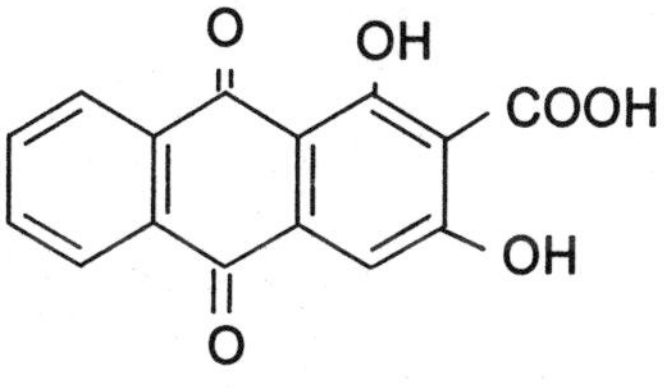

Munjistin

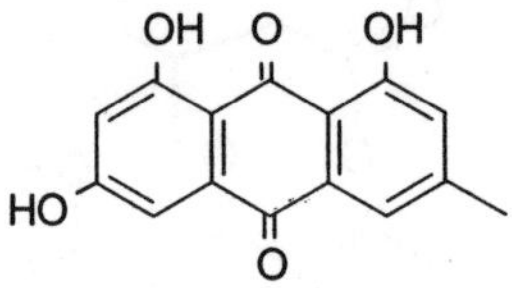

Emodin

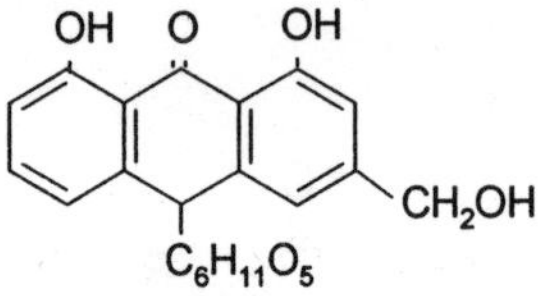

Barbaloin

d. Dianthraquinones

These are dimeric anthraquinones linked by C-C linkages. Cassiamin, cassianin and siameanin are a few members from this group reported from *Cassia.*

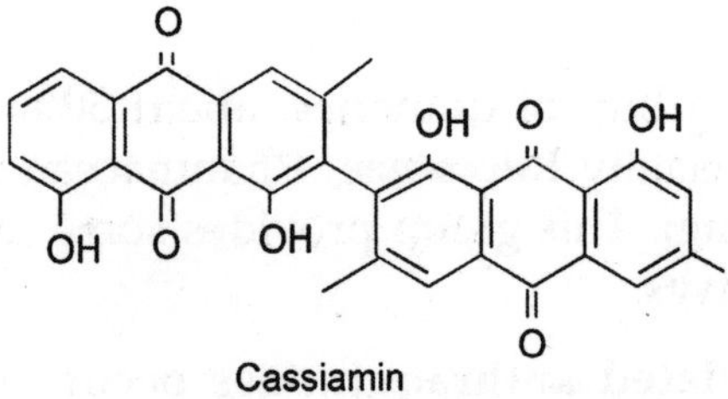

Cassiamin

e. Anthrones

Anthrone differs from other quinones in having only one keto group. Barbaloin is a C-glycoside of aloe emodin-9-anthrone. Chrysarobin (9-anthrone of chrysophanic acid), emodin anthrone and frangularoside are some other examples of this group.

f. Dianthrones

These are dimeric anthrones linked by C-C linkage; e.g. sennoside A. Dianthrones may be homodianthrones and heterodianthrones depending upon the type of monomers they possess. Fagopyrin (*Fagopyrum*) and hypericin (*Hypericum*) are derived from dianthrones by further ring closures between the two anthrone units.

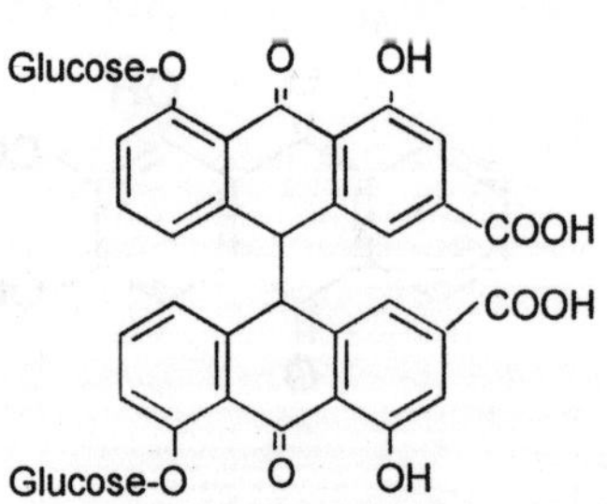

Sennidin

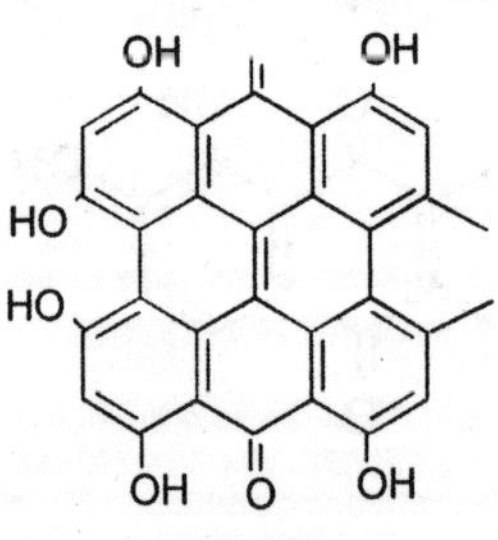

Hypericin

g. Anthranols

The anthrones on reduction give rise to anthranols. They may be formed on storage when the anthraquinones get reduced to anthranols. Emodin anthranol (*Rhamnus*) and physcion anthranol-A (*Ventilago*) are two examples of this group.

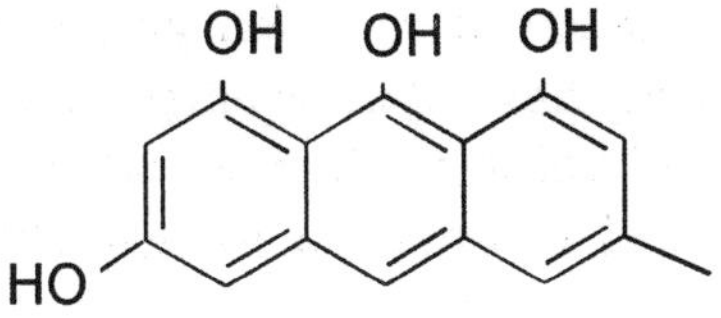

OH OH OH

HO

Emodin- anthranol

14. Tannins

Tannins are polyphenols having an astringent taste and ability to convert putrescible animal hides to hard stable leather. Two groups of tannins, hydrolysable and condensed, are recognised in plants. The former group is soluble in water, contains simple phenolic acids esterified with one or more sugar molecules and are hydrolysed by dilute acids to component compounds. Condensed tannins are polyphenols insoluble in water, which on treatment with acids yield complex products of unknown composition called "tannin reds" or "phlobaphenes".

The common phenolic components of hydrolysable tannins are gallic, digallic, and ellagic acids. The less common phenolics are hexahydroxydiphenic acid and chebulic acid. The sugar component is invariably glucose. The tannin molecules are often complex mixtures containing several different phenolic acids esterified to different positions of the sugar molecule and among themselves. Based on the phenolic nuclei involved, hydrolysable tannins are further subdivided to gallotannins and ellagitannins. Free gallic acid and the various galloyl esters of glucose constitute the 'tannic acid', of commerce, Chinese gallotannin, obtained from the aphid galls of *Rhus semialata* Murr., contains a basic structure of 1,2,3,4,6-pentagalloyl glucose with 3-5 additional galloyl groups attached by depside linkages to form chains of 2-3 *m*-galloyl groups (Robinson, 1975). Tannins from other sources differ in the type and number of phenolic acids attached to the glucose molecule. Thus tara tannin (an acidic tannin from the seeds of *Caesalpinia spinosa* Ktze) is a penta-or hexagalloyl derivative of quinic acid containing a *m*-trigalloyl group.

Acer tannin is 3,6-di-0-galloyl-1, 5-anhydro D-sorbitol. Hamamelitannin is another gallotannin of medicinal importance obtained from *Hamamelis.*

Corilagin is the simplest ellagitannin, which occurs in divi-divi (*Caesalpinia coriaria* Willd) and myrobalans (*Terminalia chebula* Retz). Other ellagitannins contain, alongwith gallic acid and ellagic acid, compounds resembling hexahydroxydiphenic acid, e.g. Chestnut tannin (*Castanea* sp) and valonia (*Quercus macrolepis* Klotschy) contain dehydrodigallic acid and valoneaic acid respectively.

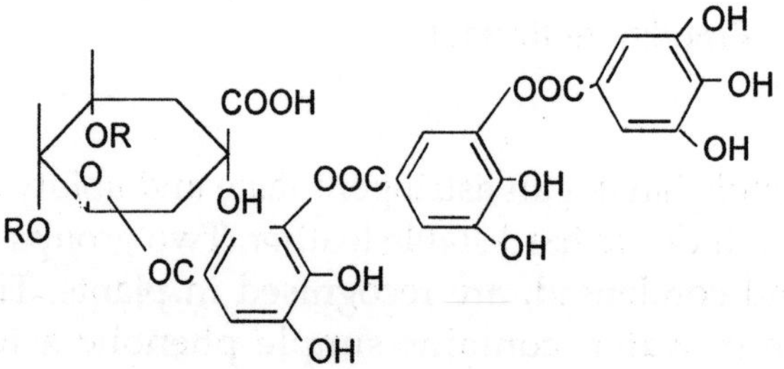

R=Gallic acid, Tara tannin

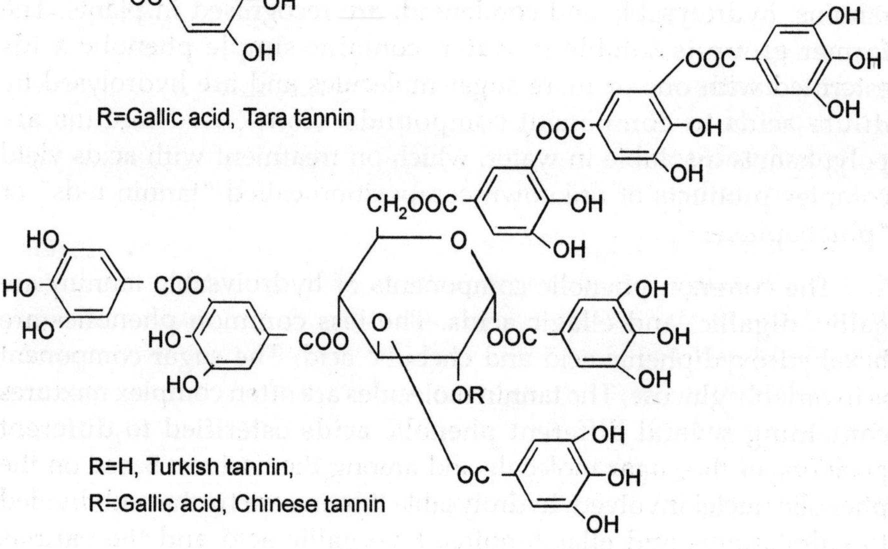

R=H, Turkish tannin,
R=Gallic acid, Chinese tannin

The condensed tannins are a complex group of compounds, the components of which are extremely difficult to isolate. Till today no complete structure of condensed tannin is known with certainty. Flavolans are undoubtedly the principal components of this group. Besides them, there exists flavonols, chalcones and flavonones, though to a limited amount. The flavolans-proanthocyanidins - yield anthocyanidins and phlobaphenes on acid hydrolysis of which the latter is of a highly complex structure. It is assumed that a molecule of catechin is linked with flavan-3, 4-diol; in a head to head, tail to tail or head to tail fashion, by -C-O- or C-C- linkage to produce a molecule of proanthocyanidin. In addition to flavolans, hydroxystilbenes (e.g. piceatannol) have the ability to tan leather and are grouped among these compounds.

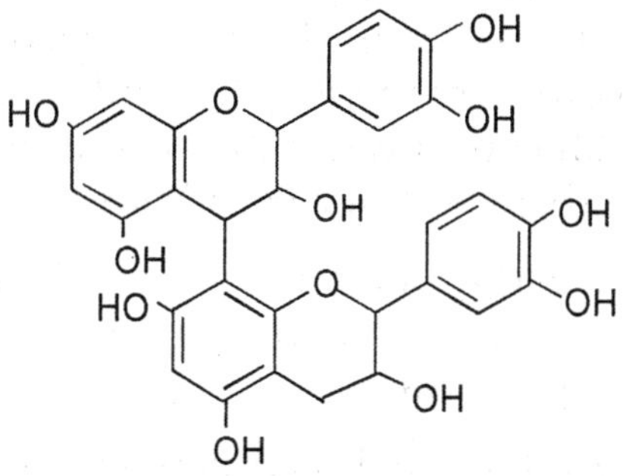

Procyanidin B4

The hydrolysable tannins are abundant in the leaves while the condensed tannins are concentrated in the wood. Usually a single species contains only one of these groups but there are instances where both types of tannins co-occur in the same plant. Thus in *Schinopsis* sp. hydrolysable tannins are predominant in leaves whereas flavolans are present in heartwood, and the sapwood contains both.

Apart from the seeds, leaves, bark and wood, commercial tannin sources include roots (*Krameria triandra* R. & P.), tubers (*Dioscorea atropurpurea* Roxb.) and rhizomes (*Rumex hymenosepalus* Torr.). Many of economically important produces like areca nut and cocoa seeds also contain appreciable amounts of tannins. The astringency caused by many unripe fruits is due to tannins.

Tannins, being fungicidal, contribute much to the resistance of the plants against the attack of fungi and other microorganisms.

Tannin is used principally for the manufacture of leather. It is also used as a deflocculent to control the viscosity and gel strength of mud in oil-well drilling; in the manufacture of gallotannate ink, for floating certain ores, as antioxidants in edible substances esp. fats, medical astringents and adhesive for plywoods and particle boards.

1.2.4. Minor components

The other compounds which are not included in alkaloids, terpenoids or phenolics are included here. They may find some similarities with the other major groups, but in those groups they may not feel at home.

1. Hydrocarbons and their derivatives

Saturated hydrocarbons occur mostly in waxes and also in volatile oil e.g. n-heptane in the major constituent of turpentine of *Pinus jeffreyi* and *P. sabiniana*. n-Nonacosane ($C_{29}H_{60}$) and n-hentriacontane ($C_{31}H_{64}$) are major constituents of candelilla wax of *Euphorbia* spp. Olive oil contains hydrocarbons of 13 to 28 carbons.

Ethylene, one of the plant hormones, is the simplest and omnipresent member of the unsaturated hydrocarbon group. Higher members are seen in plant waxes.

Aliphatic hydrocarbons and their derivatives present in waxes and cutins have a protective function in plants, against desiccation and pathogens and may be the sources of ethylene produced during ripening. These compounds are often overlooked during plant surveys. The presence of hydrocarbons in large amounts in biofuel, now keep these compounds significant in added utilization of plants. These compounds may be aliphatic, branched, unbranched, ethenoid or acetylenic with a varied number of substituents.

The normal saturated alkanes, ($CH_3(CH_2)_nCH_3$), present in plants ranges from C_7 (no lower hydrocarbons in plants !) to C_{37} and even C_{62} (stem-wax of *Leptochloa*- Graminae). Succulent plants (Crassulaceae, Opuntiaceae etc.) are invariably rich in these compounds. The lowest natural member of the group, n-heptane, occurs as a constituent of the turpentine of several species of pines. The turpentines of *Pinus jeffreyi* and *P. sabiniana* are nearly pure n- heptane and contain no other terpene hydrocarbons. Higher molecular weight hydrocarbons are often found in plant cuticle and pollen waxes. Candelilla wax (*Euphorbia antisyphilitica*) contains 50- 60% n-hentriacontane ($C_{31}H_{64}$). Cuticle waxes of several common fruits contain n-nonacosane ($C_{29}H_{60}$). The unsaponifiable fractions of fatty oils also contain these compounds. For e.g. the unsaponifiable fraction of olive oil contains hydrocarbons ranging from C_{13}-C_{28}.

Long chain aliphatic alcohols (fatty alcohols) and ketones are often seen in plant waxes. The alcohols have an even number of carbons (like fatty acids) while the ketones possess an odd number. The alcohols may occur as esters but unesterified alcohols are abundant in leaf waxes of monocots. Long chain aldehydes are seen in sugar cane wax.

Branched chain hydrocarbons (*iso* and *anteiso*) of a chain length of C_{23}-C_{35} are reported from waxes of Crassulaceae, Euphorbiaceae, and Scrophulariaceae. Olefinic hydrocarbons reported are having lesser carbon numbers, C_{23} and less, ethylene being most common. Alicyclic hydrocarbons are treated in terpenes. Aliphatic ketones like acetone, butan-2, 3-dione or 2- methyl-hept-2-en–6-one, undecan-2-one etc are reported from volatile oils of many plants.

2. *Polyacetylenes*

More than 600 polyacetylenes/polyenes are known from plants. These compounds containing one or more acetylenic bonds are particularly abundant in Asteraceae, Apiaceae, Araliaceae and Campanulaceae. They have a length of 14 to 18 carbons. In certain fungi also polyacetylenes are present, but they have a lesser carbon number, that is from 8 to 14. Some of the polyacetylenes of *Oenanthe crocata* and *Aethusa cynapium* are plant toxins while some like wyerone acid (*Vicia faba*) and safynol (*Carthamus tinctorius*) act as phytoalexins. Several highly unsaturated acetylenic conjugated compounds are similar to carotenoids in that they absorb light in the visible region and are often classed as plant pigments. Polyacetylenic alcohols reported from roots of several Umbellifers are toxic to animals and thus are defensive in nature, for e.g. cicutoxin from *Cicuta virosa*.

Brominated hydrocarbons such as laurinterol, spirolaurenone or laurencin are reported from red algae like *Laurencia*. Acetylenic amino acids, three in number, are present in the seeds of *Euphoria* (Sapindaceae): for e.g. 2-amino-4-hydroxy hept-6-ynoic acid. Acetylenic fatty acids such as tariric, stearolic, and crepenynic acids are biogenetically related to petroselinic, oleic and linoleic acids.

Thiophene derivatives of acetylenes are seen in many members of the Asteraceae such as *Ambrosia, Bidens, Spilanthes,* and *Echinops* etc. Some of the acetylenes like deca-2, 4-dien-4-methyl mercapto-6, 8-diynoic acid methyl ester from *Anthemis* contain sulphur also. Some others like capillene from *Artemisia* contain phenyl groups. Furan or pyran rings also are seen in acetylenes. Isobutyl amides or spiroketals are other variations seen in certain acetylenes. All these additions to the acetylene moiety are seen only from Asteraceae.

3. *Annonaceous acetogenins*

These acetogenins, also known as THF acetogenins, are characterized by a hydrocarbon chain of 35-37 carbons having a

γ-lactone ring on one end and one/two /three tetrahydrofuran rings in the middle. More than 350 different acetogenins have been isolated from 26 plants of the Annonaceae. The important genera yielding acetogenins are *Annona, Anomianthus, Asimina, Disepalum, Goniothalamus, Miliusa, Opyrypetalum, Rollinia, Uvaria* and *Xylopia.*

The hydrocarbon chain between γ-lactone ring and THF is known as α-chain and the other known as β-chain. The tetrahydrofuran ring close to the lactone ring is designated as A-ring and the other, if present, are B and C. the known acetogenins are divided into five classes based on the number and position of THFs. They are the following.

Adjacent bis- tetrahydrofurans: This is the second largest group having more than 40 members. Most of the acetogenins isolated from *Annona squamosa* belong to this group. All of them are characterized by a linear ten carbon skeleton that comprises of two THF rings flanked by two hydroxyl groups on either side of THF rings. A few compounds lacking one of the hydroxyl groups adjacent to ring A also are reported. These compounds are found to have the highest insecticidal potency e.g. Asimicin, Parviflorin.

Non-adjacent bis – tetrahydrofurans: Here the two rings are separated by a four-carbon chain. One of the THF rings (ring A) is flanked by one hydroxyl group which is positioned on the side of lactone ring, while the other THF ring is flanked by two hydroxyl groups in the 4 carbon chain in between the THF rings e.g. Sylvatin and Bullataticin.

Monotetrahydrofuran: This is the largest group and has only one THF ring and one or two hydroxyl groups. E.g. Gigantetrocin A., Annomontacin.

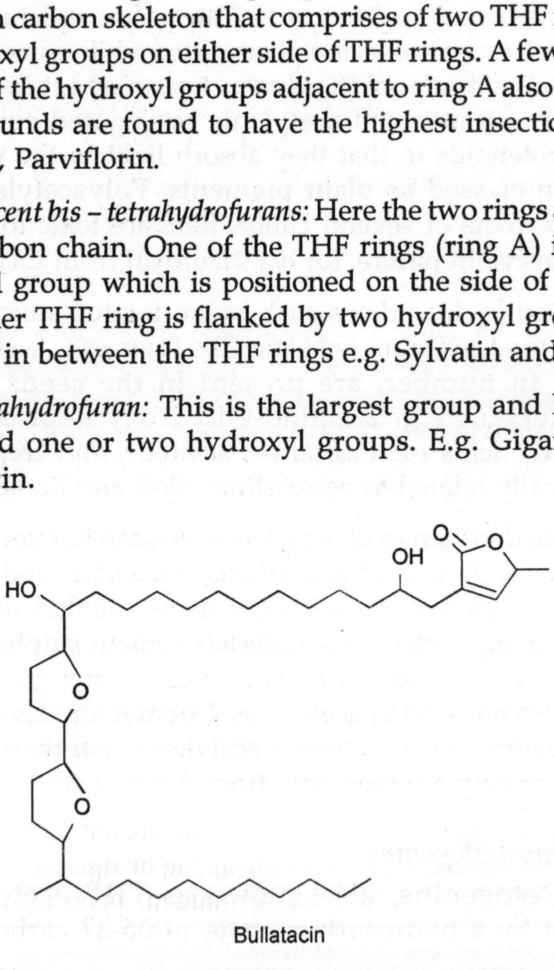

Bullatacin

Tri-tetrahydrofurans: This group has 3 THF rings adjacent to each other. E.g. goniocin.

Non-tetrahydrofurans: This group is considered as biogenetic precursors of more complex acetogenins. They have a double bond towards the center of the chain and one or two hydroxyl groups. E.g. giganin, venezenin.

Recently unusual acetogenins such as those containing a tetrahydropyran ring and a tetrohydrofuran not flanked by a hydroxyl group or a fatty acid ester (Araya, 2004), also are located.

The number of plants in the family Annonaceae is more than 2300 and all the plants are believed to contain these compounds and therefore the number of acetogenins remains to be isolated would run to thousands.

4. Pyrones

Both α–pyrones and γ- pyrones are seen in many plants. Aparajitin, a lactone of 2-methyl-4-hydroxy-pentacosanoic acid is reported from leaf wax of *Clitoria*. Kavain and methysticin (from Kava, Kava lactones) and paracotoin (Coto - bark) are the important α–pyrones located in Angiosperms. γ–Pyrones are of three types: (1) simple γ–pyrones, (2) benzo- γ- pyrones (chromones), and (3) dibenzo-γ-pyrones (xanthones). Both chromones and xanthones are explained in Phenolics. Chelidonic acid is one γ- pyrone very widely distributed in plant kingdom. Meconic acid is another example found esterified with the opium alkaloids.

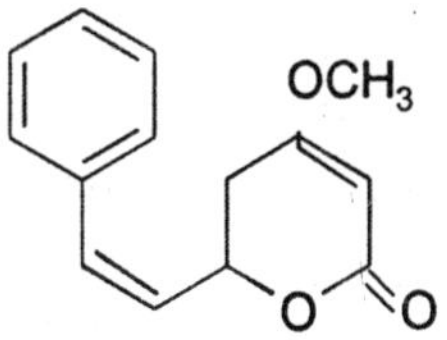

OCH_3 O O

Kawain

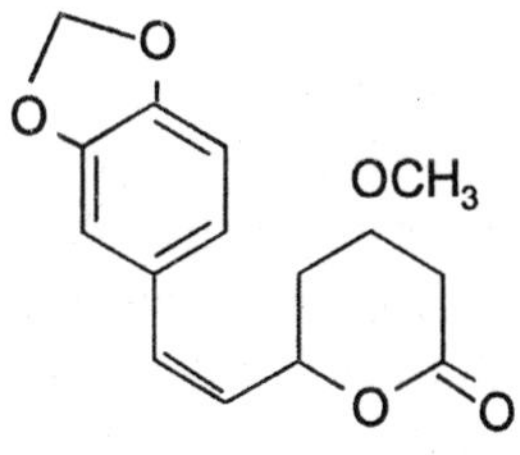

O O OCH_3 O O

Methysticin

5. Melanins

Melanin is the name given to any dark brown or black substance occurring in plants or animals. This grouping brings together a number of chemically and biosynthetically unnatural groups, and therefore the term melanin applies now to dark colored polymeric indole derivatives that involve tyrosine, dihydroxy phenyl alanine (DOPA) and dopamine and tyrosinase and oxygen.

True melanins are produced by legumes like *Vicia faba, Cytisus nigricans, Sarothamnus scoparius* and banana.

Blackening is caused by iridoids also.

6. Naphthalenes

Naphthalene itself occurs is cloves, *Iris* and rice seedlings. Hydroxy naphthalenes are reported from *Diospyros* wood and cotyledons of *Impatiens.*

7. Amines

Amines resemble alkaloids in nature and properties. Many of the amines are produced by the decarboxylation of the corresponding amino acids. Of the 100 amines isolated so far, the amines commonly located are methylamine, dimethylamine, trimethylamine, isopentyl amine, ethanolamine, putrescine, cadavarine, β-phenylethylamine, tyramine, histamine and isobutyl amine. All these amines are volatile and possess an offensive odor. The N- methyl derivatives of some of the amines are choline, hordenine and candicine. The stinging effect of *Urtica* spp. (stinging nettles) is due to the presence of histamine, serotonin and acetylcholine. Histamine is a depressor. Choline is a component of phospholipids in plants. Putrescine and other polyamines have both stimulatory and inhibitory effect depending on the tissues involved. Betaine is an antimicrobial agent in plants.

8. Cyanogenic glycosides.

More than sixty cyanogenic glycosides are reported from about 2600 taxa belonging to the Rosaceae, Fabaceae, Poaceae and Araceae. Upon wounding, by herbivores or other organisms, the cellular compartmentation breaks down and the glycosides come in contact with α- glucosidase to release 2-hydroxy Nitriles (cyanohydrins) which is further cleaved to HCN or corresponding aldehyde/ketone by a hydroxynitrile lyase.

The most common cyanogenic glycosides are linamarin and lotaustralin, usually found together in plants such as flax (*Linum usitatissimum*), clover (*Trifolium repens*) or *Lotus corniculatus.* Amygdalin is the toxic principle of bitter almonds and prunasin of *Prunus* seeds. Both of them are glycosides of mandelonitrile. Dhurrin is the glucoside of *p*-hydroxy mandelonitrile present in Sorghum. The role of these compounds in plants is protection against herbivores. Cyanide poisoning of the livestock is due to these compounds.

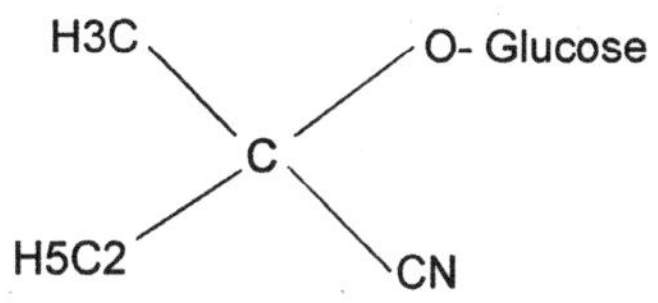

H3C, O- Glucose, C, H5C2, CN

Lotaustralin

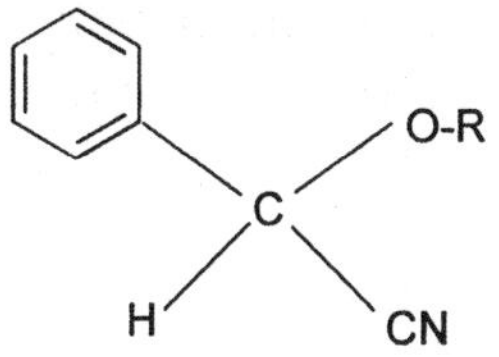

O-R, C, H, CN

R=Gentiobiose, Amygdalin
R=Glucose, Prunasin

9. Isobutylamides

A number of isobutylamides are present in certain plants which often combine with various unsaturated fatty acids by an amide linkage. Several plants like *Anacyclus pyrethrum* (Pellitory) owe their effectiveness as insecticides to these isobutyl amides. Pellitorine isolated from *Anacyclus* is a mixture of isobutylamides. The most abundant fatty acid is 2, 4-decadienoic acid.

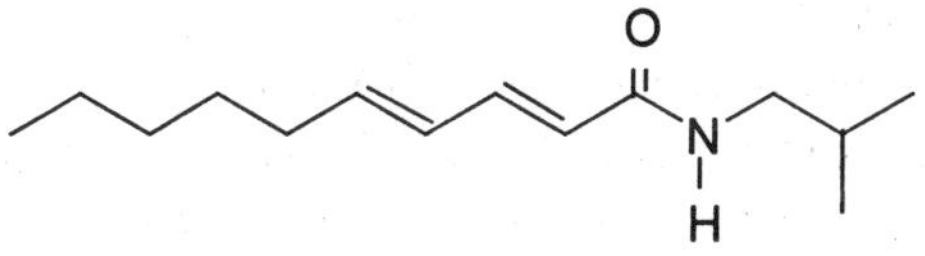

O, N, H

Pellitorine

10. Porphyrins

Porphyrins are the compounds having a cyclic tetrapyrrole-porphin- nucleus. This group is widely distributed in both animal and plant kingdoms. Porphyrins serve as part of the oxygen transport (hemoglobin) and storage system of vertebrates, in the respiratory chains of most cells (cytochromes) and as prosthetic group of certain enzymes in plant and animals. A modified porphyrin nucleus having an additional 5- membered ring forms the basic structure of chlorophylls and related pigments of green plants. They also contain magnesium in the centre of the porphyrin ring.

Chlorophylls and related compounds

Chlorophyll *a* and *b* are the green pigments of higher plants. They exist in the ratio 3:1 in the leaves. Other chlorophylls exist only in algae. Chlorophyll *b* differs from chl. *a* in having a formyl group instead of a methyl group at position 3. A diterpene alcohol -phytol- is found etherified to one of the carboxyl groups. The various chlorophyll derivatives which occur naturally or as artifacts when working with chlorophyll involve removal of magnesium and hydrolysis of one or both ester groups.

11. Betalains

Betalains (betacyanins) are a group of nitrogenous compounds responsible for the red and yellow color of plants belonging to the Centrospermae,. They possess an indole-2- carboxylic acid / proline N-linked to a pyridine dicarboxylic acid group through an acetyl group. Where indolic moiety is there, the compounds are red in color and are designated betacyanins. The second group where proline is present, the pigments are yellow in color and known as betaxanthins. Betalains always are mutually exclusive to the other red pigments, anthocyanins. Both these groups of red pigments have similar absorption maxima between 520-550 mm.

Similar to anthocyanins, betacyanins also occur as are glycosides. More than 50 betacyanins are reported and almost all of them are glycosides of betanidin or isobetanidin. At times, even phenolic acids (like *p*-coumaric acid or ferulic acid) also are found esterified to betacyanins (similar to anthocyanins). But in contrast with the anthocyanins, betacyanins are unstable in acid medium and give yellow color with alkalies. (Anthocyanin salts are stable and give blue color with bases). E. g. Amarantin (*Amaranthus*), Bougainvilleins, 1-XVI (*Bougainvillea*) and Gomphrenins 1-VIII (*Gomphrena*). Betaxanthins are few in number. e. g. Indicaxanthin (*Opuntia*), Miraxanthins 1-VI (*Mirabilis*) etc.

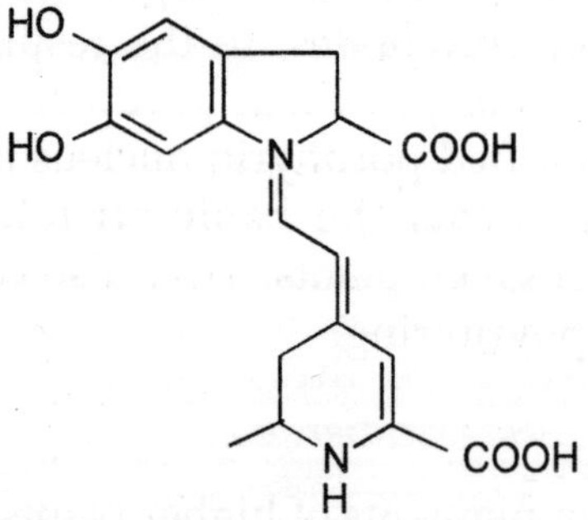

Betanidin

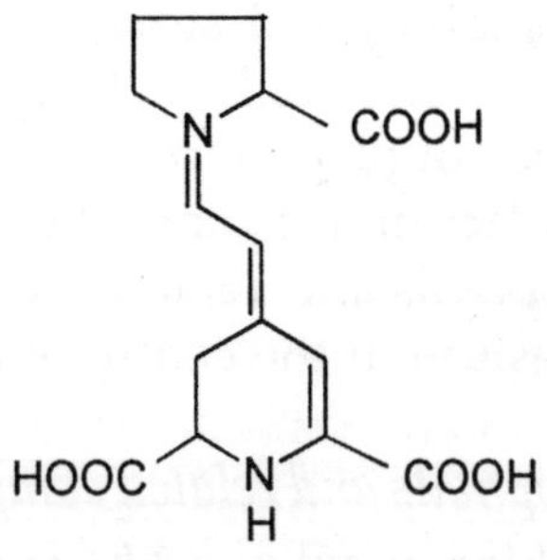

Indicaxanthin

12. Sulphur containing compounds

Sulphur exists in many different combinations in plants. The different compounds in this group are (1) thiols (RSH), (2) sulphides (RSR′), (3) disulphides (RSSR′), (4) polysulphides (RS_nR') (5) sulphoxides (RSOR′) and (6) methylsulphonium compounds. All of them are volatile, possess offensive odour and are derived from amino acids.

Onion, garlic and Asparagus are some examples of the common plants containing sulphides/ disulphides. A large number of sulphur compounds are now known from roots of the family Asteraceae. Almost all of them are linked to the polyacetylenes of the same family. They are thioketones, sulphides and cyclic sulphides. α-Terthienyl and 5-(3- buten-1-ynyl) - 2, 2′-bithienyl are two examples of cyclic sulphides. Allicin is an example of sulphoxide, but this on distillation yields diallyl disulphide, diallyl trisulphide and allyl propyl disulphide.

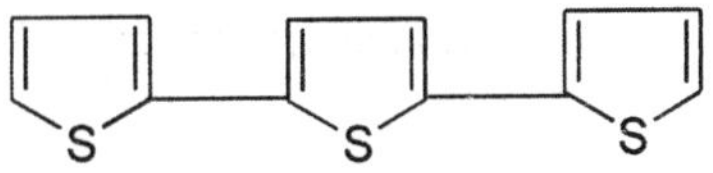

alpha-Terthienyl (*Tagetes*)

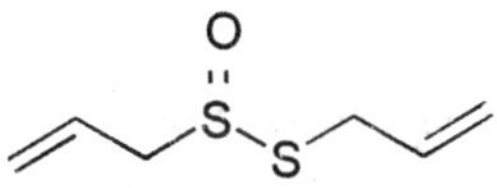

Allicin

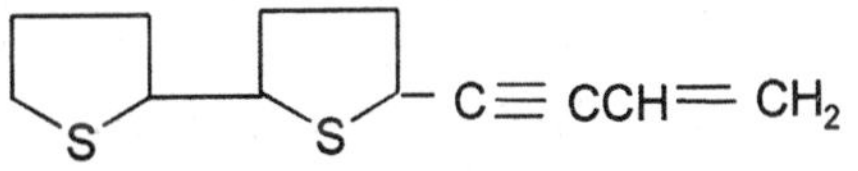

$-C\equiv CCH=CH_2$

5,2,2'-Bithienyl

Glucosinolates (Mustard oil glycosides)

These are glycosides found present in almost all the members of Brassicaceae, Capparaceae and related taxa. In the plant an enzyme myrosinase occurs spatially separated from these compounds. But on crushing, the enzyme acts on the glucosinolates and liberate volatile isothiocyanates and glucose. The isothiocyanates have a general formula R N = C= S and various members of this group differ in R-group. More than 80 different isothiocyanates are reported and due to their pungency and offensive odor, they protect the plants containing them from microbes, pests and herbivores.

Isothiocyanates are responsible for the pungent taste and characteristic flavor of mustard, horse-raddish etc. Alkyl-, alkenyl-, methyl thio alkyl- and benzyl isothiocyanates are lipids in nature and thus can penetrate biomembranes and interact with epidermal and mucosal tissues. They are also antibiotic in nature. Sinigrin present in mustard yielding allyl isothiocyanate and glucotropaeolin from *Tropaeolum majus* yielding benzyl isothiocyanate are two common glucosinolates present in plant kingdom.

$$R-C(=N-O-SO_3)(S\text{-}Glucose) \xrightarrow{\text{Myrosinase}} R-N=C=S + Glucose + H_2O$$

Glucosinolate Isothiocyanate

NOSO$_3^+$ S- Glucose → N=C=S

Glucotropaeolin Benzyl isothiocyanate

1.3. *Analytical Methods*

Since the phytochemist has to deal with very complex organisms, the plants, which manufacture millions of compounds at a time, his works are much more tedious than that of an organic chemist. He expects the unexpected and has to isolate certain compounds from among myriads of similar or dissimilar compounds that also sometimes present in miniscule amounts. Thanks to the different fractionation mechanisms developed, especially chromatography, we can isolate any compound, occurring in any amounts in pure form. The analytical methods developed in recent years would help us to identify the compounds even in microgram quantities.

Selection and preparation of plant material, to some extent, decide the success of a plant chemist. Unless otherwise needed, the plant material should be collected from healthy plants, washed free of the dirt and other impurities and dried immediately. For very accurate results, the plant material is immersed in boiling alcohol to prevent the enzymatic breakdown of compounds taking place immediately after the detachment of the plant part from the parent plant. For most classes of compounds a slow drying (preferably shade-drying or drying in an oven at 40-60°C) is recommended. The dried plant material is powdered and used for extraction. When extraction of fresh plant material is needed, the plant part is sliced well and macerated with solvent in a mixer and filtered clear of debris.

1.3.1. Extraction

Extraction methods are as important as analytical methods in the study of plant chemistry. The methods to be adopted for the extraction of a compound or a group of compounds should ensure that the complete extraction of the desired compounds take place without them undergoing decomposition, isomerisation or polymerization. Cold extraction is recommended for thermolabile compounds like volatile oils, alkaloids and proteins. In this process, the finely powdered plant material (pulverized/ blended in case of fresh material) is kept covered with the solvent with occasional stirring (magnetic stirrer) for a long time ranging from 30 min to 72h. The solvent is then decanted / filtered and the residue is extracted further by another aliquot of fresh solvent. The material may be extracted repeatedly till all the compounds are extracted. The extracts are pooled and distilled in vacuum to get a concentrated solution.

Hot extraction is more efficient, where the plant material is boiled in a solvent for a stipulated period. When low-boiling and inflammable organic solvents are used, refluxing is done in which the boiling flask is fitted with an air/water condenser. The vapours of the boiling solvent get condensed in the condenser and fall back into the extraction liquid. This prevents loss of solvents and accidental fire due to escaping vapours. Once the boiling is stopped, the flask is cooled, condenser is removed and the contents filtered. Boiling with fresh solvent is repeated if the extraction is not completed. Soxhlet's extractor is the ideal choice for hot extraction where with minimum solvent, complete extraction is effected.

A judicious selection of solvent is a prerequisite for complete extraction. An ideal solvent dissolves out the desired compounds leaving other plant constituents which may 'contaminate' the extract. 'Like dissolves like' is the main principle here. For extracting lipidic compounds, non-polar solvents are used and for extracting polar compounds, polar solvents are used. The "eluotropic series" in which the solvents are arranged based on their polarity, is a good guide here. The solvents in the beginning of this series are best suited for extracting lipids while those near the end are used for hydrophilic compounds. A generally accepted eluotropic series is given in table 5.

Table 5. Eluotropic series

1. Pentane	13. Dimethyl formamide
2. Petroleum ether (low-boiling 40-60°)	14. Ethyl acetate
3. Petroleum ether (high-boiling 60-80°)	15. Pyridine
4. Cyclohexane	16. Acetone
5. Carbon tetrachloride	17. Butanol
6. Trichloroethylene	18. n-Propanol
7. Toluene	19. Ethyl alcohol
8. Benzene	20. Methanol
9. Dichloromethane	21. Water
10. Carbon disulphide	22. Organic acids
11. Chloroform	23. Mixtures of acids with-bases, water, alcohols or pyridine
12. Diethyl ether	

Though the eluotropic series arranges a number of solvents, the most common solvents used for extraction procedures are (1) petroleum, (2) chloroform (3) methanol/ethanol 4) water and 5) DMSO. A knowledge on the compounds soluble in these solvents will help to use them to the maximum advantage.

1. Petroleum ether

Low boiling petroleum ether (40-60°) is best suited for cold extraction while the high boiling petroleum ether (60-80°) is used for hot extraction. The advantage with the former solvent is that it can be concentrated/distilled at lower temperatures so that the thermolabile compounds remain unchanged. Some of the high boiling petroleum samples may not boil when placed in a water bath. Petroleum ether is ideal for extracting all lipidic compounds such as volatile oils, fixed oils, carotenoids, chlorophyll, most of the resins (except gum resins), steroids, aglycones, highly methoxylated compounds etc. The concentrated petroleum extract may be fractionated further by saponifying the residue when the saponifiable lipids (esters) go into aqueous solution while the non-saponifiable compounds remain undissolved and can be taken out dissolved in solvent ether or petroleum ether. The classes of compounds which do not get dissolved in petroleum ether are glycosides, carbohydrates, amino acids, proteins etc. Alkaloids can be extracted in distinctly alkaline (e.g., 10% ammoniacal) petroleum ether.

2. Chloroform

Chloroform is similar to petroleum ether in its solubility properties. Almost all the compounds soluble in petroleum ether can be extracted with chloroform. But the solubility of lipids is much lower in this solvent. Chloroform dissolves moderately polar compounds also. It is extensively used in the extraction of alkaloids.

3. Methanol/Ethanol

Methanol may be called a 'super solvent' in that, except for proteins and polysaccharides, almost all other products are soluble (at least to some extent) in it. Most of the the lipids, glycosides and the aglycones are soluble in methanol. Extracting the plant material with methanol after extracting it with petroleum ether will bring out all the glycosides in solution. Since methanol dissolves both lipidic and non-lipidic components of plants, these two groups can be separated by distilling off the methanol and adding water to the residue. The lipids remain undissolved as precipitates; can be removed by filtration or by the addition of petroleum ether with the aqueous solution. On shaking petroleum ether with the aqueous solution, the lipids pass on to the petroleum ether fraction while the glycosides and other polar compounds remain in solution in water. Extraction of plant material with 70% methanol is more effective because this solution combines the properties both of methanol and water.

4. Water

Water is the best solvent to extract glycosides, sugars, amino acids, peptides, a number of proteins and many polysaccharides. The plant material may be boiled in water and filtered to get a solution rich in the above said compounds. Adding methanol to this solution precipitates proteins and polysaccharides out. Acidic water (dilute acid) dissolves all alkaloids while basic water (dilute alkali) extracts all the phenolics.

5. DMSO

Dimethyl sulphoxide (DMSO) is the solvent of choice these days. It is a very hygroscopic dipolar solvent having no odour or color and is slightly bitter with sweet aftertaste. It is soluble in water, ethanol, acetone, ether, benzene and chloroform. It dissolves many polar as well as non-polar organic compounds like fats, carbohydrates, dyes, resins etc. It is non-poisonous and used to cryopreserve and store cultured cells.

Concentrating the extracts needs careful considerations. The volatile components may escape or get isomerised at higher temperatures. The glycosides may get hydrolysed to their aglycones and sugars. Many of the alkaloids decompose on heating. The extracts of all these compounds are concentrated using a **rotary flash evaporator** in which the vacuum applied keep the solution boiling at lower temperatures. For all other compounds, the extract can be distilled to remove the solvent. Distillation is usually done using a water bath so that the temperature may not rise above 100° and this prevents accidental charring of the residue once the solvent is distilled off. All the solvents having boiling points below 80°C boil when placed in a water bath. Using a water bath reduces the fire hazards due to the escaping vapours of inflammable solvents. For those solvents having boiling points above 80°, a heating mantle, hot plate or a sand bath may be used.

Once the extract is reduced to a comfortable volume, fractionation is easy and analytical methods such as chromatography are resorted to isolate the compounds in a pure state. The structural elucidation of compounds is done using spectrometric methods. A number of chromatographic methods such as column, high performance (HPLC), paper (PC), thin layer (TLC), high pressure thin layer (HPTLC), gas (GC), ion exchange, gel and affinity chromatography are available. PC and TLC are the most useful methods for a preliminary separation of compounds. The isolated compounds are then studied by spectrometric methods such as ultraviolet/visible (UV/V), infrared (IR), Raman spectra, fluorescence and phosphorescence, nuclear magnetic resonance (NMR), electron spin resonance (ESR), mass, optical rotatory dispersion (ORD) and circular dichroism (CD). For a detailed account of the various chromatographic and spectrometric methods, the reader is referred to Daniel (1989).

1.3.2. Isolation and characterization

General methods of separation and analysis are available in Harborne (1973) and a large number of simple experiments on specific products are given by Daniel (1991). An excellent handbook to be of immense help in analysis of a large number of drugs is prepared by Wagner *et al.*, (1984). A few experiments on the separation and identification of certain representative groups of plant products taken from Daniel (1991) are described below.

1.3.2.1. Sugars

Sugars present in any plant material can be separated using paper or thin layer chromatography. The separated sugars are visualized by spraying chromogenic sprays based on Molisch Test.

a) Paper Chromatography of sugars

Sugars are extracted from plant materials by boiling the macerated pulp with 95% ethanol. The ethanol extract is filtered, concentrated to remove the alcohol and the aqueous solution remaining is directly spotted on chromatographic papers. Standard solutions of sugars (sucrose, glucose, fructose etc. 100 mg dissolved in 1-2 ml of methanol containing one drop of water) are spotted on the sides of the plant sample. Duplicate papers may be prepared with varying concentrations of the samples. The papers are developed in any one of the solvent systems mentioned below. When the run is over, the papers are taken out, dried and sprayed or dipped in any of the spray reagents. The sprayed papers are then heated over a hot plate (or in an oven) at 110° C for 5-10 min. The sugars are identified by comparing the color and Rf values of the spots with those of the standard sugars. The positions and colors of some common sugars are presented in Fig.1

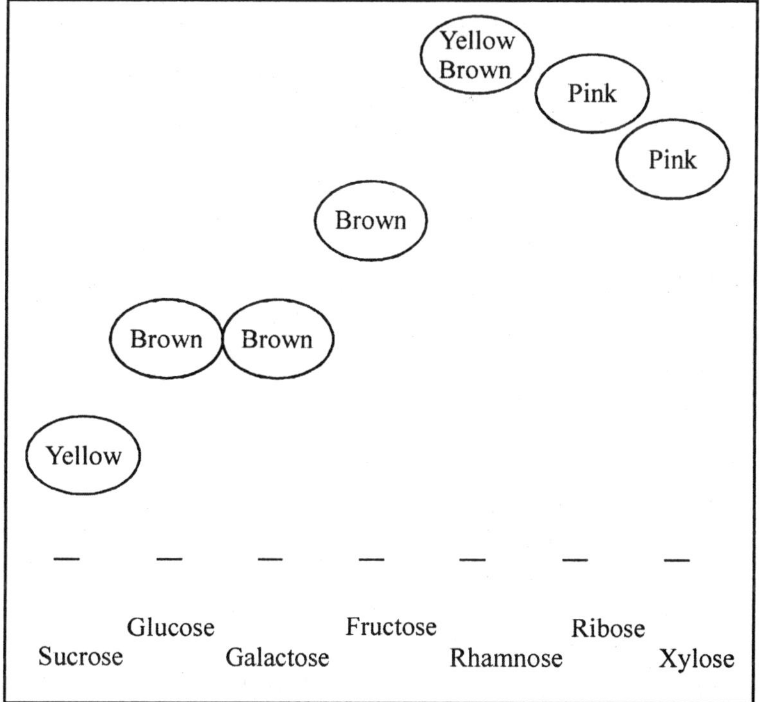

Fig. 1. Positions and colors of some common sugars in PC

Solvent systems

1. *BAW (n-Butanol-acetic acid-water, 4: 1: 5)*: While the upper organic layer is used as the developing solvent, the lower aqueous layer is used to saturate the chromatographic chamber. In this solvent system sugars exhibit low Rf values in winter. But in summer high Rf values are seen and therefore is the recommended solvent for the latter season.
2. *iso-Propanol-pyridine-acetic acid*: H_2O (8: 8: 1 : 4). This solvent system in which sugars possess higher Rf values is suitable in winter or in air conditioned laboratories.

Spray reagents

1. *Aniline hydrogen phthalate:* 100 ml n-Butanol is saturated with water by slow addition of small aliquots and shaking. Approximately 10 ml of water goes into Butanol. To 100 ml of this solution 0.91 ml (9.30mg) of aniline and 1.6 g phthalic acid is added and stirred well.
2. *p-Anisidine-HCl:* Prepared by dissolving 12.3 g of *p*-anisidine in 8.3 ml of HCl. 3g of this salt is dissolved in 100 ml of n-Butanol.
3. *α-Naphthol:* 1% alcoholic α-Naphthol - phosphoric acid (10:1)

After spraying with these regents, the chromatograms are heated at 110° for 2-5 min when distinctive colors develop.

b) TLC of Sugars

Adsorbent – silica gel G

Solvent systems

1. 65 ml Ethyl acetate + 35ml of a mixture of iso-propanol-water (2 :1)
2. n-Butanol-acetic acid-diethyl ether-water (9 : 6 : 3 : 1)

Spray regents

1. *Anisaldehyde-sulphuric acid:* Prepared by mixing 18 ml ethanol (95%), 1ml of conc. sulphuric acid, 1ml anisaldehyde and two drops of glacial acetic acid. The plates, sprayed with this reagent are heated on a hot plate at 90-100°C for 5-10 min. Sugars develop distinctive colors. (Table 2)
2. *Potassium permanganate:* 0.5N $KMnO_4$ in 1N NaOH. Sugars appear as pale yellow spots on a purple background.

Table 6. Rf values and color reactions of some sugars in TLC

Sugar	Rf in Solvents system*	Color with Anisaldehyde-Sulphuric acid
1. Lactose	0.04	Green
2. Sucrose	0.08	Violet
3. Glucose	0.17	Light Blue
4. Fructose	0.25	Grey
5. Xylose	0.39	Green
6. Rhamnose	0.62	Blue
7. Ribose	0.49	Yellow-green
8. Arabinose	0.28	Green
9. Galactose	0.18	Grey-green
10. Maltose	0.06	Violet
11. Sorbose	0.26	Violet

* Solvent system- 65 ml Ethyl acetate + 35 ml of a mixture of iso-propanol: H_2O (2:1).

1.3.2.2. Sugar alcohols

The sugar alcohols commonly present in higher plants are sorbitol (fruits of the Rosaceae members), dulcitol (members of the Celastraceae) and mannitol (exudates of *Fraxinus,* bark of *Randia* and cinnamon). These compounds do not respond to the typical sugar reagents but are readily detected by alkaline silver nitrate solution. Since all the sugars give the same reaction, duplicate chromatograms of the plant extract are sprayed with aniline hydrogen phthalate to make sure that the spots developed by $AgNO_3$ are of sugar alcohols and not of sugars.

Procedure

The powdered plant material is boiled with 80% methanol for 30 min and the filtered extract is concentrated. The concentrated extracts are spotted on Whatman No. 1 papers alongwith standard solutions of common sugars (glucose, fructose, sucrose etc.) and sugar alcohols (mannitol and sorbitol). The solvent system used for sugar alcohols is n-propanol-ethyl acetate-water (7:1:2). The developed chromatograms are sprayed with aniline hydrogen phthalate and heated to locate sugars in the plant extract. Duplicate chromatograms are dipped first in a solution of silver nitrate in acetone (prepared by adding 1 ml of a saturated solution of $AgNO_3$ in water to 200 ml acetone). Some of the sugars develop brown colors immediately and others develop colors on standing. When the chromatograms are dry, they are dipped in

0.5% aqueous NaOH when the spots develop instantly. The Rf values are noted immediately since the paper turns black on standing. The spots which do not give color with aniline hydrogen phthalate but react with alkaline $AgNO_3$ are the sugar alcohols which are identified on comparison with the standards.

1.3.2.3. Volatile oils

Volatile oils are extracted by (1) steam distillation, (2) expression (3) *enfleurage* or (4) solvent extraction. In steam distillation, steam is allowed to pass through the plant material (macerated and suspended in water), the oil gets vaporized and escape with the steam. When the steam condenses, oil also condenses and remains as a layer on water. But during higher temperatures certain oxygenated monoterpenes undergo irreversible changes and therefore steam distillation is not recommended if the components of volatile oil are to be studied or isolated. Expression is resorted to certain plants parts such as orange rind, where the oils are present in special cavities in large quantities. *Enfleurage* is the method adapted in perfumery, where the oils are present in very small quantities (e.g. rose petals). In this method an odorless fixed oil / fat is spread as a thin layer on glass plates or cloth and fresh petals are placed over it for few hours when the fat absorbs the volatile oils. Extraction of volatile oils using organic solvents such as petroleum ether or benzene, by far, is the ideal method. But in this case the solvents extract out other lipidic components such as waxes, fats, pigments etc. also.

All the spices belonging to Apiaceae (Umbelliferous fruits and their leaves) and the Lamiaceae (leafy spices) are rich in volatile oils. Since the oils are lost on cooking, these spices are mostly added as condiments.

a) Extraction of volatile oils

Steam distillation: 10 g of powdered plant material and 250 ml of water are taken in a distillation apparatus. The contents of the flask are heated to boiling. The boiling is controlled so that the rate of distillation is not more than 2-3 ml per min. the distillation is continued for 3-4h. The volatile oil which distils over condenses alongwith the water vapour and remains as a layer above the water in the receiver. When the distillation is over the contents of the receiver is shaken with 50 ml of ether. The ether layer is collected and the aqueous fraction is re-extracted with 20 ml of ether. The ether fractions combined, the solvent is allowed to evaporate in a dessicator and the oils remaining are weighed, calculated to percentage.

Solvent extraction Organic solvents such as petroleum ether, toluene, chloroform etc. are used to extract the oil. 10 g of the plant material is extracted with 100 ml solvent, at room temperature with occasional shaking, for 1 h. The extract is filtered through filter paper and the residue is put back into the beaker and re-extracted with 50 ml more of solvent for 30 min. After filtration, the filter paper and the residue are thoroughly washed with solvent and the washings added to the oil extract. The extract is distilled under reduced pressure and the oil in the flask is quantitatively transferred to a pre-weighed beaker by washing the flask 2-3 times with ether and adding the washing to the beaker. The ether is allowed to evaporate at room temperature to a constant weight. The increase in weight due to oil is calculated to percentage.

b) Analysis of volatile oils

5g Of dry powdered plant material or spices, is kept covered with 50 ml solvent ether in a conical flask for 30 min. with frequent shaking. The clear supernatant (mostly colored) is decanted into a 100 ml beaker and the extract is kept below a ceiling fan to concentrate. When the extract is reduced to ^{2}ml, it is spotted on TLC (silica gel) plates. The plates are developed in benzene or benzene-chloroform (1: 1). (In benzene the monoterpenes have higher Rf values and therefore this solvent is preferred over the second system in winter. In benzene-chloroform the monoterpenes have lower Rf values and therefore it is recommended for summer). The chromatograms are dried and sprayed with vanillin-sulphuric acid (spraying the plates first with a solution of 5% ethanolic sulphuric acid, followed by an overspray of 1% ethanolic vanillin). The plates are then warmed over a hot plate for. 3-5 min (or by keeping in an oven at 110° C for 5 min) when the monoterpenes appear as variously colored sports. A chromatogram of monoterpenes from some common spices is given in Fig. 2. The principal monoterpenes of the some essential oils are presented in Table 7.

1.3.2.4. Flavonoids.

Flavonoids are a group of polyphenols present in a large number of plants. Tea is one of the best sources of dietary polyphenols and contains all the three flavonols quercetin, kaempferol and myricetin.

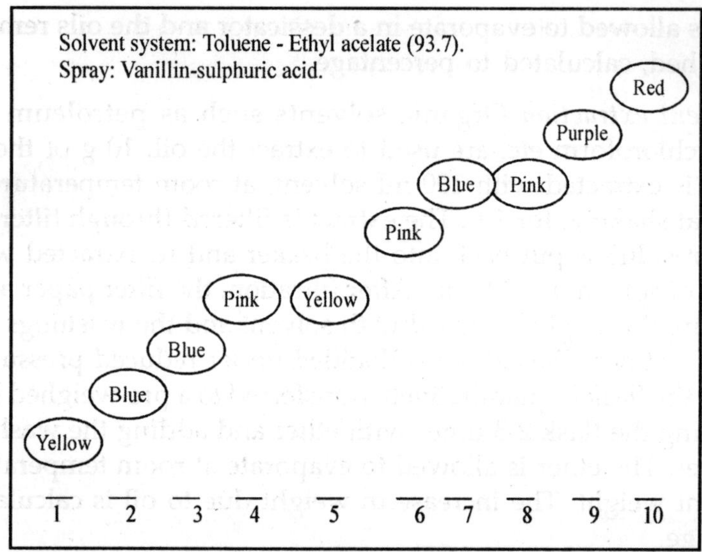

1. Geraniol, 2. Linalool, 3. Cineole, 4. Carvone, 5. Eugenol, 6. Thymol, 7. Citronellol, 8. Apiole, 9. Myristicin, 10. Anethole

Fig. 2. Thin layer chromatogram of some common monoterpenes

Table 7. The principal monoterpene components of some volatile oils

Common name	Botanical name	Principal monoterpene/s
1. Cinnamon	*Cinnamomum zeylanicum*	Cinnamic aldehyde
2. Clove	*Syzygium aromaticum*	Eugenol
3. Star anise	*Illicium verum*	Anethole
4. Caraway	*Carum carvi*	Carvone, Limonene
5. Coriander	*Coriandrum sativum*	Linalool
6. Cumin	*Cuminum cyminum*	Cuminaldehyde
7. Dill	*Anethum graveolens*	Carvone, Limonene
8. Fennel	*Foeniculum vulgare*	Anethole, Fenchone
9. Basil	*Ocimum basilicum*	Methyl cinnamate
10. Peppermint	*Mentha piperita*	Menthol
11. Rosemary	*Rosemarinus officinalis*	Bornyl acetate
12. Spearmint	*Mentha spicata*	Carvone
13. Thyme	*Thymus vulgaris*	Thymol
14. Camphor	*Cinnamomum camphora*	Camphor
15. Eucalyptus	*Eucalyptus* spp.	Cineole
16. Lemon-grass	*Cymbopogon flexuosus*	Citral
17. Rose	*Rosa damascena*	Citronellol, Geraniol, Linalool, Nerol
18. Tuberose	*Polyanthus tuberosa*	Nerol, Geraniol
19. Orange oil	*Citrus reticulata*	Limonene
20. Turpentine	*Pinus longifolia*	Δ^3 Carene

Procedure

Five grams of powdered tea leaves are boiled with 200 ml of distilled water for 30 min. The extract is filtered, filtrate transferred to a 500 ml R. B. flask, 14 ml of conc. HCI added to make the solution 7%, and refluxed for 1 h in a water bath. The flask is cooled and the aglycones precipitated are extracted in 50 ml solvent ether by shaking the hydrolysed extract with ether in a separating funnel. The yellow ether layer is taken in 100 ml beaker and the ether is allowed to evaporate. The residue is dissolved in ^{1}ml of acetone and banded in four chromatographic papers. The chromatograms are developed in Forestal solvent system (conc., HCl - acetic acid –water; 3: 30: 10) or 30 % acetic acid. Three yellow bands separate between the Rf values 0.2-0.7. The papers are dried and seen in UV (356nm) light. Flavonols appear as yellow fluorescent bands. These bands are marked with pencil and one chromatogram is sprayed with 10% sodium carbonate solution. All the three flavonols turn yellow at first, but the lowest band turns green immediately. This is myricetin. The second band is quercetin which turns to yellow brown on standing. The third band (highest Rf) which remains yellow is kaempferol.

The yellow bands marked by pencil in unsprayed chromatograms are cut out and taken in clean test tubes. The compounds are eluted by dripping spectroscopic methanol slowly from the upper end of the paper strips. Methanol flows down slowly wetting the paper and in the process elute the flavonols in the paper. The elution is continued till the paper becomes colorless. The absorption spectra of these solutions of pure flavonoids are determined in the range 200-400 nm using a spectrophotometer. The color properties of the three flavonols and their absorption maxima are presented in Table 8.

Table 8. The colors and absorption maxima of the flavonols of tea

		Colors		λmax
	Visible	UV	$Na_2 Co_3$	MeOH
1. Kaempferol	Yellow	Yellow	Yellow	267, 294, 322, 367
2. Quercetin	Yellow	Yellow	Yellow-brown	255, 269, 301, 370
3. Myricetin	Yellow	Yellow	Green	255, 272, 301, 374

1.3.2.5. Anthocyanins

Anthocyanins generally occur as glycosides and are easily analysed first for their aglycones, the anthocyanidins. They are easily extracted from flowers. Fresh flowers (*Hibiscus,* rose etc. which are distinctly red, purple, pink or blue) are cut into small pieces and macerated into a pulpy mass. This is transferred to a beaker and about 200 ml 2N HCI is added and stirred and then warmed on a water bath for one hour. The red solution is filtered through cheese cloth (when the plant material is mucilaginous) or through filter paper. The filtrate is cooled and shaken with 50 ml ether (to remove the flavonoids) and the yellow colored ether layer is discarded. The red aqueous layer is further extracted with 50 ml of amyl/isoamyl alcohol. Anthocyanidins pass on to the alcohol layer which then becomes deep red/pink. This layer is separated and concentrated to dryness by better in a flash evaporator or by keeping in a sand bath and later on a water bath. The residue is dissolved in a few drops of alcohol containing one drop of 2N HCI and banded on chromatographic papers. The compounds are separated using **Forestal** solvent system (con. HCI: acetic acid: water, 3:30:10) or 30% acetic acid. Since the anthocyanidins tend to fade in presence of light, the chromatography is conducted in a dark room and the red bands are marked with pencil. A portion of the chromatogram is cut lengthwise and sprayed with 10% sodium carbonate solution where the color of the bands (red) turns blue. The red color may be regained by dipping the papers again in 2N HCI. The bands marked in untreated chromatograms are cut out and eluted with spectroscopic methanol containing 1% HCI. The absorption maxima of the compounds are measured by scanning the samples in the region between 450-600 nm. The anthocyanidins are identified by their colors, R_f values and spectral maxima given in Table 5.

Table 9. The color, Rf values and spectral maxima of common anthocyanidins

Name	*Color*	*Rf*		*λmax*
		Forestal	*30% HOAc*	*nm/MeOH*
Pelargonidin	Red	.68	.60	520
Cyanidin	Purple	.49	.40	535
Peonidin	Purple	.63	.58	532
Rosinidin	Purple	.76	.68	524
Delphinidin	Pink	.32	.28	546
Petunidin	Pink	.46	.40	543
Malvidin	Pink	.60	.54	542

1.3.2.6. Phenolic acids.

Phenolic acids are the precursors (!) of lignin and are abundant in leaves. Healthy leaves are collected and cut to fine pieces and mashed in a mortar. The mashed material is taken in R.B flasks, covered by 2N

HCI and hydrolysed by refluxing in a boiling water bath for 1 h. The flask is taken out, cooled and extract filtered through cotton. The filtrate is taken in a separating funnel, diluted by adding 50 ml of water and extracted with 2 x 25 ml ether. The ether layer is collected, concentrated and spotted on one corner of square Whatman No. 1 papers for two dimensional chromatography. The spots have to be concentrated and can come upto 2 cm diameter because the spray reagents are not sensitive to low concentrations. Benzene: acetic acid: water (6:7:3, upper layer) in the first direction and sodium formate: formic acid: water (10: 1: 200) in the second direction are used as the solvent systems. Chromatograms are dried, viewed in UV-365, fluorescent areas are marked and then sprayed with (1) diazotized p-nitraniline/ sulphanilic acid and (2) 10% aqueous sodium carbonate solution as an over-spray. The phenolic acids give distinctive colors with the two diazotised reagents and are identified based on the color in UV, reagents and Rf values (Fig. 3. Table, 6.).

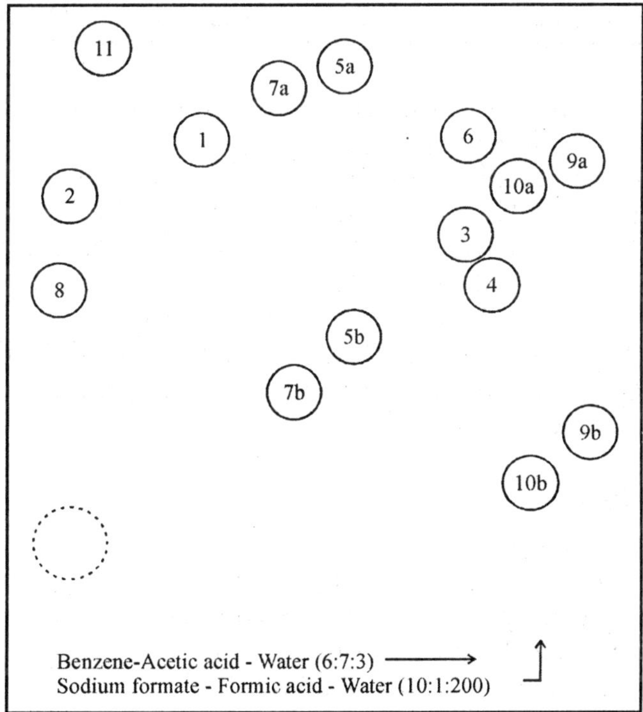

1. *p*-Hydroxy benzoic, 2. Protocatechuic, 3. Vanillic, 4. Syringic, 5a. *cis-o*-Coumaric, 5b. *trans-o*-Coumaric, 6. Melilotic, 7a. *cis-p*-Coumaric, 7b. *trans-p*-Coumaric, 8. Caffeic, 9a. *cis*-Ferulic, 9b. *trans*-Ferulic, 10a. *cis*-Sinapic, 10b. *trans*-Sinapic, 11. Chlorogenic

Fig. 3. 2-D Chromatogram (PC) showing the positions of phenolic acids

Table 10. The color reactions of phenolic acids

Sr. No.	Phenolic acid	Colors: UV	Colors: Diazotised *p*-nitraniline	Colors: Diazotised sulphanilic acid
1.	*p*-Hydroxy benzoic acid	-	Pink	Yellow
2.	Protocatechuic acid	-	Brownish Violet	Pink
3.	Vanillic acid	-	Purple	Orange
4.	Syringic acid	-	Blue	Red
5a.	*cis-o* Coumaric acid	Blue	Purple	Orange
5b.	*trans-o*-Coumaric acid	" "	" "	" "
6.	Melilotic acid	—	Purple	Orange-yellow
7a.	*cis-p* Coumaric acid	Blue	Blue	Pink
7b.	*trans-p*-Coumaric acid	" "	" "	" "
8.	Caffeic acid	Blue	Brown	Pink
9a.	*cis*-Ferulic acid	Blue	Bluish-green	Purple
9b.	*trans*-Ferulic acid	" "	" "	" "
10a.	*cis*-Sinapic acid	Green	Blue-green	Pink
10b.	*trans*-Sinapic acid	" "	" "	" "
11.	Chlorogenic acid	Blue	Blue-purple	Orange-Yellow

For preparation of diazotised reagents, stock solutions of *p*-nitraniline and sulphanilic acid are prepared by dissolving 0.7 g of *p*-nitraniline/sulphanilic acid in 9 ml of concentrated hydrochloric acid and making the volume to 100 ml. 4 ml of this stock solution is added slowly to 5 ml of 1% sodium nitrite solution in a volumetric flask kept cooled in ice. Since above 4° the diazotised compounds decompose, the nitraniline/sulphanilic acid also should be cooled before the addition. The volume is made upto 100 ml by ice-cold distilled water. After spraying this reagent the wet papers are sprayed with sodium carbonate solution.

1.3.2.7. Alkaloids

a. Tests for alkaloids

Of the many alkaloidal reagents, the three common reagents which are generally used for the detection of alkaloids are Mayer's, Dragendorff's and Wagner's reagents. Most of the other reagents are specific for a particular group of alkaloids.

Mayer's reagent

$HgCl_2$ (1.36 g) is dissolved in 60 ml distilled water (solution A) and 5g of KI in 10 ml water (solution B). Both the solutions are mixed and diluted to 100 ml with distilled water. Since this reagent reacts only with the salts of alkaloids, the alkaloidal solution should be made distinctly acidic with HCI or H_2SO_4 before adding the reagent. Only one or two drops of Mayer's reagent should be added in the beginning because the precipitates of some alkaloids are soluble in excess of the reagent. If no precipitate is obtained with the first two drops of the reagent, more reagent is added. A white precipitate indicates the presence of alkaloids. Mayer's reagent is the most useful reagent since it gives precipitate with hydrochlorides of most alkaloids.

Dragendorff's reagent

This is prepared by dissolving 0.85g of $Bi(NO_3)_3$ $5H_2O$ (Bismuth sub nitrite) in a mixture of 40 ml of water and 10ml of glacial acetic acid followed by the addition of 8g KI dissolved in 20ml of water. This reagent gives orange-red precipitate with most of the alkaloids.

Wagner's reagent

Iodine (1.27 g) and potassium iodide (2g) are dissolved in 5 ml of distilled water and the solution is made upto 100 ml with distilled water. This reagent gives brown flocculent precipitate with many alkaloids.

Only if an unknown sample gives precipitate with all the three reagents, the presence of alkaloids is confirmed. If the plant extract is to be tested, it has to be centrifuged to remove proteins and other macromolecules which also incidentally give precipitate with the above mentioned reagents.

b) Testing the plant samples for alkaloids

About 5g of dried plant material is kept in 50 ml of 10% ammoniacal ethanol for 48 h at room temperature with occasional shaking. The extract is filtered and distilled, preferably under vacuum. The dried/concentrated extract is acidified with 10 ml of 0.1 N H_2SO_4. The acid extract is centrifuged and the clear supernatant is tested with Mayer's, Wagner's and Dragendorff's.

Alkaloids may be tested by treating 1% HCl extracts or 10% acetic acid extracts of the fresh plant materials with the alkaloidal reagents. But this is not a confirmatory test as only those plants rich in alkaloids respond to this method.

1.3.2.8. Saponins

Saponins of *Smilax* roots, *Asparagus* root tubers and soap nuts are extracted in 70% or 90% ethanol. 10 g of powdered material is refluxed with 100 ml of 70% ethanol for 30 min. the extract is filtered, filtrate concentrated to 10 ml and applied to Silica gel TLC plates. The chromatograms are developed in chloroform: methanol: water (65: 50: 10). Saponins are visualized using any of the following spray reagents.

Blood reagent. Two milliliters of a mixture of sodium citrate and blood (prepared by adding 1 ml of 3.6% sodium citrate to 9 ml of fresh bovine blood) is mixed with 30 ml phosphate buffer of pH 7.4 (25ml of 2.7% aqueous potassium dihydrogen phosphate prepared in double distilled water mixed with 39.34 ml of 0.1 M sodium hydroxide and volume made to 100 ml). On spraying this mixture on chromatograms, the saponins appear as white zones against a reddish background. The plate is warmed if the zones are not visible immediately.

Vanillin-sulphuric acid. Plates sprayed heated to 100° for a few minutes; blue violet or yellow spots.

Anisaldehyde-sulphuric acid. Sprayed plates heated to 100°; saponins develop brown, purple, blue or green color (due to the sugars present in them).

Antimony trichloride reagent (20% in $CHCl_3$). Plates heated to 100°; red-violet spots (Antimony trichloride forms white precipitate with traces of water and therefore the containers should be thoroughly water-free).

Isolation of sapogenins

Since saponins are water soluble, to get sapogenins, they may be extracted in water and hydrolysed with 2N HCI. The precipitated aglycones are taken up in ether and analysed further.

100 g of fresh plant material (20 g of dry material) is boiled with 500 ml water. The extract is filtered and about 82 ml of con. HCI added (to make the solution 2N HCI) and the mixture is refluxed for 3h. The mixture is cooled and extracted with solvent ether. The concentrated ether fraction is spotted on TLC plates (silica gel) and developed using solvent systems (1) chloroform-ethyl acetate (1: 1) or (2) hexane-acetone (4:1). The developed plates are dried and sprayed

with 50% H_2SO_4 or Lieberman-Burchard reagent (2 ml con. H_2SO_4 + 40 ml acetic anhydride +100 ml chloroform). On heating the plates for 5-10 min. sapogenins develop purple violet colors. The common sapogenins and their relative positions in thin layer chromatograms (silica gel) are presented in Fig.4. Water can be used as a non-destructive spray which when sprayed smoothly to moisten the silica gel, leaves the sapogenins as white spots on a wet translucent background.

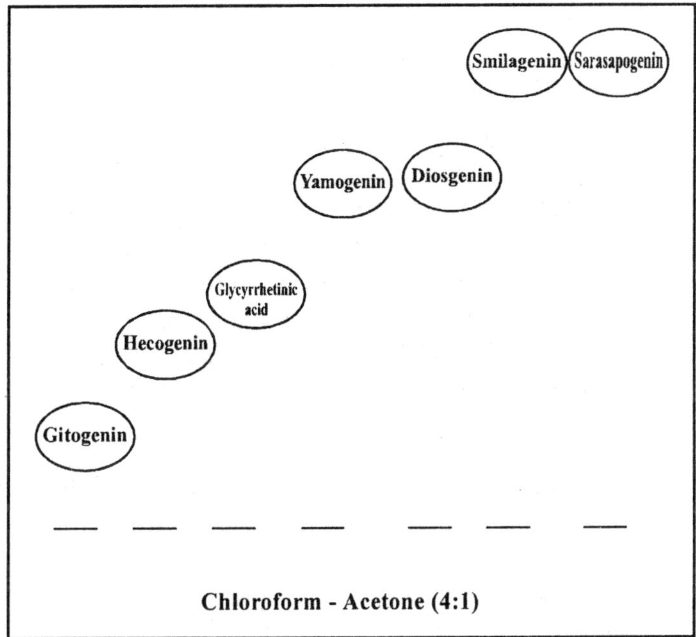

Fig. 4. TLC (silica gel) of some sapogenins

1.3.2.9. Cardiac glycosides.

The cardiac glycosides of *Digitalis* are extracted by boiling dry leaf powder (5g) in 100ml of 80% ethanol under reflux. The extract is filtered and the filtrate is treated with 50ml of 10% lead acetate to remove phenols especially flavonoids. The yellow lead phenolate, which is precipitated, is filtered off. The clear filtrate is acidified with 2ml glacial acetic acid and this solution is carefully extracted successively with 3x50ml dichloromethane. (The shaking process should be smooth lest emulsion is formed). The combined dichloromethane fractions are filtered through Buchner funnel lined with filter papers over which a layer of anhydrous sodium sulphate is placed and the solvent is evaporated to 5ml. This solution is spotted on TLC (silica gel) plates and developed in ethyl acetate- methanol-water (81:11:8).

Spray regents

Kedde reagent: This reagent is specific to cardenolides. It is prepared by mixing 10 ml of freshly prepared 3% ethanolic 3, 5-dinitrobenzoic acid and 10 ml of 2M NaOH. This mixture is always prepared fresh before spraying. Cardenolides give pink or blue colors. The color fades after a few minutes, but can be regained by repeated spraying.

Concentrated sulphuric acid, Antimony trichloride reagent and Lieberman-Burchard regent are other reagents which can be sprayed to locate the cardiac glycosides.

Digitalis purpurea leaves show a prominent blue violet spot of purpurea glycoside A (Rf about 0.3) with Kedde reagent. A number of light purple bands above this spot denote other glycosides present.

Digitalis lanata leaves show a blue violet spot of lanatoside A (Rf about 0.35) with Kedde reagent. *Nerium* leaves show at least 13 spots (purplish violet with Kedde reagent and pinkish red with sulphuric acid). The major spot nearest to the solvent front is oleandrin.

1.3.2.10. Biocrude

'Biocrude' refers to the total lipids isolated from the plants and which can be converted to gasoline or other petroleum products. A general method for the evaluation and identification of the hydrocarbon and rubber producing crops, prescribed by Buchanan *et al.* (1978), is as follows:

Dry plant material (500 g) ground to fine powder, is extracted with acetone for 24 h. The extract is concentrated, dissolved in 90% aqueous methanol (MeOH-H_2O, 9:1) or 87% ethanol (EtOH-H_2O, 7:1) and this solution is shaken with hexane. The hexane fraction contains the oils and methanol fraction contains polyphenols and complex lipids. The oil fraction (hexane solubles) is concentrated and transferred to a preweighed beaker, the solvent evaporated and the increase in weight is calculated. This is the biocrude. The biocrude is fractionated after saponification. The biocrude is saponified and the saponified extract is taken in 50% aqueous ethanol and shaken with hexane. The unsaponifiable matter (phytosterols, terpenes, wax components) may be quantified (by weighing the dry matter after the removal of solvent). The saponified fraction (ethanol fraction) is acidified to pH 6 and the precipitated lipids are extracted in hexane. The hexane fraction contains the fatty acids, rosin acids and some organic acids while the aqueous ethanol fraction contains glycerol.

The plant material left after acetone extraction is extracted further with benzene-hexane (2:3) or cyclohexane. Isoprene polymers, waxes and other terpenoids get extracted in this solvent/mixture. This fraction also is quantified and added to the biocrude to get the total biocrude from plant.

In a modified method Calvin *et al.*, (1981) recommend extraction of dry plant material directly with heptane to isolate the biocrude.

The plant material left after extraction with acetone and benzene-hexane/heptane is the oil and hydrocarbon-free residue. This residue contains cellulose, lignin, proteins, polysaccharides and inorganic components. Each one of these plant products may be quantified using appropriate methods. The scheme of extraction is presented in Fig. 5.

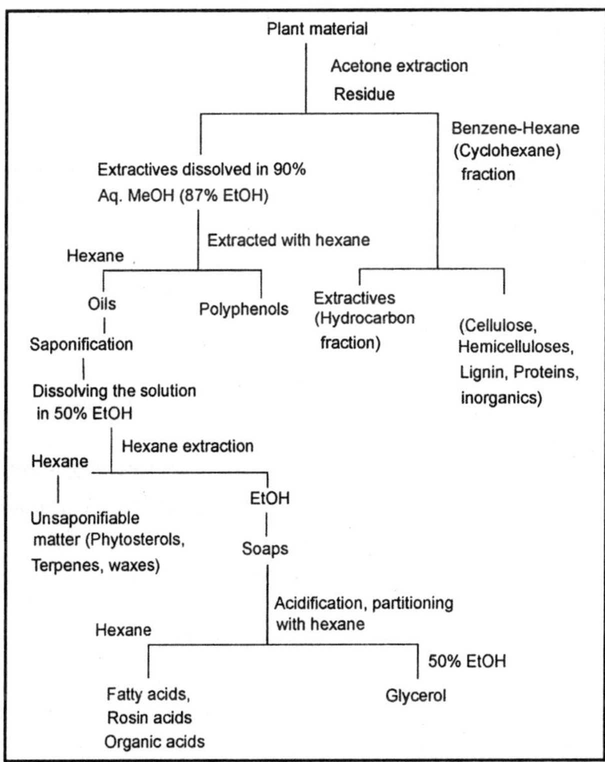

Fig. 5. A scheme of extraction to isolate biocrude from plants (Buchanan *et al.*, 1978)

□□□

2

Medicinal Plants

Medicinal plants and herbal medicines is an area where India can become the No. 1 country in the world. The well acclaimed systems of medicine like Ayurveda and Sidha and the large number of medicinal plants (more than 2500) which are in practice and a lot many which are to be discovered (lying hidden in many of the inaccessible forest regions) India can offer the world "panacea" for all diseases. China, which commands 45% of the world market of herbal medicines, does it only with the help of six acclaimed plants. India can have a much higher chance of dominating the world of herbs which will, in turn, alleviate the sufferings of countless millions, but also bring heavy revenues to the country which can be used for the overall improvement of people.

The business of herbs or herbal medicines is one area in which all parts of the country (including all geographic and climatic regions) as well as people from all strata can participate and get benefited both by health and wealth. The rural areas can cultivate the herbs, harvest them and send to the urban areas for extraction, formulation and marketing. Each village can grow the medicinal plant of that region and this brings additional revenue and empowerment for rural people. A herbal revolution similar to the "White Revolution" (of the famous Varghese Kurien in Anand - **Amul**) can transform the villages into rich repositories of wealth and will give jobs to a large member of people. A small extraction unit which would extract the active principles or standardized extracts in every village can give employment to another set of people. This will reduce the problems of storing large amount of raw materials and even save in transport in that instead of transporting a heavy load say, 100 tonnes of herbs, one need to transport only 5-10 tonnes of extract.

The picture presented here is absolutely rosy. But there are huge impediments on the way, to be overtaken; some of which are listed in one of my recent papers (Daniel, 2004). These problem-areas are short-listed below.

2.1. Identity of Medicinal Plants

In Indian systems of Medicine the medicinal plants are known by the various attributes or properties of the plant or the effect they produce in man or animals. The name **Brahmi** (giving the mental power of Brahma, the God), **Sarpagandha** (used as a medicinal for snake bites), **Ashwagandha** (giving the power of horses), **Amruta** (like legendary Amrut which gives eternal life to the consumer), **Punarnava** (giving a new life) and **Putranjiva** (good for the health of son) etc. denote the effects they produce on man. **Lakshmana** (having a lakshan – identifying mark), **Dudheli** (having milky juice), **Duralaba** (having thorns or stringing hairs), **Gojiva** (leaves similar to the tongue of cattle), **Hamsapadi** (resembling the feet of swan) etc. denote the properties of the plant which would help in identifying them. The ayurvedic physicians used to identify these plants based on the training they have received from their peers, or on the availability of substitutes resembling the same in the vicinity. Thus we have four plants known as shankhapushpi i.e. *Clitoria ternatea* (also know as aparajita), *Convolvulus prostratus, Evolvulus alsinoides* (also known as Vishnukranta) and *Canscora decussata,* recognized and used in various pasts of India. In case of Brahmi, two plants *Bacopa monnieri* (also know as neer-brahmi) and *Centella asiatica* (also known as mandukparni) are recognized. Same is the case with sariva (4 plants; *Hemidesmus indicus, Ichnocarpus frutescens, Cryptolepis bucchanani* and *Decalepis hamiltoni*); Jivanti (3 plants, *Holostemma adakodien, Leptadenia reticulata* and *Flickeringia nodosa*); Kachnar (various species of *Bauhinia*), Asoka (two plants, *Saraca asoka* and *Polyalthia longifolia*), etc are some of the large number of drugs for which two or more plants are used. The clinical efficacy of only a few plants has been studied, though all these plants have been frequently used in many parts of India.

It is to be remembered that not all plants used as drug are chemically alike in that they contain same active principles. Many a times the two plants used as a medicine are completely different. Let us take the case of Punarnava. Two plants, *Boerhavia diffusa* (along with other species of *Boerhavia* like *B. erecta* and *B. verticillata*) and *Trianthema portulacastram* are used as this drug. *Boerhavia diffusa* is found to contain alkaloids (Punarnavine 1 & 2), hypoxanthine, panarnavoside, rotenones and lignans while *Trianthema* contains ecdysone and an alkaloid (trianthemine). Similarly *Bacopa monnieri,* used as brahmi, contains saponins such as monnierin, hersaponin, bacosides and alkaloids like nicotine, brahmine, herpestine etc. whereas the other brahmi (*Centella asiatica*) contains some other saponins like asiaticoside, indocentelloside, thankuniside etc along with another alkaloid hydrocotylin. Same is the case with the plants used as shankhapushpi where four different plants are used for the same purpose. *Clitoria ternatea,* the first plant, is found to yield steroids and flavonols. *Convovulus prostratus* and *Evolvulus alsinoides* yield clavine alkaloids (convolvine, convolvamine, phyllalbine, convolvidine, and contoline etc. in the former and evolvine in the latter). *Canscora decussata,* the fourth plant used as shankhapushpi, contains an entirely different group of compounds, xanthones. All these examples clearly indicate that the chemical compounds present in different plants used as one drug may differ greatly.

The change in chemical constitution indicates change in efficacy of these medicines also. It is therefore conclusive that every drug plant should be addressed by its scientific name (Botanical name) and not by common or colloquial names. More over, it is mandatory in all developed countries that all herb based medicines should denote the scientific name of the source plant.

2.1.1. Scientific name, Botanical name or the legal name of the drug plant.

The scientific name/botanical name is the legal name that will be recognized in a court of law whenever there are disputes over any drug. This name is always frowned by the herbalists/Ayurvedic physicians of our country. Well, this is the case with "Botanists" too. For all these people, botanical name is "Greek and Latin". Yes, of course, it is Latin. It is because the botanical name is universal name. It has to be in a language practised (or known) by a large section of people. Since organized nomenclature (naming) was initiated by European botanists, that also by the first international Botanical congress meeting in Paris in August 1867, where more than 150 European and American Botanists attended, the language selected was Latin (from which most of the European languages originated) known universally by scholars and students throughout the world. (If a congress of Indian Botanists had begun a system of nomenclature, then Sanskrit, as practiced in Ayurveda, would have been selected as the language of nomenclature). This Latin is not the classical Latin of Cicero or Horace, but is the "Lingua franca" (a mixture of Italian with French, Greek, Arabic and Spanish) normally used as a medium between different nations whose own languages are not the same, providing mutual understanding) spoken and written by scholars during the middle Ages. This Latin was a development of the popular or "vulgar" (incorrect) Latin spoken by ordinary people in classical times and was written by all but a handful of poets and literary men. Latin is specific and exact in its meaning. By its preciseness and conciseness, it is particularly pertinent to the needs of descriptive phases of the natural sciences. Latin is written using the Roman alphabet.

The **binomial system of nomenclature,** in which the name of every plant consists of two words, the first designate the **genus** of which the plant belongs and the second, a particular **species** of that genus. This second word of a plant's name is a specific epithet, the two words in combination comprise of the plant's name and form a binomial and the binomial is a binary epithet.

The difficulty in understanding a botanical name stems from the lack of knowledge of its derivation or its meaning. Every **generic name** is a noun and starts with a capital letter. These names are given following the four methods by which the name is given. They are the following;

(1) Based on the name of a scientist:

(a) *Einstenia* in honour of Albert Einstein
(b) *Theophrasta* in honour of Theophrastus
(c) *Hookera* in honour of J.D.Hooker
(d) *Linnaea* in honour of Linnaeus

(2) Based on some features of the plant:

a. *Callistemon* (calli = beautiful; stemon = stamen)
b. *Calliandra* (calli = beautiful; andra = androecium)
c. *Chloroxylon* (chloro=green; xylon = wood)
d. *Haematoxylon* (Haemato=red; xylon = wood)
e. *Anacardium* (ana = inverted; cardium = heart; fruit is similar to inverted heart)

f. *Hygrophila* (Hygro = water; phila = loving - plant growing near water)

(3) Based on mythological or poetic concepts

a. *Nymphaea* (Similar to beautiful nymphs)

b. *Theobroma* (Theo=God; broma=food ; food of Gods)

(4)Based on the local name of the plants origin

a. *Saccharum* (based on sacchar = sugar - Sanskrit name)

b. *Adhatoda* (from Malayalam name 'adalodakam')

c. *Mangifera* (based on 'manga' - Malayalam name of mango)

d. *Ginkgo* (Chinese name)

e. *Nelumbo* (Sinhalese name)

Species name is the second word in a botanical name. It is always adjective in nature and begins with a small letter. This word has many sources of origin.

(1)Based on a morphological feature, abundance, habitat, distribution, utility or other traits

a) Relative abundance e.g. *vulgaris* (common), *rara* (rare)

b) Color of the flower/fruit e.g. *nigra* (black), *rubra* (red), *viridis* (green), *flava* (yellow) etc.

c) Habitat of the plant, e.g. *aquatica* (in water), *arvensis* (in fields), *muralis* (on walls), *palustris* (in swamps), *sylvatica* (in woods)

d) Region where the plant was found, e.g. *anglicus* (English), *canadensis* (Canada), *gallica* (France), *mexicana* (Mexican), *nepalensis* (Nepalese), *zeylanica* (Ceylon).

e) Size, shape or habit of the plant, e.g. *alta* (tall), *crassa* (thick), *gigantea* (giant), *natans* (swimming), *pendula* (hanging), *repens* (creeping), *scandens* (climbing), *tenuis*(thin).

f) Utility of the plant e.g. *edulis* / *esculenta* (edible), *hortensis* (in gardens), *officinalis* (sold in apothecaries shops), *arvensis* (sown for crops)

g) Other traits, e.g. *autumnalis* (of autumn), *biennis* (biennial), *foetida* (foul-smelling), *hiberna* (of winter), *religiosa* (used for religious purposes), *spinosa* (spiny), *toxicaria* (poisonous) etc.

(2) Species name made in honor of a scientist e.g. *mulleriana* (Muller), *agharkii* (Agharkar), *roxburghii* (Roxburgh), *wallichii* (Wallich) etc.

(3) Species name indicating similarity with some other species e.g. *amaranthoides* (like *Amaranthus*), *acanthoides* (like *Acanthus*), *bignonioides* (like *Bignonia*), *alliaceum* (like *Allium*) etc.

(4) Based on the Latin word of the plant e.g. *Capsella bursa–pastoris* (Bursa - pastoris was the original classical name of shepherd's purse), *Alisma plantago–aquatica* (Plantago–aquatica was the Latin name for water plantain), *Pyrus malus* (Malus was the ancient name for apple) etc.

(5) Formed from joining two words explaining some distinctive features of the plant e.g. *latifolia* (broad - leaved), *angustifolia* (narrow leaved), *cordifolia* (heart shaped leaves), *grandiflora* (large flowered) etc.

A binomial (a name of two words) is almost always followed by the **author's name,** mostly abridged. Author is the first person who named the plant or legalized its name. Thus Linn. or L. denotes Linnaeus; DC, De Candolle; Hook., Hooker etc. this author citation is a must because in case two different plants are given the same name by two different taxonomists (one not knowing that the name is already given by somebody else to another plant), the two plants can still be traced and the confusion can be clarified.

2.1.2. Change of names

This aspect of change of names causes much confusion and concern to many a non -taxonomist person interested in plants. When the same plant is known by different names at different places, according to the laws of **International Code of Botanical Nomenclature (ICBN),** the first validly published name should be accepted as the correct / legal name. This particular rule is the **rule of priority** and every plant scientist has to obey these rules. For example Cadamb plant was identified first as *Anthocephalus indicus* by A. Rich. in 1837. Without knowing this, Korth named the same plant *Anthocephalus morindae* in 1839. Not knowing both these names, Miq in 1856 identified it as *Anthocephalus cadamba*. Ignorant of these names, Roxburgh identified this plant as *Nauclea cadamba*. According to the rules of ICBN the first validily published name is *Anthocephalus indicus*

A Rich. and this is the correct name and all other names are *synonyms* or *illegitimate names*. Similarly *Randia dumetorum* was named thus by Lamarck in 1793. De Candolle identified this plant as *Randia longispina*. It was also named *Xeromphis spinosa* by Keay. But now it is found that the very same plant had been identified by Van Rheed, in his book 'Hortus Malabaricus' (which was recently republished) as *Catunaregam spinosa* way back in 15th century. Therefore the correct name (legal name) of this plant is *Catunaregam spinosa*. Therefore, the change in names is **the correction of wrong name** according to the pertinent international rules. Such correct names have great significance in the days to come when the plants are patented and in a court of law, the earlier name is the valid (correct) name and a claim on an incorrect name will be rejected. The correct (latest) names of a number of Indian medicinal plants along with their more common illegal names (synonyms) are presented in Table.

Table 1. The correct botanical names of certain medicinal plants

	Correct name	Synonyms	Common name
1	*Abelmoschus esculentus* Moench	*Hibiscus esculentus* Linn.	**Okhra/Bhindi, Ladies finger.**
2	*Abutilon hirtum* Sweet	*Sida hirta* Lam.	**Atibala, Country mallow.**
3	*Acacia nilotica* subsp. *indica* Brenan	*A. arabica* **Willd.** var. *indica* Benth	**Babul, Indian gum Arabic**
4	*Acacia senegal* Willd	*A. verek* Guillem. &Perr.	**Gum arabic tree**
5	*Actaea racemosa* Linn.	*Cimigifuga racemosa* Nutt.	**Black snakeroot, Black cohosh**
5	*Adhatoda zeylanica* Medic.	*A. vasica* Nees	**Vasaka, Malabar nut**
6	*Alangium salvifolium* Wang	*A. lamarkii* Thw. *A .decapetalum* Lamk.	**Ankol, Alangium**
7	*Alcea rosea* Linn.	*Althea rosea* Cav.	**Hollyhock**
8	*Aleurites moluccana* Willd	*A. triloba* J. R. & G. Forst.	**Candle nut tee**
9	*Alhagi pseudalhagi* Desv.	*A. camelorum* Fisch. *A. maurorum* Medic.	**Jawasa** **Camel thorn**
10	*Aloe barbadensis* Mill	*A.vera* Linn	**Kumari, Indian Aloe**
11	*Anethum sowa* Roxb.	*A. graveolens* DC, *Peucedanum sowa* Roxb.	**Sowa, Indian Dill**
12	*Anethum graveolens* Linn.	*Peucedanum graveolens* Linn.	**European Dill**
13	*Armoracia rusticana* Gaertn.	*A. lapathifolia* Gilib	**Horse-raddish**
14	*Bacopa monnieri* Wettst.	*Herpestris monnieri* H.B. & K.	**Brahmi,**

15	*Balanites roxburghii* Planch.	*B. aegyptiaca* Delile	**Hingoli, Desert Date**
16	*Baliospermum solanifolium Suresh*	*B. montanum* Muell. Arg *B. axillare* Blume	**Danti, Baliospermum**
17	*Bambusa arundinacea* Roxb.	.*B.bambos* Druce	**Vansa, Bamboo**
18	*Basella alba* Linn.	*B._rubra* Linn.	**Upodika/Poi, Red Vine spinach.**
19	*Benincasa hispida* Cogn.	*B. cerifera* Savi.	**Koshmanda, Ash gourd**
20	*Bergenia ciliata* Sternb.	*B. ligulata* Engl.	**Pashaanbheda, Megaseas**
22	*Boerhavia diffusa* Linn.	*B. repens* Linn.	**Punarnava, Spreading hogweed.**
23	*Brucea javanica* Merril	*B. amarissima* Desv., *B. antidysenterica* Merril	**Ya-Danzi, Brucea**
24	*Butea monsperma* Taub.	*B. frondosa* Konig.	**Palas, Bengal Kino**
25	*Caesalpinia crista* Linn.	*C. nuga* Ait	**Bonduc nut / Fever nut**
26	*Capsicum annum* Linn.	*C. frutescens* C.B.Clarke, *C. purpureum* Roxb. *C. minimum* Roxb.	**Lalmirich, Red Chilli.**
27	*Catharanthus roseus* G. Don	*Vinca rosea* Linn. *V. pusilla* Hook.f. *Lochnera rosea* Reichb.	**Baramasi, Vinca/ Periwinkle**
28	*Catunaregam spinosa* Tirveng	*Randia spinosa* Poir. *R. dumetorum* Poir. *Xeromphis spinosa* Keay.	**Madana, Emetic nut**
29	*Centella asiatica* Urban.	*Hydrocotyle asiatica* Linn.	**Manduk parni / Brahmi, Gotu kola**
30	*Cephaelis ipecacuanha* A. Rich.	*Psychotria ipecacuanha* Stokes	**Ipecac**
31	*Cerbera manghas* Linn.	*C.odollum* Gaertn	**Odollum**
32	*Chamaesyce hirta* Millsp.	*Euphorbia hirta* Linn *E. pilulifera* Linn.	**Dugdhika**
33	*Chrysanthemum parthenium* Pers.	*Pyrethrum parthenium* Sm. *Tanacetum parthenium* Pers.	**Featherfew, Feverfew**
34	*Cinnamomum aromaticum* Nees	*C. cassia* Blue	**Cassia**
35	*Cissus quadrangula* Linn	*Vitis quadrangula* Wall.	**Vajravalli, Bone-Setter**
36	*Citrullus colocynthis* Schard.	*Colocynthis vulgaris* Schard.	**Indravaruni, Colocynth**
37	*Coleus amboinicus* Lour.	*C. aromaticus* Benth	**Karpurvalli/ Patjarchur, Indian Borage**
38	*Clerodendrum indicum* Kuntze	*C. siphonanthus.* C.B. Clarke	**Bharangi**

39	*Convolvulus pluricaulis* Chois.	*Convolvulus microphyllus* Sieb. *C. prostratus* Forsk..	**Shankhapushpi**
40	*Crataegus laevigata* DC	*C. oxycantha* L.	**Hawthorn**
41	*Crataeva nurvala* Buch-Ham	*C. religiosa* Hook.f	**Varuna, Crataeva**
42	*Curculigo orchioides* Gaertn.	*C .malabarica* Wight *Hypoxis orchioides* Kurz,	**Musali**
43	*Curcuma zedoaria* Roscoe	*Ammomum zedoria* Christm. *Curcuma zerumbet* Roxb.	**Karcurah, Zedoary**
44	*Derris indica* Bennet	*Pongamia pinnata* Vent *P. glabra* Pierre	**Karanj**
45	*Desmodium gangeticum* DC.	*Hedysarium gangeticum* Linn.	**Prasniparni, Desmodium**
46	*Drimia indica* Jessop	*Urginea indica* Kunth.	**Junglypiaz, Indian squill**
47	*Drypetes roxburghii* Hurus	*Putranjiva roxburghii* Wall	**Putranjiva**
48	*Eclipta alba* Hassk.	*E. erecta* Linn. *E. prostrata* Linn	**Bhringaraj, Bhangra**
49	*Emblica officinalis* Gaertn.	*Phyllanthus emblica* Linn	**Amalaki/Amla, Indian gooseberry**
50	*Enicostemma hyssopifolium* Verdoon	*E. littorale* Blue.	**Chotta kirirayata, vavding**
51	*Erythrina variegata* Linn.	*Erythrina indica* Lamk.	**Paribhadrah, Coral tree**
52	*Euphorbia nivulia* Buch. - Ham.	*E. neriifolia* Linn.	**Dugdhika**
53	*Euphorbia thymifolia* Linn.	*E. prostrata* Ait	**Dugdhika**
54	*Fagonia cretica* Linn.	*Fagonia arabica* Linn	**Dasparsha/Duralabha**
55	*Feronia limonia Swingle*	*F. elephantum Corr.*	**Kota, Wood apple**
56	*Ficus racemosa* Linn	*F. glomerata* Roxb.	**Udumber**
57	*Garcinia indica Choiss.*	*G.purpurea Roxb.* Brindonia indica ***Dup.***	**Kokam, Brindon.**
58	*Gardenia resinifera* Roth.	*G. lucida* Roxb	**Dikamali**
59	*Glycine max* Merril	*G. soja* Maxim, *Soja hispida* Moench.	**Bhat, soybean.**
60	*Hedera nepalensis* Koch	*H. helix* Linn. **H. rhombea S.& Z.**	**Ivy**
61	*Hemidesmus indicus* R.Br.	*Periploca indica* Linn.	**Anantamool/Sariva**
62	*Holarrhena pubescens* Wall.	*H. antidysenterica* DC.	**Kurchi, Holarrhena**
63	*Hygrophila auriculata* Heine	*H.spinosa* T.anders. *Asteracantha longifolia* Nees.	**Kokilaksha/ Talmakhana, Hygrophila.**

64	*Ipomoea marginata* Verdc.	*I.sepiaria* Koenig.	**Lakshmana**
65	*Ipomoea mauritiana* Jacq.	*I. digitata* Linn.	**Vidarikanda/Vidari**
66	*Ipomoea nil* Roth.	*I.hederacea* Jacq.	**Krishnabijah.**
67	*Moringa oleifera* Lam.	*M. pterygosperma* Gaertn	**Sigru, Drumstick**
68	*Mucuna pruriens* DC.	*M. prurita* Hook.	**Atmagupta/Kavach, Mucuna.**
69	*Nelumbo nucifera* Gaertn	*N. nelumbo* Druce *N. speciosum* Willd	**Kamal, Lotus**
70	*Nerium oleander* Linn.	*Nerium indicum* Miller, *N. odorum* Soland.	**Karavirah/Oleander**
71	*Ocimum tenuiflorum* Linn	*O. sanctum* Linn.	**Tulsi, Basil**
72	*Operculina turpethum* Silva Manso	*Ipomoea turpethum* R.Br.	**Trivert, Indian Jalap**
73	*Oroxylum indicum* Vent.	*Bignonia indica* Linn.	**Shyonaka, Oroxylum**
74	*Pergularia daemia Choiv.*	*Daemia extensa* R.Br. *Pergularia extensa* N.E.Br.	**Kurutakah, Pergularia**
75	*Phyllanthus amarus* Schum. & Thonn.	*Phyllanthus niruri* Linn.	**Bhoomyamalki/ Bhuiavla**
76	*Plantago ovata* Forsk.	*Plantago ispaghula* Roxb.	**Isabgol, Psyllium**
77	*Plumbago indica Linn.*	*P . rosea Linn.* *P. coccinea* Boiss.	**Chitrak, Plumbago**
78	*Pogostemon cablin* Benth.	*P.patchouli* Hook.f.	**Patchouli**
79	*Premna corymbosa* Rottl.	*P.serratifolia* Linn *P. integrifolia* Linn. *P. obtusifolia* R.Br.	**Agnimanthah, Premna**
80	*Saraca asoca* De Willde	*Saraca indica* Baker	**Asoka**
81	*Serenoa repens* Small	*Sabal serrulata* Linn.	**Saw palmetto**
82	*Solanum americanum* Linn.	*S. incertum* Dunal *S. nigrum* Linn.	**Kakmachi/Makoi.**
83	*Solanum surattense* Burm. f.	*S. xanthocarpum* Schrad & Wendl.	**Kantakari**
84	*Syzygium aromaticum* Merr. & Perry	*Eugenia caryophyllata* Thaub. *E. aromatica* Bail	**Laving, Clove**
85	*Syzygium cumini* Skeels	*S. jambolanum* DC. *Eugenia jambolana* Lamk.	**Jamun/Jambuh**
86	*Tabernaemontana divaricata* Roem. &Schult.	*T.coronaria* Willd. *Ervatamia divaricata* Burkill.	**Nandyavartah/Tagar.**
87	*Tanacetum vulgare* Linn.		
88	*Taxus brevifolia* Nutt.	*T. baccata* subsp. *brevifolia* Pilger	**Pacific Yew**
89	*Thevetia peruviana* Merril	*T. neriifolia* Juss.	**Ashvaghna, yellow oleander**

90	*Toddalia asiatica* Lam	*T.aculeata* Pers.	**Kanchano, Lopez-tree**
91	*Trianthema portulacatrum* Linn	*T. monogyna* Linn.	**Punarnava**
92	*Tylophora indica* Merrill.	*T. asthmatica* W.& A.	**Anantamul, Tylophora**
93	*Vernonia cinerea* Less.	*Conyza cinerea* Linn.	**Sahadevi**
94	*Woodfordia floribunda* Salisb.	*W. fruticosa* Kurz.	**Dhataki**
95	*Wrightia arborea* Mabb.	*W. tomentosa* Roem. & Shult	**Kutaja, Pala Indigo plant**

2.2. Bioprospecting and Chemoprospecting

Bioprospecting is the search of specific compound/s in taxonomically related species of the original source plant. Since the plants are grouped/classified based on its similarities in many characters, all the plants of a taxon whether it is a subgenus/genus/tribe/family may have similar biosynthetic pathways and may produce same or similar compounds. So the search for a new or a substitute source should be based on the correct taxonomic relationships.

A correct taxonomic treatise of the taxon in question is the prerequisite to programmes of bioprospecting. Earlier classifications were mainly based on the morphologic similarities. But the recent systems of classification are based on all the characters of the plants such as morphology, embryology, palynology, pollen morphology, cytology, anatomy and phytochemistry. Since even molecular biology is used in assessing inter-relationships among plants, taxonomically related plants will possess the same or similar genes. Chemotaxonomic treatments of families in question will be of immense help here. Thus due to similarity in genetic sequences, biosynthetic pathways also are likely to be more or less the same. This means the taxonomically related plants may produce the same compound or due to the variations/adaptations which make it a different plant (different identity) may produce slightly different compounds. A classic example of bioprospecting is accomplished in case of Indian yew, *Taxus baccata.* The well known taxol, a highly potent antineoplastic agent, was first detected in the bark of pacific yew, *Taxus brevifolia.* Since the source material was the bark of the plant, the whole tree had to be uprooted for separating bark from which taxol had to be extracted. But it was found that a related species *Taxus baccata* contained baccatin in the leaves which could be semisynthetically converted to taxol. The presence of baccatin in the leaves guaranteed that the whole plant need not be uprooted for isolating the desired compound. The plant could be defoliated and the raw material could be collected without harming the source tree. Another example is the discovery of camptothecine, an anticancerous alkaloid. This compound was first discovered from a Chinese plant *Camptotheca acuminata* belonging to the family Nyssaceae of the order Cornales. Now the very same compound is extracted from *Mappia foetida* (Olacaceae) of nearby order Santalales.

The studies on chemotaxonomy reveal a number of plant groups possessing a large array of chemical compounds widely distributed or specific in them. For example, the acetogenins, the well known biopesticidal compounds, seem to be specific to the Annonaceae. Only about 230 different acetogenins have been isolated from 26 plants of this family (Keinanetal, 1997). These compounds also exhibit remarkable cytotoxic, antitumour, antimalarial and immuno-suppressive activities. When we know that the family Annonaceae contains about 2300 species (of which only 26 are studied so for!) we can only wonder on the huge dearth of data awaiting to be discovered and utilized for human welfare from the remaining 2275 plants (99%). It is almost criminal that we do not use the chemical libraries of the acetogenins provided to us by nature. Similarly the limonoids and quassinoids which also are antifeedants, pesticidal, cytotoxic etc. are restricted to the Rutales *sensu stricto* but widely distributed within. Only about 100 plants belonging to this order have been studied so far. But this order contains about 2200 plants (Rutaceae =1500 spp., Meliaceae = 550 spp., Zygophyllaceae=150 spp. and Cneoraceae = 3 spp.) and this indicates that 95% of plants are never studied for their compounds. This is the case with every plant group which contains useful herbs.

The compounds widely distributed in members of particular taxa are many; to cite a few;

1. Benzyl isoquinolines in Ranales and Papaverales,
2. Ergot alkaloids in Convolvulaceae,
3. Lupinane alkaloids in Fabaceae,
4. Tropane alkaloids in the Solanaceae,
5. Tropolones in Liliaceae,
6. Amaryllidaceous alkaloids in the Amaryllidaceae,
7. Biflavones in Coniferales and Clusiaceae,

8. Isoflavonoids and rotenoids in Fabaceae,
9. Non-protein amino acids in Fabaceae,
10. Saponins in Sapindaceae
11. Erucic acid in the Brassicaceae (Cruciferae),
12. Petroselinic acid in the Apiaceae (Umbelliferae),
13. Polyacetylenes in the Asteraceae (Compositae)
14. Sulphides in the Liliaceae,
15. Indole alkaloids in Rubiaceae, Apocynaceae and Asclepiadaceae,
16. Cardiac glycosides in Rubiaceae, Apocynaceae and Asclepiadaceae,
17. Hydnocarpic and chaulmoogric acid in the Flacourtiaceae,
18. Hypoglycin in the Sapindaceae,
19. Xanthones in the Gentianaceae,
20. Peptide alkaloids in the Rhamnaceae etc.

Bioprospecting can be done on the taxonomically related plants of a drug plant found in another geographical region, for e.g. *Echinacea* is a well known medicine in Europe used for cough and colds. In India we have *Echinops* which is grouped in the same tribe and possesses similar properties.

In contrast to the bioprospecting where the new sources of known compounds are searched in the biologically related groups, **chemoprospecting** involves looking for biologically active related molecules having some pharmacological activity, of course, from biologically unrelated plants. Betulinic acid, a triterpene first isolated from *Ziziphus mauritiana* long back, is recently found to induce apoptosis and cause cell death in melanoma cells without impacting the healthy tissue and is now moving towards clinical trials for humans. This discovery prompted studies on other related terpenoids and many triterpenes which are subjected to the bioevaluation are found to possess similar activities (Taraphdar *et al.*, 2001). Another example of chemoprospection based discovery is that of phytoestrogens. The sterility caused by sweet clover (*Trifolium subterraneum*) on the sheep grazing on this plant was found to be due to the isoflavonoids which function as estrogen antagonists. Recently it has been found that some of these isoflavonoids stimulate a signal transduction pathway leading to apoptosis and now feature in the cancer chemoprevention and hormone replacement therapy.

2.2.1. Bioassay derived pharmacophores

A number of different types of molecules from widely different sources are found to exert the same pharmacological activity. Otteliones from *Ottelia alismoides* were found to have taxol like properties as also the cholestane glycosides from *Ornithogalum saundersieae.* Epithilones from the myxobacterium *Sorangium cellulosum* and allamandin (an iridoid) from *Allamanda cathartica* are found to have anticancerous properties. All these compounds are discovered by the bio-assay oriented approach. In this method, the compounds in a plant are isolated and characterized and then subjected to high throughput screening (HTS) technology. In this technology each compound is screened against disease-based targets like enzyme inhibitors, cancer cell-lines, anti–metabolites etc. Once a compound is found to be positively active in a test–system, then it may be taken up further for clinical trials.

2.2.2. Leads from ethnomedicine

Ethnomedicine is another great repository of pharmacophores. The data collected from the tribals are to be verified by bioactivity based fractionation and when a single molecule, a group of molecules or a particular fraction is found to possess activity, they may further be studied by clinical trails. A number of plants have been discovered thus. Some of such plants are (1) Arogyapacha, *Trichopus zeylanicus* (2) asmol /a mixture of biorationally selected herbs of asthma (3) Vijayasar – *Pterocarpus marsupium* – antidiabetic (4) Salai guggul – *Boswellia serrata*-anti-inflammatory (5) *Janakia arayalpathra*-antitumour, (6) *Phyllanthus urinaria*- antitumour, and (7) Guggul-*Commiphora wightii* - hypocholesteremic etc.

2.3. Biomarkers

Biomarkers are of immense importance in the quality control procedures. Knowledge on the quality and quantity of biomarkers in a sample of raw material, extract or formulation is a prerequisite for marketing the products. It is also useful in judging the amount of active components in a sample, on the genuineness of the drug and the bio-efficacy of a drug. The absence of a biomarker indicates that the drug is completely adulterated. Lesser quantities of biomarkers indicate poor quality of the raw material or the sample/preparation is adulterated. Compounds other than the biomarkers when present indicate that the drug contains some other plant/drug.

In India mostly the drugs obtained in the market are of very poor quality. This is the main reason by which the efficacy of the drug is doubted many times. Most of the raw materials are adulterated. The adulteration of the drug materials may be due to the ignorance of the part of the plant collector on the correct botanical identity of the drug or due to the poor availability of the same plant in a particular area. All these lead to the production of substandard medicines and alternately it is the consumer who is hit below the belt.

Biomarkers are the compounds or the cells/tissues/cellular contents specific to a particular plant. They indicate the presence and the availability of the plant in a preparation. Two types of biomarkers are generally recognized.

1. Chemical biomarker,
2. Pharmacognostic biomarker.

2.3.1. Chemical biomarkers

A chemical marker is any compound, whether it is a primary or secondary metabolite, or an assortment of compounds which are seen in plant in detectable concentration. An ideal biomarker should be stable, easy to isolate and characterize and should be immune to ecological changes. Any compound occurring in appreciable concentration can be a reliable biomarker. In cases where two related/ unrelated plants possess the same compound, a second compound present in any one of the two in combination with the first compound form the biomarkers. The two biomarkers in a plant need not be of the same chemical class. They can be widely different. For example if two plants contain the same alkaloid (say berberine), a coumarin/flavone/ monoterpene present in any one of the two plants will form a biomarker along with berberine. A number of biomarkers and their methods of analysis are given in *'Indian Herbal Pharmacopoeia'* volumes jointly brought out by RRL, Jammu and IDMA, Mumbai as well as the volumes of *'Quality Standards of Indian Medicinal Plants'* published by ICMR.

A number of medicinal plants and their major compounds which may serve as the biomarkers are given in the section of active principles.

2.3.2. Pharmacognostic markers

These are the tissues, cells or cell inclusions characteristic to a particular drug/ plant. They are of immense use in finding out the purity and genuineness of raw material in the whole form or in the powder form. Even in extracts, due to the common methods of filtration (through clothes or filters), a few cell debris will be available at the base of the container which may be examined to diagnose the drug. A number of books on Pharmacognosy, both by Indian and foreign authors, are to be referred to get a basic understanding of this subject. But even now data on a large number of plants are not available and a concerted effort should be made to document all the pharmacognostic markers of all the drug plants, their substitutes and their adulterants.

Pharmacognostic markers can pinpoint the identity of the source plant, but are not of much use in analyzing the chemical constituents. Though the books on Pharmacognosy include TLC and PC methods in identifying drug samples, here we consider these methods separate in phytochemistry. Some generalizations on the analysis of a plant powder are explained here.

1. The presence of palisade and stomata in large number indicates that the drug is a leaf. The location of bark cells, more vessels and tracheids, indicate that the drug is adulterated.
2. Presence of parenchyma with chloroplasts (chlorenchyma) or hairs in a root sample is a sure indication of the sample adulterated with stem bark or leaf.
3. Flower samples will always contain pollen grains.
4. Parquetry cells and vittae indicate that the drug is an Umbellifer fruit.
5. Presence of aerenchyma indicates that the sample is an aquatic species.
6. Large number of starch grains is a sure indication that the drug is a storage organ like root, seed, bark etc.
7. Bark cells denote that the source is a stem or root bark.

Starch grains from different sources have different morphology and gelatinization patterns. Raphides, sphaeraphides, cystoliths and other crystals vary with the sources. The dimensions of cells, vessels, tracheids and sclerenchyma also vary from plant to plant. Another important distinguishing factor is the hairs which present a wide variety and the variations are much specific to the sources. Various forms of glands and stomata also aid in identification of the source materials.

2.4. Active Principles

Every medicine owes its activity to a single or a group of pharmacologically active compounds (pharmacophore/s). In many cases, the active principle is the major compound which occurs in appreciable quantities but in many cases the major compound need not be the pharmacophore. The famous *Vinca* (*Catharanthus*) alkaloids never exhibited any antineoplastic property. It was only when the alkaloid extract was fractionated and each individual alkaloid was subjected to studies, the anticancerous properties of vincristine and vinblastine (which occur in very low concentrations in roots) were discovered. The major alkaloids in *Catharanthus* are ajmalicine and serpentine and the antineoplastic alkaloids form a very minor fraction of the total alkaloids. Similarly *Allamanda cathartica* is known to contain triterpenes esters like plumericin, isoplumericin, plumieride etc. which are supposed to exhibit the cathartic action of the drug. But it was allamandin, an iridoid lactone, which was found to exhibit anti-leukemic properties.

In many herbal medicines the active component is not clearly defined/pinpointed. The existing data only indicate that the aqueous extract/lipid extract exhibits various pharmacological actions. In many cases the different components of a plant drug are found to exert widely different properties. In the case of Punarnava (*Boerhavia diffusa*), the roots are found to contain punarnavoside, rotenoids such as boeravinones A,B,C,D &E, lignans such as liriodendrin and syringaresinol, flavones, sterols, boeravine (an isofuroxanthone) and hypoxanthine-9-L-arabinofuranoside. Pharmacological studies proved that punarnavoside is the antifibrinolytic agent. Liriodendrin and hypoxanthine-9-L-arabinofuranoside are found to be antihypertensive and the former is a Ca++ channel antagonist but the whole plant extract exhibit anti-inflammatory, diuretic and hepatoprotective' activities probably due to the other compounds present in the plant. This clearly proves that the compounds other than the so-called active principle also are active in the healing processes attributed to the drug. Similarly in *Withania somnifera*, the total alkaloidal fraction exhibits hypotensive, bradicardic and respiratory stimulant activities while the other major group of components, withanolides possess antiarthritic, immunosuppressive, antitumour and antibacterial properties.

The so called minor components, which are normally phenolics, are never attributed with any activity. It is in this context, Duke's(1997) observations are interesting. He describes ferulic acid, gentisic acid, kaempferol glycosides and salicylic acid as pain relievers while ascorbic acid, cinnamic acid, coumarin, myricetin, quercetin and resveratrol are explained anti-inflammatory. Even the variety of chemicals and their richness (concentration) in a medicinal herb is of great value in assessing its property. Duke's data base states that both coriander and liquorice contain 20 chemicals with antibacterial action; Oregano and rosemary has 19; ginger 17; nutmeg 15; cinnamon and cumin 11; black pepper 14; bay 10 and garlic 13. Quantity wise liquorice contains up to 33% bactericidal compounds (dry weight basis), thyme 21%, Oregano 8.8% & rosemary 4.8%, coriander 2.2% and fennel, 1.5%. Needless to say the above data explains the drug action of the said plants.

Active principles of a number of medicinal plants along with some of their important uses are given below. The plants are arranged according to the chemical groups as **Secondary metabolites** (alkaloids, terpenoids, phenolics etc.) and **Primary metabolites.** For a detailed treatise on the active principles the reader is referred to Daniel (2006)

2.4.1. Secondary metabolites

2.4.1.1.Alkaloids

Alkaloidal amines

1) Bala (*Sida cordifolia* Linn. -Malvaceae)

Roots contain ß-phenethylamines like ß-phenethylamine, ephedrine and pseudoephedrine; carboxylated tryptamines and quinazolines like vasicinone, and sitoindoside-X, an acylsteryl glycoside. Bala is one of the important general tonics (rasayana drugs) and is especially useful in rheumatism. Sitoindoside-X exhibits adaptogenic and immunostimulant properties.

2) Jawasa, Camel thorn (*Alhagi pseudalhagi* Desv.- Fabaceae)

The aerial parts yields ß-phenylethyl amine; it's N-methyl derivative, N-methyl tyramine, hordeine, etc. The drug is used as a laxative, diuretic and expectorant. Its infusion is said to be diaphoretic.

3) Khat, Abyssinian Tea. (*Catha edulis* Forsk.-Celastraceae)

Leaves yield phenylpropanoid amines such as cathinone and *nor*-ψ-and *nor*ephedrines and other alkaloids. Cathinone is a major CNS-active compound and the leaves are used against hypertension. The leaves are chewed or consumed as tea, often as a religious ritual, to attain a state of euphoria and stimulation. This plant is considered a hallucinogen.

4) Ma-Huang, Ephedra (*Ephedra sinica* Stapf.- Ephedraceae)

The stems contain ephedrine and pseudoephedrine, which provides relief in asthma and hay fever. Ephedrine and pseudoephedrine are cardiac stimulants also.

5) Stinging nettle.(*Urtica dioica* Linn.-Urticaceae)

The stinging hairs are found to contain acetyl choline, histamine and 5-hydroxytryptamine.

Due to the irritation caused by the herb , it is used to excite activity in paralyzed limbs, arthritis by beating the paralysed body parts with the plant and taken internally for treatment of haemoptysis and haemorrhages.

Diterpenoid alkaloids

1) Ativish/Bachnag, Aconite.(*Aconitum napellus* Linn. -Ranunculaceae)

The roots contain aconitines (aconitine) and atisines. It is used as a local analgesic in liniments.

Imidazole alkaloids

1) Jaborandi (*Pilocarpus jaborandi* Holmes.-Rutaceae)

The leaves yield pilocarpine, isopilocarpine, pilosine and isopilosine used in ophthalmic practices to contract the pupil of the eye.

Indole alkaloids

1) American Yellow jasmine (*Gelsemium sempervirens* Ait.-Loganiaceae)

The dried rhizomes and roots contain gelsemine, gelsemicine, sempervirine, gelsedine and gelsevirine. The drug, being a drastic CNS depressant, is used with very great care.

2) Atmagupta/Kavach, Mucuna. (*Mucuna pruriens* DC.- Fabaceae)

Seeds and roots yield physostigmine, mucunine, mucunadine and nicotine. Roots are used as a diuretic and for dropsy. Seeds are powerful aphrodisiacs and are known to promote sexual vigour and cure impotence.

3) Baramasi, Vinca/Periwinkle (*Catharanthus roseus* G. Don -Apocynaceae)

The roots and leaves possess more than 110 indole alkaloids, important of which are indole-indoline dimeric alkaloids such as vinblastine (vincaleukoblastine, VLB), vincristine etc which are used as oncolytic alkaloids effective against leukaemia cells.

4) Bhadra, Aerva (*Aerva lanata* Juss. -Amaranthaceae)

The plant is found to contain six alkaloids, canthin-6-one, 10-OMe canthin-6-one, 10-OH canthin-6-one, ß-carboline-1-propionic acid and its 6-OMe deriv. and glucooxycanth-6-one. The plant is anthelmintic, demulcent and used against kidney and urinary stones and diabetes. It is also used in malaria and skin diseases and as an expectorant and for indigestion and wounds.

5) Calabar Bean (*Physostigma venenosum* Bal.-Fabaceae)

The main alkaloidal constituents of seeds are physostigmine (eserine), eseramine, isophysostigmine and physovenine. Physostigmine being a reversible inhibitor of cholinesterase, used extensively in opthalmology for contracting the pupil of the eye.

6) Cat's claw (*Uncaria guianensis* Gmeliu & *U. tomentosa* DC. Rubiaceae)

Stem/root yields 17 pentacyclic/tetracyclic oxindole alkaloids among which uncarine is the major component. The pentacyclic alkaloids show immune strengthening and anti-inflammatory activities while the tetracyclic alkaloids affect CNS (and thus are toxic).

7) Ergot. (*Claviceps purpurea* Tul. -Clavicipitaceae)

The sclerotia produced by *Claviceps* on the ovary of rye, contain water soluble ergometrine (ergonovine) group (ergometrine and ergometrinine) and the water insoluble ergotamine group (ergotamine, ergotaminine, ergosine and ergosinine). Ergot is used as an oxytocic ("quick delivery") agent, a property due to ergometrine. Ergotamine and ergotoxine are specific analgesics for migraine.

8) Harmal, Peganum (*Peganum harmala* Linn.- Zygophyllaceae)

The seeds contain ß-carbolines like harmine (the main constituent), harmalol and harman. The active hallucinogen is harmine, which exhibits much more activity than mescaline. The seeds find use as a spice and a valuable aphrodisiac.

9) Karaskara/Kuchla, Nux vomica (*Strychnos nux-vomica* Linn. -Loganiaceae).

The seeds contain strychnine and brucine (dimethyl strychnine) Strychnine is extremely toxic but a competent central stimulant, and therefore form an important tool in physiological and neuranatomica research.

10) Krishnabija (*Ipomoea nil* Roth.-Convolvulaceae)

Seeds are found to yield indole alkaloids (lysergic acid derivatives) chanoclavine, penniclavine, isopenniclavine and elymoclavine, and lysergol. Seeds are purgatives, anthelmintics, blood purifiers and used in rheumatism.

11) Krishna kamal, Passionflower (*Passiflora incarnata Linn.-Passifloraceae)*

The plant contains alkaloids such as harmine, harman, harmaline, harmol and harmalol. The aerial parts are used for their tranquilizing action in anxiety neurosis, insomnia and for improving concentration.

12) Murraya (*Murraya exotica* Blance - Rutaceae)

Bark yields a series of alkaloids based on carbazole skeleton, prominent among them being murrayanine and girinimbine. The bark is used as a stimulative tonic and possesses antidiarrhoeal properties.

13) Prasniparni, Desmodium (*Desmodium gangeticum* DC. -Fabaceae)

Aerial parts contain tryptamine derivatives and alkaloids such as Nb-Me-tetrahydroharman and 6-OMe-2-Me-ß-carbolium cation. The roots are good cardiotonics and are used in cardiac disorders.

14) Sarpagandha, Rauwolfia (*Rauwoflia serpentina* Benth.-Apocynaceae)

The roots yield about 0.7-1.5% alkaloids consisting of reserpine, rescinnamine and deserpidine. They exhibit antihypertensive and tranquilising action. Reserpine is used effectively against hypertension.

15) Saptaparna, Dita bark tree (*Alstonia scholaris* R. Br.-Apocynaceae)

The stem bark contains alkaloids (upto 2.7%) with echitamine as the major component and echitamidine, tubotaiwine, picrinine, akuammidine, ditamine and echitenine as minor components. It is used against chronic diarrhoea, dysentery and bowel complaints. Echitamine and also the total alkaloids cause a sharp fall in blood pressure and are hypotensives. These compounds are found to be active against human sarcoma.

16) Shankhapushpi (*Convolvulus pluricaulis* Chois. - Convolvulaceae)

The plant contains clavine alkaloids such as convolvine, convolamine, phyllabine, convolidine, contoline, convoline, subhirsine and convosine. The drug is used as hypotensive, sedative, in anxiety neurosis and for improving mental faculties.

17) Vishnukranta (*Evolvulus alsinoides* Linn.-Convolvulaceae)

The plant is found to contain an alkaloid evolvine. It is used as a remedy for all kinds of fevers. It is also an aphrodisiac, and cures insanity, nervous debility and dysentery.

Isoquinoline alkaloids

1) Abuta, False pareira (*Cissampelos pariera* Linn.-Menispermaceae)

The roots contain bisbenzyl isoquinoline alkaloids hayatine (± bebeerine/curine/chondocurine), hayatinine (4"-methyl bebeerine), hayatidine, *l*-bebeerine (pelosine / cissampeline) etc. The roots possess diuretic, antilithic, analgesic, antipyretic and emmanogogue properties and prevent threatened miscarriage.

2) Afim, Opium poppy (*Papaver somniferum* Linn.-Papaveraceae)

The latex contains morphine, codeine, and thebaine. Opium and morphine are used as pain relievers and as hypnotics. Codeine is a milder sedative and finds use in allaying coughing.

3) Amruta, Tinospora (*Tinospora cordifolia* Miers-Menispermaceae)

Stem, which forms the drug, contains jatrorrhizine, palmatine, berberine and timbeterine. Amruta is one of the rejuvenating drugs useful in general debility, rheumatism, diabetes, and improvement of intellect and for fever. It also exhibits immunosuppressive (Pendse *et al.*, 1997), antiinflammatory, analgesic and antihepatotoxic properties (Singh *et al.*, 1984).

4) Ankol, Alangium (*Alangium salvifolium* Wang-Alangiaceae)

The alkaloids of root bark are alangine A & B, desmethyl psychotrine, marckine, markidine, lamarkinine, psychotrine, tubulosine, isotubulosine, cephaeline and emetine. Root and root bark are astringent, anthelmintic, purgative, emetic and useful in skin diseases. It is a reputed single drug in Ayurveda for treatment of rabies.

5) Berberis (*Mahonia aquifolium* Nuttal - Berberidaceae)

The dried rhizomes and roots yield berberine, oxycanthine, berbamine, isotetrandrine and magniflorine. Berberis is employed as a bitter tonic, alternative, stomachic and a curative for piles. Berberine is an effective antipyretic and antiperiodic. It is also used as a medicine for erectile dysfunction.

6) Bloodroot (*Sanguinaria canadensis* Linn.-Papaveraceae)

The rhizomes yield sanguinarine, chelerythrine, protopine and allocryptopine. Bloodroot is a stimulating expectorant and emetic. Sanguinarine causes doubling of chromosome in dividing cells, a property similar to colchicine.

7) Calumba root (*Jateorhiza palmata* Linn.-Menispermaceae)

The roots yield protoberberines like palmatine, jatrorrhizine, calumbamine and a dimer, *bis*-jatrorrhizine. Calumba root is used as a bitter tonic, analgesic and for diarrhoea and dysentery.

8) Curare (*Chondrodendron tomentosum* R & P.-Menispermaceae; *Strychnos castelnaea* -Loganiaceae)

Curare contains a variety of alkaloids and quaternary compounds that differ depending on the source plants. Menispermaceous curare contains (+) tubocurarine chloride – a quaternary compound containing a bisbenzyl isoquinoline structure – as the principal alkaloid and four non-quaternary bases – isochondodendrine, isochondodendrine dimethyl ether, bebeerine and chondocurine. Loganiaceous curare contains mostly C_{40} compounds of dimeric strychnine type (indolic) designated as toxiferines and C-curarines. Crude extract exhibits a paralyzing effect on voluntary muscle and is toxic to blood vessels. Tubocurarine chloride is used as a skeletal muscle relaxant and to control convulsions caused by strychnine poisoning and tetanus.

9) Daruharidra, Barberry.(*Berberis aristata* DC,-Berberidaceae)

The root and stem bark yields berberine, berbamine, aromoline, karachine, palmatine, oxycanthine, oxyberberine and taxilamine. The total alkaloid extract is a bitter tonic, used as a cholagogue, laxative, diaphoretic and applied for eye affections, ulcers and hemorrhoids.

10) Hydrastis, Golden seal (*Hydrastis canadensis* Linn.-Ranunculaceae)

The rhizome and the numerous slender roots contain hydrastine, berberine, canadine and berberastine (5-hydroxy berberine). Hydrastis finds extensive use in checking uterine haemorrhage.

11) Ipecac (*Cephaelis ipecacuanha* A. Rich.- Rubiaceae)

The roots contain emetine, cephaeline and psychotrine. Ipecac is used as an expectorant, emetic and for amoebic dysentery. Emetine has more expectorant activity and cephaeline more emetic.

12) Ishwari, Indian Birthwort. (*Aristolochia indica* Linn. -Aristolochiaceae)

The roots contain aristolochic acid, *l*-curine (aristolochine), aristolactam, aristolic acid and aristolamide. Roots possess emmenagogue, antiarthritic and emetic properties and used also against leprosy. Aristolochic acid is found to be anticancerous but also causes renal damage. Aristolic acid showed antifertility effect in rabbits.

13) Ivy (*Hedera nepalensis* Koch.. Araliaceae)

The whole plant yields emetine. The plant, in various combinations, is used in cancer and indurations of different kinds.

14) Kamal, Lotus (*Nelumbo nucifera* Gaertn.-Nymphaeaceae)

The rhizomes contain three aporphine alkaloids, nuciferine, roemerine and nornuciferine. Another alkaloid nelumbine is isolated from petioles, pedicel and embryo. Rhizome is given for diarrhoea, dysentery and dyspepsia of children.

15) Stephania (*Stephania glabra* Miers.-Menispermaceae)

Roots yield gindarine (palmatine), gindaricine, gindarinine (all protoberberines), pronuciferine (benzyl isoquinoline), dicentrine (aporphine type), homostephanoline and hasubanonine (both morphine type). Roots are extensively used in pulmonary tuberculosis, asthma, dysentery and fever.

16) Swernakshiri, Mexican poppy (*Argemone mexicana* Linn.-Papaveraceae)

The leaves and stem contain a number of alkaloids like berberine, protopine, retundine, munitagine, argemonine, *nor*-argemonine, codeine, morphine, toddaline, sanguinarine etc. The bitter yellow juice of leaves and stems is used in chronic opthalmia.

17) Tiliacora (*Tiliacora racemosa* Cole.-Menispermaceae)

Leaves contain the alkaloids, tiliacorine (a bisbenzyl isoquinoline with a diphenyl grouping), tiliacorinine (the diasteroisomer of tiliacroine), *nor*-tiliacorinine and tiliacoridine. Root bark contains tiliarine–a stereoisomer of nor-tiliacorine. Leaves are used as an antidote for snakebites and for filariasis. Roots are used against skin infections. Total alkaloidal fraction showed marked hypotensive activity.

Monoterpenoid alkaloids

1) Chotta chirayata/ Enicostemma. (*Enicostemma hyssopifolium* Verdoon-Gentianaceae)

Aerial parts yield erythrocentaurine and enicoflavine. The plant is restorative, carminative and antidiabetic. Plant extract is useful in fever and snakebite.

2) Tecoma (*Tecoma stans* Griseb.-Bignoniaceae)

This plant possesses tecomanine, tecomine, tecostanine and tecostidine. The bark of *Tecoma* is used against snake and scorpion bites.

3) Valerian (*Valeriana officinalis* Linn.-Valerianaceae)

Rhizomes and roots contain valerine, chatinine and valeriana alkaloid XI. Valerian is used as a carminative, antispasmodic, in hysteria and other nervous disorders.

Phenanthro – indolizidine alkaloids

1) Anantamul, Tylophora (*Tylophora indica* Merrill.-Asclepiadaceae)

All the parts of plant possess tylophorine, tylocebrine and tylophorinine. Tylophorine possess powerful vescicant properties, so used effectively in fighting asthma. Tylophorine has a paralyzing action on the heart muscle, but a stimulating action on the muscles of blood vessels. All the three principal alkaloids possess anticancer properties.

Pyridine-piperidine alkaloids

1) Anar, Pomegranate (*Punica granatum* Linn.-Punicaceae)

The volatile liquid alkaloids of the root and stem bark are pelletierine, isopelletierine, methyl isopelletierine and a solid alkaloid, pseudopelletierine. The root and stem bark are used as anthelmintic and taenifuge. The pericarp is a reputed anthelmintic for children.

2) Kalimirich, Pepper (*Piper nigrum* Linn.- Piperaceae)

Chavicine, piperine ($C_{17}H_{19}O_3N$), piperettine ($C_{19}H_{21}O_3N$) and piperidine are the alkaloids contained in the fruits. Pepper is a popular condiment, and is used medicinally as a stimulant and irritant. Pepper exhibits anticonvulsant properties and enhances the bioavailability of antitubercular drugs when given together. Piperine shows CNS depressant, antipyretic, analgesic, anti-inflammatory, antioxidant and hepatoprotective properties.

3) Lobelia (*Lobelia inflata* Linn.-Lobeliaceae)

The plant yields lobeline, lobinaline, lobelidine, lobelanine, lobelanidine and isolobenaline. Lobelia is used as an expectorant in chornic bronchitis and asthma.

4) Pipal, Long pepper (*Piper longum* Linn.)

The entire unripe dried spikes are used as spices and in medicine. They, in addition to the alkaloids found in *P. nigrum,* contain one more pyridine alkaloid-piperlongumine. Long pepper is a rasayana (tonic) drug useful in improving memory power and intelligence.

5) Poison Hemlock (*Conium maculatum* Linn.-Apiaceae)

The dry unripe fruits contain 1-2.5% of the liquid alkaloid coniine, along with other alkaloids like N-methyl coniine, conhydrine, pseudoconhydrine and γ-coniceine. The unripe dry fruits – the conium of commerce – are used as antispasmodic, sedative, anodyne, in epilepsy, acute mania, and whooping cough and for treating obstruction of lymphatic system. In homeopathy it is used to prevent cataract.

6) Supari, Areca nut (*Areca catechu* Linn.-Arecaceae)

The seed contains arecoline, arecaine (N-methyl guvacine), arecolidine (methyl ester of arecaine), guvacine (demethyl arecaidine), guvacoline and isoguvacine. The nuts are used as anthelmintics especially against tapeworm.

Quinazoline alkaloids

1) Vasaka, Malabar nut (*Adhatoda zeylanica* Medic.-Acanthaceae)

The major alkaloids of the leaves are vasicine (peganine, upto 1.1%), vasicinone and vasicinol. Vasicine and vasicinone are well-known bronchodilators, respiratory stimulants, hypotensives and cardiac depressants. Vasicine exhibits oxytocic and abortificient activity.

Quinoline alkaloids

1) Asvasakhotah, Glycosmis (*Glycosmis pentaphylla* DC.-Rutaceae)

Root bark contains skimmianine, γ-fagarine, dictamine, arborine, carbazole, 3-methoxy carbazole, homoglycosolone, homoglycolone, glycolone, glycozoline and glycozolicine. This plant is used for cough, rheumatism, anaemia and jaundice. Leaves are useful in fever, liver complaints, and eczema and skin diseases.

2) Cinchona (*Cinchona officinalis* L -Rubiaceae)

The bark of roots and stems yield quinine, quinidine, cinochonine and cinchonidine. Galenicals of cinchona are used as bitter tonics and stomachics. Quinine and its derivatives are extensively used to fight malaria. Quinidine finds use for prophylaxis of cardiac arrhythmias and for the treatment of atrial fibrillation.

3) Kanchano, Lopez-tree (*Toddalia asiatica* Lam.-Rutaceae)

The root bark contains chelerythrine (toddaline), dihydro-chelerythrine (toddalinine) and skimmianine. Root bark is diaphoretic, stomachic and antipyretic and is similar to *Cinchona* in its action against malarial parasite.

Quinolizidines/Lupinane alkaloids

1) Punarnava, Spreading hogweed (*Boerhavia diffusa* Linn.-Nyctaginaceae)

The whole plant, the drug punarnava, contains two quinolizidine alkaloids punarnavine 1 & 2. Punarnava, more specifically the roots and leaves, are used best for its diuretic and antiinflammatory properties and therefore opted for treating inflammatory renal diseases, nephritic syndrome, oedema and ascites resulting from early cirrhosis of the liver and chromic peritonitis.

2) Scoparius, Broom tops (*Cytisus scoparius* Link. -Fabaceae)

The dried tops yield sparteine, sarothamnine and genisteine. Sparteine is employed as an oxytocic.

Sesquiterpene alkaloids

1) Malkangani / Jyotishmati, Intellect tree (*Celastrus paniculatus* Willd.-Celastraceae)

Seeds yield three sesquiterpene alkaloids, celapanin, celapanigin and celapagin. The seed is tranquilizing sedative, emetic, diaphoretic, and febrifugal and is used for sharpening memory.

Steroidal Alkaloids

1) Hellebore (*Veratrum viride* Aiton-Liliaceae)

Rhizome contains (1) jerveratrum alkaloids and (2) ceveratrum alkaloids. Jerveratrum alkaloids are jervine, veratramine (Veratramine+glucose) and veratrosine (veratramine + glucose) and ceveratrum alkaloids are germine & protoverine (free alkamines), germinitrine (germine+acetic acid+tiglic acid+angelic acid) and protoveratrine A. These alkaloids, in mild doses, are employed as hypotensives, cardiac depressants and sedatives. They exhibit a strong insecticidal action too.

2) Kakmachi/Makoi. (*Solanum americanum* Linn.-Solanaceae)

All parts of the plant contain solanine and solasodine glycosides. The plant, especially fruits, is used as expectorant, antiinflammatory and diuretic. The plant extracts exhibit hepatoprotective and antitumour properties.

3) Kantakari (*Solanum surattense* Burm. F.)

The whole plant contains solasodine, solamargine, β-solamargine and solasonine. The drug is a well-known expectorant, antipyretic and useful in bronchial asthma and other respiratory diseases. Solasodine exhibits antispermatogenic, hypocholestrolaemic and antiatherosclerotic activities (Kanwar *et al.*, 1990; Dixit *et al.*, 1992).

4) Kurchi, Holarrhena (*Holarrhena pubescens* Wall.-Apocynaceae)

Bark contains conessine, 7-α-hydroxy conessine, holonamine, dihydroisoconessimine, 3-aminoconan-5-ene and kurcholessine. Bark and seeds are used in dropsy and dysentery (conessine is the antidysenteric principle), stomach disorders, abdominal and glandular tumours. Fruits are anticancerous and hypoglycemic and used to regulate menstruation.

5) Kutaja, Pala Indigo plant (*Wrightia arborea* Mabb.-Apocynaceae)

The bark contains conessine, holarrhine, kurchicine and conkurchine. The root bark and seeds, often substituted for Holarrhena, are used for chronic diarrhoea and dysentery. Other uses include curing piles, hemorrhage, cardiac diseases, fever and colic. In Sidha system this drug is used for psoriasis and other skin diseases.

Tropane alkaloids

1) Belladona (*Atropa belladonna* Linn. -Solanaceae)

The leaves yield l-hyoscyamine and atropine along with a small amount of volatile bases such as pyridine and N-methyl pyrroline. Leaves are commonly administered in the form of belladonna tincture as sedatives and to check excessive secretions. Hyoscyamine and atropine, in pure form, act as anticholinergic. Atropine produces a stimulant action on Central Nervous System. The property of atropine and hyoscine (scopolamine) to dilate the pupil of the eye is frequently employed in ophthalmic practices.

2) Coca (*Erythroxylum coca* Lam. -Erythroxylaceae)

The leaves yield ecgonines such as cocaine (methyl benzyl ecgonine), cinnamyl cocaine, methyl ecognine, α- and β-truxillines; tropanes like tropeine, pseudotropeine, dihydroxy tropeine, tropacocaine and benzyl tropane and hygrines such as hygrine, hygroline and cuscohygrine. Coca leaves are chewed to resist mental and physical fatigue and to get great stimulation. Due to the power of coca to enhance meditation and to produce trance like states, it is known as a "divine" plant. Cocaine is employed as a local anesthetic.

3) Corkwood tree (*Duboisia myoporoides* R. Br. & *D. leichhardtii* F. Muell-Solanaceae)

These plants are some of the chief sources of atropine and hyoscine.

4) Dhatura, Stramonium/ thorn apple (*Datura stamonium* Linn. -Solanaceae)

The alkaloids present are hyoscyamine and hyoscine in the ratio 2:1. Stramonium is often used as a substitute of belladonna. The vapour produced at the time of burning of the powdered seed relieves asthma. On ingestion the drug produces narcotic effects. Both hyoscine and hyosyamine are anticholinergic.

5) Egyptian henbane (*Hyoscyamus muticus* Linn.- Solanaceae)

This plant, a commercial source of hyoscyamine, yields about 1.5% total alkaloids, composed largely this alkaloid.

6) Hiranpag, Small Bindweed. (*Convolvulus arvensis* Linn.-Convolvulaceae)

Roots contain the alkaloid, cuscohygrine. The plant is used as a hypotensive and it increases coronary circulation.

7) Kala Dhatura, Indian Dhatura (*Datura metel* Linn.)

Leaves contain scopolamine. Used in place of *D.stramonium*

8) Khurasaniyajvayan, Henbane (*Hyoscyamus niger* Linn.-Solanaceae)

The dried leaves contain hyoscyamine and scopolamine. The action of henbane is similar to belladonna but weaker. Often used to relieve spasms of urinary tract and with strong purgatives, to prevent griping.

9) Luckmuna, Mandrake/May-apple (*Mandragora officinarum* Linn.-Solanaceae)

The leaves and roots yield the alkaloids atropine, hyoscine, mandragorine and *nor*-hyoscyamine. Mandrake is used in place of belladona. As a poultice this is used for rheumatic pains, ulcers and for skin diseases.

10) Scopalia (*Scopolia carniolica* Jacq.-Solanaceae)

The leaves and roots contain hyoscyamine and hyoscine in high concentrations alongwith cuscohygrine, 3'-α-tigloyloxy tropane, pseudotropine and tropine. Scopolia is used extensively as a source of hyoscyamine.

Tropolone alkaloids

1) Bachnag, Gloriosa (*Gloriosa superba* Linn.-Liliaceae)

Roots yield up to 0.37% colchicine and other derivatives of colchicine. Roots are used in rheumatic fever, leprosy, skin diseases, colic and snake-bite.

2) Colchicum (*Colchicum autumnale* Linn.-Liliaceae)

The alkaloids in colchicum are non-basic like colchicine, colchiceine and demecolcine. Colchicine is used to disperse the tumours and so is used in the treatment of various cancerous diseases. It is also used to relieve gout.

2.4.1.2.Terpenoids

Monoterpenoids

1) American wormseed. (*Chenopodium anthelminticum* Linn.)

The amount of oil is upto 2% in this plant and the oil contains 60-80% ascaridole.

2) Angelica (*Angelica archangelica* Linn.-Apiaceae)

The oil from roots consists mainly of α-phellandrene. Roots are used as diaphoretics, expectorants, and diuretics and in flatulent colic.

3) Bangra, Caesulia (*Caesulia axillaris* Roxb.-Asteraceae)

The volatile oil from this plant consists of γ-asarone (48%), limonene (34%), cineole (4%) and pinene (3%). The plant is used for treating baldness and goiter and as a diuretic. The young leaves are used as a potherb.

4) Bari-elachi, Large cardamom (*Amomum subulatum* Roxb.-Zingiberaceae)

The volatile oil from seeds (up to 3%) consists of cineole (65%), terpeneol, terpenene and pinenes. Seeds are used as a substitute for cardamom. They are stimulants and stomachics and are used against indigestion, abdominal pains and rectal diseases.

5) Barakulanjar, Greater Galangal (*Alpinia galanga* Willd.-Zingiberaceae)

Rhizomes contain a volatile oil containing methylcinnamate (48%), cineole (30%), camphor and pinene. Leaves also yield a volatile oil rich in methyl cinnamate.This is a well-known medicine for rheumatism and catarrh. It is also a tonic, carminative and stimulant.

6) Calendula (*Calendula officinalis* Linn.-Asteraceae)

The flower heads contain 0.1 to 0.4% volatile oil, composed of pedunculatine, α-, and β-ionones, *trans*-caryophyllene oxide, carvone, geranyl lactone, *n*-tricosane, terpene alcohols and lactones. The ray florets form the drug *Calendula*, which is said to have tonic, sudorific, febrifuge, carminative, antiemetic and anthelmintic properties.

7) Celery (*Apium graveolens* Linn.-Apiaceae)

The volatile oil from fruits is composed of limonene (50%), selinene (15%), sedanolide and sedanonic anhydride. Celery seeds (fruits) are extensively used as spices and possess stimulant and carminative properties. They are used as nervine sedatives and tonics.

8) Chinese Cassia (*Cinnamomum aromaticum* Nees.-Lauraceae)

The vol. oil from the bark consists of cinnamaldehyde (82%), eugenol, cinnamic acid, cinnamyl acetate etc. The leaves also yield an essential oil (0.4%) containing 71% cinnamaldehyde and 10% eugenol. The flowers and unripe fruits also yield similar volatile oils. Chinese cassia is used as diaphoretic, antipyretic, analgesic, purgative, cardiac stimulant, diuretic and sedative.

9) Dhanya/Kottmir, Coriander.(*Coriandrum sativum* Linn.-Apiaceae)

The volatile oil from leaves contain linalool as major component, α- and β-pinenes, limonene, γ-terpenene, *p*-cymene, borneol, citronellol, camphor, geraniol and geranyl acetate. Coriander is a well-known spice. It has got stomachic, spasmolytic and carminative properties. The essential oil is antifungal in nature.

10) Dong quai, Female ginseng (*Angelica sinensis* Diels.-Apiaceae)

The roots yield a volatile oil containing ligustilide, safrole, *p*-cymene, butylidene dinaphthide and sesquiterpenes. They are useful in relieving menopausal symptoms and for other menstrual disorders and dysmenorrhoea. It is also tonic in nature and recommended for general debility and anaemia.

11) European Dill (*Anethum graveolens* Linn.-Apiaceae)

The fruits contain a volatile oil containing carvone and *d*-limonene. The volatile oil as well as its emulsion in water are stomachics, diuretics, anthelmintics and antiflatulent and form an important ingredient in gripe water preparations for children.

12) European pennyroyal (*Mentha pulegium* Linn).

The oil from this plant contains pulegone as its principal constituent.

13) Ganugatulsi, Hyptis (*Hyptis suaveolens* Poit.-Lamiaceae)

The volatile oil from all parts of the plant consists of sabinene, β-pinene, limonene, β-caryophyllene, 1,8-cineole, spathulenol, caryophyllene and bicyclogermacrene. This plant is a stimulant, carminative, antiseptic, sudorific and galactogogue.

14) Garden Rue (*Ruta graveolens* Linn.-Rutaceae)

The principal component of this strongly odoriferous and pungent oil is methylnonyl ketone (80-90%) accompanied by methyl heptyl ketone , α-pinene, limonene, cineole etc. Rutin amounts to 2% of the plant. Rue oil is anthelmintic, antispasmodic, antiepileptic, rubefacient and emmenagogue. Coumarins are responsible for the antispasmolytic activity.

15) German chamomile (*Matricaria recutita* Rausch.-Asteraceae)

The whole plant yields an essential oil consisting chiefly of azulene. Inhaling the vapourised oil is recommended by aromatherapists to relieve nervous disorders such as depression and anxiety associated with menopause.

16) Hyssop ***(Hyssopus officinalis Linn.-Lamiaceae)***

The whole plant yields a volatile oil consisting of cineole, β-pinene and a variety of bicyclic monoterpene derivatives (L-pinocamphene, isopinocamphone, pinocarvone etc). Hyssop (all parts) is an excellent nerve tonic and help recovery after an illness.

17) Japanese peppermint (*Mentha arvensis* Linn.)

This plant yields a volatile oil rich in menthol. Used as a carminative.

18) Jatiphala, Nutmeg (*Myristica fragrans* Houtt.-Myristicaceae)

Nutmeg yields volatile oil (5-15%), consisting of 80% terpenes like pinene, camphene and sabinene, 8% dipentene, 4%myristicin, 2% elemicin and isoelemicin; up to 1% safrole and 1% other constituents like eugenol, methyl eugenol, isoeugenol etc. This is a remedy for indigestion. Nutmeg, in doses larger than 10-15 gm (one teaspoonful), is found to produce intoxications and hallucinations-similar to LSD-after a delay of 3-5 hrs in many individuals.

19) Indian wormseed. (*Chenopodium ambrosioides* Linn.-Chenopodiaceae)

The volatile oil (0.25-0.45%) from flowering and fruiting shoots contain ascaridole (50%), xylene, *p*-cymene, tetradecanol, caproic and caprylic acids. The whole plant is used as a tonic, emmenagogue, antispasmodic and in treating nervous affections.

20) Kala zira, Caraway. (*Carum carvi* Linn. -Apiaceae)

The fruits contain a volatile oil (3-8%) consisting of carvone (55-60%) and limonene. Minor components are α-pinene, β-pinene and *p*-cymene. Caraway is a condiment and are prescribed in flatulent colic and stomach problems. Alcoholic extract is antispasmodic.

21) Kapoor, Camphor (*Cinnamomum camphora* Presl.-Lauraceae)

All the parts of this tree yield a volatile oil rich in camphor. The minor components are safrole, cineole, borneole, camphene, dipentene and terpeneol. The different varieties of this species produce volatile oil rich in compounds like safrole, linalool, cineole, etc. Camphor is used as an anodyne, rubefacient, in rheumatism, as external application on sprains or as a liniment.

22) Karpurvalli/Patjarchur,Indian Borage (*Coleus amboinicus* Lour. Lamiaceae)

Leaves contain volatile oil composed of thymol (40%) and carvacrol. This is a common remedy for cough, cold and fever of children.

23) Karcurah, Zedoary (*Curcuma zedoaria* Roscoe-Zingiberaceae)

Sesquiterpenes are the major constituents of the volatile oil obtained from rhizomes. They are zedoarone and curzerenone, curcumol, curcolene, procurcumenol, isocurcumenol etc. The rhizomes are used against cough, bronchitis, asthma, skin diseases, wounds and spleenic disorders.

24) Laving, Clove (*Syzygium aromaticum* Merr. & Perry-Myrtaceae)

70-90% of volatile oil is eugenol. Also present in the oil are minor quantities of eugenol acetate, caryophyllene, methyl-n-amyl ketone and some esters. Cloves are used for flavouring foods, as a stomachic and carminative in medicine.

25) Lavender (*Lavandula officinalis* Choix.-Lamiaceae)

Fresh flower spikes and leaves provide 0.8-2% volatile oil, comprising of chiefly *l*-linalylacetate; geraniol, linalool, limonene, cineole, esters of butyric and valeric acids. Lavender is a flavoring agent and an expectorant.

26) Lemon balm (*Melissa officinalis* Linn.-Lamiaceae)

The vol. oil distilled from the whole plant contains more than 50 compounds of which citronellal, β-caryophyllene, neral, geranial, citronellol and geraniol amount to 70%. This plant is known for its ability to calm nerves and is a common medicine for insomnia. It is also found useful in fevers, colds and spasms. Of late, this plant is found to exhibit antiviral and anti herpes properties (Dimitrova *et al.*, 1993; Wohlbling and Leonhardt, 1994).

27) Lesser Galangal (*Alpinia officinarum* Hance)

The components of the volatile oil from the rhizome are 1,8-cineole, α-pinene, β-pinene, methyl cinnamate etc. Lesser galangal is used as a spice and carminative.

28) Lilichai, Lemon grass (*Cymbopogon citratus* Stapf. -Poaceae)

Leaves yield a volatile oil consisting of citral (78-82%), *d*-limonene, citronellol, geraniol and myrcene. Lemon grass is used in elephantiasis, as CNS depressant, analgesic, antirheumatic, antipyretic and expectorant. Lemongrass oil is anticarcinogenic in that *d*-limonene and geraniol induce activity of glutathione transferase an enzyme that inhibit cancer development.

29) Madhurika /Saunf, Fennel (*Foeniculum vulgare* Mill.-Apiaceae)

The fruits contain oil (12%), the principal component of which is a phenolic ether, anethole (60%) followed by the ketone, fenchone (10-30%). Other components are methyl chavicol, limonene, α-pinene etc. The fruits are anabolic, estrogenic, analgesic, antipyretic, and antimicrobial agents. They reduce flatulence in infants.

30) Nilgiri, Eucalyptus (*Eucalyptus globulus* Labill -*Myrtaceae*)

The leaves provide a volatile oil (up to 1.5%) consisting of cineole (65-75%), caryophyllene, α-pinene, limonene etc. This is used as an expectorant and antiseptic.

31) Pudina, Peppermint.(*Mentha piperita* Linn.-Lamiaceae)

The volatile oil extracted from the whole plant have menthol, menthane, menthyl acetate, menthofuran, cineole, limonene, neomenthol, isomenthol and a sesquiterpene, viridofloral. Peppermint is spasmolytic, smooth muscle relaxant and anti-inflammatory. It helps in production of bile and is a diuretic and mild sedative.

32) Rosemary (*Rosemarinus officinalis* Linn.-Lamiaceae)

Leaves yield vol. oil composed of bornyl acetate, borneol, cineole and minor quantities of camphene and α-pinene. Rosemary, known as the "herb of remembrance", is used to enhance memory and therefore recommended in aromatherapy for patients suffering from Alzheimer's disease.

33) Spearmint (*Mentha spicata* Linn.)

The oil from leaves contains *l*-carvone, limonene, phellandrene and some esters. Spearmint is a flavouring agent used in stews, sauces, beverages, chewing gum, dentifrices etc.

34) Sowa, Indian Dill (*Anethum sowa* Roxb.)

The fruits yield a volatile oil consisting of carvone, dihydrocarvone, *d*-limonene, *d*-phellandrene, myristicin, apiole and dill-apiole. The tender plant also yields a volatile oil rich in *d*-phellandrene. Sowa is used as a carminative, antipyretic and anthelmintic and is a component of gripe water.

35) Tamalpathra, Indian Cassia Lignea (*Cinnamomum tamala* Nees & Eberm.)

The leaves yield an essential oil of varying composition. The eugenol type contains eugenol (13%), eugenol acetate (12.5%), cinnamic aldehyde (41%), linalool (15%), ß-caryophyllene, benzaldehyde, camphor and cadinene whereas the cinnamic aldehyde type yields linalool (50%), cinnamic aldehyde (12%), α-and ß-pinenes, *p*-cymene, geraniol etc. A sample from Assam contained 80-85% eugenol. The leaves are used as spice and in colic, diarrhoea and rheumatism. Leaf powder is said to be hypoglycemic.

36) Tambulah/Pan, Betel. (*Piper betel* Linn. -Piperaceae)

The bright yellow aromatic oil from the leaves (2.6%) contains chavibetol (an isomer of eugenol) and/or eugenol (25-40%), chavicol, eugenol methyl ether, cineole, caryophyllene and cadinene. Betel is a carminative, stimulant, anthelmintic, expectorant and antiseptic and cures dyspepsia, fever, flatulence and filariasis.

37) Tansy (*Tanacetum vulgare* Linn.-Asteraceae)

Flowers yield a volatile oil rich in thujone and camphor. Tansy is used for killing intestinal worms. It is also useful in reducing menopausal problems.

38)Thyme (*Thymus vulgaris* Linn.-Lamiaceae)

Leaves and flowering tops yield a volatile oil containing thymol, carvacrol, *p*-cymene, *l*-borneol, linalool etc. The oil is an antiseptic and is used in perfumery, toothpastes, as fungicides and in internal medicines.

39) Tulsi, Basil (*Ocimum tenuiflorum* Linn -Lamiaceae)

The principal components of vol. oil from leaves are eugenol (content of which is maximum in spring, up to 20%) and caryophyllene (38%). Tulsi is a known for its hypoglycaemic, immunomodulatory, antistress, analgesic, antipyretic, antiinflammatory, antiulcerogenic, antihypertensive, CNS depressant, radioprotective and antitumour properties.

40) Vacha, Sweet flag (*Acorus calamus* Linn.-Araceae)

All parts of the plant, especially the rhizomes contain an essential oil with asarone (80%), asaraldehyde, acoradin, phenylindane, a phenyl propene [1-(p-hydroxy phenol)-1-(O-acetyl) prop-2-ene], calamenol, calamene and sesquiterpene ketones. Calamus is used for improving memory power (esp. for loss of memory) and intellect and is employed in treating epilepsy and other mental disorders.

Iridoids

1) Gentian (*Gentiana lutea* Linn.-Gentianaceae)

The drug consisting of fermented rhizomes and roots yields more than 2% of a bitter iridoid principle known as gentiopicrin (which on hydrolysis yields a lactone gentiogenin and glucose), related glycosides amaropanin, amarogentin and amaroswerin. Gentian is extensively used for lack of appetite and stomach disorders and as a weak febrifuge.

2) Katuka, Picrorhiza (*Picrorhiza kurroa* Linn.-Scrophulariaceae)

Two iridoid glycosides, picroside 1 and kutkoside form the major constituents. Other iridoids present are picroside III, veronicoside and minecoside. Kutkin as well as the whole extract exhibit hepato-protective activity. It is also an immunostimulant of both cell-mediated and humoral immunity (Ansari *et al.*,1988).

3) Nirgundi, Vitex (*Vitex negundo* Linn.-Verbenaceae)

The leaves contain aucubin and agnuside. The leaves are used as a paste over inflammatory swellings of the joints formed due to rheumatism, hydrocele and spleenic enlargement.

4) Sahacharah, Barleria (*Barleria prionitis* Linn.-Acanthaceae)

The leaves/stems contain five iridoid glucosides, barlerin, acetyl barlerin, shanzhiside methylester, 5,6-ß-epoxy, 7ß-hydroxy, 8ß-methyl-1ß-rhamnosidal iridoid, gentioside, 4-carbomethoxy, 5,6-dehydro, 7,8-dihydroxy, 8-methyl-1ß-D-glucopyranosidal iridoid (Taneja & Tiwari, 1975). This drug is found very effective against neurological disorders like paraplegia, sciatica etc.

Sesquiterpenes

1) Damar, Absinthe, Wormwood (*Artemisia absinthium* Linn)

The vol. oil from the plant contains thujyl alcohol, phellandrene, pinene, *s*-guaiazulene and chamzulene. The oil is applied externally to swellings and is antimicrobial in nature.

2) Featherfew, Feverfew (*Chrysanthemum parthenium* Pers.-Asteraceae)

Feverfew contains sesquiterpenes parthenolide (a lactone, major component), santamarine and esters of parthenolide such as reynosin, artemorin, hydroxy costunolide, 8-hydroxyestatiatin etc. Also present is a volatile oil (containing pinene and its derivatives along with bornylacetate, costic acid etc), acetylenes and flavonoid glycosides. The whole plant found use as a carminative tonic, emmenagogue and employed in hysterical complaints, nervousness and lowness of spirits. The synthesis of leucotrienes, prostaglandins and thromboxanes, the compounds causing inflammations, is reported to be inhibited by this plant. It is also a primary remedy for migraine.

3) Gojiva, Elephantopus (*Elephantopus scaber* Linn. - Asteraceae)

Gojiva (roots) contains hydroxylated germacrolides, molephantin, molephantinin, phantomolin etc. The drug is used in fever, cardiac problems, insomnia, diabetes, and urethral discharges, rheumatism and for filariasis. Molephantin and molephantinin exhibited cytotoxic and antitumour properties, the latter shows antileukemic properties also. Phantomolin and its *cis*-epoxide exerted potent inhibitory action on Ehlich ascitis carcinoma and Walker 256 carcinosarcoma cells.

4) Inula (*Inula helenium* Linn.-Asteraceae)

The roots of this plant are used similarly as *I. racemosa* in Europe. They yield alantolactones whereas the aerial parts contain scopoletin and umbelliferone.

5) Jatamansi (*Nardostachys jatamansi* DC.-Valerianaceae)

The rhizome contains a sesquiterpene, jatamansone (valeranone) alongwith spirojatamol, patchouli alcohol, norseychelanone and α-, & β-patchoulenenes. Jatamol, jatamanshic acid, terpenoid coumarins oroselol and jatamansin and neolignans such as l-hydroxy pinoresinol and virolin are the other compounds present (Bagchi, 1991a, 1991b; Rucker *et al.*, 1993). Jatamansi is used for treatment of epilepsy and hysteria, insomnia and exhibits negative inotropic and chronotropic effects (Sajid *et al.*, 1996).

6) Kakamari, Fish berries (*Anamirta cocculus* W & A. - Menispermaceae)

Seeds contain a bitter principle picrotoxin (cocculin - approximately 1.5%) consisting of equal amounts of highly oxygenated sesquiterpenes picrotoxinin and picrotin. Picrotoxin is a potent long acting respiratory stimulant - used to treat overdoses of CNS depressants such as barbiturates and narcotics.

7) Kasani, Chicory (*Cichorium intybus* Linn. -Asteraceae)

The roots contain sesquiterpene lactones such as lactucin, 8-deoxylactucin, lactucopicrin, crepidiaside, sonchusides, cichoriolides and cichoriosides. The root is effective in jaundice, liver enlargement, gout and rheumatism. The whole plant is used in fevers, diarrhoea and as emmenagogue and alexiferic.

8) Kirmala, Wormseed (*Artemisia cina* Berg.-Asteraceae)

The dried unexpanded flower heads yields two crystalline sesquiterpene lactones, santonin and artemisin and a volatile oil. Both the wormseed and santonin are good anthelmintics specific for roundworms and threadworms.

9) Kushta, Costus (*Saussurea lappa* C.B.Clarke -Asteraceae)

Roots yield a volatile oil consisting of costunolide (a sesquitepene lactone-main component), dehydrocostuslactone, dihydrocostuslactone, 12-methoxy dihydrocostunolide, costol and α-ionone. Kushta is used as an incense, insecticide and used in asthma and chronic bronchitis.

10) Musta/Nagar Moth. (*Cyperus rotundus* Linn. - Cyperaceae)

The essential oil from tubers contains sesquiterpenes such as patchoulenone, cypernone, α- and β-retunol, cyperotundone and *nor*-sesquiterpenes, kobusone and isokobusone. Musta is a medicine for indigestion, diarrhoea, sprue and other stomach problems. It is also useful for epilepsy, increasing memory power and a tonic.

11) Patchouli (*Pogostemon cablin* Benth. -Lamiaceae)

The volatile oil from leaves consists of sesquiterpenes, patchouli alcohol (major compound), β-patchoulene, α-guaiene, α-bulnesene and benzaldehyde. Leaves and tops are antirheumatic and said to be good for menstrual problems.

12) Puskarmulam, Inula (*Inula racemosa* Hook.f. -Asteraceae)

The roots possessed an essential oil yielding a germacranolide *i.e.* inunolide, alantolactones such as alantolactone and isoalantolactone and β-sitosterol. Roots are used as a general tonic, aphrodisiac, diuretic and febrifuge.

13) Safed Chandan, Sandalwood (*Santalum album* Linn. -Santalaceae)

The wood yields a vol. oil (4-5%) consisting of a sesquiterpene alcohol, santalol (90%). Santalol exists both in α- and β-forms, the former being predominant. Also present in the oil are santene (C_9H_{14}), α- and β-santalenes and santalone. Sandalwood oil is one of the very important raw materials in perfumery, soaps, face creams, powders etc.

14) Sahadevi, Goat weed. (*Ageratum conyzoides* Linn.-Asteraceae)

The essential oil from leaves contains ageratochromene (75%), 7-methoxy-2, 2-dimethyl chromene, monoterpenes such as sabinene, β-pinene, 1, 8-cineole, β-phellandrene and limonene and sesquiterpenes, sesquiphellandrene, caryophyllene and caryophyllene oxide. The plant is used internally as a stimulating tonic and against diarrhoea, dysentery, colic, rheumatism and fever.

15) Swetadurva, Millfoil (*Achillea millefolium* Linn.-Asteraceae)

The aerial parts contain desacetylmatricarin and two sesquiterpene lactones of 3-oxyguaianolide type, 8-acetyl egelolide and 8-angeloloyl egelolide. The oil, maximum in heads, contains three antitumour sesquiterpenoids achimillic acids A, B & C. Millfoil is astringent, tonic, diaphoretic and anti-spasmodic and is useful in treating cold, colic, hysteria, epilepsy and rheumatism. Infusion of the plant is useful in preventing pre-mature aging of skin.

16) True Chamomile (*Anthemis nobilis* Linn.-Asteraceae)

Chamomile flowers yield a volatile oil containing α-bisacolol and chamazulene, a bitter principle (anthemic acid) and tannins. The flower heads as well as the whole plant are used as a tonic, anodyne, antispasmodic, and emmenagogue and as a remedy for hysterical and nervous affections in women at the period of menopause. Chamomile is employed as a tea for fever and insomnia.

Glucosinolates and sulphides

1) Garden Nasturtium (*Tropaeolum majus* Linn. - Tropaeolaceae)

The plant yields a water-soluble essential oil consisting mostly of benzyl isothiocyanate. The herb is employed in cystitis and for inflammation of kidneys.

2) Halim, Garden Cress. (*Lepidium sativum* Linn. - Brassicaceae)

The pungent volatile oil (Cress oil, 0.115%) from leaves contains benzyl isothiocyanate and benzyl cyanide. Garden cress is used for treating asthma, cough and bleeding piles. Leaves are also used as a diuretic, for scorbutic diseases and liver complaints.

3) Hing, Asafoetida (*Ferula asafoeida* Bios. - Apiaceae)

Asafoetida contains a volatile oil (4-20%), consisting chiefly of isobutyl propanyl disulphide and related compounds. Medicinally it is used as a carminative, expectorant, antispasmodic, laxative and against asthma, whooping cough, chronic bronchitis, hysteria, epilepsy and cholera.

4) Horse-radish (*Armoracia rusticana* Gaertn. - Brassicaceae)

The thick roots contain allyl isothiocyanate occurring free or as the glucoside, sinigrin.

5) Lashuna, Garlic (*Allium sativum* Linn. -Liliaceae)

The essential oil consists of allyl propyl disulphide, methylallyl disulphide, diallyl disulphide, dimethyl trisulphide, methyl allyl trisulphide and diallyl trisulphide. Garlic is a stimulant, diaphoretic, expectorant, diuretic, and tonic. It is found to increase fibrinolytic activity in patients with coronary artery disease. Ajoene, a derivative of allicin, is antileukemic in nature.

6) Mumiri, Shepherd's purse (*Capsella bursa – pastoris* Medic. - Brassicaceae)

The plant contains isothiocyanates sinigrin and α-methylsulfinylnonyl and 10-methylsulfinyl decylglucosinolates, thiocyanic acid and s-methyl-L-cysteine sulphoxide and a volatile oil rich in camphor. The herb is used as a bitter, pungent, astringent, antiscorbutic, febrifuge and emmenagogue.

7) Piyaz, Onion (*Allium cepa* Linn)

A. cepa yields an essential oil rich in allylpropyl disulphide. Onion juice is used for treatment of ophthalmia, ear ache and jaundice.

8) Sigru, Drumstick (*Moringa oleifera* Lam. -Moringaceae)

Root bark yields glucosinolate pterygospermin (which yields benzyl isothocyanate on hydrolysis) and three alkaloids moringine

(benzylamine), moringinine and spirochin. Roots and root bark are used against fever, asthma, rheumatism, inflammation, epilepsy and hysteria.

Diterpenes

1) Danti, Baliospermum (*Baliospermum solanifolium* Suresh-Euphorbiaceae)

Roots contain phorbol derivatives such as montanin, baliospermin, 12-deoxyphorbol palmitate, 12-deoxy-16-hydroxy phorbol palmitate and 12-deoxy-5β-hydroxy phorbol myristate. The roots are considered pungent, anthelmintic, diuretic and useful for piles, enlarged spleen and skin diseases.

2) Dravanti, Bellyache bush.(*Jatropha gossypifolia* Linn.- Euphorbiaceae)

Roots contain macrocyclic diterpenes, jatrophone and jatropholones A & B. Roots and leaves are used in dysentery, anaemia, biliousness, fistula and ulcer.

3) Ginkgo, Maiden hair tree (*Ginkgo biloba* Linn. -Ginkgoaceae)

Leaves contain both diterpene lactones (6%) and flavonols (24%) as the major constituents. The terpene lactones are ginkgolides A,B and C and a sesquiterpene, bilobalide A. The flavonols present are kaempferol, quercetin and isorhamnetin as mono-, bi- and triglycosides and/or as esters with coumaric acids. Seeds yield bilobalide, ginnol, bilobals, cardenols and phenolic lipids. In traditional Chinese medicine, seeds are held in high esteem whereas in western countries leaves are considered more important. Ginkgo is used generally for diseases appearing due to old age. Leaves are used in improving memory, removing depression and used in cardiovascular disorders, Alzheimer's and Parkinson's diseases, cerebral ischemia and liver fibrosis. It is vasodilatory, aphrodisiac and modulate cerebral energy metabolism due to antioxidant properties of flavonoids. The flavonoids and ginkgolide B act as platelet activating factor (PAF) antagonists, strengthen blood capillaries in brain and ensure that all parts of brain get proper oxygen through blood. This and the antioxidant, free radical scavenger properties of flavonoids maintain the cerebral functioning properly. Since PAF is linked with many cardiovascular, renal respiratory and CNS disorders, all these diseases are prevented by this drug. It is also useful in premenstrual syndrome, in acute allergic inflammations and increases blood flow to retina and reduces retinal deterioration and hearing loss. Seeds are used as an expectorant, for

bladder ailments, spermatorrhoea, menorrhoea, uterine fluxes and cardiovascular ailments.

4) Himalayan yew (*Taxus baccata* Wall-Taxaceae)

This plant yields diterpenoids like 10-deacetylbaccatin, from needles and twigs, which can be semi-synthetically converted to taxol.

5) Pacific Yew (*Taxus brevifolia* Nutt. -Taxaceae)

The bark yields a group of diterpenes (taxanes), alkaloids (taxine and taxagitine), lignans (isotaxiresinol), flavonoids (quercetin, rhamnetin and sciadopitysin) and glycosides. Taxanes isolated include taxol (an amide), baccatin, brevifoliol, cephalomanine and many derivatives of taxol. Taxol (also known as paclitaxel) is one of the most potent anticancer compounds today. It disrupts microtubule formation by binding to tubulin to form abnormal mitotic spindles and thus stops cell division.

6) Kairata. (*Swertia chirata* Buch. -Ham. - Gentianaceae)

Bitter principles amarogentin (0.04%) and amaroswerin (0.03%) form the major components of the whole plant which is used as a drug. Kairata is a febrifuge and used as antihepatotoxic, antiinflammatory, antimalarial and hypoglycemic.

7) Kalmegh, Creat. (*Andrographis paniculata* Wall. - Acanthaceae)

The plant yields bitter diterpenoid lactones such as andrographolide (0.6%), 1, 4-deoxy-11-oxoandrographolide and 14-deoxy-11, 12-didehydroandrographolide. This herb is extensively used as a bitter tonic and febrifuge.

8) Kuberakshi, Bonduc nut / Fever nut (*Caesalpinia bonduc* Roxb. - Caesalpiniaceae)

Seeds contain α-, β, γ, δ- and ε-caesalpins, caesalpin F and a homoisoflavone, bonducellin. They are used as antiperiodic, diuretic, antipyretics, and tonic, in malaria, diarrhoea and for treating hydrocele.

9) Nagchampa, Alexandrian Laurel (*Calophyllum inophyllum* Linn. - Clusiaceae)

Seeds yield oil containing calophyllolide, calcustralin, inophyllolide, apetalolide and calophyllic, inophyllic, inophenic, calophynic and pseudobasinic acids. The oil and seed paste are useful for painful joints, rheumatism and gout. Calophyllolide, inophyllolide and calophyllic acid are reputed remedies of leprosy and therefore the oil is intramuscularly injected in leprosy patients.

10) Shallaki/Salai,Frankincense (*Boswellia serrata* Roxb. - Burseraceae)

The oleo-gum-resin contains 8-9% volatile oil, 40-45% resin and 30-35% gum. The volatile oil consists of α-thujene (50%), and *p*-cymene (14%) as major components. The resin consists of a diterpene alcohol, serratol and eight triterpene acids such as derivatives of tirucallenoic acids (four) and four boswellic acids The gum is an acidic polysaccharide consisting of uronic acid (30%), galactose (46%), D-arabinose, xylose and mannose. Frankincense is very effective in osteoarthritis, juvenile rheumatoid arthritis and spondylitis.

11) Tung Oil Tree (*Aleurites fordii* Hemsl. - Euphorbiaceae)

Seeds yield 2 diterpene esters 13-O-acetyl-16-hydroxy phorbol and 12-O-palmityl-13-O-acetyl-16-hydroxy phorbol and a non-protein amino acid, L-3-carboxyl-1, 2, 3, 4-tetrahydro-β-carboline. Seed oil is used for treating boils, ulcers, swellings and burns. Phorbol derivatives are anticancerous.

Triterpenes and steroids

1) Apamargah, Prickly-Chaff flower (*Achyranthes aspera* Linn, - Amaranthaceae)

The plant contains ecdysterone and two alkaloids achyranthine and betaine. The whole plant is pungent, diuretic and emmanagogue.

2) Asthisandana / Vajravalli, Bone – Setter (*Cissus quadrangula* Linn. - Vitaceae)

Onocer-7-ene-3α, 21β-diol, onocer-7-ene-3β, 21α-diol, 7-oxo-onocer-8-ene-3β-diol, alkanes, taraxasterol, friedelan-3-one, β-sitosterol, δ-amyrin and δ-amyrone are the characteristic compounds of this drug. The stem is useful in piles, asthma, and irregular menstrual periods and is excellent in healing bone fracture because of the two unidentified anabolic steroids I and II, which facilitate early regeneration of all connective tissues, involved in healing and quicker mineralisation of the callus. The aqueous extract of the plant is often given topically or as intramuscularly injection.

3) Ashwagadha, Indian ginseng (*Withania somnifera* Dunal - Solanaceae)

The roots contain steroidal lactones the withanolides such as withanone, withaferin A, withanolides I, II, III, A, B, C, D, E F, G, H, I, J, K, L, M, WS-1, P and S, withasomnidienone and withanolide C.

Also present are alkaloids like tropane, pseudotropane, hygrine, isopelletierine etc. Ashwagadha is a rejuvenating drug useful in rheumatism and hypertension and shows antitumour, bradycardic, respiratory stimulant and antispasmodic properties. Withanolides exhibit antitumour, antiarthritic and immunosuppressive properties (Sudhir *et al.*, 1986).

4) Aswattah (*Ficus religiosa* Linn.)

Bark contains bergenin, lupin-3-one, methyl oleanolate and is alterative, haemostatic, laxative and is used in diabetes, diarrhoea, leucorrhoea, menorrhagia, nervous disorders and vaginal diseases.

5) Campanilla (*Allamanda cathartica* Linn. - Apocynaceae)

The stem and leaves yield triterpene esters, plumericin, isoplumericin, plumieride, ursolic acid, ß-amyrin and ß-sitosterol. The roots contain an iridoid lactone allamandin, allamandicin, allamdin and triterpene lactones such as fluvoplumeirin, plumericin, isoplumericin and plumieride. The leaves are used as cathartics. The extracts of roots and leaves cause hypotension in male cats. Allamandin is found to possess anti-leukemic properties.

6) Dugdhika (*Chamaesyce hirta* Millsp. - Euphorbiaceae)

The plant yields cycloartenol, friedlin, taraxerol, aphyldienol, ingenol triacetate, euphorbol hexacosanate, rutin, an alkaloid xanthorhamnine and dimeric tannins euphorbins A, B&C. Latex contains diesters of 12-deoxy phorbol-4ß-phorbol (Okuda *et al.*, 1993). This plant, used as a diuretic, aphrodisiac, and galactogogue, exhibited anticancer properties in that the aqueous extract reduced release of prostaglandins I_2, E_2 and D_2 and exhibited an inhibitory effect on platelet aggregation and carrageenan-induced rat paw oedema.

7) Guggul (*Commiphora wightii* Arnott. – Burseraceae)

Guggul, the oleo-gum-resin, contains about 0.37% essential oil (consisting chiefly of myrcene, dimyrcene and polymyrcene), sterols like guggulusterol I, II, & III, cholesterol, guggulusterones Z & E, diterpenes such as α-camphorene, cembrene and mukulol and long chain aliphatic tetrols and sesamin. Guggul is a medicine for rheumatism, obesity, and neurological and urinary disorders. Guggulusterols exhibit anti-inflammatory and hypocholesteremic properties.

8) Indravaruni, Colocynth (*Citrullus colocynthis* Schard. - Cucurbitaceae)

The characteristic markers of this drug are a number of bitter principles generally known as cucurbitacins consisting of α-elaterin (cucurbitacin E), elatericin B (cucurbitacin I), dihydroelatericin B (cucurbitacin L), citrullin, citrullene, citrullic acid and citrullol. It is a hydragogue, cathartic and drastic purgative.

9) Karpasa, Cotton tree (*Gossypium herbaceum* Linn. - Malvaceae)

The dimeric naphthalene derivative, gossypol is the main compound in cotton seed and root bark. Gossypol is used in treating endometriosis and uterine bleeding. Gossypol is considered a safe antifertility drug for males.

10) Madhunashi, Gymnema (*Gymnema sylvestre* R.Br. - Asclepiadaceae)

The leaves yield gymnemic acids (with gymnemic acid being the main component), 3β-glucuronides of acylated gymnemagenins (hexa hydroxy olean-12-one), gymnestrogenin and conduritol A as well as saponins gymnemasides l-VIll and gypenosides. The leaves and stems are used for treating diabetes.

11) Mrigashinga, East Indian screw tree (*Helicteres isora* Linn. - Sterculiaceae)

Roots contain a triterpenoid, isorin and 3β-27-acetoxy-lup-20(29)-en-28-oic methyl ester. Fruits contain α- and β-amyrins, lupeol and its acetate, friedelin, epifriedelinol, bauerenol acetate and taraxeron. Roots are used in cough and asthma. Leaf paste is effective against eczema and skin ailments. Pods are antidysenteric and used for colic, flatulence and stomach pain

12) Myrrh (*Commiphora myrrha* Engl. and related spp.-Burseraceae)

The oleo-gum-resin, myrrh contains 5-10% volatile oil (consisting of eugenol, cuminaldehyde, mono- and some sesquiterpenes), 25-40% resin (containing α-, β- and γ-commiphoric acids; α- and β-heerabomyrrholic acids) and 55-60% gum. Myrrh is an incense and in medicine as a tonic, stimulant and antiseptic.

13) Nyagrodhah / Vad, Baniyan Tree (*Ficus benghalensis* Linn. - Moraceae)

The stem bark contains bengalenoside, procyanidins and β-sitosterol while leaves yield friedelin and β-sitosterol, quercetin and rutin and the heartwood is found to contain ψ-taraxasterol. All the parts of the tree are used for skin diseases and the bark is used in menorrhagia, leucorrhoea and other vaginal disorders.

14) Punarnava (*Trianthema portulacastrum* Linn. - Aizoaceae)

The roots contain ecdysterone and an alkaloid, trianthemine. Leaves yield sesuvin. Roots are considered punarnava (in place of *Boerhavia diffusa*), a rejuvenative drug specific to general weakness, jaundice, oedema etc.

15) Putranjiva (*Drypetes roxburghii* Hurus. - Euphorbiaceae)

Leaves yield β-amyrin, putrone, putrol, putranjivic acid, methyl putranjivate, stigmasterol, roxburghonic acid, saponins A-D and biflavones. Seed coat contains triterpenoid saponins putrainjivoside and putranosides A-D. Leaves and fruits are used in rheumatism (for swollen joints).

16) Putikaranjah, Holoptelia (*Holoptelia integrifolia* Planch. - Ulmaceae)

Bark yields triterpenoidal fatty acid esters, holoptelin A (*epi*-friedelinol palmitate) and B(*epi*-friedelinol stearate). Leaves contain β-sitosterol, β-amyrin, hexa- and octacosanols. Bark and leaves are used as laxative, anthelmintic, for oedema,leprosy, skin diseases and diabetes. The drug exhibits lipolytic action and therefore helps in obesity.

17) Sahadevi (*Vernonia cinerea* Less. - Asteraceae)

β-Amyrin, lupeol, β-sitosterol, stigmasterol, α-spinasterol and phenols are the compounds in Sahadevi. This is a reputed drug for all types of fevers.

18) Sariva, Indian sarasaparilla (*Hemidesmus indicus* R.Br. - Asclepiadaceae)

The essential oil present in the roots consists of *p*-methoxy salicylic aldehydes. Coumarino-lignoids, hemidesminine ($C_{23}H_{22}O_8$) and hemidesmin 1 & 2 and steroids such as hemidesmol, hemidosterol, lupeol octacosanate and β-amyrin acetate are other compounds here. Sariva is a rejuvenating drug which is cooling and blood purifying.

19) Udumber (*Ficus racemosa* Linn.-Moraceae)

Bark yields ceryl behenate, lupeol, α-amyrin and β-sitosterol and gluanol acetate and is used in ulcers, skin diseases and vaginal diseases.

Limonoids and Quassinoids

1) Ardusa, Tree of heaven (*Ailanthus excelsa* Roxb. - Simaroubaceae)

The root bark contains ailantheone and glaucarubinone and useful as anthelmintic and possesses anti-spasmodic and expectorant properties. These quassinoids are anti-tumour and cytotoxic against the P-388 lymphocyclic leukemia and KB test systems.

2) Nimba, Neem (*Azadirachta indica* A. Juss. - Meliaceae)

The limonoid bitter principles of the stem and root bark are nimbin and nimbidin. Leaves yield nimbin, nimbinene, 6-desacetylnimbinene, nimbandiol and nimbolide. The fruits contain gedunin, azadirachtin, 7-deacetoxy-7α-hydroxy gedunin, azadiradione, azadiradone, 17-β-hydroxy azadiradione, 17-*epi* azadiradione and nimbiol whereas seeds yield oil containing nimbidin, nimbin, nimbinin, meliantriol as major components. All parts are antimicrobial, anthelmintic and useful against fever (especially of malaria) and skin diseases.

3) Quassia (*Picrasma excelsa* Planch. - Simaroubaceae)

The wood contains quassin (chief constituent), neo-quassin and picrasmin. Quassia is used as a bitter tonic and anthelmintic.

4) Ya-Danzi, Brucea (*Brucea javanica* Merril - Simaroubaceae)

Stem bark quassinoids are bruceantin, bruceantinol, dihydrobruceantinol, bruceine B, bruceanols D, E, F, G & H, yadanzigan, yadanziolide A and brucicanthinoside C (a glycoside). Fruits contain an alkaloid 4-ethoxycarbonyl-2-quinolone, quassinoid bruceoside C, quercetin, luteolin and vanillic acid. The plant, especially the fruit, is traditionally used in China and Indonesia for the treatment of malaria, amoebic dysentery, cancer and parasitic diseases. Bruceantin is active against B16 melanoma, colon 38 and L1210 and P388 leukemia and favoured apoptosis.

3.6c. Saponins

1) Black snakeroot, Black cohosh (*Actaea racemosa* Linn. - Ranunculaceae)

Roots and rhizomes contain triterpene glycosides actein, acetylacetal, 27-deoxyactein, cimigenol and cimifugoside. Roots and rhizomes are widely used in many gynecologic disorders esp. menopausal and menstrual disfunction.

2) Brahmi, Bacopa (*Bacopa monnieri* Wettst. - Scrophulariaceae)

The whole plant contains saponins such as monnierin, hersaponin bacoside A, A3 and bacoside B. The plant is a very well known brain tonic and is found to be effective in anxiety neurosis and to revitalize intellectual faculty.

3) Bupleurum (*Bupleurum falcatum* Linn. - Apiaceae)

Roots are found to yield a number of saponins – saikosaponins A, C, & D being the principal ones. The sapogenins are of oleanane type, e.g., saikogenin F. Roots are used as anti-inflammatory, diaphoretic and often prescribed for liver troubles.

4) Canda, Costus (*Costus speciosus* Smith. - Zingiberaceae)

Rhizomes contain saponins dioscin and gracilin based on diosgenin (2.12%), tigogenin and ß-sitosterol. The rhizome is bitter, astringent and used for dyspepsia, fever, cough and other respiratory diseases, diabetes, oedema, blood diseases and skin ailments and improves complexion.

5) Dasparsha/Duralabha (*Fagonia cretica* Linn. - Zygophyllaceae)

Aerial parts contain at least six saponins based on nahagenin, hederagenin and oleanolic acid, diterpenes fagonone, 16-O-acetyl fagonone and 7ß-fagonone and an alkaloid harmine. Duralabha is a blood purifier and cures fevers, asthma, diarrhoea, phantom tumour and dermatitis. It possesses stimulant, laxative and alterative properties.

6) Horse chestnut (*Aesculus hippocastanum* Linn. -Hippocastanaceae)

Seeds yield over 30 pentacyclic saponins based on the sapogenins proaescigenin and barringtenol, combined with angelic, tiglic or acetic acids. Seeds are useful in varicose veins and to prevent thrombosis, thrombophlebitis, calf cramping, edema and hemorrhoids.

7) Ginseng (*Panax ginseng* A. Mey and related *spp.* - Araliaceae)

The dry roots of 3-6 year old plants form the ginseng of commerce. Korean ginseng contains a number of saponins, ginsenosides R_x (x = o, a, b_1, b_2, b_3, c, d, e, f, 20-gluco-f, g_1 and g_2 – approximately 30). All these glycosides (except R_o) are based on sapogenins 20-S-protopanaxodiol and 20-S-protopanaxotriol, dammarane tetracyclic triterpenes. Ginsenoside Ro possesses oleanolic acid as the sapogenin. In American ginseng (*P. quinquefolius*) ginsenoside $R_{g\text{-}1}$ is absent and $R_{b\text{-}1}$ is dominant. Ginseng is widely used (especially in Chinese system) as a tonic, stimulant, diuretic and carminative. It is employed in anemia, insomnia and as an aphrodisiac.

8) Ginseng, Siberian (*Eleutherococcus senticosus* Maxim. -Araliaceae)

Siberian ginseng, which is mainly used in Russia, contains saponins eleutheroside I, K, L & M and steroid glycosides such as eleutheroside A.

9) Goksura/Gokhru, Tribulus (*Tribulus terrestris* Linn. - Zygophyllaceae)

The fruits contain steroidal saponins terrestrosins A-E, desgalactotigonin, F-gitonin, desglucolanatigonin and gitonin based on sapogenins diosgenin, hecogenin and neotigogenin. Gokhru is a well known diuretic and antiurolithiatic. It is also nephroprotective and cardiac stimulant.

10) Gunja, Crab's eye, Indian liquorice. (*Abrus precatorius* Linn. - Fabaceae)

The leaves, which are sweet in taste, contain up to 10% glycyrrhizin, triterpene glycosides such as abrusosides A, B, and C (which are highly sweet) and three glycosides based on cycloartane-type aglycone, abrutogenin. The roots also contain glycyrrhizin and alkaloids like abrasine and precasine besides abrine and related bases. The seeds yield lectins, the principal ones being abrins (upto 0.15% of seed). Lectins are both toxic (abrin) and non-toxic (*Abrus* agglutinin). Abrins are denoted by abrin a, b, c & d and consists of one large β-polypeptide chain (MW. 35,000) and a short α-polypeptide chain (MW. 32,000) joined by a disulphide bond. Agglutinin consists of 4 polypeptide chains; the chains are similar to those of abrins.

The roots and leaves are used as a substitute of liquorice in coughs and catarrhal affections. Abrin is extensivelly studied for its antitumour activity, where it was found to suppress tumour growth in rats and mice. Abrin is found to inhibit protein synthesis by inactivating ribosome dependant GTPase and exhibit no bone marrow suppressing affects as other cytostatic agents.

11) Hingoli, Desert Date (*Balanites roxburghii* Planch. - Simaroubaceae)

The fruit pulp contains several steroidal saponins balanitisins A-E, based on the sapogenin diosgenin, yamogenin and cryptogenin and a furostanol saponin, balanitoside along with balanitin-3 and 6-methyl diosgenin. The seed kernels yield balanitins 4, 5, 6 & 7 and balanitisin F. Yamogenin is the major component of certain samples of fruit epicarp. Almost all parts of the plant are found to be anthelmintic and purgative.

12) Indian senega (*Glinus oppositifolius* DC.-Molluginaceae)

Its roots contain oleanolic acid saponins.

13) Kanphul, Dandelion (*Taraxacum officinale* Weber - Asteraceae)

Roots contain steroids such as taraxasterol, taraxacerin, taraxacin, gum and rubber. Roots and leaves are alterative, cholagogue, diuretic, lithotropic, tonic and used for a variety of ailments connected with menopause, like mood swings.

14) Kantala, American Aloe (*Agave americana* Linn. - Agavaceae)

The leaves yield saponins such as agavasaponins A-I. The leaves are used as laxatives and emmenagogue and in scurvy.

15) Madana, Emetic nut (*Catunaregam spinosa* Tirveng. - Rubiaceae)

The bark contains saponins based on randialic acid A & B, mannitol (up to 6%), 10-methyl lixoside (an iridoid) and scopoletin. The fresh fruits contain saponins (10% dry wt. 10% dry wt. mostly concentrated in pulp) such as dumetorins A, B, C, D, E and F (based on oleanolic acid), randioside A (on oleanolic acid), ursosaponin (on ursolic acid), two triterpenoid glycosides based on oleanolic acid and 10-methyl lixoside. Seeds contain saponins based on oleanolic acid. The bark is used in diarrhoea, dysentery and is abortifacient, anthelminitic, antipyretic and emetic. The roots are used for treating gonorrhoea and for poisoning fish. The fruit pulp, dried and powdered, is emetic and substituted for ipecac.

16) Manduk parni /Brahmi, Gotu kola (*Centella asiatica* Urban. - Apiaceae)

The triterpenoid glycosides present in this plant are asiaticoside, indocentelloside, brahmoside, brahminoside, thankuniside and isothankuniside. These saponins are based on asiatic, indocentoic, brahmic, thankunic and isothankunic acids as aglycones. (Samples from Europe are found to contain madecassocide-aglycone madecassic acid). Asiatic, indocentoic and brahmic acid along with isobrahmic and betulic acids occur in the free form also. Gotu kola is used as **Brahmi** (*Bacopa monnieri*) at many places. This is one plant recommended for improving memory, reducing anxiety neurosis and hypertension and for treatment of leprosy and skin diseases.

17) Methika, Fenugreek (*Trigonella foenum-graecum* Linn. - Fabaceae)

Seeds contain saponins based on diosgenin, tigogenin and gitogenin. Seeds are mainly used as a galactogogue. They are carminative, tonic and as a poultice, are useful in ulcers, and abscesses.

18) Milkwort or Snakeroot (*Polygala senega* Linn. - Polygalaceae)

The dry rootstock and roots contain a number of triterpenoid saponins that on hydrolysis yield the sapogenins, senegenin (a chlorinated sapogenin), presenegenin, senegenic acid, polygalic acid (hydroxy senegenin) and glucose. Senega is used as a cure for snakebite. It is a known expectorant, emetic and stimulant. The plants are given to cattle to increase the production of milk.

19) Musali (*Curculigo orchioides* Gaertn. - Amaryllidaceae)

Tubers contain a glycoside of yuccagenin. The tubers (musali) are a well known rejuvenative (rasayana) drug and an aphrodisiac.

20) Mu-T'ung/Mokutsu (*Akebia quinata* DC - Lardizabalaceae)

The saponins present in the stem bark are akeboside St-a, St-b, St-c, St-d, St-f, St-h, St-j and St-k based on hederagenin and oleanolic acid sapogenins. The seeds of this plant also yield saponins based on hederagenin. The drug is used as an anti-inflammatory agent, diuretic and menses stimulant.

21) Platycodon (*Platycodon grandiflorum* DC - Campanulaceae)

Of the 8 or more saponins – platicodins A-H- known from the roots, platicodin-D forms the major one. The crude saponin extract on hydrolysis yields platicodigenin as the main sapogenin and polygalacic acid as well as platycodigenic acids as minor sapogenins.The whole saponin extract is employed as an expectorant and an antitussive agent.

22) Safed musali (*Chlorophytum arundinaceum* Baker- Liliaceae)

The tubers are found to contain a bibenzylxyloside, 2', 4, 4'-trihydroxy-2-xylopyranosyl bibenzyl, and steroidal sapogenins such as sarasapogenin, tigogenin, neotigogenin and tokorogenin, a disubstituted tetrahydrofuran, stigmasterol and its glucoside. This is used as a general tonic and aphrodisiac.

23) Sarasaparilla (*Smilax aristolochiaefolia* Miller and related spp.-Liliaceae)

The drug contains a number of saponins based on smilagenin and its isomer, sarasapogenin. Sarasaponin, one of the principal saponins, yields sarasapogenin, two molecules of glucose and one molecule of rhamnose on hydrolysis. Smilonin, another saponin, yields smilagenin and as many as five molecules of sugars. Sarasaparilla is used in skin diseases, rheumatism and as a flavouring agent. The sapogenins are used as the starting material for the synthesis of cortisone and other steroid hormones.

24) Saw palmetto (*Serenoa repens* Small - Arecaceae)

Fruits yield steroidal saponins. A concentrated extract of berries is useful in maintaining a healthy prostrate function.

25) Shataveri, Asparagus (*Asparagus racemosus* Willd. - Liliaceae)

The tubers contain saponins, named shatavarins I,II,III,IV, glycoside AR-4, tridecaacetyl shatavarin-I, and tetradeca-O-methyl shatavarin-I ($C_{65}H_{114}O_{23}$). Shatavarin-IV is based on sarasapogenin. Asparagus is a tonic, demulcent and aphrodisiac and used for hyperacidity, diarrhoea and dysentery. It is a well-known galactogogue and beneficial to menstrual troubles and nervous break down.

26)Swetamusali/Safedmusali(*Chlorophytum borivilianum* Sant. & Fernandes. - Liliaceae)

The tubers contain saponins (2-3%) based on hecogenin, protein (8%) and carbohydrates (40%) mainly as mucilage. Safed musali is much valued for its aphrodisiac properties. It is a valuable nervine and general tonic.

27) Yashtimadhuh, Liquorice (*Glycyrrhiza glabra* Linn. - Fabaceae)

The roots contain saponin glycyrrhizin (2-9%) and a mixture of potassium and calcium salts of glycyrrhizinic acid as the major constituents as well as some other triterpenoid saponins such as glabranin A& B , glycyrrhetol, glabrolide and isoglabrolide; The drug exhibits demulcent, expectorant, antimicrobial, antihepatotoxic and anti-inflammatory properties due to glycyrrhizin. It is also useful in treating peptic ulcer.

28) Yiian-chi / Onji (*Polygala tenuifolia* Willd. - Polygalaceae)

The roots contain onjisaponins A-F (onjisaponin A and B are identical with senegin IV and III respectively of *P. Senega*). It is used as a sedative and to strengthen the nervous system.

29) Ziziphus (*Ziziphus jujuba* var. *spinosa* Hu - Rhmnaceae)

The seeds, which form the drug, contain two saponins jujubosides A & B. Both on hydrolysis yield ebelin lactone. The seeds are used for strengthening the nervous system and for insomnia.

Cardiac glycosides

1) Alkarka, Dead Sea Apple (*Calotropis procera* Ait.f.-Asclepiadaceae)

The root bark yields cardiac glycosides calotropin, uscharin, calotoxin, calactin, uscharidin (all based on the genin calotropagenin), voruscharin, proceroside and two genins, uzarigenin and syriogenin. *C. procera* is found to possess almost all medicinal properties of *C. gigantea* and used likewise.

2) Arkah, Milkweed (*Calotropis gigantea* Ait.f. - Asclepiadaceae)

The root bark yields cardiac glycosides such as gigantin, giganteol, isogiganteol and calotroposides A-G. The latex, found in all parts of the plant contain gigantin, calotropin, uscharin, calotoxin, calactin and uscharin (all based on the genin calotropagenin). Root bark is a substitute for ipecac and in small doses it acts as a diaphoretic and expectorant. It is useful in leprosy and eczema.

3) Ashvaghna, yellow oleander (*Thevetia peruviana* Merril - Apocynaceae)

All parts of the plant contain cardiac glycosides; seeds being the richest source of these compounds (containing up to 7 times) than the other parts. The various compounds isolated are cereberoside (thevetin B), 2'-O-acetyl cerberoside and neriifolin, 2'-O-neriifolin (cerberin) all based on the genin, digitoxigenin; thevetin A (19-oxocerberoside) and peruvoside (19-oxonerifolin), both based on cannogenin; thevenerrin (19-oxyneriifolin-based on cannogenol) and peruvosidic acid (based on cannogenic acid.). Bark is found to contain neriifolin and peruvoside. Roots yield thevetin and neriifolin. All the glycosides are used as cardiotonics, the most preferred ones being peruvoside and thevetin. Bark is a bitter cathartic and emetic.

4) Black hellebore (*Helleborus niger* Linn. - Ranunulaceae)

Rhizomes and parts of the aerial stem constitute the drug. Of the few glycosides it contains, hellebrin (hellebrigenin + rhamnose + glucose) shows the maximum cardiotonic activity. Hellebrin, having approximately twenty times powerful action compared to **Digitalis,** is extensively used in veterinary practices.

5) Convallaria (*Convallaria majalis* Linn. - Liliaceae)

More than 25 glycosides are isolated from the drug and the important ones are convallatoxin (strophanthidin + rhamnose), convalloside (convallatoxin + glucose) and convallatoxol (strophanthid-19-ol + rhamnose). The drug is a cardiotonic.

6) Digitalis (*Digitalis purpurea* Linn. - Scrophulariaceae)

The leaves contain more than 40 cardiac glycosides based on 4 genins, digitoxigenin, gitoxigenin, gitaloxigenin, and gitaligenin along with their formyl or acetyl derivatives. The glycosides may be: (1) monoglycosides like verodoxin and odoroside H, (2) diglycosides like gitalin and digitalinum verum, (3) triglycosides like gitoxin and gitaloxin or (4) primary or tetraglycosides like purpurea glycoside A. Digitoxin is a cardiotonic increasing the tone of cardiac muscle. Gitalin alone or with other glycosides are often used in treatment of congestive heart failure.

7) Digitalis (*Digitalis lanata* Ehrh.-Scrophulariaceae)

This species yields more than 70 glycosides based on the four genins reported from *D. purpurea* and two new ones digoxigenin and diginatigenin. Digoxin – a triglycoside and the primary glycosides, lanatosides A, B, C, D and E are major ones. Lanatosides and digoxin are valuable cardiotonics.

8) Junglypiaz, Indian squill (*Drimia indica* Jessop - Liliaceae)

Bulbs contain bufadienolides (maximum in dormant stage of the bulb) such as proscillaridin A, scillopheoside and anhydro-scilliphaosidin. This is used as an expectorant, cardiac stimulant and diuretic. It is also useful as an anticancer and hypoglycaemic drug and for skin diseases.

9) Jew's mallow (*Corchorus olitorius* Linn.- Tiliaceae)

Seeds contain cardenolides such as olitoriside A, erysimoside, coroloside, helveticoside, corchoroside A, evonoside, strophanthidol and strophanthidin. Seeds are ecbolic in nature and are used as a cardiotonic. The cardenolides are very effective in acute and chronic cardiac insufficiency, peroxystic tachycardia and tachyarythmia.

10) Kakatundi, Blood flower (*Asclepias curassavica* Linn. - Asclepiadaceae)

The leaves of Brazilian plants yield cardenolides such as clepogenin, curassavogenin, ascurogenin, carotoxigenin, uzarigenin, coroglaucigenin and a glycoside uzarin. But the leaves from Indian plants are devoid of the first four compounds but are found to contain 22 cardenolides of which calactin, calotropin, calotropagenin and asclepin are the major compounds. The roots are found to contain vincentoxin (asclepiadin). All parts of the plant are used in medicine. Root is considered emetic, cathartic, astringent and a remedy for gonorrhoea and piles. Roots are used as substitute of ipecac. The whole plant is used as emetic, styptic and purgative and the extract is found to inhibit carcinomatous cells of human nasopharynx. The plant, as powder, balm or enema, is used to destroy abdominal tumours. The latex is used against warts and corns.

11) Karaunda, Christ's thorn (*Carissa carandas* Linn. - Apocynaceae)

The roots contain cardiac glycosides based on odoroside H and terpenoids such as carissone, carindone, carinol and related compounds. All parts of the plant are attributed with medicinal properties. Roots are cardiotonic, anthelmintic and effect a prolonged blood pressure lowering effect.

12) Karavirah/Oleander (*Nerium oleander* Linn. - Apocynaceae)

The bark yields a number of cardiac glycosides designated as odoroside A,B,D,F,G,H,K-M and odorobioside K based on digitoxigenin, uzarigenin and 16-anhydrodigitoxigenin. Leaves contain oleandrin (a cardenolide based on acetyl gitoxigenin), nerium E (16-deacetyl anhydro-oleandrin) and nerium F (16-anhydrodigitoxigenin). The root, the official drug, is considered highly toxic, but in controlled doses it is a cardiotonic. It is anthelmintic, diaphoretic, carminative and used for ulcers, leprosy and skin diseases.

13) Kurutakah, Pergularia (*Pergularia daemia* Choiv. - Asclepiadaceae)

Stem and seeds contain cardenolides such as calactin, calotropin, calotropagenin (from seeds), uzarigenin and coroglaucigenin (from stems). The plant is used for vesical calculus, dysurea and anurea. Leaf juice is emetic and expectorant and used in asthma, rheumatism, menstrual disorders and diarrhoea.

14) Odollum (*Cerbera manghas* Linn. - Apocynaceae)

All parts of the plant contain cardiac glycosides. The stem contains neriifolin, seven glycosides based on digitoxigenin, two glycosides based on tanghmigenin and one based on oleagenin and free tanghinigenin. Also present are lignans, (-)- olivil, (+)-cyclo olivil, 5,5'''-*bis*-olivil and similar dimers. Iridoids such as 10- and 11- carboxy carbinal also are isolated from the stem bark. The leaves are found to contain cerleaside A and B as the major cardenolides, and 5 glycosides of digitoxigenin, 5 of tanghinigenin, neriifolin, thevetin B, thevoside and thevirioside as minor components. Seeds yield cerleaside A, 2'-O-acetyl cerleaside A, 17α-neriifolin, 17β-neriifolin, cerberin, neriifolin, thevoside, thevirioside, odollin, odollotoxin etc. The bark and leaves are cathartic and emetic. Seeds are fish poisons. Seed oil is used as hair oil and as a rubefacient to cure cold.

15) Strophanthus (*Strophanthus kombe* Oliv., *S. hispidus* DC - Apocynaceae)

The seeds yield about 8-10% strophanthin or K-strophanthin, a mixture of more than 10 glycosides. The chief constituent is K-strophanthoside (Strophoside), a triglycoside consisting of the genin, strophanthidin, cymarose and 2 molecules of glucose. Two other major glycosides are cymarin and K-strophanthoside B. The genins present are strophanthidin, alloperiplogenin, strophanthid-19-ol, strophanthidic acid and periplogenin. Strophanthus is the cardiotonic drug preferred over *Digitalis* in Europe.

16) *Strophanthus gratus* Oliv.

The seeds of this plant yields ouabain (g-strophanthin), a rhamnoside of ouabagenin. This is one of the most rapidly acting cardenolide.

Carotenoids

1) Kesar, Saffron (*Crocus sativus* Linn. - Iridaceae)

The color of saffron is due to a diterpene crocin, which is the gentiobiose ester of crocetin – a diterpene acid having a structure resembling carotenoids. Saffron is a coloring and flavouring agent. Saffron extract shows cytotoxic and antimutagenic activities and anti-tumour activities against ascites tumours in mice. Crocin is used to stabilize light sensitive drugs.

2) Papeetha, Papaya (*Carica papaya* Linn. - Caricaceae)

The ripe fruits are a good source of carotenoids like β-carotene, cryptoflavin, violaxanthin, phytoene, γ-carotene, chrysanthemaxanthin and neoxanthin. Ripe fruit is stomachic, galactogogue, diuretic and is effective in dysentery, chronic diarrhoea, bleeding piles, dyspepsia and as an emmenagogue.

3) Rakthamachi, Tomato (*Lycopersicon esculentum* Mill. - Solanaceae)

The carotenoids present here are lycopene, β-carotene, lutein and related compounds. Lycopene is highly recommended for prostrate cancer and is an antioxidant of great demand.

4) Shonapushpe, Annato (*Bixa orellana* Linn. - Bixaceae)

The red pulpy seed coat contain a group of pigments, with bixin, a sesterpene (having a carotenoid skeleton) amounting to 70-80%. Other pigments present are orellin, methyl bixin, ß-carotene, cryptoxanthin, lutein and zeaxanthin. Isobixin is the *trans*- form of the unstable naturally occurring *cis*-bixin. Annato dye is a food color widely used. The pulp surrounding the seed is used as a haemostatic, antidysenteric, diuretic and febrifuge.

2.4.1.3. Phenolics

Simple phenols

1) Aspidium/ Male fern (*Dryopteris filix-mas* Schott. - Polypodiaceae)

The rhizome, frond bases and apical buds yields 6-15% oleo-resin consisting of mono-, bi-, tri- and tetracyclic phloroglucinol derivatives, which are formed by the condensation of a molecule of butyric acid with one (aspidinol), two (albaspidine) or three (filicic acid) molecules of phloroglucinol. The drug is an effective taenicide against tapeworm. It is also used in cancerous tumours, as a contraceptive and a bactericide.

2) Benzoin (*Styrax benzoin* Dryand and related spp. - Styraceae)

Benzoin contains up to 20% free balasmic acids in which cinnamic acid amounts to 10% and benzoic acid 6%. Benzoin is a stimulant, expectorant and used in the preparation of perfumes, soaps, tooth powders and cosmetic lotions.

3) Bhilawa, Marking Nut (*Semecarpus anacardium* Linn. - Anacardiaceae)

The nut contains a variety of phenols like bhilawanol (a catechol derivative with a $C_{15}H_{37}$ side chain), anacardic acid, cardol, catechol, anacardol and semecarpol. The resinous juice extracted from the nut is used to remove rheumatic pains, aches, sprains and for any sort of veneral complaint or leprosy.

4) Haridra/Haldi, Turmeric (*Curcuma longa* Linn. - Zingiberaceae)

The rhizome contains curcuminoids of which curcumin amounts to 60%, with desmethoxy curcumin, *bis*desmethoxy curcumin and dihydrocurcumin forming the rest. *Curcuma* shows excellent antiinflammatory properties due to curcumin and deriv. It is also antihepatotoxic, antiulcer and anticancerous. It is a good antioxidant too.

5) Hops (*Humulus lupulus* Linn. - Cannabinaceae)

Hops contain bitter phloroglucinol derivatives with isoprenoid side chains like lupulones (iso-α and β-acids, lupulin, principal component), humulinones (γ-acids), hulupones (Δ-acids), 4-deoxyhumulones, isohumulinones, lupoxes a & b, and lupdoxes A & B and polyphenols. Hops stimulates appetite, is antibiotic and regulate menstrual periods. It exhibits sedative and hypnotic effects and is a source of antioxidant phenols.

6) Kavakava (*Piper methysticum* Forst. - Piperaceae)

The rhizome and roots yields 5-10% of a resin consisting of a number of styryl pyrones, such as dihydromethysticin, methysticin, yangonin etc. The drug is a popular beverage exhibiting sedative and hypnotic properties.

7) Lalmirich, Red Chilli (*Capsicum annum* Linn. - Solanaceae)

Chillies are found to contain an oleoresin (16%) consisting of capsainoids such as capsaicin, dehydrocapsaicin, homocapsaicin II, bishomo and trishomocapsaicin, homodihydrocapsaicin I and II, nordihydrocapsaicin, vanillyl amides of caprylic, nonylic and decylic acids and capsiamide. Chilli is a very popular spice and in small doses acts as a powerful stimulant and carminative. It is a well-known remedy in atomic and flatulent dyspepsia and dipsomania. A regular use is beneficial in varicose veins, anorexia and liver congestion.

Xanthones and coumarins

1) Bavchi, Psoralea (*Psoralea corylifolia* Linn. - Fabaceae)

Seeds yield about 1% furanocoumarins, psoralen and isopsoralen. Seeds are used as stomachic, deobstruent, in leprosy and leucoderma.

2) Bel, Bangal quince (*Aegle marmelos* Correa - Rutaceae)

The fruits contain a group of coumarins such as marmelosin, allo-imparatorin, marmalide, psoralen, scoporone, scopoletin, umbelliferone, marmasin and skimmine. The unripe fruit is an astringent, digestive and stomachic and is effective in chronic diarrhoea and dysentery.

3) Bhringaraja/Bhangra (*Eclipta alba* Hassk. - Asteraceae)

The plant yields (up to 1.6%) coumestan derivatives like wedelolactone and demethylwedelolactone. The drug is hepatoprotective (antihepatotoxic) due to wedelolactone and demethylwedelolactone. It is also antimyotoxic and antihaemorrhagic in nature.

4) Mangusta, Mangosteen (*Garcinia mangostana* Linn.- Clusiaceae)

Bark, fruit hulls and dried latex yield xanthones, mangostin, β-mangostin, and γ-mangostin (normangostin). Fruits contain gartenin, 8-deoxygartenin, normangostin and garcinones A-C. Maclurin also is reported from fruits. Mangostin and other xanthones are found to be hepatoprotective, CNS depressants, anti-inflammatory and antiulcer. Mangostin is a cardiotonic.

5) Pashaanbheda, Megaseas (*Bergenia ciliata* Sternb. - Saxifragaceae)

The dried rhizome, which forms the drug, yields an isocoumarin deriv. bergenin. Pashanbheda is considered very effective in dissolving kidney stones. The drug is a tonic, antiscorbutic and is given for pulmonary infections, ulcers, cough and fever.

Chromones

1) Visnaga, Khella (*Ammi visnaga* Lam.- Apiaceae)

The fruits yield 0.5 - 1.5% chromones consisting of khellin, visnagin, khellinin, visnaginol and visinagionol. Fruits are used for expelling kidney stones and for renal colic. Khellin is antispasmodic and a vasoditator.

Stilbenes & Lignans

1) Bakrachimyaka, Indian podophyllum (*Podophyllum emodi* Wall.)

This plant yields 6-12% resin with a higher percentage of podophyllotoxin (40%). Excepting the peltatins, the other ingredients are the same as in *P. peltatum.*

2) Bhoomyamalki/Bhuiavla (*Phyllanthus amarus* Schum. & Thonn. - Euphorbiaceae)

The plant contains lignans, phyllanthin (a diaryl butane, major constituent, 0.5%), hypophyllanthin (aryltetrahydronaphthalene, upto 0.2%), nyrphylline (lignan) and phyllnirubin (a neolignan). *P. amarus* is known for the hepatoprotective action it exerts, an action due to the lignans. The drug exhibits antiviral actions on hepatitis B.

3) Flax, Linseed (*Linum usitatissimum* Linn. - Linaceae)

The hull of the seed is rich in lignans such as secoisolariciresinol diglycoside (SDG). The lignans are found useful in prostrate and skin cancer, melanoma and diabetes. They also act as phytoestrogens useful in treating breast tenderness, sweating, vaginal dryness, hot flashes, uneven menstrual cycle and PMS symptoms. Lignans are produced by intestinal bacteria from precursors present in linseed meal.

4) Kabab Chini, Cubeb (*Piper cubeba* Linn.-Piperaceae)

Cubeb contains lignans cubebin (0.8%), hinokinin, clusin, dihydrocubebin, cubebinolide, cubebinone and 5′′methoxyhinokinin. Cubeb is a carminative, antiseptic and is used medicinally for urinogenital diseases.

5) Kasmari, Gmelina (*Gmelina arborea* Roxb. - Verbenaceae)

Heartwood yields lignans such as arborone, 7-oxodihydro-gmelinols, arboreal, gmelanone, 6′′- bromoisoarboreol, epieudesmin and gummidiol. Wood is hypoglycaemic and also used in skeletal fractures.

6) Magnolia wine, Wu-wei-zi (*Schisandra chinensis* Linn. - Schisandraceae)

The fruits contain lignans such as schisandrin, deoxyschisandrin, gomisin and pregomisin, essential oil and numerous acids. These compounds are found to regenerate liver tissue damaged by hepatitis, alcohol and other disorders.

7) Mandrake/May-apple (*Podophyllum peltatum* Linn. - Berberidaceae)

The rhizome and roots yield 2-8% of a resin consisting of lignans such as podophyllotoxin, α-, and β-peltatins, demethylpodophyllotoxin and dehydropodophyllotoxin. Mandrake is a drastic purgative. Podophyllotoxin shows excellent antimitotic action.

8) Til, Sesame (*Sesamum indicum* Linn. - Pedaliaceae)

The seed oil contains a phenol, sesamol, produced by the hydrolysis of a lignan, sesamolin. Sesame oil is found be one of the stablest oils; the stability attributed to the antioxidant properties of sesamol.

Flavonoids

1) Asanah, Indian/Malabar Kino (*Pterocarpus marsupium* Roxb. - Fabaceae)

Bark contains *l*-epicatechin. Heartwood yields liquiritigenin, isoliquiritigenin and a number of C-glycosides like pterocarposide, pteroisoauroside (isoaurone C-glucosides), marsuposide, benzo-furanone C-glucoside, pteroside (a benzofuran C-gluoside), vijayosin [1,2-bis(2,4-dihydroxy,3-C-glucopyranosyl)ethanedione] and a glucoside of 2,6-dihydroxybenzene. The heartwood is well-known for its activity against diabetes, leprosy and other skin diseases. It is also useful in diarrhoea and haemophilic disorders. The bark and leaves are used in scabies, leucoderma, leprosy and skin diseases.

2) Ashta. (*Bauhinia racemosa* Lam. - Caesalpiniaceae)

Bark contains a tetracyclic 2, 2-dimethyl chroman. The bark is highly astringent and is antidysenteric, antiinflammatory and cholagogue. Leaves are anthelminitic and used for diarrhoea.

3) Bhat, soybean (*Glycine max* Merril - Fabaceae)

Seeds contain several isoflavone glycosides genistin (glycoside of genistein), diadzin (glycitein 7-O-β-D-glucoside), glycitin, 6''-O-acetyl genistein, 6''-O-acetyl diadzin and their aglycones, and 6-O-malonylated derivatives of diadzin, glycitin and genistein. These isoflavones are estrogenic and therefore are useful in treating breast cancer. This is done through preventing effect of protein tyrosine kinase (PTK) activity. Some other effects of genistein are inhibition of topoisomerase 2 and accumulation of protein associated single strained breaks of DNA in the whole cell. Genistein is also reported to prevent leukemia and prostrate cancer by arresting cell cycle, including apoptosis, modifying cellular differentiation programme and by modulating cell-cycle events. Genistein-induced apoptosis of prostrate cancer is preceded by a specific decrease in total adhesion kinase activity. Use of genistein is found to prevent incidence of cancer of colon, lung, liver, oesophagus, oral epithelium and bone marrow.

4) Bilberry, Huckleberry (*Vaccinium myrtillus* Linn. - Vacciniaceae)

Bilberry contains at least 15 anthocyanins based on malvidin, cyanidin and delphinidin. The anthocyanin content increases as the fruit ripens mainly due to the conversion of proanthocyanidins and catechol tannins to anthocyanins. Bilberry anthocyanins are used for improving the eyesight as increased night vision, delayed cataract and other eye disorders. This is especially useful for age-related macular degradation (AMD), which is the cause of blindness in people over the age 55 and regenerate rhodopsin.

5) Buckwheat, Kotu (*Fagopyrum esculentum* Moench. - Polygonaceae)

The leaves and blossoms yield rutin (3-5%). Rutin is used in treatment of increased capillary fragility with associated hypertension sometimes resulting in the bursting of blood vessels in brain, retina etc. It also prevents weakening of capillaries caused by the administration of drugs such as salicylate, arsenical, sulphadiazine etc.

6) Dhataki, Woodfordia (*Woodfordia floribunda* Salisb. - Lythraceae)

Flowers yield glycosides of cyanidin, pelargonidin; quercetin, myricetin and chrysophanol, ellagitannins and β-sitosterol. Flowers are used as a blood purifier and specific to uterine diseases like menorrhagia and leucorrhoea. It is also used in preventing abortion, healing ulcers, dysentery and diarrhoea. Leaves are used as a tanning material.

7) Dikamali (*Gardenia gummifera* Linn. f. - Rubiaceae)

The resin and leaves contain flavones such as gardenin, dimethyl tangeretin, nevadensin, wogonin, isoscutellarien, apigenin and demethoxy sudachitin. Dikamali is a well known antiseptic and carminative and is used in dyspepsia and as an anthelmintic.

8) Dikamali (*Gardenia resinifera* Roth.)

The resin contains flavones such as gardenins, A, B, C, D, & E, desmethyl tangeretin, nevadensin, acerosin, wogonin, demethoxy woganin, trimethoxy woganin, hydroxywoganin, apigenin, demethoxy sudachitin and 5,7,3′,5′-tetrahydroxy-8,4′-dimethoxy flavone.

9) Elderberry (*Sambucus nigra* Linn. - Caprifoliaceae)

Flowers are found to contain flavonoids (rutin, isoquercitrin and kaempferol) and tannins. Fruits also contain anthocyanins and flavonoids. Berries are diaphoretic, diuretic and laxative. Leaves and flowers also possess diaphoretic property. All these parts esp. berry are recommended for colds, fevers, chronic nasal catarrh, sinusitis and rheumatism. Elderberry is recommended especially for flu viruses.

10) Hawthorn (*Crataegus laevigata* DC & *C. monogyna* Linn. - Rosaceae)

Berries contain anthocyanins and flavonoids such as rutin, vitexin and orientin and proanthocyanins. Leaves yield vitexin, hyperoside and proanthocyanins. Flowers contain flavonols and anthocyanins. Hawthorn (berries, leaves and flowers) is taken orally to relieve chronic heart conditions primarily congestive heart failure, anguina and arrhythmias. It is also used for treating anxiety and insomnia.

11) Indravalli, Ballon vine (*Cardiospermum halicacabum* Linn. - Sapindaceae)

The leaves are found to contain flavones such as apigenin, acacetin, 7-O Me apigenin, 7, 4′- diOMe apigenin and 3′, 4, - diOMe luteolin.The roots and leaves are diuretic, laxative, stomachic and alterative and are recommended for hair growth, rheumatism, nervous disorders, piles, fever etc.

12) Japa, Shoe flower (*Hibiscus rosa-sinensis* Linn. - Malvaceae)

Flowers contain mucilage, anthocyanins based on malvidin and cyanidin and flavonoids such as quercetin and gossypetin. Flowers are emmenagogue and useful in treating vaginal and uterine discharges. They are effective in treatment of arterial hypertension and have significant antifertility effect. Flower extract is both cytostatic and cytotoxic. Flowers are also considered hypoglycemic, aphrodisiac and for piles. Leaves and flowers are good for healing ulcers and for promoting the growth and color of hair.

13) Kamala, Kamila dye tree (*Mallotus philippensis* Don. - Euphorbiaceae)

The fruit hairs which form the dye, contain rottlerin (a complicated derivative of chalcone) and isorottlerin. The bark and wood contain about 0.5% bergenin (an isocoumarin). The dye is a bitter anthelmentic and cathartic.

14) Karanj, Pongam (*Derris indica* Bennet - Fabaceae)

Stem bark contains chromenoflavone, pongachromene, glabrin, glabra-I, glabra-II, karanjin, pogapin, kanugin and demethoxy kanugin. The wood also contains same compounds. Leaves yield a furanoflavone, 3'- OMe pongapin, karanjin, kanjone and other known flavonoids. Flowers contain another furanoflavone, pongaglabol, aurantiamide-OAc and β-sitosterol. The fruits yield furanoflavonoids, pongapinnol A-D and a coumestan, pongacoumestan. The seeds yield flavonoids such as dimethyl chromenoflavanone, isolonchocarpin, demethoxy kanugin, pongol, a β-diketone karanjin, pongapin, lanceolatin B etc. (Yadav *et al.*, 2004). The plant is used in leprosy and an effective remedy for all types of skin diseases like scabies, eczema and ulcers.

15) Lajjalu, Biophytum (*Biophytum sensitivum* DC. - Oxalidaceae)

The leaves are found to contain flavones such as apigenin, acacetin and luteolin and glycoflavones like methoxy orientin and its isomer, proanthocyanins and tannins. The plant is a tonic and used in insomnia, inflammations, respiratory troubles, and as a diuretic and antiseptic.

16) Milk thistle (*Silybium marianum* Gaertn. - Asteraceae)

The seeds contain three flavonolignans, silybin, silydianin and silychristin – collectively known as silymarin. Silymarin is the only drug having genuine antihepatotoxic properties *i.e.*, a protective effect on liver. It is antagonistic to α-amanitine poisoning.

17) Paranti, Ixora (*Ixora coccinea* Linn. - Rubiaceae)

The flowers contain cyanidin and flavonoids. Flowers form a component of hair oil.

18)Pushkarmula (*Iris germanica* Linn. - Iridaceae)

Rhizomes contain isoflavones, irisolone, irilone, tectoridin, homotectoridin, and a benzophenone, 2, 4, 6, 4′-tetrahydroxy benzophenone and flavones. Petals contain embinin, a glycoflavone. The roots are used for cough, dyspnoea, anaemia, cardiac disorders, skin diseases, dysmenorrhoea and dysurea. It is also an aphrodisiac, diuretic and febrifuge.

19) Sarapunkhah, Wild Indigo (*Tephrosia purpurea* Pers. - Fabaceae)

All the parts of the plant contain rotenoids. Leaves contain rutin, quercetin, 4′-OMe quercetin and phenolic acids. Seeds also contain rotenoids. The roots or even the whole plant is used against enlargement and inflammation of liver and spleen.

20) Shankpushpi, Canscora (*Canscora decussata* Roem. & Schult. - Gentianaceae)

The whole plant/aerial parts contain xanthones and glycoflavones such as vitexin and isovitexin.The plant is laxative, tonic and febrifuge. It is reported to promote memory power, cure leprosy, oedema and urinary disorders and is hypertensive.

21) Shyonaka, Oroxylum (*Oroxylum indicum* Vent. - Bignoniaceae)

The stem and root bark contain three flavones, oroxylin-A (5,7-diOH,6-OMe flavone), baicalein (5,6,7-triOH flavone) and chrysin (5,7-diOH flavone) and tannins. The root bark is useful in diarrhoea and dysentery. Bark is used in acute rheumatism and is diaphoretic. It is also useful in skin diseases, oedema, urinogenital diseases, piles and cough.

22) Tea, Chai (*Camellia sinensis* Kuntze - Theaceae)

The young leaves contain polyphenols amounting to 30%, of which about three quarters are flavanols generally referred to as catechin.

The members of this group are (+)- and (-)-catechin, (+)- and (-)-gallocatehin, (-)-gallocatechin gallate, (-)-epiafzelchin, (-) -epigallocatechin and its gallate and epicatechin and its gallate. Of these, epigallocatechin and its gallate are the prinicipal components which, during fermentation, produce the pigments characteristic to black tea. Dimeric gallates of procyanidin and prodelphinidin, dimeric flavan-3-ol gallates (theasinensins A and B) and acylated flavan-3-ols, [(-) epigallocatechin 3-O-*p*-coumaroate, (-)-epigallocatehin 3,3′-di -O-gallate and (-) epigallocatechin 3,4′-di-O-gallate] and tropolones are the other members of catechins. Flavonols such as kaempferol, quercetin, and myricetin in glycosidic forms, saponaretin, vitexin, isovitexin, theiferins A & B and several other flavone glycosides are the other polyphenolic constituients of tea. Black tea, due to the fermentation during processing, contains polymeric flavonoid complexes theaflavins and thearubigins. Theaflavins are yellow compounds. Tea polyphenols inhibit platelet aggregation and possess anticoagulant activity. They strengthen the walls of capillaries and are antiinflammatory. Green tea is found to inhibit the growth of cancer tumours in alimentary canal. Quercetin is a known antitumour agent. Catechin is useful in treating nephritis and chronic hepatitis.

Quinones

1) Alder Buckthorn bark, Frangula bark (*Rhamnus alnus* Mill. - Rhamnaceae)

The purple bark contain frangulosides A & B, frangulin B, emodin dianthrone and its glycoside, and palmidin C and its rhamnoside. The seeds contain aloe-emodin. Alder buck thorn bark is used as a cathartic.

2) Amaltas, Indian Laburnum (*Cassia fistula* Linn.-Caesalpiniaceae)

The bark contains flavonoids such as 3,5,7,3′,4′- pentahydroxy,6,8-dimethoxy flavonol, 3,5,7,4′-tetrahydroxy,6,8,3′- trimethoxy flavonol, 1,3,8- trihydroxy,3,7-dimethoxy xanthone, (+)-catechin, epicatechin, fistacacidin, fistucacidin, kaempferol, procyanidin, propelargonidin and rhein glycoside. The pulp from pods contains sennosides A and B, rhein and its glucoside barbaloin, aloin, pectin, tannin and sugars. The bark is used as a tonic and against dysentery and skin diseases. It is also useful in leprosy, jaundice, syphilis and heart diseases. The pods are well known laxatives. The pulp from fruits is a safe purgative for children and pregnant women.

3) Cascara bark (*Rhamnus purshiana* DC. - Rhamnaceae)

Cascara bark contains about 6-9% anthraquinones mostly as glycosides such as *cascarosides* (the primary glycosides: cascarosides

A and B), barbaloin and chrysaloin in the free form, emodin, emodin oxanthrone, aloe emodin, chrysophanol etc. free or as O-glycosides and homodianthrones of emodin, aloe-emodin and chrysophanol and heterodianthrones like palmidin A, B & C. Cascara is a purgative having an action very similar to that of senna.

4) Chakramarda, Sickle senna (*Cassia tora* Linn.-Caesalpiniaceae)

The stem bark and leaves contain emodin, rhein, euphol, β-sitosterol and basseol. Seed contain glycosides of 1, 8-dihydroxy-3-methyl anthraquinone, 8-OH,3-Me,1-OMe anthraquinone, two glycosides of 9,10-diOH,7-OMe,3-Me-1H naphthopyran-1-one, cassiamide, rhein, emodin, aloe emodin, physcion, rubrofusarin, nor-rubrofusarin, 8-hydroxy-3-methyl anthraquinone and chrysophanic acid. The whole plant is purgative, anthelmintic and exceedingly useful in eczema, ringworm and other skin diseases.

5) Chitrak, Plumbago (*Plumbago indica* Linn. - Plumbaginaceae)

Root bark contains a napthaquinone plumbagin (0.91%), sitosterol, and tannins.Roots are highly valued for their curative action on leucoderma and other skin diseases.

6) Chrysarobin, Goa powder (*Andira araroba* Aguiar - Fabaceae)

The wood of *A. araroba* contains 50-75% chrysarobin which is an anthrone-mixture consisting of 40% chrysophenol-anthrone, 20% monomethyl ether of emodin-anthrone and about 30% monomethyldehydroemodinanthrone. Even though chrysarobin possesses a laxative action, it is seldom used thus, but is extensively used in fighting skin diseases.

7) Hindisana, Senna (*Cassia angustifolia* Vahl. - Caesalpiniaceae)

The leaflets yield about 2% quinones consisting of aloe-emodin, rhein, and sennosides A, B, C & D and many other heterodianthrones. Senna is widely used as a purgative. It is preferred over rhubarb because it does not produce the astringent after effect of the latter.

8) Kasamarda, Coffee senna (*Cassia occidentalis* Linn. - Caesalpiniaceae)

The leaves are found to contain anthraquinones such as chrysophanol, emodin (and their glycosides), physcion, metteucinol-7-rhamnoside, jaceidin-7-rhamnoside and 4, 4′, 5, 5′-tetrahydroxy-2, 2′-dimethyl-1, 1′-bianthraquinone. Also present are tetrahydroanthracene derivatives, germichrysone and occidentalins A & B. The pods yield glycosides of 7-OMe quercetin and 3,6,3'-trimethoxy quercetin. The seeds contain anthraquinones such as 1,8-dihydroxy-2-methylanthraquinone, 1,4,5-trihydroxy-7-methoxy-3-methyl anthraquinone, physcion, rhein, aloe-emodin, chrysophanol and steroidal glycosides based on campesterol and β-sitosterol. The roots yield bitetrahydroanthracene derivatives, occidentol-I and II, pinselin (cassiolin), rhein, aloe-emodin and their glycosides, chrysophanol, physion, emodin, islandicin, helminthosporin, xanthorin, 1,8-dihydroxyanthraquinone, α-hydroxy anthraquinone and 1,8-dihydroxy-3-methoxy xanthone. All parts of this plant exhibit similar medicinal properties such as purgative, tonic, febrifugal, diuretic and for skin diseases.

9) Kumari, Indian Aloe (*Aloe barbadensis* Mill. - Liliaceae)

The leaves contain mucilage, anthraquinones and chromones. The quinones present are barbaloin and isobarbaloin (C-glycosides) and their aglycone, aloe-emodin. Fresh leaf juice is cathartic and also used for eye troubles, spleen and liver ailments, dermatitis and other skin diseases.

10) Mangishta/Indian madder (*Rubia cordifolia* Linn. - Rubiaceae)

The roots contain two anthraquinones, purpurin (trihydroxy anthraquinone) and munjistin (xanthopurpurin–2-carboxylic acid) in major amounts and xanthopurpurin (purpuroxanthin), pseudopurpurin (purpurin-3-carboxylic acid), naphthoquinones, and hexapeptides. Mangishta is one of the acclaimed blood purifiers and used as an alterative, tonic and for diseases of blood, skin and urino-genital system.

11) Patala (*Stereospermum suaveolens* DC - Bignoniaceae)

All parts contain a napthaquinone, lepachol. The wood contains lepachonone also. The roots are diuretic, cardiotonic, expectorant and effective in piles.

12) Raktachandan, Red sandalwood. (*Pterocarpus santalinus* Linn. - Fabaceae)

Wood contains a quinone, santalin (santalic acid, 16%), desoxy santalin (a napthaquinone), santal (isoflavone) and colorless pterostilbenes, pterocaspin and homopterocarpin. Raktachandan is a reputed remedy to remove pimples, scars and other skin diseases.

13) Ratanjot (*Arnebia nobilis* Rachinger - Boraginaceae)

The roots yield up to 2.5% of a red dye consisting of seven naphthaquinones, arnebins 1-7. The roots are used for dyeing clothes such as wool and silk as also as a food color. Arnebins are known to possess anticancer activity in that the crude extract/arnebin-1/arnebin-3 is found to inhibit Walker Carcino-Sarcoma tumour cells.

14) Rhubarb (*Rheum palmatum* Linn. - Polygonaceae)

The rhizome yields a variety of anthraquinones, which occur as glycosides or aglycones. They are: (1) hydroxy phenols such as chrysophanol emodin, aloe emodin and methoxylated phenols like physcion and their glycosides. (2) Carboxylated anthraquinones, e.g., rhein (cassic acid) and glucorhein and (3) Anthrones and dianthrones based on chrysophanic acid, rhein or physicon. Sennidin is a dianthrone glucoside. (4) Heterodianthrones derived from 2 different anthrones like palmidin A (aloe-emodin anthrone and emodin anthrone), palmidin B (aloe-emodin anthrone and chrysophanol anthrone) etc. The drug is used as a bitter stomachic and in the treatment of diarrhoea because of the purgative action of anthrones and the astringency of tannins.

15) Sonamukhi, Alexadrian/Tinnevelly Senna (*Cassia senna* Linn.-Caesalpiniaceae)

The leaves and the shells of fruits contain sennosides A,B,C and G, three sennidins, free anthraquinones such as rhein, chrysophanol, emodin and aloe-emodin. The leaves and shells are used as tonics in Ayurvedic and Unani Systems of medicine. It is an efficient purgative but due to the tendency to cause gripe, usually administered along with carminatives.

16) St. Johns wort. (*Hypericum perforatum* Linn. - Hypericaceae)

Aerial parts contain hyperin, rutin and quercetin. Hypericin (hexahydroxy, dimethyl-napthadianthrone), the aglycone of hyperin, is present in flowers also. The herb is a spasmolytic, diuretic and stimulates gastrointestinal secretions especially bile.

17) Vavding, Embelia (*Embelia ribes* Burm.f. - Myrsinaceae)

The fruits contain a benzoquinone, embelin (2, 5-diOH-3-undecyl-1,4- benzoquinone). Embelin and embelia are well-known anthelmintics esp. for tapeworm. The fruits are effective in cystic and abdominal tumours.

Tannins

1) Arjun/Arjun Sadad (*Terminalia arjuna* W. &A.)

Bark and fruits are found to contain tannins (both hydrolysable and condensed). The bark of Arjun tree is a well known cardiotonic. This is used in all sorts of cardiac failure and dropsy.

2) Asoka (*Saraca asoca* De Willde - Caesalpiniaceae)

The bark yields tannins (6%) and catechol. Asoka bark is widely used for treating uterine disorders such as hemoarrhages, dysmenorrhoea and menorrhagia.

3) Babul, Indian gum Arabic (*Acacia nilotica* subsp. *indica* Brenan - Mimosaseae)

Babul bark contains catechin, epicatechin, dicatechin, quercetin, gallic acid and procyanidin. The pods contain 12-19% tannin consisting of gallic acid, *m*-diagalic acid, catechin, chlorogenic acid, galloylated flavan-3, 4-diol and robidandiol. Bark is used in asthma, bronchitis, leucorrhoea and pneumonia and is hypoglycemic.

4) Hirda/haritaki, Chebulic myrobalans (*Terminalia chebula* Retz. - Combretaceae)

The fruits contain 30-40% of tannins mostly as di-O-galloyl and tri-O-galloyl glucose. Some of the component tannins isolated are chebulagic acid ($C_{41}H_{30}O_{27}.10H_2O$), chebulinic acid and corilagin. This is one of the best laxatives, and used for diabetes, fever, anaemia, cardiac disorders, diarrhoea and dysentery.

5) Jamun/Jambuh (*Syzygium cumini* Skeels - Myrtaceae)

The fruit pulp contains gallic acid, tannins and glucosides of petunidin and malvidin. Stem bark also contains tannins, myricetin, gallic acid and ellagic acid. The fruit, especially seeds, are extensively used against diabetes. Bark is used in dysentery.

6) Khair, Cutch Tree, Catechu (*Acacia catechu* Willd. - Mimosaceae)

Katha is almost pure catechin whereas cutch contains up to 60% tannins and 12-18% catechin. Kheersal is the crystalline or powdery deposits of catechin found naturally in old trees. Katha is known for its antileprotic activity and is used as an astringent and in cases of cough and diarrhoea.

7) Palas, Bengal kino tree (*Butea monsperma* Taub. - Fabaceae)

Bengal kino contains catechin, procyanidin, gallic acid and mucilage. It is a powerful astringent, used for diarrhoea, dysentery, phthisis and hemorrhage from stomach and bladder.

8) Parisa/palash, Portia tree (*Thespesia populnea* Sol. - Malvaceae)

Bark contains tannins (7%), gossypol and a coloring matter. Bark is used mainly for treating skin diseases and urinary disorders.

9) Samudraphal, Indian Oak (*Barringtonia actuangula* Gaertn. - Lecythidaceae)

The bark contains ellagitannins (16%). Bark finds use in diarrhoea and mennorrhoea and as a febrifuge.

10) Siris tree (*Albizzia lebbeck* Willd. - Mimosaceae)

The heartwood contains polyphenols such as melanoxetin, okanin, propelargonidin, lebbecacidin and melacacidin. Bark yields condensed tannins (7-11%) consisting of D-catechin, procyanidin, melacacidin and lebbecacidin. *A. lebbeck* is said to have anti-dysenteric and anti-tubercular properties. The bark is used for bronchitis, leprosy, paralysis and worms and as on expectorant.

11) Witch Hazel (*Hamamelis virginiana* Linn. - Hamamelidaceae)

Leaves contain 'hamamelitannins' (a group of three gallotannins, α-, β- and γ-hamamelitannins) and ellagitannins. The leaves are used as astringents and haemostatic and widely used for sprains, bruises, wounds and in eye lotions.

2.4.1.4. Gums and mucilages

1) Atibala, Country mallow (*Abutilon indicum* Sweet. - Malvaceae)

The leaves and seeds contain mucilage. Atibala root is considered aphrodisiac, diuretic and a nervine tonic.

2) Coneflower, Echinacea (*Echinacea angustifolia* DC - Asteraceae)

Roots and rhizome contain an arabinogalactan, inulin and a variety of phenols of which echinacoside is the principal component. Echinacea is said to have immune stimulating properties. It is also an alterative and aphrodisiac.

3) Dhawa, Gum ghatti (*Anogeissus latifolia* Wall. - Combretaceae)

Gum ghatti is highly branched polysaccharide with long chains of 1-6, linked β-D-galactose with a small proportion of 1-3-links. The side chains contain glucuronic acid, 1→2 linked with mannose and/or 1→6 linked with galactose. Gum ghatti is used as a substitute for gum *Acacia*.

4) Gum arabic tree (*Acacia senegal* Willd. - Mimosaceae)

The gum is acidic in nature and is a mixture of two polysaccharide fractions. The major sugar components are galactose (37%), arabinose

(30%), rhamnose (11%) and glucuronic acid (13%). Gum Arabic is used as a binding agent for pills and against dysentery, gonorrhoea and nodular leprosy.

5) Hollyhock (*Alcea rosea* Linn. - Malvaceae)

The mucilage is seen in all parts of the plant. The roots of *A. rosea* are used in fever and dysentery.

6) Isabgol, Psyllium (*Plantago ovata* Forsk. - Plantaginaceae)

Seeds contain mucilage consisting of weakly acidic arabinoxylans with a xylan backbone and branches of arabinose, xylose, rhamnose and galacturonic acid. Psyllium is a bulk forming laxative useful in constipation and diarrhoea. It is also found to be hypocholesteremic.

7) Karaya gum, Sterculia gum (*Sterculia* spp. – Sterculiaceae)

The gum is a partially acetylated polysaccharide of highly branched structure with residues of D-glucuronic acid (occurring as end groups), D-galacturonic acid, D-galactose and L-rhamnose and acetic acid. Karaya gum is used as a bulk laxative.

8) Khaira, Marshmallow (*Althea officinalis* Linn. - Malvaceae)

Mucilage, present in all parts of plant, is an acidic polysaccharide consisting of galacturonic acid, galactose, glucose, xylose and rhamnose. The roots, as well as the other parts of the plant, are used as demulcent, emollient and expectorant.

9) Kokilaksha/Taalmakhana (*Hygrophila auriculata* Heine - Acanthaceae)

Seeds contain mucilage and sterols. Kokilaksha (seeds) is a reputed remedy for arthritis and is a well known aphrodisiac.

10) Okhra/Bhindi, Ladies finger (*Abelmoschus esculentus* Moench. - Malvaceae)

Okra mucilage is an acidic polysaccharide consisting of galactose (80%), rhamnose (10%), galacturonic acid (6%) and arabinose (3%). Certain samples contained protein, minerals and glucosamine. The mucilage from pods is said to have emollient and demulcent properties and is useful in dysentery. Okra mucilage is used as a plasma replacement or blood volume expander.

11) Slippery Elm (*Ulmus rubra* Muhl. - Ulmaceae)

The mucilage is found to be a mixture of two or more polyuronides, mostly polygalacturonides. The mucilage is a good demulcent and possesses great nutritive properties.

2.4.2. Primary Metabolites

2.4.2.1. Organic acids

1) Amalaki/Amla, Indian gooseberry (*Emblica officinalis* Gaertn. - Euphorbiaceae)

The fruit contain vitamin C (ascorbic acid, up to 2%, richest source), free amino acids and gallotannins. Emblica exhibits antitumour, hypotensive and immunomodulatory properties.

2) Gamboge (*Garcinia combogia* Desv.-Clusiaceae)

The fruit rind contains (-)- hydroxycitric acid (about 30%). The latex yields cambogin and camboginol, two isoprenylated benzophenones. The fruit rind is used as a spice and believed to help slimming.

3) Cangeri/Indian sorrel (*Oxalis corniculata* Linn. - Oxalidaceae)

Leaves contain tartaric acid and citric acids, calcium oxalate and flavonoids. The whole plant is a household remedy for indigestion and diarrhoea.

4) Cinca, Tamarindus (*Tamarindus indica* Linn. - Caesalpiniaceae)

The brownish red pulp surrounding the seed contains pectin, tartaric acid and soluble sugars. The fruit pulp is carminative, digestive, laxative, antiscorbutic and antibilious.

5) *Citrus* spp.- (Rutaceae)

All the *Citrus* fruits are rich in vitamin C (ascorbic acid) and minerals. The constituents of fruits are citric and malic acids, their salts and sugars. All these fruits are rich in free amino acids, which amounts to about 70% of the total nitrogenous compounds. All *Citrus* fruits, being sources of vit. C, are used for treating scurvy.

6) Kokam, Brindon (*Garcinia indica* Choiss. - Clusiaceae)

The fruit contains (+) hydroxycitric acid. Fruit is antiscorbutic, cholagogue, emollient and demulcent. Oil is used in skin diseases.

7) Lonika, Purslane (*Portulaca oleracea* Linn. - Portulacaceae)

The whole plant, especially leaves, is exceptionally rich in antioxidant vitamins A, C and E. Purslane is considered a tonic, cooling, blood purifier and useful in cough, inflammations of bladder and kidney and piles.

2.4.2.2. *Fatty acids, triacylglycerols and glycolipids*

1) Erandah, Castor (*Ricinus communis* Linn. - Euphorbiaceae)

Castor oil contains mainly ricinoleic acid (up to 90%). Seed oil is purgative and used for fever, inflammation of scortum and colic.

2) Evening Primrose (*Oenothera biennis* Linn. - Onagraceae)

Seed oil is rich in essential fatty acids such as *cis*-linoleic acid, alpha-linoleic acid and gamma linoleic acid. Also present in the oil is Vit.E. Evening primrose oil is recommended for PMS (premenstrual syndrome) like irritability, breast pain, tenderness and mood changes.

3) Garudaphala, Chaulmoogra (*Hydnocarpus wightiana* Bl. - Flacourtiaceae)

The seed oil consists of a mixture of glycerides of hydnocarpic (45%), chaulmoogric (20%), gorlic (15%), oleic (12%) and palmitic acids. The unsaturated fatty acids show strong bactericidal properties towards the *Micrococcus* of leprosy and so are used for the treatment of leprosy.

4) Jalap (*Exogonium purga* Benth. - Convolvulaceae)

The resin from dried roots contains convolvulin as the principal component, which on hydrolysis yields glucose, rhodeose and convolvulinic acid. Jalap is drastic purgative and hydragogue.

5) Orizaba jalap (*Ipomoea orizabensis* Led. - Convolvulaceae)

The roots contain 6-18% glycosidal resins consisting of jalapin which yields on hydrolysis jalapinolic acid (11-hydroxy palmitic acid), glucose, rhodeose and methyl tetrose. Used as cathartics and hydragogues.

6) Trivert, Indian Jalap (*Operculina turpethum* Silva Manso - Convolvulaceae)

The resin from the roots, contain three glycosides turpethin (major component), α-turpethin and β-turpethin and a volatile oil. Root and root bark are used as a purgative in case of constipation, piles and jaundice.

2.4.3. *The changing scenario of active principles*

A large number of medicinal plants and their active principles are described above. It was customarily believed that the major compound present in a medicinal plant is the active component in its drug action. But this concept is changing these days. The major compounds need not be the active copound. A number of new compounds having pharmacological action are reported recently as also new properties are discovered for some of the known compounds. Entirely new medicinal properties also are discovered these days. Another area of interest is that a large number of polysaccharides and lectins are found to exhibit distinct pharmacological properties. Some of these aspects are described below.

a. New compounds reported from the known medicinal plants

1. *Strychnos nux-vomica* - main constituents are alkaloids. New compounds identified are iridoids such as 5-loganic acid, 6-0-acetyl loganic acid and 3-0 acetyl loganic acid (Zhang *et al.*, 2003).

2. *Cissampelos pereira* – Known for bis-benzyl isoquinoline alkaloids such as are hayatine, hayatinine, hayatidine, *l*-bebeerine etc. New compounds isolated are chalcone-flavone dimer-cissampeloflavone (Ramirez *et al.*, 2003).

3. *Apium graveolens*- Known for volatile oils, is found to contain five sesquiterpenoid glycosides (celerioside A-E) along with *nor*-carotenoid glucosides and a lignan glycoside (Kitajima *et al.*, 2003).

4. *Pimpinella anisum*- Known for volatile oils, now contains glycosides of 3- hydroxy anethole, 3-OH estragole, methyl syringate, hexane-1,5-diol and 1- deoxy –L-erythritol. (Fujimatu, *et al.*, 2003)

5. *Cuminum cyminum*- Principal components are volatile oil rich in cuminaldehyde, now found to contain water soluble sesquiterpene glycoside hydroxyguadienolide, dihydroxyeudes-menolide and alkyl glycosides (Takayanagi *et al.*, 2003).

6. *Sida spinosa*- Known to contain mucilage and alkaloids like ephedrine, now known to contain ecdysone and glyceryl-1-eicosanoate. (Darwish and Reinecke, 2003).

7. *Allium cepa*- Known for sulphides, now known for 4-substituted carboxy pyranocyanidin (Fossen and Andersen, 2003)

8. *Commiphora wightii* , Known for guggulusterols now contain an antifungal flavonone, muscanone [3-O (1″,8″14″- trimethylhexa-decanyl)- naringenin] (Fatope, *et al.*, 2003).

9. *Pterocarpus marsupium*- Known for tannins, now found to contain glycoflavones based on aurones and flavonones (Maurya *et al.*, 2004)
10. *Humulus lupulus*- Known for humulones and lupulones (Phenolics with isoprenoid side chains), is found to contain estrogenic prenylated chalcones and prenyl flavonones. (Chadwick *et al.*, 2004)
11. *Vitex negundo*- Found containing iridoids and flavones, this plant is now discovered with a phenyldihydronaphthalene type lignan (vitedoin), phenyl naphathalene type lignan alkaloid (vitedoamine A) and a new trinorlabdane type diterpene alongwith known lignans like detetrahydroconidendrin, vitofolal E, F and cedrusin(Ono *et al*, 2004)
12. *Capsicum annum*- The chillies, known for the phenolic amides capsaicins are known known to yield 19 sesquiterpenes. (Kawaguchi *et al.* 2004)
13. *Murraya koenigii* is well-known for the volatile oil and flavonoids. But now carbazole alkaloids like murrayanine and 8, 8" bis koenigine are isolated which are found to display cytotoxic activity against cultured KB cells. (Wang *et al.*, 2003)
14. *Picrorhiza kurroa*- Found to contain iridoids, picroside and kutkoside, this drug is now detected with cucurbitacin glycosides and phenol glycosides (Stuppner and Wagner, 1989)
15. *Randia spinosa* – Known to contain saponins (randialic acid A&B) and mannitol. New compounds identified are iridoids such as randinoside, galioside, deacetyl asperulosic acid methyl ester, scandoside methyl ester, geniposide and gardenoside (Hamerski *et al.*, 2003)
16. *Emblica officinalis*- The principal components are tannins and ascorbic acid. A good amount of free amino acids such as free alanine, aspartic acid, glutamic acid, lysine and proline are reported now which may have therapeutic effects.

b. New properties

Resistance modifying agents / Multidrug resistance reversal agent.

A pentasubstituted pyridine-2,6-dimethyl-4-phenyl pyridine–3,5-dicarboxylic acid diethyl ester from *Jatroplia elliptica* is found to be exhibiting antibacterial and resistance modifying activities against strains of *Staphylococcus aureus* possessing MsrA and NorA resistance efflux mechanisms. (Marquez *et al.*, 2005).

c. New types of active principles

1. Polysaccharides

1. A galacturonan isolated from *Nerium oleander* Linn. flowers has been found to possess antitumour and immunological activities. Polysaccharides from the leaves of the same plant were found to stimulate mitogen-induced T & B. lymphocyte proliferation (Miller *et al.* 1991).
2. Cell wall derived arabinogalactan and fucogalactoxyloglucans are found to be immune-stimulating principles of *Echinacea* (Steinmuller and Roesler, 1993).
3. An acetylated mannan, acemannan, isolated from *Aloe vera* is found to be immune system modulating (Peng and Normal, 1991).
4. Polysaccharide fraction of the aqueous extracts of *Viscum album* is cytotoxic and immunostimulatory.
5. Xylans from *Phyllanthus niruri* (Meltinger *et al.* 2004).
6. Hexahydrodibenzopyrans((HHDBP) like machaeriol A,B,C, & D) and secohexahydrodibenzopyrans from *Machaerium multiflorum* (Muhammad *et al.*, 2003)
7. Wall polysaccharides such as arabinan, acidic rhamnogalacturonan, homogalacturonan, galactans, xyloglucans etc. of ginseng roots are found responsible for some of the immunostimulatory activity. (Zhu *et al.*, 2005)

2. Lectins and Proteins

1. Trichosanthin, a protein produced in the roots of *Trichosanthes kirilowii* (Chinese cucumber plant) cause abortions. But now it is found to selectively inhibit replication of HIV virus *in vitro* by inhibiting ribosomal protein synthesis and cellular reproduction (Duke and Foster, 1989).
2. Aloe lectins purified from the leaf epidermis of *Aloe* are found to inhibit the growth of a fibrosarcoma in mice (Imanishi and Ishiguro, 1981).
3. Lectins from mistletoe, are found to inhibit tumour growth due to the fragmentation of DNA in cancerous cells by the lectins (Janssen *et al,* 1993).

Other compounds

1. Piperidine alkaloids such as (-)-3-O-acetyl spectaline, (-)-7-hydroxy spectaline, iso-6-spectaline and spectaline are the new cytotoxic (DNÁ- damaging) principles isolated from the flowers and green fruits of *Cassia spectabilis* (Viegas *et al.*, 2003).
2. Cyclopentatetrahydrobenzofurans such as rocaglamide from *Aglaia* spp. (*Melia*) are strongly insecticidal and show pronounced cytostatic effects against human cancer cells in vitro (Bringmann et al., 2003.
3. Proteoglycans from *Convolvulus arvensis* are found to inhibit angiogenesis and stimulate immune responses.

New activities of known compounds

1. Rhein, an anthraquinone found in *Rheum* (Rhubarb) is known for the purgative action it exhibits. But it is found to be anti neoplastic (Iosi *et al.*, 1993).
2. Kaempferol glycoside isolated from *Forsteronia refracta* (Apocynaceae) is tumour inhibitory in that it inhibits the action of cancer linked protein RSK (ribosomal S6 kinase) which plays a role in cancer proliferation by blocking the cell cycle during Gl phase. This compound is found to inhibit the proliferation of breast cancer cell line MCF-7, producing a cell cycle block in the G1 phase with an efficacy paralleling its ability to inhibit RSK in intact cells (Smith *et al.*, 2005).
3. Kaempferol–3,7,-O-(α-)- dirhamnoside isolated from *Bauhinia forticata* leaves is found to be hypoglycemic.
4. Vitexicarpin, a flavone isolated from leaves of *Vitex negundo* is found to be cytotoxic (Diaz *et al*, 2003).
5. Murrayacoumarins A,B,C and other coumarins of *Murraya siamensis* are found to be anti-tumour promoting (Ito, C. *et al.*, 2005).
6. Triterpenes such as 3β-acetoxy-12,19-dioxo-13(18)-oleanene and related compounds isolated from aerial roots of *Ficus microcarpa* are found to be cytotoxic against three human cancer cell lines (Chiang *et al.*, 2005).

d. Synergism

The case of *Hypericum perforatum*, (St. John's Wort) wherein a number of different molecules are found to have the same activity is interesting. The plant is used for treating anxiety, depression and sleep disorders. The compounds present in the plant are, lipid soluble hypericins (up to 0.75%), flavonoids (2-4%), xanthones (0.0004%), procyanidins (8%), hyperforin (2.8%) and volatile oil (0.1 –1%), of which hypericins, xanthones and hyperforin are characteristic to St. John's Wort. It was shown earlier that hypericin inhibits monoamine oxidase (MAO) and thus increases the amount of neurotransmitters in the synapse between neurons and leads to enhanced mood (Suzuki *et al*, 1984.) Recent studies have shown that flavonoids and xanthones in *Hypericum* extracts inhibit catechol-O-methyl transferase, another enzyme that catabolises neurotransmitters (Thiede *et al*, 1994). Perovic and Muller, (1995) have shown that *Hypericum* extract decreases the uptake of the neurotransmitter, serotonin, in all rat synaptosomes. Thus it appears that a number of mechanisms act synergistically to increase the neurotransmitter signal.

2.5. Cultivation of Medicinal Plants

In earlier days nobody used to cultivate the medicinal plants. Whenever needed the plants were collected from their natural habitats. The cultivation of these plants began only in the last decade when people realized that the demand is much higher than the supply and now any body and every body wants to cultivate them. It is to be remembered that not every plant can be cultivated every where. Even if some plants are able to grow in other regions, the chemistry of the plant growing in two ecologically different regions are found to be different. To take an example of 'guggul', a very important drug, it is the oleo-gum-resin of a xerophyte growing in xeric regions of Kutcch. This resin contains a large amount of guggulusterols, lignans and flavonones. Since this plant pay handsome dividends to the farmer in terms of revenue, other farmers of the nearby Kheda region (supported by NGO's and other institutions) wanted to grow them in this region. This plant could be replicated by fresh seeds (Dutt *et al*, 1987) or stem cutting; the plants were then grown as a fence to other crops. Since in a cultivated field, water was available plenty, guggul plants flourished well. But when the question of harvesting guggul came, there was no guggul exudations (normally occurring) and the guggul collected after wounding the bark was very poor in guggulusterols but contained more of lignans. From this example it is imperative that the climate and other ecological factors are very important in stimulating the secondary metabolites. In Kutcch the plant had to produce resin to retain water due to the extreme heat and for wound healing purposes. But in Kheda District where the temperature in low and climate is moist, there is no need for the plant to produce protective resins. It is well-known that the *Cannabis sativa* (bhang), the narcotic plant, produces cannabinoids only in tropical countries. In temperate climates like Europe this plant is grown for its fibre and oil. Thus it is implicit that the plants produce various secondary/metabolites as a result of their interaction with the environment. Therefore we cannot cultivate each and every medicinal plant at all places and a plant can be cultivated only in a place where the climate and other ecological factors are the same. Once the plants are selected for cultivation, their farming has to be done in the organic way. Synthetic fertilizers like NPK are to be avoided because nitrogen facilitates the speedy vegetative growth and the reproductive phase is delayed. The fast-grown plants are lean and thin and are not healthy and thus are liable to insect attack. The plants are to be allowed to grow naturally and if they are infected by pests, only biological pesticides should be used. Methods of cultivation of a number of medicinal plants are available with NBRI, Lucknow, of course, on payment basis. Farooqui and Sreeramu (2001) have provided cultivation practices of about 50 medicinal and aromatic plants recently.

2.6. Harvesting and Storage

Harvesting has to be done when the desired secondary metabolites are at their maximum in plants. The time of harvesting for many medicinal plants is recommended in standard texts. If leaves are to be collected, care is taken to avoid diseased leaves. If the whole aerial parts are to be collected, they are harvested and have to dried in shade. Sun drying is to be avoided because many of the compounds are light sensitive. The sunlight and accompanying heat will cause certain enzymatic changes also in the harvested plant material. Finally the plant material is dried preferably in an oven at 60oC and stored preferably in nitrogen.

2.6.1 *Aflatoxins*

Aflatoxins are a group of mycotoxins produced by certain fungi (*Aspergillus flavus*-from which the name is derived) growing saprophytically on stored plant materials (plant parts) or grains as well as other food preparations. These compounds were discovered when more than 100,000 young turkeys died in poultry farms of England after ingesting contaminated peanut chicken feed. These toxins produce a lot of harmful effects on animals and in man they are potent carcinogens and even cause death.

Aflatoxins are produced primarily by some strains of *A. flavus* and by most, if not all, strains of *A. parasiticus, A. nominus* and *A. niger*. There are four major aflatoxins, B1, B2, G1and G2 plus two additional metabolic products M1&M2. The aflatoxins M1and M2 were first isolated from milk of animals fed with aflatoxin preparations (and hence the designation M-for milk). The names B1 and B2 are given due to their blue fluorescence in UV light while the G designation is due to the green (yellowish green). All these toxins are highly oxygenated heterocyclic compounds of the following molecular formulae.

B1 = $C_{17}H_{12}0_6$

B2 = $C_{17}H_{14}0_6$

G1 = $C_{17}H_{12}0_7$

G2 = $C_{17}H_{14}0_7$

Aflatoxins B2 and G2 are the dihydroxy derivatives of B1 and G1, whereas M1 is 4 – hydroxy aflatoxin B1 and M2 is 4-hydroxy aflatoxin B2. *A. flavus* produces B aflatoxins while *A. parasiticus* produces both B and G aflatoxins.

Aflatoxin production is favored by water, insect damage, humidity and high temperature. In animals aflatoxins cause a disease aflatoxicosis which is a hepatic disease. Acute aflatoxicosis in man causes vomiting, abdominal pain, convulsions, coma and even death. Only aflatoxin B1 is proved to be a potent carcinogen in man causing liver cancer.

Aflatoxins can be detected by placing the powdered sample in UV (365 mm) light. If a greenish blue fluorescence is observed it will indicate the presence of aflatoxins (!). Aflatoxins can be extracted from the finely ground plant material (50g) with 200 ml chloroform. The chloroform extract is filtered, and the filtrate is then shaken with n - hexane (to remove lipids) and this hexane extract is discarded. The chloroform extract is evaporated and the residue is spotted in TLC plates (silica gel G) and developed in a solvent system Toluene: Isoamyl alcohol: Methanol (90:30:2). When the developed TLC plate is viewed in UV light (long wave length, 365 mm) two blue and two green spots will be visible. The two blue spots correspond to B1 and B2 and the two green spots are G1 and G2. On spraying the plates with 25% H_2SO_4 the fluorescence changes from blue/green to yellow.

Aflatoxins are analysed also by HPLC and immunochemical methods like Radioimmunoassay (RIA), Enzyme Linked Immuno Sorbent Assay (ELISA) and Immunoaffinity column assay (ICA)

Since aflatoxin contamination is unavoidable, numerous methods for their detoxification are being devised. In a chemical treatment, ammoniation or reaction with sodium bisulphate is considered better and safe (to humans) to degrade aflatoxins. Aflatoxins are unstable when exposed to oxidizing agents, UV light or pH below 3 or above 10. They decompose at their melting points which are between 237°C (G1) and 299°C (M1). They can be completely destroyed by autoclaving in presence of ammonia or by treatment with bleach. The action level of aflatoxins in human food is 20 ppb.

Another group of mycotoxins found in stored plant materials are produced by the fungus *Fusarium*. Zearalenone, trichothecines or moniliformin are the toxins produced by *F. moniliforme, F. oxysporum, F. culmorum, F. avenaceum, F. equiseti, F. roseum* and *F. nivale*. Other mycotoxins present in food include ergotamine, ergot or ochratoxins.

Only B1,G1 and M1 are conclusively proved to cause carcinogenicity in liver. There is a limited evidence for B2 and inadequate evidence for G2.

Aflatoxins are detected in many ayurvedic preparations. This is because of the improper drying mechanisms and unhygienic storage. The drying of plant material at constant temperature and keeping the dry powder under N_2 will prevent the fungal infection and aflatoxins in herbal drugs.

2.7. Extraction Methods

The Ayurvedic physicians used two types of extractions (1) aqueous extraction and (2) oil extraction. Taking a clue from this and with the knowledge of the chemical compounds, the various methods of extraction are devised. First and foremost method is the aqueous extraction closely followed by the alcoholic extraction and thirdly the petroleum ether/solvent ether/hexane extractions. Extracts are preferred these days because of the less area they occupy (when compared to large amount of bulky raw material) and the transportation cost of extracts is much less (almost $1/10^{th}$) as compared to raw materials. Extracts are preferred by industries also because of the easiness by which the formulations could be prepared.

Aqueous extracts

Most of the extracts available in the market are aqueous extracts. To reduce the salt content (and thus mineral content) deionised water is preferred. The water from tube wells, city pipelines and rivers are to be avoided since it contains high amount of dissolved salts which will get accumulated during concentration (water from tube well has maximum salt content). Glass equipments are the best for getting mineral free extracts.

The extract thus obtained has to be concentrated in vacuum. Spray–drying is by far the best method. Concentrating by boiling causes destruction of a good amount of compounds. The alkaloids are particularly susceptible to thermal decomposition. The concentrated extract is packed preferably in nitrogen.

Alcoholic extracts

Both methyl alcohol and ethyl alcohol are used in commercial extractions. A number of compounds which are insoluble in water are extracted in methanol/ethanol. But alcohols cannot dissolve highly non-polar compounds like fats & oils and many other lipids. Acetone, propanol and butanol are the other solvents used to extract the compounds. The solvents are then distilled off to get a concentrated extract.

Petroleum ether/cyclohexane/hexane/toluene

These solvents are generally used for extracting oils, waxes and highly non-polar lipids. The solvents are then removed by distillation and the concentrated extracts are then stored in air tight containers.

Supercritical fluid extraction

A super critical fluid is an element in a super critical environment i.e. at a temperature and pressure above the critical values (above criticial temperature of a compound, the pure gaseous component cannot be liquefied regardless of the pressure applied). In a supercritical environment only one phase exists and that is the fluid state when it is neither a liquid nor a gas and is in the intermediate state. At this state the fluid has the dissolving power as that of liquids and mobility as that of gases. The density, viscosity and the diffusivity of a fluid is in between those of gas and liquid.

The behavior of a fluid in the super critical state is described as that of a very mobile liquid. The solubility will be almost equal to the liquid while penetration into a solid matrix is facilitated by the gas like transport properties. Variation in pressure changes properties of SCFs continuously from more gas-like behavior to more liquid like behavior. As a result, the rate of extraction and phase separation are significantly faster than for conventional extraction processes. Since the extraction is dependent on the density of the fluid, the density can be manipulated by change in temperature and pressure. The dissolving power of a fluid can be increased either with the increase in density or increase in temperature. Once the extraction is over, with the change in temperature and pressure the SCF will get separated due to volatilization from the compounds extracted and there is absolutely no trace of residual solvent in the compound extracted. The SCF may then be recompressed to the extraction conditions and recycled. In SCF even the high - boiling components can be extracted at relatively low temperature and thermally labile compounds can be extracted at low temperatures.

Carbon dioxide is the most commonly used SCF, due primarily to its low critical parameters (31.1°C, 73.8 bar) low cost and nontoxicity. Superficial CO_2 is an excellent solvent to many substances and its solvent properties can be tuned over a wide range by changing the combination of temperature and pressure in the extraction chamber. For extraction of a desired chemical profile, a specific combination of temp. and pressure is applied.

Besides CO_2, *water* is the other increasingly applied solvent. One of the unique properties of water is that, above its critical point (374°C, 218 atm), it becomes an excellent solvent for organic compounds and very poor solvent for inorganic salts. This property gives the chance to use the same solvent to extract the inorganic and the organic compounds separately. Ethane (32.4°C, 48.8 bar), ethylene (0.94°C, 50.4 bar), propane (96.8°C, 42.5 bar), cyclohexane (280.5°C, 40.7 bar) etc. are some of the other solvents used as SCF.

Volatile oil of flowers and spices are generally extracted by SCF. The other compounds extracted by SCF are those of Saw palmetto (fats), Marigold, Hawthorn, Chamomile, Arnica etc.

The solubility of components in SCFs is enhanced by the addition of a substance referred to as an enantiomer or cosolvent. Ethanol is a cosolvent used with CO_2 for extraction of medicinal herbs.

2.8. Formulations

Formulations are the combinations of various drugs made to combat a particular disease. The earlier Ayurvedic tests provide all details of various combinations often exceeding 30 or more components. But recent formulations prepared by the herbal pharmacies contain comparatively fewer drugs mainly because of the difficulty in procuring all the prescribed drugs in sufficient amounts.

The present day quality control procedures demand that all plant/ drugs employed in a formulation should be assessed qualitatively and quantitatively. But the difficulty here is the non-availability of standard samples of biomarkers in the market and the absence of detailed analytical procedures for each and every drug/plant product. For many plants even the biomarkers are not defined.

There is no single analytical method for assessing the presence or the quantity of all the biomarkers in a formulation. Since it is found that almost all formulations are heterogeneous mixtures of a wide variety of drugs containing both polar and non-polar (lipidic) compounds, methods are to be devised for extracting these compounds separately from a formulation and then testing then individually. For example if a formulation contains asafoetida, salai guggul and senna, a sequential extraction in which first the formulation in extracted with petroleum ether, the second extraction by alcohol and a third extraction in water, has to be performed. Of these three extracts, petroleum ether extract contain volatile oil and steroids from salai guggul, sulphides from asafoetida and aglycone quinones from senna. The alcohol extract brings out quinone glycosides of senna, phenolic acids and coumarins from asafoetida and polar lipids from salai guggul. The third extract (aqueous) contains gum (from asafoetida) and highly glycosylated compounds of senna. All these extracts are then to be analysed by TLC/GLC for volatile oils & sulphides and by PC/TLC/HPLC for steroids and quinones. For gums, electrophoresis or hydrolysis and further analysis for sugars are to be carried out. A number of methods of analysis of natural products are available (Daniel,1991).

Not all biomarkers are available commercially. But even the standard compounds which are available are priced very high beyond the reach of a small time or a medium- sized pharmaceutical firm. In cases where markers are not available commercially, these compounds can be isolated oneself. For this, all one has to do is to find out the original research article on the isolation of the major compound or biomarker. Once the procedure, by which the said compound was isolated first, is understood, that process is repeated in the exact way described in the original paper and one can locate the desired compound, isolate the same and keep it as a standard. Thus all the standard compounds can be isolated, stored and used when needed.

2.9. Pharmacology, Pharmacokinetics and Pharmacodynamics

Pharmacology is the study of the interaction of chemical substances with living systems. The substances having medicinal properties are known as pharmaceuticals. The subject of pharmacology encompasses drug composition and properties, interaction, toxicology, therapy and medical applications and antipathogenic capabilities. Pharmacokinetics describes the behavior of these compounds in the body like its absorption, distribution in the body, metabolism and excretion. Pharmacodynamics refers to the relationship between the drug concentration of the site of action (receptor) and pharmacologic response including biochemical and physiologic effects that influence the interaction of drug with receptor.

Pharmacological studies are now conducted easily due to availability of many test systems including animal models (normally rat or mice) and cultures of diseased cells. To test the pharmacological activity, the drug is administered to these models and results are noted. In many cases the plant extracted are fractionated using different solvents of increasing polarity (say petroleum ether, chloroform, alcohol and water) and each fraction is screened for its activity using the models. When a fraction, say alcoholic fraction, is found to have activity in a test system, then the different compounds in that fraction are isolated separately and each compound is then assessed for its pharmaceutical profile. A review of pharmacology of Indian medicinal plants and natural products is available now (Dahanukar *et al.*, 2000)

Once an extract or a compound is found to be active pharmacologically, clinical trials follow. This is the administration of the drug in dog, monkeys and finally man. These are very elaborate programmes and take 8-10 years to complete. But for herbal preparations of a good history, a clinical research direct on human volunteers would suffice.

For herbal pharmaceuticals, pharmacodynamics and pharmacokinetics are generally not followed because each drug contains many compounds of varying concentration and properties.

2.10. Herbal Cosmetics

Though some products claimed to be herb-based are available in market, the quality aspects are never given a serious thought. Spurious drugs, adulterants, unscientific ways of extraction etc. are the drawbacks in this industry. The market for herbal cosmetic products is very huge. But similar to herbal medicines, this industry also is plagued by the very same problems.

A number of herbs are attributed with properties of skin care. They are often used as external applications. The plants having "blood purifying" action also can be used in skin care because a good circulatory system will take care of a number of skin problems. Many of the plants used for skin care contain antioxidant phenols. Phenols when present in an external application like creams and lotions will prevent harmful UV radiation also. Another property of phenolics is their antifungal and antibacterial properties by which the microbial attack on the skin also can be prevented, reduced or removed.

Free radical scavenging is another property of antioxidant phenols. It is found that aging is closely dependant on increase of free radicals in the body. These free radicals are produced in the body in many ways. They may leach out from mitochondria. Free radicals leaching from mitochondria double with ageing. They even damage the mitochondria resulting in more leakage of free radicals. Free radicals are also produced by UV light and pollutants (mainly pesticides) in air and foods. All these free radicals interact with the molecules around (propagating free radicals) and produce more free radicals causing a kind of cellular pollution. The targets of free radicals are nucleic acids, proteins including enzymes and mostly membranes bounding the organelles and endoplasmic reticulum. These factors slow down metabolism. This coupled with the decrease in regular cell function reduces the efficacy of existing cells and organs. Free radicals generally attack the unsaturated fatty acids present in phospholipids of the membranes by oxidative processes resulting in oxidative rancidity of lipids. The destruction of lipids causes the destruction of membranes. The free radicals attack the fatty layers below the skin also. The skin shrinks down in the areas where the underlying fat layer is lost, causing wrinkles.

Of the large number of plants used in herbal cosmetics, a majority contains quinones. Some of the quinone containing plants are various species of *Cassia.* such as *C. tora, C. occidentalis, C. fistula, Aloe, Plumbago, Embelia, Cyperus, Rubia* etc. Simple phenols, rather phenyl propanes are the active compounds of *Curcuma* and *Psoralea*. Flavonoids and tannins are the compounds in *Pterocarpus marsupium, P.santalinus, Thespesia, Pongamia, Oxalis, Ficus* spp. (*F. benghalensis* and *F. religiosa*) and *Azadirachta. Aristolochia,* and *Alangium* contain alkaloids and *Cascabela* contain cardiac glycosides. *Centella* is another plant used in skin care.

Hair care is another area where a good number of plants are found to be effective. Some of them are *Indigofera, Vitex, Eclipta, Ixora, Hibiscus,* brahmi etc.

A fairly large number of plants are used as blood purifiers and when taken internally they reduce the skin problems and produce a healthy and glowing skin. The common plants used thus are sarsaparilla (*Smilax* spp. containing saponins), Indian sarsaparilla (*Hemidesmus indicus, Ichnocarpus fruticosa, Cryptolepis buchanani* and *Decalepis hamiltonii*), Neem, *Rubia cordifolia,* and others.

Extraction of the active principle also needs careful considerations. Most of the compounds of the above plants are extracted in water except for turmeric and red sandal wood where in methanolic extraction is recommended. The concentrated extract then is mixed with a vehicle and applied. For creams, preservative added is sodium benzoate.

2.11. Antioxidant Therapy

Free radicals are a necessary evil. They are fundamental to any biochemical process and form an integral part of aerobic life and metabolism. They are generally produced by respiration, which uses oxygen and some cell immune functions. In the body, these compounds are counteracted and balanced by the natural antioxidants. According to the present knowledge, in addition to the traditional role of protecting the fats, proteins and carbohydrates, the biological antioxidants manage repair systems such as iron transport proteins (transferrin, ferrutin, caeruloplasmin etc), antioxidant enzymes and factors affecting signal transduction, vascular homeostasis and gene expression (Frankel and Mayer, 2000). But environmental pollutants such as air/water contaminants, radiation, pesticides etc. produce a large influx of free radicals in the body and tip the balance between prooxidant (free radicals) and anti-free radical (antioxidicants) in favour of the former, resulting in a cumulative damage of protein, lipid, DNA, carbohydrates and membranes leading to **oxidative stress.** The oxidative stress, in which the free radicals outweigh antioxidants in number, is suggested to be the cause of ageing and other diseases like atherosclerosis, stroke, diabetes, cancer and neuro-degenerative diseases such as Alzheimer's disease and Parkinsonism. Studies conducted worldwide have provided enough evidence stating that providing the body systems with sizable amount of antioxidants may correct the vitiated homeostasis (Tiwari, 1999; Pietta, 2000) and a proper administration of antioxidants can treat or prevent the onset of ailments.

We normally receive a good amount of antioxidants through our diet consisting of spices, vegetables, pulses and cereals which contain a large variety of these compounds and their role in preventing human diseases, including cancer, atherosclerosis, stroke, rheumatoid arthritis, neurodegeneration and diabetes are currently being followed (Fang et al., 2002). But at the time of oxidative stress, a good amount of external antioxidants are to be pumped into, and this is the basis of various antioxidant-based therapeutics.

Atherosclerosis is considered mainly due to oxidation of low-density lipoproteins (LDL) leading to the accumulation of cholesterol in the atherosclerotic lesion. In addition, the high fat diet also is found to decrease the expression of genes which produce free-radical scavenging enzymes resulting in an oxidative stress (Sreekumar et al., 2002). Recently probucol and α-tocopherol-based compounds have been designed and developed to combat atherosclerosis, which are undergoing clinical trials. Similarly therapeutic formulations from Arjuna, Abana, Shosaikoto etc having antiatherosclerotic properties are of great promise (Inoue, 2000; Wasserman et al., 2003). Stroke and cerebral thrombosis are the other related diseases receiving attention in free radical scavenging and antioxidant based therapeutics.

The increased levels of sugar in blood, caused by *diabetes* also leads to the production of free radicals leading to oxidative stress. This paves ways to further cellular damage and late diabetic complications. In this case also antioxidants are found to decrease oxidative stress and further complications (Karazu et al., 1997). Supplements of antioxidants have shown to decrease oxidative stress and diabetic induced defects in diabetic animals. Since some of the plants used for diabetes are rich sources of antioxidants, the activities of these plants are now considered based on their antioxidant principles (Scartezzi and Speroni, 2000).

Oxidative stress also is found to induce neuronal damage, leading to neuronal death and cause neurodegenerative diseases such as *Alzheimer's and Parkinson's diseases.* In the former case the characteristic histopathological alterations such as neutristic plaques composed largely of amyloid ß-peptides and neuronal aggregates of abnormally phosphorylated cytoskeletal proteins, are found to be due to the overproduction of ROS (Reactive Oxygen Species). The free radical mediated selective degeneration of dopaminergic neurons in the nigrostriatal system, which results in loss of dopaminergic influence on other structures of basal ganglia, is found to result in normal Parkinsonian symptoms. A number of studies show that antioxidants like L-DOPA, tocopherols, ascorbate etc. can block neuronal death and may have therapeutic properties.

Cancer is another disease in which antioxidants can play a beneficial role. Metabolic activation of a carcinogen is a free radical dependant reaction. It is also found out that endogenous DNA damages arise from a variety of intermediates of oxygen reduction and free radicals and this play a positive role in carcinogenesis. Misrepair of damaged DNA as also the oxidation of guanine by free radicals are other problems here. The higher amount of H_2O_2 and other reactive oxygen species in cancer cells also points to the persistent oxidative stress in cancer. Phenolics and isothiocyanates are effectively used in inducing H_2O_2 and they activate stress genes, which in turn cause large scale irreparable damage to DNA and thus inhibit proliferation and act as anticancer agents (Loo, 2003). The anticancer activities of antioxidants involve (i) trapping the carcinogen, (ii) blocking the metabolic activation of carcinogens, (iii) scavenging the free radicals and their production, (iv) block lipoxigenase / cycloxygenase pathway etc (Smith *et al.,* 1995).

From among the kinds of radicals (hydroxyl, alkoxyl, peroxyl and carbon centered) involved in oxidative stress, the peroxyl radicals are considered to be the major target radical for radical scavenging antioxidants in vivo (Niki *et al.*, 1995). Peroxyl radicals are electrophilic in nature and therefore electron-donating substituents in phenolics increase the reactivity towards these radicals. Phenolics having substituents on the ortho-position are found to be better peroxyl scavengers. In addition, flavonols are found to contain a strong nucleophilic centre that reacts with electrophilic species and thereby decreases the bioavailability of the ultimate carcinogens.

In plants phenolics form a very large group. They include, simple phenolics, phenolic acids, phenyl propanes, acetophenones, stilbenes, xanthones, flavones, flavonols, catechins, aurones, chalcones, isoflavonoids, flavans, neoflavonoids, tannins etc. All these different types of phenolics are very effective antioxidants. The large number of medicinal plants containing phenolics (as active principles) will effectively support the role of these compounds in therapy. In addition the various phenolics found in other medicinal plants (as such there is no plant without phenolics) having different active principles will ably support the curative action of the latter compounds by providing a free radical- free environment.

2.12. Aromatherapy

It is a relatively new term for an ancient practice of using fragrant essential oils to treat a wide range of physical and emotional ailments. It was a French Chemist Rene-Maurice Gattefose, who coined the term aromatherapy to describe the medicinal uses of essential oils. The oils are directly applied to the affected parts like wounds and burns or massaged into the skin so that the oil reaches bloodstream or inhaled to lungs so that the effect of smelling reaches the mind.

Since the monoterpenes are small molecules and are lipidic, they are rapidly absorbed by the skin. This property is used for chest rubs or abdominal rubs. The antimicrobial, expectorant and local irritant properties of these compounds are fairly known and therefore they are often used thus by many.

But today aromatherapy is more concerned with the physiological and emotional effects of the fragrance of essential oils than with their antimicrobial and healing qualities. How scents can affect mood, state of mind, and health by sending messages directly to the brain, are subjects to an increasing number of research studies.

The part of the brain, that receives and processes smells, is known as the **limbic system.** This is the oldest part of the brain, the core around which higher brain functions are located. It includes the primary control systems for most of our body functions including hormone secretion and the autonomic nervous system. Thus inhaling a fragrance can stimulate the release of hormones, which cause euphoria, pain relief, sexual response stimulation, calm anger or fear, induce sleep or reduce stress.

It is found out that fragrance can affect both mood and behavior both at home and workplace. People are found to cooperate more when placed in a room having a pleasant smell. The aroma of oils of rosemary, cinnamon, basil, jasmine, black pepper etc are found to cause brain-wave patterns much like those caused by caffeine. Researchers in Japan measured the brain waves of women who were receiving facials scented with orange, tangerine and peach designed to lift their spirits. They found that these patterns are similar to those achieved during deep relaxation or meditation. It is also found that the womens blood pressure also was lowered. The oils of valerian, neroli, nutmeg and mace also cause a lowering of blood pressure.

The fragrances of chamomile, marjoram, sandalwood and bergamot are found to sedate mind and body. International Fragrance and Flavour, a large perfume company, has acquired a patent to produce a product to ease stress in the workplace by using a blend of neroli, valerian and nutmeg. Adding a few drops of these oils to a warm bath also produce similar effects. In an article appeared in British Journal of Occupational Therapy, aromatherapy is described as a way to promote health and well being through massage, inhalation, baths, compresses, creams and lotions. Using fragrance in hospitals is advocated to reduce stress and depression, sedate or invigorate, stimulate sensory awareness and provide pain relief. Memorial Sloan-Kettering cancer centre in New York is already using aromatherapy to lower the stress levels not only of patients, but also their families and staff (Keville and Green, 1995).

2.13. Nutraceuticals

Nutraceuticals are foods supplemented with compounds having pharmaceutical properties. They contain a sizable amount of nutrients in the form of carbohydrates or proteins (the major compounds) and a dose of compounds having varied pharmacological properties. On developed countries the foods especially those which are canned, are fortified with a good dose of minerals, vitamins etc. and are prepared to take good care of health. Calcium enriched beverages, food bars, cereals, yoghurt and fermented foods dominate the nutraceuticals market in Japan, while in Europe it is omega 3 fatty acids along with calcium.

A number of nutritional factors are added to nutraceuticals. They are:

1. L- Lipoic acid (an antioxidant to scavenge free radicals),
2. Creatinine monohydrate (to increase muscle strength),
3. γ- Linolenic acid
4. Conjugated linolenic acid (to increase muscle mass and to reduce fat.)
5. Inulin (to increase calcium absorption),
6. L- Carnitine (as a fat burner),
7. β- Carotene (provitamin A),
8. Omega-3- fatty acids (to optimize cardiac and nervous system) and
9. Phytosterols (to reduce cholesterol)

Recently herbal medicines, in the form of plant parts (containing active components) are added to prepare nutraceuticals. Fenugreek (to prevent and treat diabetes), isoflavones (to prevent breast cancer and to reduce the incidence of osteoporosis), dietary fiber, proteins, amino acid supplements, β- glucans, antioxidants, vitamins from fruits, tocopherols from oilseeds, vegetable proteins especially texturised vegetable proteins from soybean, β- glucans from oat and barley (to reduce LDL, cholesterol and lower blood pressure), omega-3- fatty acids (precursors of prostaglandins , leukatrienes and brain lipids) anthocyanins etc. are the other components added to the foods to prepare nutraceuticals. Animals fed with flax seeds (rich in omega 3 - fatty acids), are founds to produce eggs and milk rich in omega-3- fatty acids).

Phase II enzyme inducers from *Brassica,* compounds such as sulforphane and glucoraphanin present in broccoli and similar compounds seen in many other Brassicaceae members like *Lepidium* are also introduced in nutraceuticals (Phase II enzymes are part of the body's RDS defense system. They prevent cellular damage).

Global market for nutraceuticals is 40-50 billion US dollars and this market is estimated to grow at the rate of 6% per annum. Nutraceuticals also are subjected to the regulatory systems applied to pharmaceuticals or foods. "Prevention is better than cure" is the principle of nutraceuticals.

India can produce a wide variety of nutraceuticals .Foods can be prepared with all the "rasayana" drugs. The plants like Ashwagandha, Amrut, Punarnava, Shataveri, Amla etc. are general tonics which can be mixed with food materials. They will take care of all body functions and regulate all body systems. These provide principles active in normalizing the metabolism and contain, in addition, a number of vitamins, minerals, co- factors and antioxidants. Depending on the disease to be taken care of, different foods tailor made with the drugs prescribed for that disease can be prepared and sold as nutraceuticals. The well known preparations like "Chavanprash, Dhanwanthari rasayana, Ashwagandhadi" etc. contain plant drugs with energy rich ghee, suger etc. These preparations can be recommended as nutraceuticals.

2.14. Herbal Veterinary Medicine

This is an area which is to be the thoroughly explored. Most of the plants used for man can be given for animals. Even some of the biopesticides are used to drive away the insect pests of the animals. Galactogogues are another group of plants which are to be investigated.

Regional Animal Production and Health Commission for Asia and the Pacific (APHCA) has published a number of books and treatises on the plants used for the well being of the domestic animals (Anon. 1991a, 1991b). The major disease conditions of animals are (1) Gastrointestinal disorders, (2) Respiratory problems, (3) Reproductive disorders, (4) urinary disorders, (5) Musculo-skeletal conditions, (6) skin affections, (7) Parasitism and (8) other conditions.

1. *Gastro-intestinal disorders* include bloat, flatulence, diarrhoea, indigestion, constipation, inappetence, abdominal distress etc. The plants used for bloat and flatulence are *Cocos nucifera* (expressed extract from kernels-coconut milk), *Musa sapientum* (flowers), *Carica papaya* (fruits), ginger (rhizome), garlic (bulb), *Desmodium triflorum* (whole plant), pepper (seed), mustard (seed), *Datura* (leaf), *Solanum* (whole plant), cinnamon (leaf), *Aristolochia* (root, leaf), *Vitex* (leaf) etc. For diarrhoea, the plants used are *Aegle* (fruit), guava (seed), Banana (leaf), *Mangifera* (root), bamboo(leaf), Jamun (bark extract), papaya (fruit), *Ipomoea aquatica* (leaf), Arjuna (stem bark), *Ficus glomerata* (stem bark),

Sida (root), *Basella* (leaf), *Amaranthus* (whole plant), *Lannea coromandeliana* (stem bark) etc. The plants recommended for dysentery are *Lagerstroemia speciosa* (leaf), *Achras* (bark), *Rheo discolor* (leaf), *Bauhinia malabarica* (flowers and stem bark), *Ixora* (stem bark), *Holarrhena* (stem bark), *Garcinia morella* (stem bark), *Barringtonia* (stem bark), *Emblica* (fruits), *Abutilon* (leaf) etc. For indigestion, the plants used are the leaves of cotton, *Vitex*, lemon, Bonduc and *Nigella*. Other plants used thus are garlic (leaves), *Moringa* (stem bark), *Acorus* (rhizome), betel (leaves), ginger (rhizome), shataveri (leaf), brahmi (leaves), neem (leaves), coriander (seeds), cumin (seed), fennel (seed) and curry leaves. Even pepper (seed), *Jatropha glandulifera* (root), *Cissus* (roots) and shataveri (root) also are used for the same purpose. In cases of constipation, seeds of *Bassia longifolia*, bulbs of garlic, root bark of *Moringa*, roots of *Croton tiglium* and liqorice and leaves of *Tinospora* are used. *Lantana*, *Premna* and *Coscinium* are the plants used for colic.

Respiratory problems of animals include cough, colds and nasal discharges. For these diseases the plants used are *Lantana* (flowers and leaves), *Blumea* (leaf), Bitter gourd (leaves), *Eucalyptus* (leaves), Mango (leaves), *Premna* (leaves), chillies and carrot seeds, *Dregea* (roots), garlic, calamus, pepper, ginger, *Ficus religiosa & F. racemosa* (leaves) and *Cardiospermum* also are being used.

For *epistaxis*, juice from leaves of *Chromolaena*, *Tinospora rumphii*, *Tagetes* and *Artemisia vulgaris* are applied to the nasal mucous membranes. In case of asthmatic signs, *Euphorbia hirta* (whole plant), *Lantana* (leaves of flowers) and brinjal roots are administered.

Reproductive disorders

The cases of **infertility** are treated with *Acorus*, coconut, *Clitoria* (root), *Terminalia chebula* (fruit), carrot (leaves & seeds), *Datura* (flowers and seeds), *Cyperus rotundus* (roots), *Caesalpinia bonducella* (roots), *Nelumbium* (root), *Moringa* (root) and cinnamon. For the cases of difficult delivery, *Vitex*, neem, *Cassia tora* (leaves) *Feronia*, ginger, *Acorus*, *Cissampelos*, *Randia dumetorum*, *Anisomeles*, chilli roots and garlic are given to the cattle. The cases of retained placenta are treated by pineapple (juice), *Tinospora*, *Gossypium* (leaves), *Calotropis* (stem bark), *Gloriosa* (root) and seeds of *Vernonia zeylanica*, mustard, *Dolichos biflorus* and *Toddalia asiatica*. At some places roots of *Oroxylum indicum*, *Crataeva nurvala* and leaves of *Abrus* also are used. Prolapse of uterus is treated by *Adhatoda* (leaves) and *Garcinia morella* fruits.

Urinary disorders include urinary stasis (retention of urine) and urinary incontinence. The former disease is treated by the following plants: roots and seeds of rice plant and leaves of *Corchorus olitorius* and *Lagerstroemia speciosa. Desmodium triflorum* (whole plant), til oil, chillies, pepper seeds, leaves of *Tinospora, Achyranthes aspera, Vernonia zeylanica, Vitex negundo* and *Acalypha indica* and coconut water also are used for this purpose. The urinary incontinence is checked by the whole plant of *Mimosa,* leaves of *Luffa* and *Cucurbita maxima* and seeds of the later plant. Heamaturia is treated with roots of *Sida,* leaves of *Abutilon* & emblica, whole plants of *Bacopa,* stem bark of *Ficus glomerata* or seeds of *Dolichos biflorus* and *Phaseolus mungo.*

Musculo-skeletal conditions like sprain, fracture and rheumatism are treated with leaves of *Jatropha curcas, Lantana, Eucalyptus, Andropogon citratus, Blumea balsamifera* and *Ceiba pentandra* and oil of *Cymbopogon*. The swelling of foot is remedied by administering leaves of *Hibiscus, Moringa,* brinjal and *Annona squamosa.*

Skin afflictions are many in animals. Old wounds and wounds infected by maggots are treated by applying pound leaves or extracts of *Chromolaena odorata, Barringtonia racemosa, Capsicum,* brinjal, drumstick, guava, garlic or *Euphorbia neriifolia.* Fresh wounds are treated by the same plants or the leaves of *Paspalium scorbiculatum,* tobacco, banana, ginger or whole plants of *Euphorbia hirta* and *Portulaca oleracea* or leaves of *Heliotropium indicum.*

Pox and wart lesions are treated by pepper, chillies, leaves of bitter gourd, *Leucaena leucocephala, Annona muricata* and *Paspalium* or the vines of *Tinospora rumphii.* Foot rot is effectively treated by *Citrus microcarpa,* fruits, leaves, root and stem of brinjal or leaves of *Hyptis capitata.*

For wounds, *Trigonella* (seed), *Garcinia morella* (fruit), *Nigella sativa* (seed),Cumin , *Strychnos nux-vomica* (stem bark), *Canna* (roots), oils of til, tobacco, turmeric , garlic , ajowain (*Trachyspermum ammi*), *Morinda tinctoria* (stem bark), *Terminalia chebula, T. bellerica,* emblica, *Acorus* , *Glycyrrhiza,* catha , *Coscinium fenustratum* (stem), *Morinda citrifolia* (leaf), red sandalwood, *Clerodendrum infortunatum* (leaf), *Alstonia* (stem bark), *Achyranthes aspera* (leaf), *Erythrina indica* (leaf) or *Pogostemon hyneanus* alone or mixed with carbon soot or copper sulphate are applied. Castration wounds are treated with leaves of guava, betel, tulsi or seeds of *Areca.*

Boil, abscess and eczema are treated with leaves of *Lantana, Moringa, Musa, Symphytum, Gliricidia* and *Coleus* or flowers of *Hibiscus*. Scaly skin lesions and FMD and skin lesions are cured with leaves of *Gliricidia, Tinospora rumphii, Cucurbita maxima,* and guava, fruits of *Citrus,* ginger, pepper, *Acorus,* garlic or stem bark of neem and mango tree.

General conditions of the body such as *fever, inappetence* and *dehydration* also are treated with plants. When the animals get fever the following plants are given internally. They are *Vitex, Tamarindus, Coleus* and garlic. The animals are given topical (external) treatment with garlic (pound bulbs mixed with vinegar) leaf extracts of *Lantana, Symphytum* and *Bauhinia malabarica*. *Inappetence* is treated with leaves of *Cassia tora* and *Tinospora rumphii* or honey. For dehydration, the animal is given coconut water.

Parasitism

The infection of roundworms, tapeworms and coccidia are dealt by giving the seeds of *Leucaena,* papaya or leaves of papaya, *Tagetes, Persea, Leucaena,* and *Areca,* or fruits of *Momordica* or pineapple. Leaves of neem, *Aristolochia, Acorus, Citrus, Cannabis* , *Atlantia* and *Vitex,* seeds of ajowain, *Nigella,* cumin, *Embelia,* bonduc nuts and nutmeg or roots of *Aristolochia* or asafoetida also are being used for this purpose. Diseases like mange are treated with topical application of extracts of leaves of *Gliricidia,* guava, sunflower, *Chromolaena odorata, Cassia sophera,* roots of *Cassia sophera,* fruits of papaya, coconut, chillies, vines of *Tinospora rumphii* or bark of *Plumeria acuminata.* Lice, ticks and fleas are managed by applying the extracts of leaves of *Cucurbita maxima, Premna odorata, Gliricidia sepium, Tinospora rumphii* and *Vitex negundo.*

A number of other diseases also are treated by plants. For *eye inflammation,* the root extract of *Derris elliptica* and for diseases of eyes the extracts of red sandal wood, *Strychnos* (seed) and leaves of *Citrus* are used. *Ear diseases* are treated by a mixture of extracts of *Alternanthera sessilis* (leaf), *Asparagus adsendens, Aristolochia indica* (root), cinnamon, clove, pepper, nutmeg, *Acorus* and *Cleome gynandra* mixed with margosa oil is used. *Toddalia asiatica, Luffa acutangula, Dregea volubilis* (root), *Momordica dioica, Cannabis* (leaves) or seeds of castor, cardamom, *Evodia, Ocimum, Terminalia chebula* or flowers of *Mesua ferrea* are also used for the same purpose. *Hoof diseases* are cured with stem bark of *Careya arborea, Terminalia arjuna,* roots of *Cassia fistula, Calotropis gigantea,* leaves of *Thespesia,* Tobacco or ginger, pepper, latex of *Artocarpus* or whole plant of *Eclipta alba.* Sprains are treated with extract of leaf or stem bark of *Bombax, Adenanthera, Cassia tora, Olax europea* or turmeric.

For *fractures,* the plants used are leaves of *Asparagus, Cissus, Adenanthera, Evodia, Pongamia,* stem barks of *Garcinia, Canarium, Dipterocarpus, Pongamia, Ficus religiosa, Allophylus cobbe* or *Artocarpus nobilis,* seed of *Bassia latifolia, Enteda scandens,* fennel, cumin, til, or flowers of coconut palm, *Areca* or turmeric.

Galactogogues are another area of great interest. The cattle are given the roots of *Traverniera cuneifolia,* twigs and leaves of *Tinospora cordifolia,* twigs of *Leptadenia reticulata, L. pyrotechniqua, Holostemma annulare, Marsdenia volubilis* and *Grewia oppositifolia* and pods of *Prosopis, Cineraria* and *Acacia nilotica* to increase the milk production. Also given are *Anethum, Trigonella, Nigella,* anise (seeds) and milk thistle (*Silybium marianum), Glinus lotoides* and nettle leaves for the same purpose.

The above mentioned literature indicates that many of the plants used for treating human ailments are used for animals also. In addition, a number of other plants also are used for the same purpose. These are potential candidates for new medicines for man. Since the physiology of man and animals is almost the same, it is certain that these plants effectively used in animals, can be effective for humans also.

For further information on the plants used in veterinary medicine the reader is referred to the publications of FAO/APHCA. (Anon. 1991a, 1991b).

2.15. Documentation, TKDL and DNA Fingerprinting

Documentation of all the data on herbs acquired during ethnobotanical or ethno medicinal survey is to be done to prevent biopiracy of these ideas and concepts. A number of journals, scientific and semi-scientific (as Honeybee of Sristi, Ahmedabad) are there to publish this literature. TKDL, Traditional Knowledge Digital Library is a laudable effort in this direction.

TKDL

TKDL is a bulky encyclopedia of about 30 million pages of traditional knowledge to prevent granting of patents to non-original inventions. In the absence of published information, the traditional knowledge is patented by outsiders. Therefore this is an attempt to publish all such information and make it available to patenting authorities.

Traditional Knowledge Digital Library (TKDL) is a collaborative project between NISAIR (National Institute of Science communication and Information Resources) and ISM & H (Department of Indian Systems of Medicine and Homeopathy). This is planned to give legitimacy to the existing traditional knowledge and enable protection of such information from getting patented by other (foreign) inventors acquiring patents on our traditional knowledge systems.

The project of TKDL is documenting the knowledge available in public domain (in the form of books, literature etc) by sifting and collating the information on traditional knowledge from the existing literature on Ayurveda, in digitalized format in five international languages (German, English, French, Japanese and Spanish). Traditional knowledge Resource classification (TKRC), an innovative structured classification and retrieval, has been evolved for about 5000 subgroups against one group in international patent classification, i.e. AK61K35/78 is the subgroup related to medicinal plants.

The need for TKDL arose from the need to safeguard our traditional knowledge from patenting by persons or agencies outside India. India possesses a rich traditional knowledge of treating various diseases. Some are available in texts like Ayurvedic books while some have been passed down by word of mouth from generation to generation. In the absence of documented literature made available to the patenting authorities, this wealth of knowledge can be misused for personal benefit (in the form of patents) by unscrupulous individuals or organizations. The cases of patenting of turmeric, neem and basmati rice by outside agencies are well-known. India fought successfully for the revocation of these patents due to the availability of published literature on them. Therefore the TKDL was conceived and formulated.

At present in TKDL there is an inter-disciplinary team consisting of 25 Ayurvedic experts, one patent examiner, five IT experts, three NISCAIR scientists and four Technical officers.

TKRC, the structural classification system of traditional knowledge divides the sources of indigenous knowledge to six sections such as (A) Ayurveda (B) Unani (C) Siddha, (D) Yoga (E) Naturopathy and

(F) Folklore medicine. These sections are further divided to classes. For example the section of Ayurveda is divided to classes such as:

01. Pharmaceutical preparations,
02. Personal hygiene preparations,
03. Dietary preparations,
04. Biocides and fumigatives.

The pharmaceutical preparations are then divided to the following subclasses.

01A - Based on plants,
01B - Based on animals,
01C - Based on minerals,
01D - Characterised by Roga (diseases),
01E - Characterised by Karma (action),
01F - Drug administration,
01G - Miscellaneous.

So the sub-class A0IA indicates pharmaceutical preparations based on medicinal plants.

By this method TKDL is able to include the entire traditional knowledge classified and coded in 5000 subgroups. All the data on Indian systems of Medicine are prepared in patent application format. The data available in Sanskrit, Urdu, Persian, Arabic and Tamil have been converted using TKRC and Unicode metadata software Technology to English, French, German, Spanish and Japanese.

TKDL has been created for 53,000 formulations of Ayurveda and 41,000 formulations of Unani. By December 2006, about 1, 50,000 formulations in Ayurveda, Unani and Sidha will be transcribed. TKDL of Yoga will be taken up shortly. TKRC was presented and discussed by the committee of experts in IPC union and the major decisions that were taken were to create 200 new subgroups to classify traditional knowledge subject matter particularly the medicinal plants and link TKRC with IPC. The technical standards used for the creation of TKDL have been adopted by World Intellectual Property Organization in the 5th session of the inter-governmental committee on Intellectual Property and Genetic Resources, Traditional Knowledge and Expression of Folklore held in 2003.

DNA fingerprinting (DNA profiling)

This is a technique to analyze the DNA of an individual to reveal the pattern of repetition of particular short nucleotide sequences (called Variable Number Tandem Repeats) throughout the genome. This pattern is unique to the species concerned and nowadays the DNA fingerprinting pattern is a standard criterion for confirming the identity of any medicinal herb.

Southern blot is one way to analyze genetic patterns in an organism. In this, cellular DNA is isolated and by using restriction enzymes, this DNA is cut to small pieces. The pieces of DNA are then arranged based on their size (molecular weight) by gel electrophoresis. The DNA fragments are then denatured by heating or by chemical treatment to single stranded structures. The gel containing size fractionated DNA is then applied to a sheet of nitrocellulose paper and then the DNA is baked to get attached to the sheet. This is southern blot. The next step is making a radio active probe. This is done by introducing breaks (nicks) to the DNA taken in a test tube, adding individual nucleotides, some of which are radioactive and adding DNA polymerase to this mixture. The DNA polymerase will become immediately attached to the nicks of DNA and attempts to repair the DNA by picking up the free nucleotides (some radioactive also) and new strands synthesized which are complimentary to the parent strand. The nicked DNA is then heated to denature the molecule and single stranded radioactive and non-radioactive pieces are formed. The radioactive DNA is the probe. The probe is then hybridized with denatured original DNA by dipping the sheet of nitrocellulose paper containing the gel with fragments (produced by the restriction endonuclease) in a solution containing the probe. A probe for a sequence that is repeated several times in the genome identifies a few of the thousands of DNA fragments and hybridises. These fragments are then revealed by auto radiography.

DNA fingerprints of only a few medicinal plants are available. Every medicinal herb has to be screened for its "fingerprint" and these fingerprints can be used to assess the purity of the sample (adulterants will have a different fingerprint) thus aiding in quality control.

2.16. Metabolomics: Small molecule 'Omics'

Metabolomics is the study of all the "small" molecules (except the macromolecules, proteins and nucleic acids) of an organism. Each organism has a unique assortment of both primary and secondary metabolites which gives a chemical fingerprint highly specific to that organism. The sum total of all the metabolites in an organism is the Metabolome. The mRNA gene expression data and the proteomics do not give a true picture of the metabolic happenings of the cell. But a metabolic profiling will give an instantaneous picture of the entire chemicals and the reactions associated with them. Metabolomic research has two important aspects, the first is the profiling of all the metabolites belonging to all classes and the second is summarizing and analyzing this data by multivariate methods.

The plants are metabolically much richer than the animals or microbes. The bacterial system can support the biosynthesis of just a few hundred metabolites (e.g. 580 for *Bacillus subtilis* and 500 for *Escherichia coli*). The metabolic richness comes not only just from the number of genes present (20,000 to 50,000) but also from multiple substrate specificities for many enzymes, sub cellular compartmentalization and the occurrence of non-enzymic reactions. Approximately 50,000 different compounds have been elucidated in plants (De Luca and St. Pierre, 2000) and it is predicted that the final figure for the plant kingdom will approach or even exceed 200,000 (Fiehn, 2002).

For profiling the metabolites the instruments generally used are based on mass spectrometry in association with other chromatographic and spectral techniques. GC-MS and HPLC - Photo diode array – MS are considered best for quantitative and qualitative metabolite profiling. Rapid scanning Time-Of-Flight (TOF) MS coupled with GC separation and integrated with peak deconvolution software (in which the number of metabolites detectable by GC-MS in crude plant extracts is 500-1000), as well as NMR analysis of crude extracts as also direct examination of crude extracts by MS either in the from of Quadropole (Q) TOF- MS or ultra high resolution Fourier transform ion cyclotron MS (FT-MS) are the other methods frequently used. Direct infusion of extracts into MS instruments using "soft" electrospray or atmospheric pressure chemical ionization sources is a novel way to generate "fingerprints" of the molecular ions of the metabolites.

The data thus generated are arranged in data bases and then analysed using univariate statistical methods such as hierarchical cluster analysis, partial least squares and linear discriminant analysis.

Metabolomics is of great use in monitoring and measuring the large scale physiological changes that occur in response to subtle changes in environment, thus helping to improve our monitoring of adverse drug reaction and better understanding of individual sensitivities to drugs. It will also enable rapid disease diagnosis through metabolic profiling. Metabolomics is an excellent tool in quality control procedures for medicinal plants. It can also be very useful in system biology approaches in studying the activity of medicinal plants. If compared with the present day paradigm of drug discovery using the reductionistic approach of single target, single compound, the holistic approach of systems biology has the advantage that it also may detect prodrugs and synergy of compounds in plant extracts.

2.17. Patenting

A patent is the grant of exclusive rights to an inventor issued by the Patent and Trademark Office. The term of a new patent is usually 20 years from the date on which the application for the patent is filed in the United States. The right of a patent is the right to exclude others from making, using, offering for sale or selling the invention. In other words a patent is a right to a monopoly on the use and sale of a product.

Patenting products of nature was not possible earlier, but in 1980 U.S. Supreme Court opined that a bacterium genetically engineered to break down crude oil could be patented. Following this Patent and Trademark Office (PTO) announced that "non-naturally occurring non human multicellular organisms, including animals to be patentable subject matter". Thus a genetically engineered mouse or a purified and isolated DNA sequence (cloned gene) could be patented . Rio conference of 1992 upheld the Convention on Biological Diversity (CBD). The CBD protects patent rights, but also calls for technology transfer and "fair and equitable" access to the results of biotechnology, especially to the developing countries which provided the genetic resources being utilized.

It was in this Earth Summit, it was agreed on the sovereign right of states (countries) over their bioresources and knowledge systems. Some of the CBD provisions include.

Article 3. Recognizes the sovereign right of countries over their biological resources.

Article 8(j). Underlines the need to protect TK and point to the potential benefits to be realized from such knowledge through involvement of its holders and equitable benefit-sharing.

But WTO (World Trade Organization) does not recognize any Traditional Knowledge (TK), Traditional Medicine (TM) or Traditional Resource Rights (TRR). Intellectual Property Rights (IPR) also does not recognize the informal systems of innovation of indigenous people/ national minorities.

Under the new rules of the General Agreement on Tariffs and Trade (GATT) which took effect from January 1,1995; all member countries must bring their IPR laws into conformity with certain provisions of the new agreement on Trade-Related Intellectual Property Rights (TRIPPS). This agreement obliges governments to provide for "protection of plant varieties by patents or by an effective *sui generis* (of their own kind) or by any system or by any combination there of".

At present only new plant varieties can be patented. None of the existing plant or plant product can be patented. But a new use of a plant (unknown to the world, see TKDL) or a compound, or a new compound having an activity (pharmacophore) can be patented. Many countries have allowed patents on processes and not on products.

If and when a new compound is isolated from a plant which is known to have a distinct pharmacological activity and that activity is attributed to this new compound, this can be patented. Once the patent application is filed, the compound has to undergo clinical trials. In phase I trial, the drug is tried on dogs under GLP conditions. The maximum tolerable dose (MTD) of this compound in dogs is determined, 1/60th of MTD in dogs is then given to a first man, double dose to second man and this increase in dose is continued till a toxicity symptom is developed and this is the MTD for man. Phase II trial is a double blind placebo trial involving at least 200 human volunteers to decide on the optimum dose for humans. In phase III, the drug has to go through different continents and different races and will involve thirty to forty centers, several hundred doctors and at least 3000-5000 people are involved. After this, the drug can be manufactured and sold. These trials will take 10-15 years and incur an expenditure of 10-15 million US dollars.

2.18. DISTRIBUTION OF MEDICINAL PLANTS IN INDIA

India is divided broadly into 5 regions.

1. Himalayan hills, above 2500m.
2. Lower Himalayan hills and other hills, between 1000m to 2500m
3. Tropical south-east region (Tropical southern moist forests) & coastal parts.
4. Arid parts (Rajasthan and Gujarat)
5. Plain land - below 1000m

Indian flora contains 15,000 to 17,500 species which contain 3000/ 3500 / 7500 medicinal plants. But the plants commonly used are only about 1000. The drugs very frequently used are about 500.

(1) Higher Himalayas

These regions are above 2500m up to 5,000m. The temperature ranges from 0-20°C. Nearly 15% of total medicinal plants are obtained here. The important species found here are the following.

1. *Aconitum ferox* , *A. heterophyllum* and other species
2. *Orchis latifolia*
3. *Picrorrhiza kurroa*
4. *Rheum emodi*
5. *Nardostachys jatamansi*
6. *Ephedra racemosa*
7. *Inula racemosa*
8. *Saussurea lappa*
9. *Taxus baccata*
10. *Abies webbiana*
11. *Onosma echioides*
12. *Gentiana kurroa* etc. Nearly 60 plants.

(2) Lower Himalayan hills and other hills

They are at a height of 1000 to 2500m. Temperature is 30°C and below. The vegetation is dry deciduous to semi-evergreen. About 10% of the medicinal plants are seen here.

Important plants of these areas are the following.

1. *Rubia cordifolia*
2. *Valeriana jatamansi*
3. *Berberis aristata* and other species of *Berberis*
4. *Bergenia ligulata*
5. *Hedychium spicatum*
6. *Crocus sativus*
7. *Swertia chirata* and related spp.
8. *Acorus calamus*
9. *Habinaria* sp
10. *Viola odorata* etc.

(3) Plains and lower hills below 1000m

Temperature in these areas ranges from 10° to 45°C. The vegetation is dry moist deciduous. Nearly 65% of the total medicinal plants are seen here. Few of the important species are given below.

1. *Terminalia chebula , T. arjuna, T. bellirica*
2. *Holarrhena antidysenterica*
3. *Withania somnifera*
4. *Asparagus racemosus*
5. *Phyllanthus emblica*
6. *Chlorophytum* spp.
7. *Bacopa monnieri*
8. *Oroxylum indicum*
9. *Solanum* spp.
10. *Semecarpus anacardium*
11. *Plumbago zeylanica*
12. *Stereospermum suaveolens*
13. *Pueraria tuberosa*
14. *Hemidesmus indicus*
15. *Eclipta alba*
16. *Boerhavia diffusa*
17. *Convolvulus microphyllus*
18. *Azadirachta indica* etc.

(4) Tropical South East & Coastal parts

Here the vegetation is of moist evergreen forests. Temperature ranges from 20°-35°C. This region accounts for 7-8% plants, a few of them are endemic. Important plants of this region are listed below.

1. *Piper longum*
2. *Pterocarpus marsupium, P.santalinus*
3. *Alpinia galanga*
4. *Coscinium fenustratum*
5. *Saraca asoca*
6. *Myristica fragrans*
7. *Cinnamomum zeylanicum*
8. *Piper nigrum*
9. *Elettaria cardamom*
10. *Hydnocarpus wightiana*
11. *Mesua ferrea*
12. *Calophyllum inophyllum* etc.

(5) Arid regions (Deserts)

The vegetation is mostly of thorny forests. Temperature is between 10-48°C. This region accounts for 2-3% of medicinals mostly specific to this zone. The following plants are seen here.

1. *Commiphora wightii*
2. *Acacia senegal*
3. *Aloe barbadensis*
4. *Tecomella undulata*
5. *Capparis decidua*
6. *Peganum harmala*
7. *Tribulus terrestris*
8. *Fagonia cretica* etc.

The region in which the plant grows is very significant in that the specific chemicals (active principles) of the plants are synthesized due to the interaction of the plant with its specific environment, temperature, soil types etc.

2.19. Substitutes of the Drugs / Drug Plants

Since India is a very large country with varied topography and ecological conditions, it is difficult to get the same plant growing everywhere. The same is true of the medicinal plants (Please see the distribution of medicinal plants in India). It is therefore imperative that substitute sources are sought in those places where the drugs are not available. In India one can find more substitutes than the original drugs. In some cases the substitutes are found better than the orignal drug in terms of its drug action or decreased toxicity. But in majority of cases the substitutes are poor in their pharmacological actions and are at times toxic. There are many types of substitutes. These may be (a) other species of the same genus, (b) other parts of the same plant or (c) other members of the same family or other families. For detailed information of substitutes the reader is referred to a recent book by Joshi (2006).

In the cases of *Swertia chirata* (chirata), *Berberis aristata* (Daru haridra), *Aconitum ferox* (Batsanabha) , *Solanum indicum* (Brihati), *Plumbago zeylanica* (chitrak) or *Sida cordifolia* (Bala), various other species of the same genus are used as substitutes. When the drug plant is a tree or a shrub, other parts of the same plant are used as substitute (e.g. *Stereospermum, Oroxylum* etc.) Instead of *Swertia*, species of *Andrographis, Exacum* or *Enicostemma* are used as substitutes. "Pashanabheda" is *Bergenia*, but *Aerva, Coleus, Didymocarpus, Celosia* or *Homonoia* are used thus. *Pluchea* is the rasna, but *Vanda, Alpinia or Withania* are used as a rasna in many places. The original Nagkesar is *Mesua*, but *Calophyllum* and *Ochrocarpus* also are used thus. *Hemidesmus* (Sariva) is replaced by species of *Ichnocarpus, Decalepis* or *Cryptolepis*. Parpataka is *Fumaria*, but the other plants used are *Peristrophe, Justicia* or *Oldenlandia.*

Many of these substitutes were used traditionally because they were similar in properties and were cheaper compared to the original drug. Some of the plants restricted to certain geographic zones were replaced by the locally available plants in some other places. For example *Saussurea, Bergenia, Inula, Aconitum* etc are not available in plains and south and in their places the respective substitutes used are *Costus, Aerva* and *Pistia*. In a few cases the local names of the substitutes were same as that of the original drugs (e.g. Pashanabheda is the local names of *Kalanchoe* as well as *Didymocarpus*).

The overexploitation of the original drugs resulting in their near extinction necessitates the need for legitimizing these substitutes. For this purpose, these substitutes are to be subjected to the rigorous methods adopted for the original plants. The chemistry, pharmacognosy, ecological conditions and experimental evidences are needed to make them genuine drugs.

New substitutes are to be searched in folk information, ethnobotany literature or in allied genera or species. Bioprospecting and chemical prospecting are the other methods to discover new medicinals which can be utilized as the substitutes.

3

Natural Dyes

The natural dyes gave color and beauty from the very ancient times till the first synthetic dye was manufactured in 1956. Since then these natural compounds were almost replaced by their synthetic counterparts. The role of synthetic dyes in polluting the environment and the reported carcinogenicity of certain diazo dyes make the synthetic dyes highly unpopular and poorly accepted. There is an urgent need to replace them now. But in the absence of viable alternatives, the synthetic dyestuffs are being manufactured and used as a matter of necessity. The large lacuna, if and when synthetic dyes are not available for dyeing, has to be efficiently filled by natural dyes. Whether these compounds would be in a position to do so is the daunting question frequently asked.

The clothes dyed by natural colors evidently possess a higher price tag. This is the first thing, which makes them unpopular in the market because, only the rich and savvy can afford them. They are not within the reach of common man who, therefore, shuns these clothes. To rectify this, the cost factor of naturally dyed clothes is to be brought down. This is possible only if large-scale production of dyestuffs is accomplished. This can be done only if large amounts of plant material which produce dyes are grown and harvested.

The unstability of the natural dyes is another matter of great concern. The great irony in this crisis is that natural colors are accepted because they are "biodegradable". If they are degradable, much stability cannot be achieved. Therefore, the quality of biodegradability, that keeps these dyes endearing, becomes a curse. On the other hand, if somehow, by chemical means we make the concerned compound stable (so that they cannot be broken down by the climatic conditions), the biodegradability is lost and thus the same character by which the clothes are accepted is lost in the bargain.

This is the paradox we have to tackle. On one hand the compounds, which impart the colour, decompose because they protect us from oxidative damage. These compounds scavenge the free radicals which attack our body, and thus our body is protected when these compounds decompose. This fact makes the biodegradability a must for all natural colors. One way to overcome the problems with biodegradability is to dye the clothes periodically. This is possible only if the clothes have a same uniform color and not a pattern/design. By dyeing the clothes periodically, the consumer gets a fresh dose of antioxidants, which in turn will protect the body in a better manner (than the old faint clothes devoid of antioxidants). Thus one can have a body friendly cloth armed with fresh antioxidants.

Another serious problem to handle is the non-availability of large quantities of dyes. This is because we are using a very few selected plants which are used somewhere else also. Mangishta, Raktachandan, Vijaysar, Turmeric etc. are well-known medicines to be wasted in dyeing industry. We have to search for other plant sources that are not otherwise commercially in demand. Bioprospecting is a way to achieve the same. This is done by screening the biologically related (belonging to the same genus, tribe or family) plants for the presence of the very same compounds or related compounds. Since the plants belonging to the same taxa would posses same/similar biosynthetic pathways, the related plants may contain similar compounds. For example, indigo is produced by many species of *Indigofera* (we have more than 800 species of *Indigofera* in the world) and related *Tephrosia* or other genera of the tribe Indigoferae. But only a few species of *Indigofera* are commercially utilized for the dye. A systematic search for coloring materials in wild relatives may prove to pay handsome dividends.

Proanthocyanidins and iridoids are the two other groups of compounds which can provide valuable colours. Proanthocyanidins are condensed tannins consisting of catechins, flavan-3, 4-diols and related compounds often present in wood, bark or leaves of many trees. These compounds, on treatment with acids or alkali produce anthocyanidins (25%) and very complex polymers (of unknown chemical constitution, phlobaphenes/tannin reds). The red color, produced when areca nuts (in 'paan') are chewed with 'chuna' (sodium hydroxide), is due to anthocyanins. The red/violet/brown color of tea, chocolates etc. is due to tannin reds. Cutch (*Acacia catechu*), Cotinus (*Cotinus coggygria*) and gamboge (*Garcinia hanburyi*) are the tannin sources used in dyeing. But the sources of tannins are legion. Some of them are chestnut (*Castanea dentata*), divi-divi (*Caesalpinia coriaria*), myrobalans (*Terminalia chebula, T. bellerica*), oak (*Quercus montana*), sumac (*Rhus* spp.), teri pods (*Caesalpinia digyna*), avaraam (*Cassia auriculata*), canaigre, (*Rumex hymenosepalus*), gambier (*Uncaria gambier*), hemlock (*Tsuga canadensis*), larch (*Larix decidua*), mallet bark (*Eucalyptus* spp.) etc. All these can be used as sources of dyes.

Almost all the above sources are barks of trees. But a large number of plants contain proanthocyanidins in leaves. Almost all the plants of Guttiferae, Anacardiaceae and Sapindaceae etc. are rich in leaf proanthocyanidins. These leaves, in addition, contain abundant flavonols also. Flavonols can give soft yellow to yellow-brown colors to the clothes and with ferric chloride they produce blue/green or black colours. Flavonoids are another group of phenolics, the potential of which as dyes is poorly exploited. Both proanthocyanidins and flavonols are excellent antioxidants.

Iridoids are another class of compounds having a great potential as dyes. They possess a cyclopentanopyran skeleton and occur as glycosides in many advanced dicots such as Rubiaceae, Apocynaceae, Loganiaceae, Oleaceae, Scrophulariaceae and Verbenaceae etc. They are detected by a blue/green color produced when an aqueous solution (prepared by keeping the fresh chopped plant material in water for 3 hours) is boiled with Trim-Hill reagent (10ml of glacial acetic acid, 0.5 ml of con. HCl and 1 ml of aqueous 0.2% copper sulphate solution). The formation of blue/green/red colors on treating with copper sulphate (which is often used as a mordant) indicates the ease by which this color can be produced. The fact that the leaves of *Lantana, Vitex* etc. turn black on drying denotes that these compounds can also produce black color. But till today, nobody has used iridoids as a dye and a lot of procedures are to be standardized before using them as dyes. Iridoids are excellent antimicrobials also.

Food Colors

Natural food colors are another area of great concern. It seems that everybody is content with consuming large amounts of synthetic food colors which are part of jams, jellies, stored foods etc. There is a general belief that all the food colors available in market today are harmless. The imposition of limits in the amount of food color indicates that these compounds interfere with our metabolism if taken in large amounts. The large amount of jams, jellies etc., which are part of the junk food, ensures that a person is deposited with a fairly good amount of these compounds.

On the contrary, natural colors impart a number of benefits for the food they impart color. Most of the colors like saffron, turmeric that are used in food, are good preservatives (antioxidants). I would suggest utilizing a large number of colors from sources such as tree barks in food. Most of these colors are proanthocyanidins, so are very effective antioxidants. So are the flavonoids abundant in leaves. A culinary preparation colored by natural colors will effectively protect the body of the consumer from oxidative damage. Since the free radicals are now believed to be responsible for diabetes, cardiac failure, brain damage, many types of cancer etc., the antioxidants added to food will definitely be protective in nature and will add years to the life of a consumer. These antioxidants will increase the shelf life of foods also.

Pharmaceutical Colors

The bright colors of cough syrups and tablets, by which manufacturers attract customers, ensure that the patient consumes a good amount of synthetic dyes. Natural dyes, if used, for which a separate license is required (because of the addition of another plant product to the drug) may impart the medicine, additional benefits which otherwise are not available. The use of Mangishta, Rakthachandan, Sappanwood etc. will provide the pharmaceutical preparations with a large amount of antioxidants. These antioxidants will, not only benefit the person, but will increase the shelf life of the active components, and may provide a congenial environment (free from microbes because most of the phenolics are antimicrobial) as also protecting the lipids and membrane systems at the site of activity of the drug for a better drug action. An extract of Arjuna or Sariva if used as a color, will thin the blood, reduce the agglutination of RBC and purify the blood. Flavonoids as dyes will reduce the incidence of vascular purpurea and strengthen capillary walls. All these compounds reduce the aging process because of their antioxidant properties. Similarly anthocyanins as dyes will help in increased vision and reduce the incidence of cataract.

Paintings

The paintings of Ajanta and Ellora caves and many ancient temples of India are based on natural colors. Unfortunately these secrets are almost lost or being lost. Nature offers myriads of color combinations and one would be wise to use them. The paintings using mangishta and turmeric are common. The fact they fetch a higher price in market should be encouraging to the artisans using them. These paintings will be ecofriendly and protect the surface from degradation by microbes or biological agents. Urgent and extensive researches are to be done for saving ancient secrets of natural paintings from going to oblivion. All the natural colors on suitable vehicles can be used for paintings.

Paints

Natural paints can be prepared in the same way they are prepared for paintings. A wall or a room painted with natural colors will have a soothing effect on eyes. Moreover a room protected by these antioxidants will be safe from free radicals and will have great health benefits also. However, painting of walls even by synthetic dyes also has to be done frequently and therefore natural dyes can be employed on surfaces with a distinct advantage to the people who interact with these surfaces.

Cosmetics

Great strides can be achieved in the cases of face powders and lipsticks. At present the colors used or the bases to hold the colors (waxes, jellies etc.) are synthetic and interact with the skin. Applying such lipsticks destroys the texture of the skin, which in turn creates wrinkles and makes the skin rough. A natural dye like anthocyanins, proanthocyanins or quinones will, on the other hand, refresh and rejuvenate the skin so that the skin remains fresh. A face powder containing these antioxidants will prevent aging and prove beneficial to the skin.

There are very many avenues where natural dyes can be used for the benefit of mankind. There is nothing new in this approach. We have been doing this for the past many centuries, but have forgotten the ways for the last half a century or so. What we have to do is to just revive the old tradition and culture. A greater dependence on natural dyes will result in greater demand for the source plants which are to be cultivated in a large scale. Resorting to natural dyes will reduce pollution and these dyes will also reduce human sufferings due to pollution.

The principles contained in various dyestuffs are of diverse origin. Most of them contain simple or complex phenols free or in glycosidic form. Some of the indoles, diterpenes, porphyrins and tetraterpenes also contribute much to the colour. The location of these compounds in plants may be systemic or specific to a plant organ.

The plants yielding natural dyes are arranged based on the type of dying principles they possess.

3.1 PHENOLICS

Phenols being antioxidants add to the stability of the substrate into which they are added. The range of phenols present in dyes varies from methyl gallate to phlobaphenes.

3.1.1. Simple Phenols

1. *Curcuma longa* Linn. (Zingiberaceae)

Haridra/Haldi, Turmeric

Turmeric is a native of Cochin China and E. Indies. This is cultivated everywhere for its rhizome which is used as a spice. This plant is a perennial with a short thick rhizome (dark yellowish-brown inside), simple long-petioled ovate-lanceolate leaves and pale yellow flowers in dense bracteate strobiliform spikes.

Curcuminoids are the principal coloring materials (up to 6%) of which curcumin amounts to 60%, with demethoxy curcumin, *bis*-demethoxy curcumin and dihydrocurcumin forming the rest. This dye is abundantly used to impart a yellow colour to clothes and foodstuffs. A number of derivatives such as chlorocurcumin, nitro-chloro-curcumin, amino-chloro-curcumin and acetylated curcumin also are used as dyes.

The aqueous extract of rhizome is used for dyeing silk fast yellow with equal quantity of pomegranate rind. With iron mordant the color is changed to dark grey. On wool it gives various shades of light brown with mordants and brownish green with copper. With indigo a green color is produced. The dye turns red with alkalies.

2. *Eclipta alba* Hassk. (Asteraceae)

Syn. *E. erecta* Linn.

E. prostrata Linn.

Bhringaraja/Bhangra

This is an erect/prostrate annual herb, often rooting at nodes, with stems and branches covered by white hairs. Leaves sessile, oblong-lanceolate with appressed hairs on both sides. Heads heterogamous, white and achenes cuneate and winged. A very common weed in cultivated fields.

The plant yields (up to 1.6%) coumestan derivatives like wedelolactone and demethylwedelolactone. Thiophenes present are ecliptal (a terthienyl aldehyde), 2-angeloyloxy methylene-5′-(but-3-en-1-ynyl) dithiophene and 5-isovaleryloxy methylene-2-(4-isovaleryloxy-but-3-ynyl) dithiophene. Saponins (eclalbosaponins I-1V), triterpenoids, flavone (luteolin), alkaloid (ecliptine) and polypeptides are other constituents of the plant. Roots are rich in thiophene acetylenes such as 5′-senecioyl oxymethylene-2-(4-isovaleryloxy but-3-ynyl)-dithiophene, 5′-tigloyloxymethylene-2-(isovaleryloxy but-3-ynyl)-dithiophene and 2-(3-acetoxy-4-chloro-but-1-ynyl)-5-(pent-1,3-diynyl)-thiophene.(Sikroia *et al*, 1982; Wagner *et al*;1986; Das and Chakravarthy,1991;Singh and Bhargava, 1992)

The leaves are used to dye grey hairs black and the juice is used for tattooing. The color produced is blue black. Roots are used as a black dye with salts of iron.

The leaves are boiled in oil and applied to hair to make it black It is also used as a dyeing herb for tattooing.

3. *Roccella and Lecanora* spp. - (Lichens)

Pather ka - Phool, Litmus, Orchil and Cudbear

The various species of *Roccella* and *Lecanora* are the sources of dyes like litmus, orchil and cudbear. These colouring matters themselves are not constituents of lichens but are formed from orcinol, a degradation product of lecanoric acid which is a common component of lichen genera *Roccella* and *Lecanora.* Litmus is prepared by treating orcinol with ammonia and sodium carbonate at 60-80°C and kept for a few days; after which the dye is precipitated by the addition of hydrochloric acid. Orcein is a complex mixture of colouring matters separable into 14 pure compounds having a nucleus of either 7 amino-2-phenoxazane (e.g. g-amino orceimine), 7-hydroxy-2-phenoxazone (γ-hydroxy orcien) or 7-aminophenoxazime (γ-aminoorceimine). Litmus has been found to be a polymer obtained from substances having 7-hydroxy-2-phenoxazone chromophore.

Litmus is an acid-base indicator used in the laboratories. Orcein and related colouring materials are much valued for their high quality and colour and they are considered superior to synthetic dyes due to the moth-repelling properties.

4. *Wedelia chinensis* Merril (Asteraceae)

Syn. *W. calendulacea* Less.

Bhringaraj

W. chinensis is a procumbent perennial with camphor like odour, subsessile linear oblong/lanceolate leaves and solitary bright yellow heads on long axillary peduncles. A native of tropics.

The leaves contain wedelolactone, isoflavones and an oil soluble black dye.

Similar to *Eclipta,* the leaves of this plant also are used to dye grey hairs black and the juice is used for tattooing. Roots also are used to prepare a black dye with salts of iron.

3.1.2. Benzophenones and Stilbenes

1. *Chlorophora tinctoria* Gaud. (Moraceae)

Fustic

A highly branched laticiferous, dioecious tree of 25 m with a crooked bole with yellow bark having small lumpy spines, simple toothed acuminate leaves, male flowers in long narrow spikes in male plants; female flowers in green spherical heads and a sorosis type of fruit.

Fustic is a yellowish brown dye obtained from the heartwood of this tree, native to Central South America and W. Indies. This dye contains maclurin (2, 4, 6, 3′, 4′-pentahydroxy benzophenone) as the principal colouring agent. It also contains flavonoids, 1, 3, 5, 6-tetrahydroxy xanthone and a stilbene chlorophorin (2-homogeranyl-2′, 3′, 3, 5-tetra hydroxy stilbene).

Fustic imparts yellow, brown or olive colors to leather and clothes.

2. *Garcinia cambogia* Desv. (Clusiaceae)

Cokam

This plant is a moderate sized tree resembling *G. indica,* but differing in having a grooved (7-8) berry, each vertical furrow containing an ovoid compressed seed surrounded by a white aril. Common in South Western Ghats.

The latex yields cambogin and camboginol, two isoprenylated benzophenones. The fruit rind contains (-) - hydroxycitric acid (about 30%).

This dye also is used like other dyes of *Garcinia*. The fruit rind is used as a spice and believed to help slimming. The plant is used in skeletal fractures. The dried rind is used in rickets and enlargement of spleen.

3. *Garcinia indica* Chois. (Clusiaceae)

Syn. *G .purpurea* Roxb.

Brindonia indica Dup.

Kokam, Brindon

This is a small tree with glabrous oblong-lanceolate leaves (red when young) and solitary or fascicled greenish white unisexual flowers. Male flowers are tetramerous with many stamens. Female flowers are with staminodes. Fruit is a purple berry containing 5-8 compressed seeds embedded in pulp. Grown in coastal areas of Western Ghats.

The seed is the source of kokam butter. The fruit rind contains camboginol and benzophenone derivatives, garcinol and isogarcinol. Also present in the fruit are (+) hydroxycitric acid and cyanidin glycosides.

This dye gives a purple/brownish red color which can be used as a food/pharmaceutical color. Medicinally the fruit is antiscorbutic, cholagogue, emollient and demulcent. Oil is used in skin diseases.

4. *Garcinia mangostana* Linn.

Mangusta, Mangosteen

This is a small laticiferous tree yielding the edible fruit mangosteen. It bears elliptic lanceolate glossy leaves, solitary pink flowers and globose slightly flattened reddish berries with a persistent thick calyx at the base, a thick rind and 5-8 flat seeds enclosed in a thick white jelly like pulp.

The pulp contains carbohydrates (14%) mainly invert sugar and sucrose, citric acid (0.5%), pectin and tannins (7-14%). Bark, fruit hulls and dried latex yield xanthones, mangostin, β-mangostin, and γ-mangostin (normangostin). Fruits contain gartenin, 8-

deoxygartenin, normangostin and garcinones A-C. Maclurin also is reported from fruits. (Balasubramanian and Rajgopalan,1988)

This dye also is similar to that obtained from *G. indica* in properties and uses. Mangostin and other xanthones are found to be hepatoprotective, CNS depressants, anti-inflammatory and antiulcer. Mangostin is cardiotonic.

5. *Maclura pomifera* Schneider (Moraceae)

Osage orange

A bright orange dye is obtained from the woods of this small dioecious tree, native to Central and North America. This plant possesses axillary spines, long pointed leaves, male flowers in short racemes and female flowers in small dense globular heads. The enlarged fleshy perianths enclose the drupelets and form a syncarp as large as an orange.

The principal colouring principle in the dye is maclurin. The dye contains other yellow pigments like maclura-xanthone, alvaxanthone, osagaxanthone and a stilbene 2′-hydroxyreservatrol. The bark is a rich source of tannins.

6. *Milicia excelsa* C.Berg (Moraceae)

Syn. *Chlorophora excelsa* Gaud.

Fustic

M. excelsa is a large laticiferous dioecious tree of 50 m in height (9m in girth) with a clean cylindrical bole, large broadly and obliquely oblong elliptic leaves, male flowers (apetalous) in dense long spikes, on male trees; female flowers on compact spikes (in female trees) and a multiple fruit having a flavour of vinegar. A native of tropical America and Africa.

The heartwood is brownish red in color and contains chlorophorin, 3′,4,5′-trihydroxy-4′-geranyl stilbene, 2,3′4,5′-tetrahydroxy stilbene, methoxy quercetins and hydroxy benzaldehyde. The bark contains tannins.

The bark and wood are used as a source of dye.

3.1.3. Coumarins and Xanthones

1. *Feronia limonia* Swingle (Rutaceae)

Syn. *F. elephantum* Corr.

Kota, wood apple

F. limonia is a moderate-sized armed tree with axillary thorns, pinnate leaves containing obovate leaflets, white polygamous flowers in lax panicles and a large globose berry having a woody pericarp, and numerous compressed seeds embedded in a aromatic sweet pulp. A native of Srilanka, found throughout India.

The edible pulp of fruit contains protein (7%), carbohydrates (15%) and minerals. The leaves yield an essential oil (0.73%), containing estragol (90%) as the main constituent, a flavonol, 3'OMe-quercetin and phenolic acids like *p*-hydroxy benzoic, vanillic, syringic, *p*-coumaric and ferulic acids. A gum, naturally exuding from bark consists of arabinose, xylose and galactose.

The rind of unripe fruit yields a yellow dye. The pulp is used in preparation of yellow dye in certain gesso work.

2. *Mangifera indica* Linn. (Anacardiaceae)

Am, Mango

Mango, one of the important fruit trees of India, is a large tree bearing elliptic-lanceolate leaves, polygamous reddish white flowers in terminal panicles and oblique ovoid yellow fruits containing a single stony pyrene.

The bark contains a xanthone-6-glucoside, mangiferin, and 16-20% of tannins and yields a coloring matter which produces yellow shades on cotton, silk and wool. With turmeric and lime it produces a rose-pink color.

Bark extract yields an ochre-yellow dye. The leaves yield green and yellow colors. The bark is used as an auxiliary also. The dried immature fruit is used as a mordant in dyeing with safflower and in place of acids in dyeing silk and wool. The dye is extracted in water and is used for topping turmeric dyed materials to improve fastness and with a little alum to get a greenish yellow color. It is also used to develop green color on indigo dyed materials.

3.1.4. Anthocyanins

1. *Allium cepa* Linn. (Liliaceae)

Onion

Onion is a biennial with a single large bulb formed of the swollen leaf bases, bearing hollow cylindrical fleshy leaves and a single leafless scape bearing a compound umbel of white flowers. Cultivated extensively for the fleshy edible bulbs.

The dry outer scales of the bulb which are red in colour yield glycosides of 5-carboxypyranocyanidin, cyanidin and quercetin (Fossen and Andersen, 2003).

The dry scales are used to dye clothes red and or yellow.

2. *Chukrasia tabularis* A. Juss. (Meliaceae)

Chittagong Wood

A tall tree having a straight cylindrical bole and a crown consisting of long softly pubescent paripinnate leaves (leaflets ovate/elliptic), white/pink/red flowers and ellipsoidal brown capsules containing winged seeds. A native of India.

The flowers containing anthocyanins yield a red/yellow dye. The wood contains limonoids such as chukrasins A-E and usnic acid. The bark yields melianone, scopoletin and 6-7-dimethoxy coumarin. The bark and leaves contain 15-22% tannins. Seeds contain limonoids phragmalin-3, 30-di-isobutyrate, phragmatin, 3-isobutyrate-30-propionate and their 12-μ-acetoxy derivatives and 7-deacetoxy-7-hydrogedunin. The leaves contain 5, 7-dihydroxy-2′, 4′, 5′, 6-tetramethoxy flavone and quercetin. Roots contain cedrelone.

The flowers are used as a source of dye.

3. *Clitoria ternatea* Linn (Fabaceae)

Aparajita, Clitoria

A rambling twiner with scandent stem, pinnately five-foliolate leaves (leaflets ovate/oblong), bright blue/white/papilionaceous flowers with yellow/orange center and flat beaked pods containing yellowish brown subglobose seeds. A native of India.

The color of the blue flower is due to glucosides of delphinidin, malvidin, cyanidin and kaempferol.

The flowers yield a blue dye used for coloring cakes bright blue and or dyeing white cloth and matting in Malaysia.

4. *Hibiscus rosa-sinensis* Linn. (Malvaceae)

Japa, Shoe flower

An arborescent mucilaginous shrub with long-petioled, ovate-lanceolate leaves (serrate towards the tip) basally three-veined, large axillary solitary red flowers (with epicalyx, staminal tube and reniform anthers.). A native of China now cultivated every where.

Flowers contain mucilage, anthocyanins based on malvidin and cyanidin and flavonoids such as quercetin and gossypetin. Leaves yield taraxeryl acetate, β-sitosterol and flavonoids such as acacetin, 4′-OMe vitexin, 7-OMe vitexin and phenolic acids vanillic, syringic, *p*-coumaric and ferulic acids.

The crushed flowers yield a dark purplish dye used for coloring hair, eyebrows, foods, liquors and blackening shoes. Leaves and flowers are used for preparing hair oils for promoting the growth and color of hair. Flowers are also considered hypoglycemic, aphrodisiac and for piles.

5. *Solanum americanum* Mill. (Solanaceae)

In Oman a purple dye is extracted from the fruits of this plant, which is used for dyeing clothes and skin (This plant is also referred in chapter on Biopesticides).

6. *Woodfordia floribunda* Salisb. (Lythraceae)

Syn. *W. fruticosa* Kurz.

Dhataki, Woodfordia

A straggling leafy shrub with white pubescent young branches, opposite sessile ovate-lanceolate velvety leaves having an intramarginal vein, deep orange-red flowers in 2-15 flowered cymes arising form the axils of leaf scars in leaf-less portions of stem and an ellipsoid capsule covered by the calyx enclosing many small obovoid seeds. Common on rocky slopes, river banks, all over India, Africa and S.E. Asia.

Leaves contain tannins (12-20%) and lawsone. The bark contains 20-27% tannins. Flowers yield glycosides of cyanidin, pelargonidin, quercetin, myricetin and chrysophanol, ellagitannins and α-sitosterol.

The flowers yield a red dye and are employed for dyeing fabrics or as an adjunct or mordant with *Morinda citrifolia*. The dye is extracted by steeping the flowers in cold or hot water (alum or lime at times be added as a mordant) and material to be dyed is immersed in this solution several times until a pink color of desired depth is obtained. It is also used as a food color.The twigs and leaves are also used in dyeing.

3.1.5. Flavonoids

1. *Artocarpus heterophyllus* Lam. (Moraceae)

Syn. *A. integrifolia* Hook. f.

A. integra Merril

Panasa, Jack tree

Jack tree is a large evergreen laticiferous tree of 10-15m in height with green/black bark, straight stem, obovate-elliptic leaves, solitary axillary cauliflorous (on short leafy shoots) male heads, oblong ovoid female heads and large cylindric yellow fruits (syncarp-30-100cm x 25-30cm) covered by tapering obtuse processes and enclosing a large number of oblong-ellipsoid seeds enclosed in a fragrant soft/hard pulp. A native of India, cultivated for wood & fruits.

The heartwood is bright yellow and darken on exposure. The color is due to several flavones morin, dihydromorin, cyanomaclurin, artocarpin, isoartocarpin, norartocarpin, cycloartocarpin, artocarpesin, cyclodihydroartocarpesin, artocarpetin, norartocarpetin, cycloartinone and artocarpanone. Also present is gallic acid. Bark contains betulinic acid and flavones cycloheterophyllin and isocycloheterophyllin. Also present are cycloartenol, cycloartenone and tannins. Leaves and young stems yield saponins, cycloartenol, cycloartenol, ß-sitosterol and tannins.

A yellow dye is obtained by boiling wood or saw dust in water. This dye is fixed by alum and is used to dye clothes.

2. *Artocarpus hirsutus* Lam.

Anjili, Wild Jack

This is a large evergreen tree up to 70m, with a straight bole, grey bark, elliptic/rhomboid/ovate, pinnatifid leaves, paired unisexual heads and short, cylindric, orange syncarps containing ovoid seeds. A native of South India.

The wood, golden brown in color turning to blackish brown on exposure contains all the flavonoids reported from *A. heterophyllus* except norartocarpin.

A plant grown for timber, this tree can be used as a source of stable yellow dye as jack tree.

3. *Baphia nitida* Afzel. (Fabaceae)

Cam wood

This tree, native of W. Africa, yields red dye from its wood. The wood contains isoflavonones like homopterocarpin, baphinilone (a coumaranochroman), baphiin (a glycoside of homopterocarpin) and pterocarpin (inermin-7-methyl ether).

This dye is used for clothes especially wool.

4. *Butea frondosa* Koenig. (Fabaceae)

Syn. *Butea monosperma* Taub.

Palas, Bengal kino tree

Palas is a deciduous tree with a crooked trunk, bluish grey bark, long petioled trifoliolate leaves (leaflets large, broadly obovate from a deltoid base) which are densely silky below but glabrous above, bright orange red somewhat sickle shaped flowers borne on long racemes on bare branches and silky tomentose pendulous pods enclosing a single flat reniform seed at the apex. A native of Indo-Malaya, seen in most of the forests ('Flame of forest')

The bark yields tannins (6%), 3-OH,9-OMe pterocarpan (medicarpin), isoflavones such as 5-methoxy genistein and prunetin besides lupanone, lupeol and sitosterol. Leaves contain protein (14%), proanthocyanidins and phenolic acids such as *p*-hydroxybenzoic and gentisic acids. Flowers yield chalcones and aurones such as butin, butein and their glycosides, butrin, isobutrin, palasitrin, coreopsin, isocoreopsin, sulphuretin monospermoside and isospermoside. Also present in flowers are waxes and β-sitosterol. Pods yield an imide, palsimide and 1-carbomethoxy, 2-carbamyl hydrazine. The seeds contain an imidazole alkaloid, monospermin, phytolectin and palasonin, besides protein (20%), fatty oil (15.5%), mucilage (47%), β-sitosterol and α-amyrin. Also present in seeds are enzymes like proteinase and polypeptidase as well as lipolytic enzymes.

Butea gum, which is naturally exuded from cracks or due to incisions, is ruby-red in color and translucent. This is water soluble and contains procyanidin, gallic acid and catechins. This gum is useful for dyeing and tanning.

The flowers also yield a brilliant yellow dye containing chalcones and aurones such as butin, butrin, isobutrin, palasitrin, coreopsin, isocoreopsin, sulphuretin etc. The color is not very stable. But stability is attained on the addition of alum or lime. This dye is used to color food stuffs and to dye cotton, wool and also for holi. The flower extract (dye) is also used for controlling white ants in the field.

The flowers are boiled in a weak solution of lime water. With alum it yields a yellow dye which turns to orange by the addition of a little soda. On wool this dye is more permanent than that obtained from turmeric. With silk also the dye is permanent.

Green color is produced by mixing *Butea* with indigo.

Butea superba Roxb., a related woody climber common in the forests of India, also yields a red dye from its roots. Its flowers, which contain butrin, butin and butein, also are used as sources of yellow dyes.

5. *Caesalpinia sappan* Linn. (Caesalpiniaceae)

Bakam, Sappanwood

The orange red heartwood of this tree contains a dyestuff useful in dyeing cotton, silk and wool fabrics. This small prickly tree, a native of India and Malaya, has bipinnate leaves with oblong inequilateral leaflets, small yellow flowers in panicles and woody polished compressed beaked pods containing 3-4 seeds.

The principal colouring matter in this dyestuff is another aryl chroman, brazilin. The wood contains tannins too. The pod cases contain about 40% tannin. The dye is used in the manufacture of red ink.

The dye is prepared by boiling the wood chips/powder in water till 65% (2/3) of the water has evaporated. With cochineal mixed a fast dye is produced. The color is magenta/petunia color. For purple color the wood/powder is extracted with water containing 25% (of the wt. of wood) each of washing soda and copper sulphate. This is used in *Bandhana* and wood for *gulal.*

6. *Chrozophora tinctoria* A. Juss. (Euphorbiaceae)

Tournesol

An annual prostrate herb densely clothed with stellate tomentum, thick tomentose ovate, sinuously toothed or lobed leaves, small male and female flowers and a capsule containing silvery sheened seeds. Native of Mediterranean, found cultivated or wild as a weed.

The plant contains two flavonoid glycosides and in France and Holland "Le tournesol en drapeaux" is extracted and used for coloring cheese.

7. *Cinnamomum zeylanicum* Breyn. (Lauraceae)

Thaj, Cinnamon

The bark is found to contain a flavonol, kaempferol and phenolic acids such as *p*-hydroxy benzoic, melilotic, vanillic, syringic, gallic, gentisic, ferulic and protocatechuic acids besides some tannins.

Bark extract yields a brown dye. The leaves and bark are used as dye auxiliaries.

8. *Datisca cannabina* Linn (Datiscaceae)

This is a glabrous bushy dioecious herb of 2-3m with imparipinnate leaves having 5, lanceolate, serrate, acuminate leaflets.

The roots and leaves yield a flavone glycoside, datiscin (6-10%). Datiscetin, the aglycone (a tetrahydroxy flavone), is present in the root.

The roots give a yellow dye useful in dyeing cotton and wool. With aluminum mordant the color produced is bright yellow.

9. *Didymocarpus pedicellata* R. Br. (Gesneriaceae)

This is a small herb with a reduced stem with two to three pairs of opposite roundly ovate glabrous gland-punctate leaves. Common in Western Himalayas.

The leaves have a characteristic spicy odor and a dust of reddish coloring matter. The volatile oil contains a sesquiterpene didymocarpene and related didymocarpol and didymocarpenol. The coloring principles are chalcone pedicin and related pedicellin, pedicinin and methyl pedicinin. The leaf extract yields an orange dye.

10. *Equisetum arvense Linn.*

The green shoots contain many flavones like methoxy apigenins and gossypitrin, polyphenols and many phenolic acids

An aqueous extract of the plant is used to dye silk. Silk is impregnated with the aqueous extract and treated with K_2CO_3 and $FeSO_4$ mordant to give beige and brownish green colors respectively. This is also an indicator of gold.

11. *Haematoxylon campechianum* Linn. (Fabaceae)

Patang, Haematoxylin, Lagwood

The heartwood of this small thorny tree yields this dye. (Please see Biopesticides also)

The purplish red dye obtained from the heartwood contains a neoflavonoid, haematoxylin (an arylchroman) which is responsible for the red colour. Other compounds are haematoxylol A, tetra-D-methyl haematoxylol B, myricetin, quercetin, gallic acid and considerable amount of tannins.

Haematoxylin, which is up to 10% of the heart wood, on mild oxidation give haematein- a dark violet crystalline material with a green metallic lusture. Chipped logwood is subjected to ageing process to facilitate haematein formation. Logwood extract prepared from unaged wood by aqueous extraction, is concentrated in vacuum pans to get a haematein paste.

Logwood and it extractions are used to dye wool and silk and to cotton with iron chromium mordants. They are also used in dyeing rayon, nylon, fur, jute and leather. It is also used in histological work. The heartwood is a tonic and astringent.

12. *Kaempferia rotunda* Linn.(Zingiberaceae)

Bhui-Champa

This is a stemless herb with a tuberous rootstock and many thick tuberous oblong tubers. The leaves are long (up to 40cm.), oblong, mottled green above and pale red-purple beneath. Flowers are fragrant in various shades of purple and white, borne on a spikate radical scape. Cultivated in many parts of India

Tubers containing flavonols, crotepoxide and other cyclohexane diepoxides are used as a source of dye.

13. *Kirganellia reticulata* Bail. (Euphorbiaceae)

Syn. *Phyllanthus reticulates* Poir.

A large glabrous straggling shrub having small oblong/elliptic leaves, male flowers in fascicles of 2-6, female solitary flowers and a purple fleshy berry containing 8-16 trigonous finely granulate seeds. A native of Tropical Asia and Africa

The roots are used to make a red dye.

14. *Mallotus philippensis* Don. (Euphorbiaceae)

Kamala, Kamila dye tree

Kamila dye consists of the fruit hairs of *M. philippensis*, a native of Asian and African tropics. It is a monoecious tree with ovate-elliptic leaves. Male flowers are in axillary racemes. Capsules and the seeds are covered with minute red stellate hairs and a red resinous powder.

Red glandular pubescence covering the ripe capsules contains rottlerin (a complicated derivative of chalcone) and isorottlerin which form 11% of the powder. Other constituents are homorottlerin, a dark red resin, yellow resin, wax and traces of volatile oil. The bark and wood contain about 0.5% bergenin (an isocoumarin), folic acid, α-amyrin, and sitosterol and acetyl alcur.

The red pubescence is extracted manually or by stirring the fruits vigourously in water when the dye settles at the bottom. The yield is up to 3.5% of the fresh fruit. This dye is freely soluble in alkalies, alcohol and ether. The dye is used in alkaline medium. Kamala is also an excellent antioxidant useful in ghee, margarine and foods. Rottlerin is toxic to fish. Used for dyeing silk and wool.

In another method the colored red glands of the fruit are collected as a powder. This powder is made into a paste with a little water and turkey red oil, some more water is added and kept aside. The banana leaves & leaf sheaths are burned and the ash (containing 25% alkali) collected. Equal quantities of ash and kamila powder are dissolved in 12 times water. This solution is filtered and added to the first prepared solution of kamila and boiled. Then $^1/_{32}$ part of kamila powder is added and boiled again. A drop of this solution is kept on the nail and if the nail is stained the dye is ready. Otherwise the boiling is continued till it stains the nail. The wetted silk is kept in this solution for ½ h., taken out and washed thoroughly. A yellow or flame color is produced.

Kamila is used to color ghee and oil. The leaves and bark contain tannins and are used to give a deep tan color to leather. The roots also are a source of red dye.

The dye is a bitter anthelmintic and cathartic.

15. *Mesua ferrea* Linn (Clusiaceae)

Nagkesar, Iron wood

M. ferrea is a medium–sized tree with oblong-lanceolate (red when young, glabrous above and prurinose beneath) leaves, large fragrant white tetramerous flowers borne singly (stamens, numerous yellow) and a sub woody fleshy capsule surrounded by enlarged sepals and dehiscing by 2/4 valves. Seeds are 1-4, angular, smooth and chestnut brown. A native of Tropical Asia.

The finely powdered bark is kept in cold water to extract the dye and the fabric is steeped in this solution and dried. The color produced is light yellow.

16. *Oroxylum indicum* Vent. (Bignoniaceae)

Syn. *Bignonia indica* Linn.

Shyonaka, Oroxylum

A small tree, bark soft with green juice and corky lenticels. Leaves very large (1-2m long) opposite, 2-3 pinnate. Leaflets large, ovate-elliptic and unequal. Flowers purple, large, zygomorphic with 5 stamens, 4 in didynamous condition. Fruit a long flat woody capsule containing many flat seeds, winged all around. A common tree in India.

The stem and root bark contain three flavones, oroxylin-A (5,7-diOH,6-OMe flavone), baicalein (5,6,7-triOH flavone) and chrysin (5,7-diOH flavone). Also present are sitosterol and tannins. Seeds yield a fatty oil (of oleic 80% and rest saturated acids), baicalein and its glucoside tetuin. Stem and root bark are used as dyestuffs.

One of the *Dasamula* (a formulation of 10 roots, used as a tonic), root bark useful in diarrhoea and dysentery. Bark is used in acute rheumatism and is diaphoretic. It is also useful in skin diseases, edema, urinogenital diseases, piles and cough. Leaves are useful in rheumatism and as a poultice for enlarged spleen, headache and ulcers.

17. *Polygonum hydropiper* Linn (Polygonaceae)

Syn. *P. flaccidum* Meissn

Pani maricha, water pepper

A glandular reddish annual growing in wet places with stems reaching a height of 1 m, linear lanceolate leaves, pink/red flowers borne in slender racemes and granulate finely dotted nuts. A native of Central Asia.

The plant is rich in flavonoids such as quercetin, quercitrin, rutin, hyperoside, rhamnetin, isorhamnetin, etc. Seedlings contain a deep red pigment idacein. Roots contain tannins and anthraquinones.

The plant is used to dye wool yellow.

18. *Pterocarpus dalbergioides* Roxb. (Fabaceae)

Syn. *P. indicus* Baker

Chalangada, Andaman Redwood

This is a large evergreen buttressed tree of 45m, with a clean cylindrical bole, imparipinnate leaves (leaflets ovate lanceolate, tapering to a point), golden yellow flowers in terminal panicles and sorbicular flat cringed pods (edge the pod is concave) containing 1-2 smooth shining seeds. A native of Andamans now cultivated in South India and W. Bengal.

The wood contains santalin and a flavone santal. The bark and wood contain pterostilbene. Pterocarpin, liquiritigenin, (7, 4′-dihydroxy flavone), isoliquiritigenin (2, 4, 4′-trihydroxychalcone) and homopterocarpin are the other components of wood.

This is a new source of dye.

19. *Pterocarpus indicus* Willd.

This plant is similar to *P. dalbergioides*, but distinguished by ovate leaves round to a blunt point, axillary panicles and convex edge of the pod. A native of India.

The bark yields a kino. Wood contains santalin, narrin and angolensin. Narrin is a dark red amorphous powder yielding resorcinol and phloroglucinol on treatment with alkali.

The wood and bark are used as dyes.

20. *Pterocarpus marsupium* Roxb. (Fabaceae)

Asanah, Indian/Malabar Kino

This is a large deciduous tree with yellowish grey bark, imparipinnate leaves having 5-7 elliptic pubescent, alternate leaflets with emarginated apex, yellow papilionaceous flowers having crisped margins in terminal panicles and a circular winged samaroid fruit (wing veined) containing a subreniform seed. A common tree in the forests of India and nearby countries.

Bark contains *l*-epicatechin. Heartwood yields liquiritigenin and isoliquiritigenin. Other compounds isolated are a number of C-glycosides like pterocarposide, pteroisoauroside (isoaurone C-glucosides), marsuposide, benzofuranone C-glucoside, pteroside (a benzofuran C-glucoside), vijayosin [1,2-bis(2,4-dihydroxy,3-C-glucopyranosyl)ethanedione], a glucoside of 2,6-dihydroxybenzene and a sesquiterpene. Kino contains kinotannic acid, phlobaphenes, protocatechuic acid, resin and gallic acid. (Handa *et al.*, 2000; Maurya, *et al.*,2004)

Bark and wood dyes are used for dyeing clothes. This can be used as a food or pharmaceutical dye also.

The heartwood is well-known for its activity against diabetes, leprosy and other skin diseases. It is also useful in diarrhoea and hemophilic disorders. The bark and leaves are used in scabies, leucoderma, leprosy and skin diseases. The gum-resin "Malabar Kino" is useful in diarrhoea. Clinically the wood is found to be an effective hypoglycemic and hypocholestremic.

21. *Quercus velutina* Lam. (Fagaceae)

Quercitron

This bright yellow dye is obtained from the powdered/crushed bark of *Q. velutina*, a native of U.S.A. This tree has pinnately lobed obovate leaves, male flowers in catkins and female flowers in clusters of 2-5 and an obovoid acorn.

The colouring principles are various glycosides of quercetin and related flavonols. The bark contains proanthocyanidins also.

Quercitron is used for leather and clothes.

22. *Reseda luteola* Linn. (Resedaceae)

Weld, Dyer's Rocket

This is a biennial glabrous herb of 1-2m height, with sessile narrowly oblanceolate rosette leaves, narrowly oblong stem leaves, yellowish green flowers in terminal spike like racemes and 3-lobed ovoid capsules. A native of Western Europe now cultivated at many places.

The leaves and the whole plant are rich in flavones like luteolin (principal component), apigenin and their glycosides. Also present is a glucosinolate, glucobarbarin.

The plant yields an excellent deep yellow dye used for coloring silk and wool.

23. *Tamarindus indica* Linn. (Caesalpiniaceae)

Cinca, Tamarindus

A large tree with paripinnate leaves having 15-17 pairs of small subsessile oblong leaflets, creamy yellow (pink dotted) zygomorphic flowers in axillary racemes (stamens only 3, monadelphous) and an oblong incurved thick torulose pod containing black shining obovate seeds enveloped by a fleshy sweet acidic mesocarp. A native of tropical Africa now cultivated widely in tropics.

The brownish red pulp surrounding the seed contains soluble sugar (30%), pectin and tartaric acid (12-15%). More than half of the tartaric acid present is in the form of potassium hydrogen tartarate. Leaves contain tartaric and malic acids, flavones like apigenin, acacetin and luteolin, glycoflavones such as the vitexin, isovitexin, orientin and isoorientin and phenolic acids vanillic, *p*-coumaric, *p*-hydroxy benzoic and gallic acids. Also present are prodelphinidins (Daniel, unpublished). The bark contains an alkaloid hordenin. The red color of fruit pulp is due to chrysanthemin.

The leaves yield a reddish yellow dye used in coloring wool and silk. The flowers also are used as auxiliaries in dyeing.

24. *Trigonella foenum-graecum* Linn. (Fabaceae)

Methika, Fenugreek

Fenugreek is a herbaceous annual with pinnately trifoliolate leaves having long oblanceolate dentate leaflets, white/yellow flowers singly or in pairs from axils and long (up to 15cm) cylindrical pods containing 10-20 yellow, grooved trapezoid seeds. A native of Mediterranean, grown for seeds and as pot-herb.

Seeds contain flavonoids such as quercetin and luteolin (free and as glycosides), protein (26%), fat (6%), carbohydrates (44%), saponins based on diosgenin, tigogenin and gitogenin, trigonelline and minerals. A galactomannan is the main component of carbohydrates. Leaves yield proteins (41%), carbohydrates (6%) and free amino acids. Also present in leaves are saponins (similar to those of seeds), flavonoids such as kaempferol and 7,3'-diOMe quercetin (Daniel, 1989).

The seeds yield a very stable yellow dye suitable as a coloring material in foods and pharmaceuticals. Seeds are used as a galactogogue, carminative, tonic and as a poultice, are useful in ulcers and abscesses. They are used as emollient for inflammations in intestinal tract. The mucilage is used a tablet binder. The leaves are a pot-herb much in demand.

3.1.6. Quinones

a. Benzoquinones

1. *Embelia ribes* Burm.f. (Myrsinaceae)

Vavding, Embelia

E. ribes is a large woody scandent shrub with long internodes, bark studded with lenticels and coriaceous, elliptic-lanceolate leaves (mature leaves covered with minute reddish sunken glands), small greenish-yellow flowers in lax paniculate racemes and small globose

1-seeded fruit tipped with persistent style. Seen throughout India in forests.

The fruits contain a benzoquinone, embelin (2,5-diOH-3-undecyl-1,4- benzoquinone).

Embelia fruit extract is dark red in color and is used in coloring clothes. Embelin and *Embelia* are well-known anthelmintics esp. for tapeworm. The fruits are effective in cystic and abdominal tumours. With *Piper longum* and asafoetida, it forms a well-known contraceptive. *Embelia* is a male contraceptive plant in that embelin impaired spermatogenesis to the level of infertility (effect is reversible). Embelin exhibited significant anti-implantation and post-coital antifertility activity also in animals. With leaves of *Ficus religiosa* and borax in cow's milk, is a quick aborticide.

b. Naphthaquinones

1. *Alkanna benthamii* Johnston. (Boraginaceae)

Syn. *Macrotomia benthamii* A. DC.

Ratanjoth, Kashmiri Gozuban

This is an erect perennial, reaching a height of 1m with large radical narrow lanceolate and small cauline leaves, pink/purple flowers in dense terminal spikes and ovoid spikelets. Found in Himalayas.

The flowering shoots are extensively used in giving a pleasing red color to foodstuffs, oils, fats and medicinal preparations. It is also used for dyeing silk and wool.

2. *Alkanna euchroma* Johnston

Syn. *Macrotomia perennis* Boiss.

Ratanjot

This is an erect caespitose perennial with thick purple roots, many stems arising from the base, leaves hairy (lower lanceolate, upper sessile), flowers in subglobular terminal racemes (bracts leafy) and coarsely tuberculate nutlets. A native of Central and Western Asia.

The plants contain acetyl shikonin, 4 related compounds and alkannin-β, β-di methyl acrylate.

The roots are used as food colors and for dyeing silk. They also exhibit anti-pyretic and anti-inflammatory properties.

3. *Alkanna tinctoria* Tausch.

Alkanna

The roots of *A. tinctoria,* a native of Mediterranean, yield this red violet dye. This dark green prostrate herb has reddish purple roots, spathulate leaves, blue flowers and beaked nutlets.

The roots contain two closely related naphthaquinones alkannan and alkannin. Alkannan is 1, 2-naphthaquinone while alkannin (hydroxy alkannan-alkanna red) may occur esterified with angelic acid.

This dye is used for colouring foodstuffs, medicines and in histological works.

4. *Jatropha curcas* Linn. (Euphorbiaceae)

Ratanjot

The latex dries to a bright reddish brown brittle substance staining linen. Bark yields a dark blue dye used in Philippines for coloring clothes, fishing nets and lines. A yellow dye is extracted from leaves and tender stems also. This imparts different shades of tan and brown on cotton.

5. *Jatropha glandulifera* Roxb. (Euphorbiaceae)

Ratanjot

A small evergreen laticiferous tree with dichotomously branched stem, palmately 3-5 lobed long-petioled leaves having serrate margins and each serrature ending in a gland-tipped bristle (stipules very long, divided into many filiform segments, each with a glandular head), male flowers having a staminal column, female flowers without petals and a 3-lobed capsule containing 3 ellipsoid shining seeds. A native of Tropical Africa now grown as a weed.

The aerial parts contain naphthaquinones, 3, 3-dimethyl acrylylshikonin and isohexenylnaphthazarin.

The dry stem and latex produce a red dye

6. *Juglans regia* Linn. (Juglandaceae)

Akhrot, Walnut

This native of Iran is a monoecious deciduous tree having imparipinnate leaves, small yellowish flowers in catkins and an ellipsoid drupe with a coriaceous pericarp and a 2-valved endocarp enclosing a 4-lobed corrugated oily seed.

All parts of the tree contain a naphthaquinone, juglone. Root bark contains cyclotrijuglone, bisjuglone and α-sitosterol. Leaves contain hyperoside, and kaempferol and quercetin glycosides.

The green walnut hulls, shells, bark and leaves, containing juglone and related naphthaquinones, are used for dyeing and tanning. The bark is used for cleaning teeth and for chewing to redden the lips. Green walnut shells have been used (as an oily extract or alcoholic extract with alum) as a hair dye. Juglone imparts a brownish yellow on mordanted wool and a rose color on mordanted cotton.

7. *Lawsonia inermis* L. (Lythraceae)

Mehndi, Henna

Henna is an orange dye obtained from the leaves and young shoots a small tree, a native of Central and S. Asia. The leaves are small elliptic-lanceolate in shape and flowers are white fragrant borne in corymbose panicles.

The leaves contain 1.0%-1.4% (dry wt.) of a naphthoquinone lawsone, 5% tannin, flavones such as 4'OMe luteolin and 7-OMe apigenin and gallic acid. The flowers yield vol. oil consisting of α-and β-ionones.

For preparing the dye, the leaves are dried and ground to paste. Henna is a fast dye used for clothes, leather and dyeing the body parts. Henna (dry leaf powder) is used for dyeing hair, silk and wool. Henna leaves act as prophylactic against skin diseases.

The powdered leaves are beaten up with catechu paste yielding a copper red dye. This is used to give designs to body parts and coloring hair. Leather also is dyed with henna, but as the plant contains no tannin, the color is not permanent. Leaves yield a gold-green color. Roots also yield a red dye.

8. *Plumbago indica* Linn. (Plumbaginaceae)

Syn . *P . rosea* Linn.

P. coccinea Boiss.

Chitrak, Plumbago

P. indica is an erect branched shrub with long tuberous roots, ovate-oblong short-petioled leaves, rose red flowers in long terminal spikes (calyx is glandular hairy outside) and a membraneous capsular fruit enclosed in the persistent calyx. This is a native of Sikkim, cultivated throughout for medicinal purposes and as an ornamental.

Root bark contain a napthaquinone, plumbagin (0.91%), sitosterol, and tannins.

Chitrak roots are used for dyeing clothes. Roots are highly valued for their curative action on leucoderma and other skin diseases. They are digestive stimulants, diuretic, germicidal, vesicant and abortifacient. Chitrak is also used for haemorrhoidal inflammation of anus, diabetes, diarrhoea and elephantiasis.

Plumbago zeylanica Linn., a perennial herb with ovate leaves having an amplexicaul leaf base and white flowers, is used as a substitute of chitrak.

c. Anthraquinones

1. *Aloe barbadensis* Mill. (Liliaceae)

Syn. *A. vera* Linn

Kumari, Indian Aloe

A. barbadensis is a succulent perennial with a short stem, fleshy narrow-lanceolate leaves having a dentate margin and yellow flowers in racemes borne on scapes.

The leaves contain mucilage, anthraquinones and chromones. Mucilage on hydrolysis yields glucose, galactose, mannose and galacturonic acid. The quinones present are barbaloin and isobarbaloin (C-glycosides) and their aglycone, aloe-emodin. Aloesone (7-hydroxy chromone) and its C-glycoside- aloesin are the chromones located. Also present are glycoproteins, alocutin A and B

Aloe gives a yellow stable dye useful for dyeing clothes and pharmaceuticals.

2. *Andira araroba* Aguiar. (Fabaceae)

Chrysarobin, Goa powder

The wood of *A. araroba*, a native of Brazil, but extensively cultivated in Goa, provides chrysarobin. This plant is a large tree with glabrous branches, odd pinnate leaves having oval leaflets, white flowers in racemes and an oblong pod.

The yellowish brown powder, found in large lacunae of the stem wood, contains 50-75% chrysarobin, 7% bitter substances and about 2% resin. Chrysarobin is an anthrone-mixture consisting of 40% chrysophanol-anthrone, 20% monomethyl ether of emodin-anthrone and about 30% monomethyldehydroemodinanthrone.

The dye is yellowish red in color and is used to dye clothes. Even though chrysarobin possesses a laxative action, it is seldom used thus. It is extensively used in fighting skin diseases.

3. *Carthamus tinctorius* Linn. (Asteraceae)

Kusumbha, Safflower

This is a much branched slender annual growing up to 50 cm to 1m in height with lanceolate spinulose-serrate leaves, orange red flower heads having spiny involucrate bracts, and obovoid 4-angled achenes. A native of Europe now cultivated throughout.

The flowers yield two coloring matters, a water soluble yellow (carthamin) and an insoluble red (carthamone). The yellow color is worthless and the red color, though beautiful, fades quickly because it has no affinity to any known mordant. Also present in flowers are glycosides of luteolin and kaempferol, steroids and polyacetylenes. Seeds yield edible oil, safflower oil.

The flowers, dried in the shade, rubbed between the hands and kept in a basket filter and pure steam or acidulated water is poured over. This removes most of the yellow dye. When the water filtering down is clear, the florets are partially dried, pressed into cakes and dried. From these cakes the red dye is extracted using sodium carbonate solution. The dye can then be precipitated with dilute acids.

The dye is weak and gives red, pink, rose, crimson or scarlet shades – used for cotton and silk.

***Cassia* Linn. (Caesalpiniaceae)**

A number of species of *Cassia* give stable yellow or yellowish red colors. Many of them contain quinones or tannins or both and exhibit a wide variety of properties. The important species are the following

4. *Cassia angustifolia* Vahl (Caesalpiniaceae)

Hindisana, Senn

C. angustifolia Vahl, a native of Arabia and India, constitute senna It is a subglabrous erect shrub with stipulate pinnate leaves having 8

12 obovate oblong leaflets, pale yellow flowers in narrow peduncled cymes and a flat glabrous recurved pod containing 6-12 seeds.

The leaflets yield about 2% quinones consisting of aloe-emodin, rhein, sennosides A, B, C & D and many other heterodianthrones. Also present in the drug are flavonols like kaempferol, sterols, mucilage and resins.

Senna is a tanning and dyeing agent. It is widely used as a purgative also. It is preferred over rhubarb because it does not produce the astringent after effect of the latter.

5. ***Cassia fistula*** Linn

Amaltas, Indian Laburnum

C. fistula is a deciduous medium-sized tree with a grey smooth bark, large pinnately compound leaves bearing 4-8 pairs of stalked ovate leaflets having a silvery pubescence, bright yellow flowers in axillary lax pendulous racemes and pendulous cylindrical black pods (up to 60 cm length) containing hard brown biconcave seeds embedded in sweetish pulp.

The bark contains flavonoids such as 3,5,7,3′,4′- pentahydroxy,6,8-dimethoxy flavonol, 3,5,7,4′-tetrahydroxy,6,8,3′- trimethoxy flavonol,1,3,8- trihydroxy,3,7-dimethoxy xanthone, (+)-catechin, epicatechin, fistacacidin, fistucacidin, kaempferol, procyanidin, propelargonidin, rhein glycoside and steroids like lupeol and β-sitosterol .The leaves yield sennosides A and B, chrysophanol, physcion, rhein, kaempferol, quercetin, epiafzelchin and epicatechin and their polymers; gallic, ellagic and protocatechuic acids. The roots are found to contain 7-methyl physcion, oxyanthraquinones, betulinic acid, α-sitosterol fistucacidin, tannins, and phlobaphenes. Flowers contain kaempferol, propelargonidin tetramer, rhein, fistulin rhamnoside, stigmasterol, 28-isofucosterol and methyleugenol. The pulp from pods is rich in protein (up to 20%) and carbohydrates (26%), and contains sennosides A and B, rhein and its glucoside barbaloin, aloin, pectin, tannin and sugars. Pods yield procyanin B-2, epicatechin dimers, epiafzelchin dimers and epicatechin-epiafzelchin dimers. The seeds yield a gum and a fixed oil containing cyclopropenoid fatty acids such as vernolic, malvalic and sterculic acids. (Gupta *et al.*, 1989; Asseleiah *et al.*, 1990)

The bark is used in tanning heavy leather and for dyeing fish nets. The bark of twigs gives much darker color than the tree bark. The bark is used as a tonic and against dysentery and skin diseases. It is also useful in leprosy, jaundice, syphilis and heart diseases. The pods are well known laxatives. The shell showed anti-fertility activity in mice and is found to cause abortion and expulsion of placenta. The pulp from fruits is a safe purgative for children and pregnant women. It is also an analgesic, antipyretic (against malaria), anthrax, dysentery, leprosy and diabetes. Seeds and leaves are used in preparation of a drink, which is carminative, laxative and antipyretic. The roots are useful in rheumatism, hemorrhages, ulcers, skin diseases and leprosy. The flowers are said to be purgative, febrifugal and anti-bilious.

6. *Cassia occidentalis* Linn.

Kasamarda, Coffee senna

This is an erect undershrub reaching a height of 1.5 m with reddish stems, pinnate leaves having 3 pairs of glaucous membraneous lanceolate leaflets, yellow flowers in short racemes and recurved, compressed glabrous pods containing olive green, ovoid, compressed seeds. A common weed throughout India.

The leaves are found to contain anthraquinones such as chrysophanol, emodin (and their glycosides), physcion, metteucinol-7-rhamnoside, jaceidin-7-rhamnoside and 4,4′,5,5′-tetrahydroxy-2,2′-dimethyl-1,1′-bianthraquinone. Also present are tetrahydroanthracene derivatives, germichrysone and occidentalins A & B. The pods yield glycosides of 7-OMe quercetin and 3,6,3′-trimethoxy quercetin. The seeds contain a gum (galactomannan), a fixed oil and anthraquinones such as 1,8-dihydroxy-2-methylanthraquinone, 1,4,5-trihydroxy-7-methoxy-3-methyl anthra-quinone, physcion, rhein, aloe-emodin, chrysophanol and steroidal glycosides based on campesterol and α-sitosterol. The roots yield bitetrahydroanthracene derivatives, occidentol-I and II, pinselin (cassiolin), rhein, aloe-emodin and their glycosides, chrysophanol, physion, emodin, islandicin, helminthosporin, xanthorin, 1,8-dihydroxyanthraquinone, α-hydroxy anthraquinone, 1,8-dihydroxy-3-methoxy xanthone, quercetin, sitosterol, campesterol and stigmasterol. Also present in the roots are pinselin, questin, germichrysone, methyl germitorosone and singuenol.

All parts of this plant provide similar colors and are used as tanning and dyeing agents.

7. *Cassia senna* Linn

Sonamukhi, Alexadrian/Tinnevelly Senna

An erect branching shrub of 1.5 m ht. with pinnate leaves having lanceolate elliptic leaflets in 3-9 pairs, brilliant yellow flowers in terminal racemes and dark brown flattened pubescent pods containing 5-7 obovate dark brown seeds.

The leaves and the shells of fruits contain sennosides and related anthraquinones. Shells contain 3-5% of these compounds whereas the leaves have 2.5 - 4%. The various sennosides are sennosides A,B,C and G as well as three sennidins. Also present are free anthraquinones such as rhein, chrysophanol, emodin and aloe-emodin. Leaves contain monoglucosides of rhein and aloe-emodin whereas the pods contain monoglucosides of chrysophanol, rhein, aloe-emodin and 6-hydroxymusizin. The flavonoids present are isorhamnetin and kaempferol as glycosides. Leaves contain 7% mucilage also. Roots are found to contain physcion, physcionin, chrysophanein and sennidin C.

Senna can be used as a dyeing agent giving yellow or yellowish brown color. The leaves and shells are used as tonics in Ayurvedic and Unani Systems of medicine. It is an efficient purgative but due to the tendency to cause gripe, usually administered along with carminatives; other uses of senna include as a febrifuge, in spleenic enlargements, anemia, jaundice, rheumatism, tumours, and bronchitis and against skin diseases.

8. *Cassia tora* Linn.

Chakramarda, Sickle senna

A foetid annual shrub with pinnate leaves bearing 3 pairs of ovate oblong leaflets (two pairs of leaflets have glands at their bases, all show sleeping movements) bright yellow flowers in pairs and stout long pods containing 25-30 green seeds. A very common weed throughout India.

The stem bark contains emodin, rhein, euphol, β-sitosterol and basseol. The leaves also contain emodin, kaempferol, *d*-mannitol stachydrine and choline. The immature seeds yield brassinosteroids, brassinolide, castasterone, typhasterol, teasterone and 28-norcastasterone and the monoglycerides, monopalmitin and mono-olein. Seed is rich in protein (24%), carbohydrates (49%), oil (5.4%) and contain glycosides of 1,8-dihydroxy-3-methyl anthraquinone, 8-OH,3-Me,1-OMe anthraquinone, two glycosides of 9,10-diOH,7-OMe,3-Me-1H naphthopyran-1-one, cassiamide, rhein, emodin, aloe emodin, physcion, rubrofusarin, nor-rubrofusarin, 8-hydroxy-3-methyl anthraquinone, chrysophanic acid, obtusin, toralactone, trigonelline, stachydrine and choline. Seed oil contains linoleic (45%), oleic (22%) and palmitic (22%) acids, 24-ethyl cholest-5-en-3β-ol, 24-ethyl cholest-5,22-dien-3β-ol and 24-ethyl cholest-7-en-3β-ol. The carbohydrates include a gum (galactomannan - 7.6 %) also. Roots yield choline, an anthraquinone, propelargonidin and β-sitosterol.

Chakramarda can be used as a dyeing agent giving yellow or yellowish brown color. The whole plant is purgative, anthelmintic and exceeding useful in eczema, ringworm and other skin diseases. Mature leaves, sometimes adulterated with senna, are anthelmintic and antiperiodic. Seeds are substituted for coffee beans, and are well-known tonics and stomachics. It is used as a blood purifier also.

Senna pods. The fruits of *C. angustifolia* and *C. acutifolia* Delile also provide stable colors. The active principles are very similar to those of the leaf.

9. *Hedyotis umbellata* Lam. (Rubiaceae)

Syn. *Oldenlandia umbellata* Linn.

This is a diffuse woody annual/biennial of 30 cm, with angular rough branches, small sessile linear leaves fascicled at nodes, small white flowers in 3-7 flowered umbellate cymes and small globose capsules. A native of S. E. Asia.

The roots are similar to madder in containing alizarin, rubichloric acid and ruberythric acid. Also present are alizarin α-methyl ether, acetyl alizarin, anthragallol dimethyl ethers, hystazarin and *m*-hydroxy anthraquinone.

The root bark of two year old plants is the source of chayroot dye used in imparting red color to wool, silk and calico fabrics.

10. *Heliotropium indicum* Linn. (Boraginaceae)

Indian Heliotrope

Seeds contain a pyrrolizidine alkaloid heliotrine and indicine-N-oxide (an antitumour principle).

The leaves yield an impermanent black dye.

***Morinda* Linn.**

Al, Surangi

The roots of many species of *Morinda* such as *M. citrifolia, M. coreia* (main sources), M. *bracteata, M. tomentosa, M. umbellata* and M.

angustifolia are used as sources of dyes. The roots of *M. angustifolia* Roxb. yield a yellow dye, while that of *M. citrifolia* gives a red dye.

11. *Morinda citrifolia* Linn.

This is a small tree with a straight trunk having broadly elliptic glabrous leaves, white flowers in dense ovoid heads and ovoid glossy drupes containing winged compressed pyrenes. A native of Tropics.

The roots of 3-4 year old trees yield maximum dye. The dye consists of morindin, a glycoside of morindone (trihydroxy methyl anthraquninone, the principal component), rubichloric acid, alizarin, rubiadin, morindadiol and soranjidiol.

The root bark as well as stem bark is used with mordants to give a red color to cotton, silk and wool. The various colors, red, purple and chocolate, are fast to soap. In acidic medium this dye turns yellow.

From the roots of *Morinda angustifolia a* yellow dye is extracted which is used for dyeing cotton yarn and cloth. With mordants the dye gives shades of red.

12. *Morinda coreia* Buch-Ham. (Syn. *M. tinctoria*).

The root bark of this tree contains morindin. The wood contains morindone, damnacanthal and *nor*-damnacanthal.

13. *Morinda umbellata* Linn.

Root bark contains morindone, morindin, rubichloric acid (principal components), morindanigrin and other anthraquinones.

For extracting the dye of *Morinda persicaefolia,* small pieces of roots are boiled in water till the solution turns yellow. The yarn is dyed by immersing it in this solution and stirring rapidly. To produce red colored dye, the yarn is first soaked in a mixture of mustard oil and water. The yarn is dried and soaked for three days in water containing the powder.

14. *Onosma echioides* Linn. (Boraginaceae)

Gozuban

The root yields a red dye which is extracted in alkaline medium. With different mordants the dye gives shades of grey. Roots are used for coloring wool, hair, and wood. It is also used to color medicines.

The roots of *O. emodi* Wall (*Maharanga emodi* Wall) also are used for coloring wool and silk. This plant is more common in Eastern Himalayas. The dye obtained from the roots of *O. hookeri* C. B. Clarke (a native of Sikkim and Bhutan) is used for coloring medicinal preparations - and is an expectorant and purgative.

15. *Onosma hispidum* Wall (Boraginaceae)

Syn. *O. echioides* C. B. Clarke

Ratanjot

This is an erect perennial herb found in Western Himalayas having a tapering purplish red root. Leaves are of two types. The radical leaves are narrowly oblanceolate and cauline leaves are linear-oblong. Leaves are without visible veins and are densely covered with coarse hairs having discoid bases. Flowers are fragrant, white/pale yellow borne on terminal cymes. Nutlets are smooth and lustrous.

Roots contain a red coloring principle containing naphthaquinones alkannan and alkannin, similar to *Alkanna tinctoria*.

The roots are used for dyeing wool and medicinally as an anti inflammatory.

16. *Pterocarpus santalinus* Linn. (Fabaceae)

Raktachandan, Red sandalwood

A tall tree with 3-5 foliolate imparipinnate leaves (leaflets large, ovate -orbicular, apex emarginated and alternate on rachis), small papilionaceous flowers in axillary racemes and a broad orbicular winged pod containing a single subreniform seed. Common in Western Ghats.

The wood is extremely hard and dark purple. Wood contains a quinone, santalin (santalic acid, 16%), desoxy santalin (a napthaquinone), santal (isoflavone), two aurone glycosides 6–hydroxy, 5–methyl, 3′,4′,5′- trimethoxy aurone 4-O-α-L-rhamnopyranoside and 6,4′-dihydroxy aurone-4-O-rutinoside (Kesari *et al.* 2004) and colorless pterostilbenes, pterocaspin and homopterocarpin.

The dye is extracted in water containing alum and safflower oil. It can also be extracted in acetone or alcohol. This is used for coloring sauces and pharmaceutical preparations. The wood is used for dyeing wool, cotton and leather. Raktachandan is a reputed remedy to remove pimples and scars and for treating other skin diseases.

Rhamnus spp. (Rhamnaceae)

Various spp. of *Rhamnus* are used as sources of dyes in different areas. The bark of *R. utilis* Deone (Lokao/Chinese green) native to China, the fruits of *R. infectoria* L. (Persian berries) and *R. cathartica* (sap green) both natives of Europe, yield green or yellow dyes useful in dyeing clothes.

All these dyes are rich in anthraquinones like chrysophanic acid, emodin and their various glycosides.

The presence of anthraquinones makes these dyes active cathartics, used in medicine.

17. *Rhamnus alnus* Mill (Rhamnaceae)

Alder Buckthorn bark, Frangula bark

The bark of Alder buckthorn, a native Europe & W. Asia, provides the drug. The plant is a deciduous tree attaining a height of 5 m with broadly ovate shining leaves, greenish yellow flowers in sessile glabrous umbels and reddish black drupes.

Glycosides form the major group of the 2-5% anthraquinones present in the purple bark. The principal compound in the dye is a pair of isomers frangulosides A & B, together known as frangulin or rhamnoside franguloside. The minor compounds are frangulin B, emodin dianthrone and its glycoside, and palmidin C and its rhamnoside. The seeds contain aloe-emodin.

Alder buckthorn bark is used mainly as a cathartic, but this can be an excellent dye.

18. *Rhamnus purshiana* DC. (Rhamnaceae)

Cascara bark

Cascara bark is derived from a deciduous tree native of U.S.A. and Canada. The leaves of this plant are elliptic or oblong tufted at the tips of branches and flowers are borne in pubescent umbels. The fruits are black drupes.

For better preparations of the dye, the bark is to be stored for at least one year, during which period a slow hydrolysis of the constituents is believed to take place. The drug contains about 6-9% anthraquinones mostly as glycosides. The C-glycosides amounts 70-80% of the total glycosides and O-glycosides 10-20%. The different types of principles present are: (1) *cascarosides* – The primary glycosides: cascarosides A and B are the most important compounds in this dye. They are the O-glycosides of barbaloin – a C-glycoside. Cascarosides C & D are also active compounds and are O-glycosides of chrysaloin. (2) Barbaloin and chrysaloin in the free form. (3) Emodin, emodin oxanthrone, aloe emodin, chrysophanol etc. free or as O-glycosides and (4) Homodianthrones of emodin, aloe-emodin and chrysophanol and heterodianthrones like palmidin A, B & C.

This plant also is used mainly as a cathartic, but this can be an excellent dye. Aloe-emodin isolated from the seeds of *R. frangula* Linn. exhibited anticancerous property in that, it showed significant inhibitory activity when tested in mice against the P-388 lymphocytic leukemia.

19. *Rheum emodi* D. Don (Polygonaceae)

Gandhit, Indian Rhubarb

This is a robust perennial herb of 1-m with long thick roots, orbicular large leaves, small dark purple flowers and angled purple fruits.

It contains up to 2.2% anthraquinones. Emodin and chrysophanol alongwtih their derivatives form the major constituents. Also present are rhein, aloe-emodin, physcion and their glycosides as also hydrolysable and condensed tannins.

R. emodi Wall., *R. webbianum* Royle, or other species of *Rheum* native to India are known in commerce as Indian Rhubarb. They contain also emodin and chrysophanol alongwith their derivatives as the major constituents. The roots are used, along with madder, and potash for dyeing fabrics red. *R. moorcroftianum* Royle is a herb at high attitudes of Himalayas, the roots of which are used for dyeing woollens yellow.

20. *Rheum palmatum* Linn. (Polygonaceae)

Rhubarb

Rhubarb is the dried rhizome, deprived of bark, of *R. palmatum* and many other species of *Rheum*. Rhubarb is a stout perennial herb with clumps of large broadly cordate and deeply palmately lobed radical leaves and green, short pedicelled flowers in leafy panicles borne on flowering shoots. These plants, natives of Hong Kong, are widely cultivated in China.

The drug yields a variety of anthraquinones, which occur as glycosides or aglycones. They are: (1) dihydroxy phenols such as chrysophenol; trihydroxy phenols like emodin, aloe emodin and methoxylated phenols like physcion and their glycosides. (2) carboxylated anthraquinones, e.g., rhein (cassic acid) and glucorhein, (3) anthrones and dianthrones based on chrysophanic acid, rhein or physcion. Sennidin is a dianthrone glucoside. (4) heterodianthrones derived from 2 different anthrones like palmidin A (aloe-emodin anthrone and emodin anthrone), palmidin B (aloe-emodin anthrone and chrysophanol anthrone) etc. Rhubarb contains a large amount of both hydrolysable and condensed tannins, starch and calcium oxalate also.

The presence of anthraquinones and tannins make rhubarb potential dyes. But the plant is mainly used as a drug in the treatment of diarrhoea because of the purgative action of anthrones and the astringency of tannins and as a as a bitter stomachic.

21. *Rhizophora mucronata* Linn. (Rhizophoraceae)

Syn. *R. mangle* Roxb.

Mangrove

R.. mucronata is an evergreen tree of 12-15m, with reddish black bark, broad elliptic or oblong coriaceous leaves, white/yellow flowers in cymes and conical-ovoid fruits containing germinated (hypocotyle protruding) seeds. A native of tropics.

Bark containing tannins (25-35%) is used as a dyeing and tanning agent.

22. *Rhus chinensis* Mill. (Anacardiaceae)

Syn. *R. javanica* Linn.

R. semialata Murr.

Tatri, Nutgall Tree

This is a small deciduous tree with pinnately compound leaves having 7-13 brownish-pubescent toothed ovate leaflets and winged rachis and petioles, creamy white flowers in large terminal panicles and a large glandular hairy 1-seeded drupe. A native of China and temperate Asia.

Galls in leaves of this plant, caused by species of *Aphis, Melaphis* and *Eriphyes,* (Chinese tannin) contain 50-80% tannins. They are used for dyeing and tanning.

23. *Rubia cordifolia* Linn. (Rubiaceae)

Mangishta/Indian madder

R. cordifolia is a scabrid hispid perennial climbing herb with 4-angled rough stem, very long cylindric red roots, numerous scandent divaricate or deflexed branchlets, ovate acute scabrid leaves in whorls of four (one pair of each whorl larger, as also the lower leaves are large, margins with minute white prickles, petioles also with sharp recurved prickles), greenish small flowers in axillary or terminal cymes and small globose purple fruits containing two seeds. Common in hills throughout India, Africa, Malaysia and Japan.

The roots contain a number of anthraquinones such as purpurin (trihydroxy anthraquinone), munjishtin (xanthopurpurin-2-carboxylic acid), xanthopurpurin, pseudopurpurin, 3-carboxymethyl-1-hydroxy anthraquinone, 1,4-dihydroxy-2-methyl anthraquinone, 1-hydroxy-2-methyl anthraquinone, tectoquinone, 1-hydroxy-2- hydroxymethyl anthraquinone & 1,3-dihydroxy anthraquinone; three naphthaquinones such as dehydro-α-lepachone, 2-carbonyl-3-methoxy-1,4-naphthoquinone and 2-carbomyl-3-hydroxy-1,4-naphthoquinone; an iridoid glycoside (6-methoxy geniposide), two pentacyclic triterpenes (rubicoumaric acid and rubifolic acid) and a number of antitumour bicyclic hexapeptides RAs (RA-1—RAXIV, RAI-RA3, RA-XV and RA-XVI) (Koyama et al., 1992, Takeya et al., 1993).

Purpurin is the main coloring principle (purple) while munjishtin is an orange dye. With alum, purpurin dyes cotton fabrics a bright scarlet red. The dye is extracted by boiling the stems/roots in water or by steeping in water for a few hours. Alum or myrobalan is used as a mordant. It imparts scarlet, coffee-brown or mauve to cotton. This is used to color medicines and oils.

Mangishta is one of the acclaimed blood purifiers and used as an alterative, tonic and for diseases of blood, skin and urino-genital system. It improves complexion, improves the microcirculation within the body and cures jaundice, paralysis, freckles etc.

24. ***Rubia sikkimensis*** **Kurz.**

Naga-Madder

This species is differentiated in having all the four leaves in a whorl of equal size, yellow flowers and absence of prickles. Common in Eastern Himalayan ranges and contains the same constituents as madder.

This is the main dye yielding species in Eastern India. The roots are fluted in nature and used for dyeing wool, hair, cane/bamboo etc.

25. ***Rubia tinctoria*** **Linn. (Rubiaceae)**

Madder

The roots of madder, a native of Mediterranean, yield a scarlet dye-turkey red. This herb is a climbing perennial with long fleshy roots, simple lanceolate leaves in whorls of 4, greenish yellow flowers in terminal panicled racemes and a small fleshy one-seeded black fruit.

Roots contain glycosides of alizarin and purpurin such as ruberythric acid (alizarin-2-ß-primeveroside), galiosin (pseudopurpurin primeveroside) and rabiadin (2-methyl purpuroxanthin-3-glucoside) and free pseudopurpurin, purpuroxanthin and munjishtin. Also present are asperuloside, sugars and organic acids.

Madder yields the famous Turkey Red color with aluminum and calcium as mordants.

Used for giving a red color to clothes, rugs, food products etc. This dye, now prepared synthetically, is used for clothes.

26. ***Rumex acetosa*** **Linn. (Polygonaceae)**

Garden sorrel

This is an erect perennial with fibrous roots, deeply grooved stem, ovate oblong radical leaves having a hastate base, unisexual (dioecious) flowers in long terminal panicles and a trigonous shining black nut. Common in Western Himalayas.

Leaves contain flavonol glycosides rutin, hyperin, a glycoflavone (vitexin) and oxymethyl anthraquinone. Roots contain chrysophanic acid and 1,8-dihydroxy naphthalene. Fruits also contain hyperin and quercetin.

Fruits and roots are sources of dyes.

27. *Rumex dentatus* Linn.

An erect deep-rooted annual of 1m with red colored roots, oblong leaves and green bisexual flowers in verticillate clusters. Nut acutely trigonous, brown. A native of S.E.Asia.

Roots containing chrysophanic acid and emodin are used as dyes.

28. *Rumex nepalensis* Spreng.

Roots contain chrysophanic acid and nepodin (a derivative of protocatechuic acid) and are used as yellow dyes.

29. *Tectona grandis* Linn.f. (Verbenaceae)

Sag, Teak

A tall tree with quadrangular branchlets, large broadly elliptical leaves (coriaceous stellately tomentose below with minute red glandular dots, young leaves red in color), small white flowers in terminal panicles and a hard bony globose fruit, enveloped by a bladder like persistent calyx, containing 1-3 seeds. A native of India.

The wood contains a large number of naphtha- and anthraquinones, triterpenoids, hemiterpenoids and tectol. The anthraquinones reported are tectoquinone (2-methyl anthraquinone), 2-hydroxymethyl anthraquinone, 3-methyl anthraquinone, 3-hydroxy, 2-methyl anthraquinone, 4, 5-dihydroxy methyl anthraquinone, anthraquinone-2-carboxylic acid etc. Leaves contain tannin and anthraquinones such as tectoleafquinone (1,4,5,8-tetra hydroxy-2-iso-pentadienyl anthraquinone).

The leaves yield a red dye extracted by heating chopped leaves in alkaline water (with 0.25% sodium carbonate). The best yield of dye is obtained during May-August. The dye is used as such for wool and silk but for cotton a mordant is required. The color varies from lilac to grey depending on mordants. Without mordants, on wool and silk gives a Caledonian-brown color and with 5% potassium dichromate or stannous chloride, a flesh color is developed. Root bark is used to color mats yellow.

30. *Ventilago maderaspatana* Gaertn (Rhamnaceae)

V. maderaspatana is a large much branched woody climber having pubescent young branches, oblong-lanceolate puberulous (beneath) leaves, small yellow fragrant flowers in drooping terminal panicles (sepals pubescent, petals, clawed cucullate, enveloping the stamens and a 5-angled pubescent disc) and a dry indehiscent fruit, the apex of which is prolonged to a coriaceous wing.

Root and bark are extracted in alkaline water. The dye gives shades of maroon and brown with different mordants. This is also used to improve / modify the colors produced by Indian madder and *Ceriops.*

3.1.7. Tannins

1. ***Acacia catechu*** **Willd (Mimosaceae)**

Khair, Cutch Tree, Catechu

A. catechu is a small tree of 3m ht, with exfoliating dark grey bark, deciduous bipinnate leaves having a pair of recurved prickles as stipules, yellow flowers in long cylindrical spikes and glabrous pods. This is a common tree inhabiting the drier regions of India.

Catechu is a resin extracted by boiling the red heartwood in water. Katha (pale catechu) is prepared by keeping the concentrated extract for a few days when the catechin settles (crystallizes) at the bottom. These crystals collected are moulded to various shapes and is the katha. The mother liquor on concentration yields Cutch (dark catechu). Katha is almost pure catechin whereas cutch contains up to 60% tannins and 12-18% catchin. Kheersal is the crystalline or powdery deposits of catechin found naturally in old trees. Khair gum is naturally exuding and occurs as pale yellow tears. It is sweet in taste, dissolves in water to form strong mucilage and is a polysaccharide consisting of galactose, arabinose, rhamnose & aldobiouronic acid. Leaves contain up to 12% crude protein. Seeds possess protein (44%) and other extractives (40%).

This dye gives brown, yellow, grey and black color. The brown color is obtained by using catechu with iron sulphate and myrobalan. Black color is produced by mixing cutch with castor oil and turmeric.

Katha is known for its antileprotic activity and is used as an astringent and in cases of cough and diarrhoea. It is extensively used in pan preparations. Katha is a highly efficient anti-oxidant. Kheersal is used for treatment of cough. Gum is a good substitute of gum Arabic.

2. ***Acacia leucophloea*** **Willd. (Mimosaceae)**

Armida/Babul/Fragrant Acacia

This is a deciduous thorny tree with a crooked and gnarled trunk, black bark, reddish brown wood, bipinnate leaves, bright yellow fragrant flowers in globose heads and strap-shaped flat coriaceous dehiscent fruits containing 15-20 ovoid compressed seeds. A native of tropical America naturalized in India.

The bark and pods yield a black dye.

3. *Adenanthera pavonina* Linn. (Mimosaceae)

Kundhandana, Coral wood

This is a tall tree with bipinnate leaves having 4-8 elliptic-oblong alternate leaflets, yellow fragrant flowers in racemes and long pods enclosing scarlet biconvex seeds. Common in India.

Seeds contain a fat (25-30%), protein (30%), stigmasterol, dulcitol and non-protein amino acids such a γ-methylene glutamic acid, γ-methylene glutamine and γ-ethylidine glutamic acid. The wood contains 2,4-dihydroxy benzoic acid, robinetin, chalcone butein and dihydroxyricetin. Leaves yield octacosanol, dulcitol, β-sitosterol and stigmasterol. Bark contains saponins based on oleanolic and echinocystic acids.

The wood yields a dye, which is used as a substitute for red sandalwood. Seeds are used for treatment of cholera and paralysis. Seed extract is useful in pulmonary affections and chronic opthalmia. A decoction of leaves/bark is used against rheumatism and gout. Root extract is hypotensive.

4. *Albizzia lebbeck* Willd. (Mimosaceae)

Siris tree

This is a large erect deciduous tree with dark brown bark, bipinnate leaves having 8-18 leaflets, greenish yellow flowers in globose heads and yellowish brown strap- shaped fruits containing 6-10 seeds. Planted in many parts of India as a social forestry tree yielding timber.

The heartwood contains polyphenols such as melanoxetin, okanin, propelargonidin, lebbecacidin, melacacidin, pinitol and a saponin, lebbekanin E, based on a sapogenin acacic acid. Bark yields condensed tannins (7-11%) consisting of D-catechin, procyanidin, melacacidin and lebbecacidin, a saponin lebbekanin D (sapogenin acacic acid) and steroids friedlin and β-sitosterol. Bark yields a reddish brown gum also. Leaves contain alkaloids, kaempferol, and quercetin and caffeic acid. Flowers yield a volatile oil consisting of *p*-nitro benzoate, benzyl alcohol and benzoic acid and the saponin, lebbekanin D. Also present in the flowers are lupeol, α-and β-amyrins and a coloring pigment similar to crocetin. Seeds yield protein (40%), and saponins lebbekanin A &B based on sapogenins echinocystic and oleanolic acids.

The wood extract is used in tanning and dyeing. *A. lebbeck* is said to have anti-dysenteric and anti-tubercular properties. The bark is used for bronchitis, leprosy, paralysis and worms and as on expectorant.

5. *Albizzia procera* Benth.

Safed siris

A. procera also is a large tree with pinnate leaves having 4-12 leaflets and a few flowered small heads in terminal panicles. Pods are thin containing 8-10 seeds.

Roots are found to contain a saponin, based on oleanolic acid methyl ester, and α-spinasterol .The root bark yields gum, tannins (12-17%) and β-sitosterol. α-Spinasterol and oleanolic acid are the compounds present in roots while the leaves contain the former compound only. Stem bark contained β-sitosterol and isoflavones such as biochanin A, formonetin, genistein and diadzein. Stem wood also contains all these isoflavones besides a pterocarpan. Seeds are found to possess two saponins, proceranin (based on the aglycone machaerinic acid) and proceranin A (based on proceragenin A and B). An oleanolic acid saponin also is reported from seeds.

The stem and root bark are used in tanning and dyeing. All the parts of the plant exhibit anti-cancer properties. The plant is used for intestinal diseases.

6. *Albizzia odoratissima* Benth.

Kala siris

This tree also yields a gum in the form of rounded tears. The stem and leaves contain tannin and seeds, a saponin odoratissimin based on echinocystic acid and machaerinic acid. The bark is considered tonic and effective in leprosy and ulcers. Leaves are used as expectorants and the bark yields a brown dye.

7. *Areca catechu* Linn. (Arecaceae)

Supari, Areca nut

Areca nut is the ripe dried seed of a tall palm, a native of Malaya. It is a slender, monoecious, unbranched palm attaining a height of 25 m or more with a ringed stem and a crown shaft. The petiole of the long pinnate leaves are broad, and stem clasping, and the numerous pinnae are more than 1 m long and long acuminate with upper ones often confluent. The much-branched inflorescences are intrafoliar and bear many staminate flowers at the top and a few pistillate flowers at the base. The drupe is one seeded, orange scarlet in colour with a fibrous pericarp.

The seeds contain tannins which vary from 14-26% and consist of gallotannic acid and catechins. The other components of the seed are 0.29-0.7% pyridine alkaloids in which arecoline-the principal alkaloid-amounts to 0.1-0.5%. The accompanying alkaloids are arecaine (N-methyl guvacine), arecolidine (methyl ester of arecaine), guvacine (demethyl arecaidine), guvacoline and isoguvacine. Also present in the seed are 13-17% fats similar to coconut oil in composition (consisting of 19.5% lauric acid, 46.2% myristic acid, 12.7% palmitic acid and small amounts of oleic, linoleic and hexadecanoic acid), 47-80% carbohydrates including saccharose, reducing sugars, galactans and mannans; 5-10% proteins, up to 5% carotenes and traces of sitosterol. The husk of the fruit contains 47-50% cellulose.

The extract, prepared by boiling seeds in water, provides a rich brown dye range.

In earlier days betel (areca) nuts were often chewed with betel pepper (*Piper betle*) leaves and lime to redden lips of the ladies. This color is due to the anthocyanidins which are released because of the alkaline hydrolysis of the condensed tannins of areca nut. The nuts are also used as anthelmintics especially against tapeworm.

8. *Azadirachta indica* A. Juss. (Meliaceae)

Nimba, Neem

Neem is a large evergreen tree with imparipinnate leaves having ovate lanceolate, serrate unequal-sided leaflets, small white scented flowers in axillary panicles and green drupes containing a large seed. Found all over India cultivated or naturally grown. This plant is a very well-known biopesticide and the chemistry is given there.

Bark extract yields condensed tannins including procyanidins and other proanthocyanins. The dye produced gives a range of beige to brown tones.

9. *Bauhinia purpurea* Linn. (Caesalpiniaceae)

Kanchanar, Butterfly Tree

A medium-sized evergreen tree with bifid coriaceous, ovate leaves shallowly cordate at the base, pink/white flowers in terminal panicles and long flat distinctly reticulate beaked pods containing 12-15 seeds. Cultivated throughout India for the flowers and beautiful foliage.

Leaves contain quercetin, quercitrin, rutin and apigenin. Flowers are found to contain astragalin, isoquercitrin, quercetin and glycosides of pelargonidin. Seeds yield 16% fat, 27% protein, 15% carbohydrates and 15% oil, with linoleic (49%), palmitic (18.5%), stearic (18%) and oleic (11%) acids, besides the chalcones butein and 3,4 dihydroxy chalcone and DOPA (2%). A phytohemagglutinin, trypsin inhibitor and an alkaloid also are reported.

Seeds and the bark yield a reddish brown dye. Stem bark is used extensively in glandular diseases and as an antidote to poison. It is also anthelmintic and cures ulcers, swellings, leprosy, menstrual disorders and prolapse of rectum. Leaves are hypoglycemic. Flower buds, which are used as a vegetable and also pickled, are considered laxative and anthelmintic.

10. *Bauhinia racemosa* Lam.

Ashta

This yellow flowered bauhinia is a small bushy tree with drooping branches. Leaves bifid, broader than long and pods are falcate and glabrous.

Bark contains a tetracyclic 2,2-dimethyl chroman, de-O-methyl racemosal, β-amyrin, β-sitosterol and octacosane. Seeds contain protein (10%), mucilage (4%) and oil (Prabhakar, *et al.*, 1994).

The bark yields a reddish brown dye. It is also highly astringent and is antidysenteric, anti-inflammatory and cholagogue. Leaves are anthelminitic and used for diarrhoea.

11. *Bauhinia tomentosa* Linn

Kachnar, Buddhist Bauhinia

B. tomentosa is a medium sized deciduous tree with grey bark, bifid broad leaves, variously colored flowers in short peduncled corymbs and 10-15 seeded long hard flat pods. A common garden tree found all over India.

The flowers contain flavonoids such as isoquercitrin (6%), rutin (4.6%) and traces of quercetin. Seeds yield proteins (19%), mucilage (7%) and saponins.

Leaves are used as dyes. The fruit is used as a diuretic and seeds, a tonic. Root bark is found useful in liver troubles and as vermifuge.

12. ***Bauhinia variegata*** **Linn.**

Kovidara, Mountain Ebony

This also is a medium sized deciduous tree, with sub-coriaceous bifid broad leaves, variously colored flowers borne in short peduncled corymbs and long hard glabrous flat 1pods containing 10-15 seeds. Cultivated in gardens for the showy flowers.

Stem contains β-sitosterol, lupeol and glycosides of kaempferol and 5,7- dimethoxy flavonone. Stem bark yields tannins (9.6%). Violet flowers are found to contain cyanidin, malvidin and peonidin while white flowers yield kaempferol. Seeds yield protein (44%), pentosan (11%), a fatty oil containing oleic (32%), linoleic (36%), palmitic (17%) and stearic (13%) acids. Mucilage and phytohemagglutinins also are reported from seeds.

The bark is used for extracting a yellow dye which is used as a dye auxiliary and for tanning. The pods also yield a beige to brown dye.

Root is used in dyspepsia, flatulence and to prevent obesity. Bark is antidysentric, tonic and anthelmintic. Flowers and flower buds are used as vegetables and in the treatment of diarrhoea, dysentery, piles, tumours, heamaturia, and menorrhagia and as a laxative.

13. ***Bischofia javanica*** **Blume (Euphorbiaceae)**

Bhillar, Bishop wood

This is an evergreen tree of 20m with dark brown bark exfoliating in angular scales, trifoliolate leaves having ovate-oblong crenate, bronze green leaflets, greenish yellow unisexual flowers in slender racemes and reddish brown to blue-black berries containing 3-4 dark brown shining seeds embedded in an edible pulp. Common in forests of India.

The heartwood is red or dark reddish brown. The bark contains tannin and steroids such as epi-friedelanol, friedlin, betulinic acid and β-sitosterol. Leaves also contain ellagitannins and steroids similar to those of stem.

The bark is a source of a red dye used to dye rattan baskets.

14. ***Ceriops decandra*** **Ding Hou. (Rhizophoraceae)**

Syn. *C. roxburghiana* Arn.

Goran, Ceriops

This is a small mangrove tree of 6m, with obovate or elliptic oblong leaves, greenish white flowers 3-6 together in axillary condensed capitate inflorescences and ovoid conical fruits. Common in the mangroves of India, forming more than half of the total mangroves.

The bark contains 40% tannins while the tannin yield of wood is 10%. Leaves contain tannins and steroids (α - amyrin, lupeol, oleanolic acid etc.).

The bark and leaves are important sources of tannin. The bark extract is used for dyeing jute and by the addition of mordants various colors can be obtained.

The bark, rich in tannin, is boiled with a small quantity of soda till the solution is pink. The yarn is boiled in this solution till it is dyed.

15. ***Cotinus coggygria*** **Scop. (Anacardiaceae)**

Syn. *Rhus cotinus* Linn.

Tunga

The leaves of this plant, a native of Mediterranean and China, yield a vol. oil containing α-pinene (40%), limonene (20%), β-pinene (12%) and β-phellandrene (4%).The twigs yield a yellowish brown dye. Fisetin is the principal coloring matter, occurring as a salt of tannic acid. Also present are methyl gallate and tannins like leucofisetinidin (a flavan-3, 4-diol). It is used for dyeing wool and gives an orange or scarlet color to leather. Also used for coloring foodstuffs.

16. ***Cocos nucifera*** **Linn. (Arecaceae)**

Nariyal, Coconut Palm

Coconut palm is a native of Malay Archipelago, widely cultivated in coastal region of India. This palm, often attaining a height of 30m, possesses an unbranched bole and a heavy crown of many drooping pinnate leaves. The leaves are 5-7m long with a large number of 1m long linear leaflets. The staminate flowers are borne on the upper part of the interfoliar branching spadices coming out from a boat-like spathe and pistillate flowers on the lower parts. The fruit is a large sub-globose, one-seeded drupe having a fibrous mesocarp.

The dye is extracted from the fibrous mesocarp by boiling it in water containing a small quantity of turmeric till the solution turns brown. The yarn is soaked in this for a few hours. A dark brown color is obtained.

17. *Diospyros kaki* Linn.f. (Ebenaceae)

Halwatendu, Kaki Persimmon

This native of India is a small deciduous tree with opposite leaves, unisexual flowers and orange red berries having an enlarged calyx at the base.

The fruits (immature) contain tannins which include both gallotannins and proanthocyanidins. The ripe fruit is rich in sugars (up to 15%), mannans, carotenoids and phlobaphenes.

The tannins present in the unripe fruit are used for dyeing.

D. montana contains diospyrin (6, 2-bis-7-methyl juglone) in its bark and wood.

18. *Emblica officinalis* Gaertn. (Euphorbiaceae)

Syn. *Phyllanthus emblica* Linn

Amalaki/Amla, Indian gooseberry

E. officinalis is tall deciduous tree with small narrowly oblong leaves arranged distichously on branches so that the young branches resemble a pinnately compound leaf, small green unisexual flowers and depressed globose, obscurely six lobed fleshy fruits containing six (sometimes only one) trigonous seeds. It is a native to India, now seen wild and cultivated.

The fruit contain vitamin C (ascorbic acid, up to 2%, richest source) gallotannins (5%), carbohydrates (14%), phenolic acids, alkaloids, pectin and minerals. The phenolic acids include gallic, ellagic and phyllemblic acids and emblicol. Also present are a good amount of free amino acids such as alanine, aspartic acid, glutamic acid, lysine and proline. The alkaloids present are phyllantidine and phyllantine. Curcuminoids also are reported from certain samples.

Leaves and fruits yield a dye useful for giving a brown color to silk. Fruits are used for hair dyes. *Emblica* exhibits antitumour, hypotensive and immunomodulatory properties. Extracts are anticytotoxic, antigenotoxic, anticlastrogenic, and antimutagenic in nature. This is one of the richest and stablest sources of vitamin C because of the presence of antioxidant gallotannins in the fruit. The drug is a rasayana drug. It is used in the treatment of peptic ulcer and dyspepsia.

19. *Ficus benghalensis* Linn. (Moraceae)

Nyagrodhah / Vad, Baniyan Tree

This is a very large laticiferous tree with many aerial roots from the branches which root on the ground and thus extending the growth of the tree infinitely. Leaves are large broadly elliptic, glabrous above and minutely pubescent beneath. Figs monoecious, axillary depressed, globose, 2 cm in diam, brick red when ripe with 3 broad rounded basal bracts. Male flowers numerous near the mouth, female flowers at the base of receptacle and neuter flowers in the middle. Achenes, globose ellipsoid. A native of India cultivated or found wild through out India.

The stem bark contains bengalenoside, procyanidin, β-sitosterol, *meso*-inositol and ketones. Leaves yield friedelin and β-sitosterol, quercetin and rutin. Heartwood contains ψ-taraxasterol.

The bark provides a reddish brown dye which can be used for dyeing clothes and coloring medicines. All the parts of the tree are used in medicine especially for skin diseases. The bark is useful in menorrhagia, leucorrhoea and other vaginal disorders and is found to cure diarrhoea and dysentery. Bengalenoside is found to be hypoglycemic. Latex and young buds are used in urinogenital diseases. The roots also are useful in haemoptysis, menorrhagia and ulcers. Young roots are found to cure pimples.

20. *Ficus religiosa* Linn

Aswattah

A large glabrous laticiferous tree with long petioled broadly ovate rotund leaf having a long lanceolate cuspidate tip and monoecious paired dark purple figs. A highly venerated tree found throughout India.

Bark contains bergenin, lupin-3-one, methyl oleanolate, lanosterol, n-octacosanol, β-sitosterol, stigmasterol and caffeic acid (Swami *et al.*, 1989).

Bark extracts yield beige to buff dye. Both bark and leaves yield red/brown dyes.

The bark is alterative, haemostatic, laxative and is used in diabetes, diarrhoea, leucorrhoea, menorrhagia, nervous disorders and vaginal diseases. It is also an excellent remedy for skin diseases and is anthelmintic. Leaves are also used in skin diseases. Seeds have the power to sterilize women.

21. *Garcinia hanburyi* Hook f. (Guttiferae)

Gamboge

Gamboge is a gum resin obtained as an exudate from *G. hanburyi* and related species; natives of Malaya. This dye which possesses a beautiful yellow colour contains flavonoids and related tannins, is used by artists and in medicine (as a violent cathartic).

22. *Lannea coromandelica* Merr. (Anacardiaceae)

Syn. *Odina woodier* Roxb.

Jingini, Wodier

This is a deciduous tree of 20m, with stout branches, exfoliating ash colored bark, stellately puberulous young branches, imparipinnate leaves (leaflets 7-11, ovate-oblong, pink when young) crowded about the ends of branches, unisexual purplish flowers appearing when the tree is bare of leaves (male flowers in panicles, female flowers in simple racemes, tetramerous), and small compressed reniform red drupes containing a single stone.

The bark contains tannin, gum and a good amount of potassium carbonate.

The fruit on boiling in water containing a little salt, gives a grey dye. The bark, used in combination with *Phyllanthus emblica* and *Artocarpus* (in equal proportions) when steeped in water for 15-20 days yields a orange dye, which gives the yarn a charcoal grey color. The bark of *Lannea* is also used in combination with barks of *Careya arborea* and *Eugenia jambolana* in a similar way.

23. *Madhuca indica* Gmel. (Sapotaceae)

Syn. Bassia longifolia Koenig

B. latifolia Roxb.

Mawra

Mahua is a large laticiferous tree with broad lanceolate leaves crowded at the tips of branches, clusters of creamy white flowers having fleshy sweet caducous corolla and fleshy ovoid fruits containing 1-3 ellipsoidal compressed seeds. A native of India.

The kernels yield 55-65% of whitish yellow oil consisting of glycerides of palmitic 23-26%, stearic 19-25%, oleic 41-45% and linoleic acid 9-13%. In some samples, myristic acid replaces stearic acid. Besides fatty oil, the seeds contain 4.6% of a bitter toxic saponin, mowrin, which on hydrolysis yields the sapogenin, bassic acid.

The bark contains 17% tannin and is used for dyeing and tanning.

24. *Manilkara hexandra* Dubard (Sapotaceae)

Syn. *Mimusops hexandra* Roxb.

Khirin

This is a medium sized evergreen tree with a spreading crown and a massive straight bole, elliptic-obovate leaves, white fragrant flowers (corolla + stamens fall off from flowers) and ovoid or ellipsoid reddish yellow fruits containing 1 or two shining ovoid reddish brown seeds. A native of India.

The bark contains 10% tannin. The wood is red to light purplish brown with darker lines.

The bark of this plant and also of *M. littoralis,* found in Andamans, yield a red dye.

25. *Peltophorum pterocarpum* Baker (Caesalpiniaceae)

Syn. P. roxburghii Degener.

P. inerme Naves

P. ferrugineum Benth.

Copper pod

This is a tall handsome tree (up to 25m. high) with grey smooth bark (red inside), bipinnate leaves bearing many small, obliquely oblong leaflets, yellow fragrant flowers (stamens 7) in large erect terminal panicles and flat winged reddish brown pods containing 1-5 oblong flattened brownish seeds. A native of Andamans, now cultivated as an ornamental tree.

The bark, leaves and wood contain condensed tannins. Bark yields a dye as also a tanning material.

26. *Peristrophe bivalvis* Merr. (Acanthaceae)

Syn. *P. tinctoria* Nees

Bhatia-rang

P. bivalvis is an erect spreading herb of 60cm with ovate/sub-acute leaves, rose/purple flowers and ellipsoid capsules containing ovoid/ellipsoid seeds. A native of S. E. Asia, cultivated for the dye.

The twigs yield a yellow orange to deep red dye. The dye is extracted by boiling the cut branches in water, used to dye sticks (for mats) and clothes to get a dull red color.

27. *Pithecellobium dulce* Benth. (Mimosaceae)

Jungle jilebi, Quamachil, Manila tamarind

This is a medium-sized evergreen spiny tree (up to 18m. in ht.) with smooth bark, bipinnate leaves (pinnae 2, pinnules 2 in each pinna, leaflets obliquely ovate-oblong), small white flowers in dense heads and twisted/curved, constricted, reddish brown pods containing 6-10 shining black seeds enveloped in pinkish white pulpy edible aril. A native of Tropical America, now cultivated for the aril.

The bark contains tannins (up to 37%) and a gum. Seeds yield an edible oil (17%), starch (41%) protein (18%) and saponins.

The bark yields a yellow dye.

28. *Psidium guajava* Linn. (Myrtaceae)

Amrud, Common Guava

This is a small tree with grayish white bark (bark peeling off as scales), finely pubescent and chartaceous leaves, white fragrant flowers and green/yellow fruits containing many white seeds enveloped in a creamy white flesh. A native of Tropical America, now naturalized in India.

Bark and leaves contain considerable amount of hydrolysable and condensed tannins. The leaves yield vol. oil containing d- and dl-limonenes, β-caryophyllene and flavonols like quercetin and saponins.

Bark is used for tanning and dyeing. The volatile oil is antibacterial.

29. *Pterospermum acerifolium* Willd. (Sterculiaceae)

Hathipaila

This is an evergreen tree growing up to 25m in height, with leaves covered by dark brown tomentum. Leaves are very variable, entire or palmately lobed or peltate and the flowers are white fragrant large, solitary or in pairs. The capsules are 5-angled oblong and woody containing winged seeds. A native of India.

The bark yields a purple dye.

30. *Pterospermum canescens* Roxb.

Syn. *P. diversifolium* Blume

Muchkand

A medium sized tree found in South India having broadly elliptic coriaceous leaves, white fragrant flowers singly or in pairs and ovoid-oblong woody capsules containing winged seeds.

The bark is chewed with betel leaves as a masticatory. It is cut and boiled to extract a red dye useful in dyeing fish nets and clothes. The bark serves as a fish-poison also.

31. *Pterospermum lanceaefolium* Roxb.

A medium-sized tree with lanceolate/oblong leaves found in sub-Himalayan tracts. The leaves are chewed to redden the lips.

P. rubiginosum Heyne also yields a red wood.

P. semisagittatum Buch-Ham. The bark is used as a mastictory.

32. *Pyrus communis* Linn.(Rosaceae)

Pear

The leaves of pear containing tannin and flavonoids are used to dye mordanted wool to yellow and brown shades.

33. *Quercus infectoria* Olivier (Fagaceae)

Manjuphal, Dyer's Oak

This is a small tree of 5m, native to Mediterranean countries with spinous toothed rigid long leaves, long male spikes, short female spikes and cylindrical acorns. Oak galls are spherical or pear-shaped excrescences on young twigs caused by the deposition of eggs by an insect *Adleria gallae-tinctoriae.*

Galls, spherical or pear-shaped structures arising on the young twigs due to the deposition of egg by a small hymenopterous insect, contain tannic acid (gallotannic acid- 50-70%), gallic acid, ellagic acid, gum and essential oil.

Galls are extensively used in tanning, dyeing, mordanting and in the manufacture of ink.

All the species of *Quercus* are sources of tannins.

34. *Sapindus mukorossi* Gaertn. (Sapindaceae)

Ritha, Soapnut tree

S. mukorossi is a handsome tomentose tree with large pinnate leaves (leaflets 10-16, lanceolate-oblong) crowded near the ends of branches, white/purple ciliated flowers (petals clawed, having two woolly scales on either side at the top of claw) in panicles and fleshy yellow saponaceous fruits. Common in India, China and Japan.

The saponin mukorosside (sapogenin-hederagenin), present in fruit rind, is used as a level-dyeing agent in dye baths.

35. *Sapindus trifoliatus* Linn.

The saponin present in the fruits, emarginatoside (a hederagenin glycoside), is used in the preparation of certain fungicides and insecticides. Fruit and root bark are used as fish poison.

Seeds yield an oil (45%) rich in oleic (55%) and arachidic (22%) acids. The hard black shell of seeds yields a dye used for dyeing light shades on mordanted cotton, wool and silk. It is also used for dyeing leather and for coloring wood.

36. *Shorea robusta* Gaertn.f. (Dipterocarpaceae)

Sal, Shorea

This is a large deciduous tree of 40m with reddish brown bark, large ovate oblong coriaceous leaves, small yellow flowers in lax axillary and terminal panicles and an ovoid reddish one seeded fruit having five wing-like persistent sepals. A native of Tropical Asia, this is one of the important timber trees of India.

The heartwood is pale brown, turning to dark reddish brown with age. The wood contains 12% tannin. Bark and leaves contain 12% and 20% respectively.

37. *Symplocos cochinchinensis* Moore (Symplocaceae)

Lodhra, Symplocos

Lodhra is a medium-sized tree having elliptic-obovate, crenate leaves, small white fragrant flowers (corolla gamopetalous, stamens

12 or more, ovary inferior) in tomentose spikes and a small globose drupe having an apical ring, enclosing oblong seeds. A native of South East Asia, common in forests.

The bark is used for dyeing. To extract the dye, bark and/or leaves are boiled in water for many hours and alum is added with lac and this gives a permanent red to silk. With alum it gives a madder purple color and with iron, a black color is obtained. This dye is suitable for cotton, silk.

The bark is an acclaimed remedy for menstrual disorders, uterine problems and urinogenital diseases. It is a specific uterotonic and is useful in menorrhagia, leucorrhoea, diarrhoea, dysentery, fever, cough and gum diseases.

38. *Symplocos paniculata* Miq.

The bark and leaves yield a yellow dye used as a mordant with madder.

39. *Symplocos racemosa* Roxb.

Lodhra

S. racemosa is an evergreen tree of 8m with elliptic-oblong, orbicular coriaceous leaves, white (turning yellow) flowers in simple or compound racemes, and purplish black subcylindric drupes containing 1-3 seeds. A native of S.E. Asia.

Bark contains propelargonidin type tannins and alkaloids such as harman, loturidine and collaturine.

Leaves and bark yield a yellow dye used as a mordant and as a color in Holi.

40. *Symplocos sumunita* Buch-Ham

A native of Himalayas, the bark and leaves of this plant yield a yellow dye with hot water, turning red with sodium carbonate - used as a mordant.

41. *Syzygium cumini* Skeels (Myrtaceae)

Syn. *S. jambolanum* DC.

Eugenia jambolana Lamk.

Jamun/Jambuh

A large tree with exfoliating bark, elliptic oblong leaves (with intramarginal vein), dull white tetramerous flowers crowded in panicled cymes from branches below the leaves (stamens many, ovary inferior), and a dark purple elliptic berry containing a globose seed. Common as a cultivated avenue tree or in forests of India, S.E.Asia and Australia.

The edible pulp from fruits contains sugars, gallic acid, malic acid, tannins, glucosides of petunidin and malvidin and oleanolic acid. Seeds contain protein (8.5%), tannins (19%), ellagic and gallic acids, and a glycoside, jamboline. Flowers yield oleanolic acid and Eugenia-triterpenoid A&B along with isoquercitrin, quercetin, kaempferol and myricetin. Stem bark contains betulinic acid, β-sitosterol, friedlin, tannins, myricetin, gallic acid and ellagic acid.

The bark gives a red dye while the red fruit color may be used in foods and pharmaceuticals. The fruit, especially seeds, are extensively used against diabetes. Bark is used in dysentery. Also used in cough, asthma, dyspepsia and hemorrhages. Seed extract is used as a male contraceptive also. (Sinha *et al.*,1986).

42. *Tamarix dioica* Roxb (Tamaricaceae)

T. dioica is a small tree of 2-3m in height commonly seen in river-beds and near sea coasts. The branches are drooping and the leaves are minute obliquely truncate and acuminate and closely adpressed to stem forming a sheath. Flowers are pink borne in dense peduncled spikes and capsules are oblong/conical. A native of temperate Europe and Asia.

The galls which are formed on the flowering spikes yield 50% tannin. They are used as a mordant in dyeing and tanning. With salts of iron, they produce shades of grey and black.

43. *Terminalia arjuna* W. &A. (Combretaceae)

Arjun/Arjun Sadad

This is a large tree with a thick trunk, horizontally spreading branches and greenish white bark flaking off in large flat pieces. Leaves are large, subopposite, oblong, crenate –serrate and have 1-3 glands at the top of petiole. Flowers are apetalous, yellowish green in axillary spikes. Fruit a 5-winged drupe with veins of wings curved upwards. Common in Western India.

Bark is found to contain tannins (both hydrolysable and condensed, 20-40%), arjunic acid, a trihydroxy triterpene monocarboxylic acid, ($C_{39}H_{48}O_5$) and its glucoside arjunetin, β-sitosterol, friedlin and ellagic acid. Also present in the bark are CaO, MgO and Al_2O_3. Fruits contain 7-20% of tannin, arjunic acid, β-sitosterol, mannitol and KCl.

Bark extract yields a red dye. The bark of Arjuna tree is a well known cardiotonic. This is used in all sorts of cardiac failure and dropsy. It is also a good diuretic, febrifuge, tonic and antidysenteric.

44. *Terminalia bellirica* Roxb. (Combretaceae)

Vibhitaki/Behada, Belliric myrobalan

A large deciduous tree with opposite broadly elliptic leaves crowded at the end of branches. Flowers, both unisexual and bisexual, are apetalous, yellowish green borne in axillary spikes (male flowers towards the top). Fruit is a globose drupe, tomentose containing a single seed. Seen throughout India.

Fruits are found to contain tannins, β-sitosterol, ellagic acid, gallic acid, chebulagic acid, ethyl gallate and sugars. Tannins present are both hydrolysable and condensed type. Seeds yield a fatty oil and protein.

This also is a source of tannin and dye. The rind of the plant is one of the 'triphala', a general tonic. It is used for anaemia, cough, fever and is laxative and germicidal. It is considered to be useful in treating leprosy, tuberculosis, diarrhoea, inflammations and problems of eye, nose and throat.

45. *Terminalia catappa* Linn. (Combretaceae)

Desi-badam, Bengal Almond

This is a small buttressed tree with branches borne in horizontal whorls, large obovate leaves, small yellowish white apetalous flowers in spikes and an ellipsoid fruit keeled all around. A native of Moluccas, this tree is cultivated for edible fruits

The bark and leaves yield a brownish yellow dye used in coloring cotton, silk and wool. The leaves and fruits contain corilagin, gallic acid, ellagic acid, brevifolin etc. whereas the bark and wood contain catechins, epicatechin and procyanidin.

46. *Terminalia chebula* Retz. (Combretaceae)

Hirda/Haritaki, Chebulic myrobalan

This is a moderate-sized much branched tree with opposite elliptic oblong large leaves (pubescent when young, have two glands at the top of petiole), apetalous cream colored flowers in axillary spikes and a glabrous obovoid , faintly five-ribbed fruit containing a bony stone. Abundant in Western Ghats.

The fruits contain 30-40% of tannins mostly as di-O-galloyl and tri-O-galloyl glucose. Some of the component tannins isolated are chebulagic acid ($C_{41}H_{30}O_{27}.10H_2O$), chebulinic acid and corilagin. Also present are the breakdown products (partial) such as chebulic acid, 3,6-digalloyl glucose, gallic acid, ellagic acid, terchebin (1,3,6-trigalloyl glucose) and 1,2,3,4,6-pentagalloyl glucose. Also present in the fruit are sugars (glucose and sorbitol), amino acids, and organic acids (succinic, quinic and shikimic acids).

The powdered rind of the fruit with alum yields a fast yellow color. With iron sulphates black color is produced. The fruit is extensively used as a dye mordant. Flowers on extraction with water yield a yellow dye and khaki color is produced with mordants. The powdered flower is used for producing a yellow dye and to top indigo dye materials to develop green color. It is also used to print yellow and green colors to leather.

Haritaki (the fruit rind) is a component of Triphala (a group of three fruits *T. chebula, T. bellirica* and *Emblica officinalis*) which is used as a 'rasayana ' (rejuvenative) drug. Haritaki is antiseptic, laxative, diuretic and carminative. This is one of the best laxatives, and used for diabetes, fever, anemia, cardiac disorders, diarrhoea and dysentery.

Haritaki is one of the chief sources of tannin in India.

3.2.*Alkaloids*

1. ***Berberis aristata*** **DC (Berberidaceae)**

Daruharidra, Barberry

B. aristata is a spiny shrub with obovate/elliptic toothed leaves, sharp spines from the nodes, yellow flowers in corymbose racemes and oblong-ovoid red berries. It is a native of Nepal, widely cultivated or grown as an escape in many parts of Western India.

The root bark, as well as stem bark, yields (up to 4%) isoquinoline alkaloids, such as berberine, berbamine, aromoline, karachine, palmatine, oxycanthine, oxyberberine and taxilamine.

Berberine is one of the few alkaloids which are yellow in color. The roots and stem of barberry yield a yellow dye useful for coloring leather and clothes. The stems are used in Tibet to stir boiling butter to impart a fresh golden color to ghee. The total alkaloid extract is a bitter tonic, used as a cholagogue, laxative, diaphoretic and applied for eye affections, ulcers and hemorrhoids. It is also an emmanagogue and is useful in the treatment of jaundice. The fresh berries are laxative, antiscorbutic and useful in piles and eye diseases esp. conjunctivitis.

B. asiatica Roxb., common in Himalayas, also yields similar alkaloids (berberine amounts of 2% in root bark) and used in place of barberry. Related species such as *B. chitria, B. coriaria, B. floribunda* etc also are used as substitutes.

2. *Punica granatum* Linn. (Punicaceae)

Anar, Pomegranate

Pomegranate is a small tree, with simple ovate shining leaves, red solitary flowers and a round berry crowned with calyx. The plant is a native of Central Asia, now cultivated everywhere for the very refreshing edible fruits.

The yield of volatile liquid alkaloids from the root and stem bark is about 0.5-1%.The principal alkaloid is pelletierine which is accompanied by isopelletierine, methyl isopelletierine and a solid alkaloid, pseudopelletierine. All these alkaloids occur in the form of tannates. The tannin content of the root bark amounts to 22%. Fruit rind contains 26% tannins used to tan leather.

The fruit rind containing 26% tannins along with stem bark (tannins 10-25%), root bark (26%) and leaves (11%) are used for tanning and for dyeing leather. This gives yellowish brown to khaki shades to wool and silk. Flowers containing pelargonidin glycoside, also are used to dye clothes to a light red color.

The dye is extracted by boiling bark or fruit rind in water. Washing soda is the mordant. Pond green, olive or yellow ochre colors are obtained by mixing pomegranate extract with indigo, turmeric or sulphate of iron and alum. Black color is obtained by boiling silk in a solution of pomegranate and iron fillings. A *kishmishi* or raisin color is obtained with pomegranate rind, cotton flower and alum.

The root and stem bark are used as anthelmintic and taenifuge. The pericarp is a reputed anthelmintic for children.

3.3. *Indoles*

1. *Indigofera* spp. (Fabaceae)

Nil, Indigo

Indigo is obtained from several species of *Indigofera* especially *I. tinctorius* L. (Asia) and *I. suffructicosa* Mill. (America). These plants are small woody shrubs with pinnately compounds leaves and apical racemes of small pink or blue flowers and long pods.

The dye is present as a colourless soluble glycoside 'Indican' in the leaves. This glycoside gets hydrolysed on fermentation, when kept in water, to an insoluble indoxyl (indigo) and glucose.

Indigo

Indigo is insoluble in water and therefore needs an alkaline medium and subsequent oxidation to yield its blue color. Two methods, wet process and dry-leaf process are followed for extraction. The wet process consists of steeping the leaves in water in large steeping vats for 12-14h. The yellowish green liquor is drained to a lower beating vat. After beating, done by hand or wheel, the liquor is allowed to stand until 'indigo' is settled leaving a clear red supernatant liquor. After boiling, the solution is filtered through clothes and the residue is pressed and dried into cakes. In the dry-leaf process, the leaves are dried and are steeped in water and allowed to ferment. The fine green liquor thus obtained is drawn off to the beating vat where the matured indigo is precipitated, concentrated and filtered. The inferior quality of indigo (20%) is used to dye cotton fabrics while the superior quality is preferred for dyeing silk fabrics.

Indigo mixed with sulphate of iron and myrobalan produces a black color. Indigo with oak galls give grey color and with safflower it gives a purple color. Green color is produced by mixing indigo with flowers of *Butea* and *Nyctanthes.*

The blue colour imparted by this dye is the stablest among dyestuffs. Presently natural indigo is replaced by synthetic ones.

2. *Isatis tinctoria* Linn. (Brassicaceae)

Woad

Woad is a small biennial, the leaves of which yield a blue dye. This plant, a native of Europe, bears oblong radical leaves, auricled stem leaves, small yellow flowers in panicles and small 1-seeded indehiscent samaroid fruits.

The fermented and dried leaves contain a compound isatan-B which on hydrolysis yields indoxyl and 5-ketogluconic acid.

Used as a substitute for indigo.

3. *Musa paradisiaca* Linn. (Musaceae)

Kela, Banana

This is a tall herb native to India and Malaya, having a large spherical corm, a pseudoaerial stem consisting of sheathing leaf bases, a crown of large oblong pinnate parallel leaves and an inflorescence consisting of bunches of compact white flowers surrounded by purple spathes. The fruits are seedless berries found in spirally arranged bunches on an elongated axis.

The watery juice oozing out of the cut ends of leaves and pseudostem may leave a permanent black color on clothes. The diluted paste of pseudostem is used to finish textiles. Ripe peel, dried and triturated, is used as a potash-rich mordant to enrich yellows. Sap contains tannin. Ash prepared from leaves and bark's outer layer yields a red dye.

4. *Wrightia arborea* Mabb. (Apocynaceae)

Syn. *W. tomentosa* Roem. & Shult.

Kutaja, Pala Indigo plant

Kutaja is a small deciduous tree with abundant yellow latex, tomentose elliptic leaves, white fragrant flowers (with an orange corona of numerous fimbriate scales in 2-3 series) in lax terminal cymes and a pair of pendulous cylindrical grooved follicles rough with tubercles, containing slender seeds comose at the base. This tree is common in deciduous forest of India and nearby countries.

The bark contains alkaloids, conessine, holarrhine, kurchicine and conkurchine. Also present are β-sitosterol, lupeol and α-amyrin. Seeds yield an oil similar to that of *Holarrhena*. Leaves yield quercetin, kaempferol, *iso*-orientin, procyanidin and phenolic acids such as vanillic, syringic, *p*-OH-benzoic and sinapic acids (Daniel and Sabnis,1978). Roots, leaves and seeds are shown to contain an indigo-yielding glycoside.

Leaves yield a blue dye used for dyeing clothes. The yellow latex, on dilution with water, yields a yellow dye, which is used for dyeing cotton. The root bark and seeds, often substituted for *Holarrhena*, are used for chronic diarrhoea and dysentery. Other uses include curing piles, hemorrhage, cardiac diseases, fever and colic.

W. tinctoria R.Br. is a related tree used as a substitute. Its bark is poor in alkaloids but contains β-amyrin, lupeol, β-sitosterol, α-amyrin, ursolic acid and oleanolic acid. The leaves are source of indigo, a blue dye named Mysore Pala Indigo. Seeds are used as an adjuvant to dyeing materials.

3.4. Betacyanins

1. *Beta vulgaris* Linn. (Chenopodiaceae)

Beet root

This is a succulent biennial herb with a large swollen spherical fleshy root, basal rosette of ovate / oblong leaves and greenish flowers in panicles. A native of Mediterranean region.

The blood-red color of the beetroot is due to betalains, a group of indole derived pigments. The betalains consist of red-violet betacyanins and yellow betaxanthins. Betanin is the major component of the former group (75-95%) and vulgaxanthin I is the principal component of later group. The other betalains present are isobetanin, isobetanidin, prebetanin, isoprebetanin and vulgaxanthin II.

The content of betacyanins is increased 5-7 times by fermentation of the extracted juice with *Torulopsis utilis*. Betacyanins are used to color foods and pharmaceutical products. Though the color is not very stable, it is successfully used in meat substitutes, summer sausage, gelatin desserts and diary products. The root juice is used for intensifying the color of tomato products like tomato juice, puree and catsup. Betacyanin gives best color at pH 6.3-6.6 and turns yellow at basic pH.

2. *Iresine herbstii* Hook. f. (Amaranthaceae)

Syn. *Achyranthes verschaffelti* Lem.

This is a small, much branched perennial of 1m in height, with bright crimson or maroon colored leaves. The leaves are employed in Java for coloring agar-agar jelly.

3. *Rivina humilis* Linn. (Phytolaccaceae)

Bloodberry

R. humilis, a native of Tropical America, is a small beautiful herb with ovate leaves (having wavy margins), small white flowers in apical racemes and small blood-red one-seeded berries. This plant is widely cultivated as an ornamental plant.

Berries containing betacyanins is used as a source of textile dyes and food color.

3.5. Terpenoids

3.5.1. Iridoids

1. *Catunaregam spinosa* Tirveng. (Rubiaceae)

Syn. *Randia spinosa* Poir.

R. dumetorum Poir.

Xeromphis spinosa Keay.

Madana, Emetic nut

A thorny tree with dark brown bark, obtuse deciduous leaves narrowed into a short petiole, axillary thorns, solitary fragrant white flowers turning yellow and globose yellow berries enclosing flat angular seeds. Common as undergrowth in sal forests.

The bark contains saponins based on randialic acid A & B, mannitol (up to 6%), 10-methyl lixoside (an iridoid) and scopoletin. Stems yield iridoids such as randinoside, galioside, deacetylasperulosidic acid methyl ester, scandoside methyl ester, geniposide and gardenoside. Roots also contain mannitol, scopoletin and a yellow dye. The leaves yield flavonoids, kaempferol, 4-OMe kaempferol and quercetin and phenolic acids, *p*-hydroxy benzoic, gentisic, melilotic, vanillic and syringic acids (Thomas, 1989; Hamerski *et al.*, 2003).

The fresh fruits contain 2-3% saponins (10% dry wt.) which mostly are concentrated in pulp. They are dumetorins A, B, C, D, E and F (based on oleanolic acid), randioside A (on oleanolic acid), ursosaponin (on ursolic acid), two triterpenoid glycosides based on oleanolic acid and 10-methyl lixoside. Seeds contain protein (14%) and saponins based on oleanolic acid.

The leaves and fruits give a blue dye. The bark is used in diarrhoea, dysentery and is abortifacient, anthelminitic, antipyretic and emetic.

2. *Holarrhena pubescens* Wall. (Apocynaceae)

Syn. *H. antidysenterica* DC.

Kurchi, Holarrhena

A small tree with broadly elliptic leaves, white flowers (corona absent) in terminal corymbose cymes and a pair of long cylindric (often dotted with white spots) follicles enclosing numerous linear-oblong seeds tipped with a spreading deciduous corona of brown hairs. A native of India.

Leaves contain alkaloids such as kurchiphylline, kurchiphyllamine, kurchaline, holadysine and holadysamine; aminoglycosteroids like holantosines A,B,C & D and holarosine A and aminodeoxyglycosteroids holarosine B,E and F, flavonoids such as kaempferol and quercetin and phenolic acids (*p*-coumaric, vanillic, syringic,α-resorcylic and *p*-hydroxy benzoic acids). Seeds also contain a good amount of conessine. (Bhutani *et al.*, 1990; Siddiqui *et al.*, 1994; Daniel and Sabnis, 1978).

Leaves yield a dye similar to indigo. Bark and seeds are used in dropsy and dysentery (conessine is the antidysenteric principle), stomach disorders, abdominal and glandular tumours.

3. *Vitex negundo* Linn. (Verbenaceae)

Nirgundi, Vitex

Nirgundi is a small tree with grey pubescent branches, 3-5 foliolate leaves (leaflets lanceolate , white tomentose beneath, petioluled, lateral leaflets small, lanceolate and terminal leaflets larger), small blue bilabiate flowers (stamens 4-didynamous) and ovoid black drupes containing 4 obovate seeds. Common in India, Afghanistan, Philippines and Ceylon.

The leaves contain, aucubin, agnuside; alkaloids nishindine ($C_{15}H_{21}ON$) and hydrocotylene ($C_{22}H_{33}O_8N$), glycoflavones orientin and isoorientin and 5-hydroxy, 3,6,7,3′,4′-penta methoxy flavone.

Oil prepared with leaves is useful for the blackening and growth of hair and increases the functions of brain. The leaves, as a paste, are used for inflammatory swellings of the joints formed due to rheumatism, hydrocele and splenic enlargement. They are also used in skin diseases, nervous disorders, leprosy and as a vermifuge.

Vitex trifolia Linn, a related species with 3-foliolate, sessile, obovate leaflets and blue flowers are used as a substitute for *V. negundo.*

3.5.2. Carotenoids

1. *Bixa orellana* Linn. (Bixaceae)

Batkan, Annatto

The scarlet aril surrounding the seeds of a small tree, a native of tropical America, yields this bright reddish yellow dye. This tree possesses large slender petioled ovate leaves, pink showy flowers in terminal panicles and red-brown spiny two valved capsules enclosing numerous arillate seeds.

The principal coloring matter is a sesterpene bixin which amounts to 70-80% of total pigments. Also present are orellin (a water soluble yellow dye), methyl bixin, β-carotene, cryptoxanthin, lutein and zeaxanthin. Bixin occurs in *cis*-form, an unstable compound, in plants, which changes to *trans*-form, called isobixin. Bixin on saponification gives a water soluble nor-bixin. At low pH bixin may turn pink.

For extracting the dye, the seeds are bruised and the pulp macerated with hot water, soaked for several days till the coloring matter forms a fine suspension. The seeds are then removed and the *brei* which contains the pigment is allowed to ferment for a week when the dye settles at the bottom. The dye is extracted also by boiling the seeds with sodium carbonate solution, filtering and then acidifying the filtrate. The dye is soluble in oil and also ethyl acetate.

In another method of extraction, the seeds are boiled in a vessel of water containing a small amount of turmeric paste and lime along with the yarn for 3 days. This process is repeated for darkening the color. Treatment with copper sulphate improves fastness for light.

Annatto is extensively used in dairy industry for coloring butter, ghee, cheese, margarines, ice cream, chocolate, meat and sausages. It is also used in nail-gloss, hair-oils, shoe polishes, soap, pharmaceuticals and leather. Annatto is employed to paint bodies of dancers for folk dances and as mosquito repellant. Annatto is abundantly used for colouring wool, paints and soaps.

2. *Capsicum annum* Linn. (Solanaceae)

Syn. C. *frutescens* C.B.Clarke,

C. *purpureum* Roxb.

C. *minimum* Roxb.

Lalmirich, Red Chilli

A herbaceous perennial with oblong glabrous thin leaves, solitary white flowers from internodes, long green berries changing to orange to red containing pungent white/yellow circular seeds. A native of Tropical America now cultivated throughout the world. A number of varieties and cultivators are available.

Chillies are found to contain an oleoresin (16%), ascorbic acid, protein (16% dry wt.), fat (6%, mainly in seeds), carbohydrates (32%), carotenoids, a volatile oil and vit. E. The oleoresin, yield varies from 15-18%, extracted by solvents, consists of capsainoids and some carotenoids. The most important component of capsainoids is a phenylamide – capsaicin- that is generally responsible for the pungency of chilli. Other capsainoids are dehydrocapsaicin, homocapsaicin II, bishomo-and trishomocapsaicin, homodihydrocapsaicin and I II, nordihydrocapsaicin, vanillyl amides of caprylic, nonylic and decylic acids and capsianide and I. Capsaicin is maximum before the ripening stage of the fruit. Vitamin C is more in large chilies. The volatile oil, which varies from 0.1 to 2.6%, consists of 2-methoxy-3-isobutylpyrazine, *trans*-β-ocimene, limonene, methyl salicylate, linalool etc. Fruit pericarp yielded glycosides of luteolin, quercetin, and apigenin and *p*-coumaric and ferulic acids (Materska *et al.*, 2003). Seeds yield a fatty oil rich in linoleic acid and many steroids such as 4α-methyl-5α-cholest-8(14)-en-3β-ol, fucosterol, isofucosterol, citrostadienol and derivatives of lanosterol. Seeds contain free sugars and capsicoside.

The green fruits contain many polyphenols (which are found to be absent in mature ones). They are petunidin, luteolin and quercetin glycosides, esters of *p*-coumaric and quinic acids, acyclic diterpene glycosides such as capsiansids I to V and A-E. All parts except flowers and stem contain solanine and solanidine.

The color of red chillies is due to carotenoids such as capsanthin and capsorubin (main components), capsanthin-5, 6-epoxide and capsochrome. Zeaxanthin, cryptoxanthin, lutein, neoxanthin, violaxanthin, β-carotene are the other carotenoids present. (The green peppers have lutein as the major pigment). Some cultivars contain petunidin glycosides also. This is used to color foods and pharmaceuticals.

3. ***Crocus sativus*** **Linn. (Iridaceae)**

Kesar, Saffron

The dried stigmas and tops of styles of this perennial herb constitute the saffron of commerce. The plant has small globose corms bearing numerous, radical, linear, ciliate leaves. The lilac/purple flowers are borne in 2-valved spathes. The styles of flowers are branched and blood red in colour.

The color of saffron is due to a diterpene crocin, which is the gentiobiose ester of crocetin – a diterpene acid having a structure resembling carotenoids. Also present are lycopene, β-carotene, γ-carotene and zeaxanthin. The bitter taste of saffron is attributed to picrocrocin – a glycoside yielding glucose and safranal on hydrolysis. Safranal is dihydro β-cyclocitral and is the main odoriferous constituent. The bulbs contain four isolectins. Tepals yield myricetin, quercetin, kaempferol, delphinidin, petunidin, astragalin and helichrysoside. Pollen is found to contain crosatosides A and B and kaempferol glycoside.

Saffron is a coloring and flavouring agent. Saffron extract shows cytotoxic and antimutagenic activities and anti-tumour activities against ascites tumours in mice. It is also used for its mild hypotensive properties, as an emmanagogue, in amenorrhea, dismenorrhagia, hysteria and to prevent premature ejaculation. Crocin is used to stabilize light sensitive drugs.

4. ***Gardenia grandiflora*** **Lour. (Rubiaceae)**

A Chinese tree, the fruit of which is employed in dyeing the yellow robes of the Mandarins, contain crocin. *Gardenia campanulata* fruits also are used to dye clothes.

5. ***Gardenia jasminoides*** **J. Ellis (Rubiaceae)**

Cape jasmine, Common Gardenia

This is an evergreen shrub of 1-2m. with dark-brown glossy ovate leaves, creamy white highly fragrant large flowers (petals double) and an oblong fruit (2-3cm long) bearing persistent calyx at the top. A native of China often cultivated for the fragrant flowers.

A number of iridoids such as genipin, its glycosides, geniposide, acetyl geniposide, geniposidic acid, deacetyl aspelurosidic acid, scandioside, gardoside etc, crocin (and crocetin), gardenin (a flavonoid) and gardenolic acid & β-sitosterol are reported from fruits.

In China, this is used as a food coloring and natural dye of long shelf life. A blue dye is produced by hydrolyzing geniposide and reacting the resultant genipin with a primary amine.

6. *Lycopersicon esculentum* Mill. (Solanaceae)

Rakthamachi, Tomato

The color of the tomato is due to β-carotene and lycopene along with minor components neolycopene, prolycopene, neo-β-carotene, phytofluene etc. These pigments are ideal for coloring foods and as a pharmaceutical dye. Lycopene is a an antioxidant well- sought after and has a curative and preventive action for prostrate problems including prostrate cancer.

7. *Nyctanthes arbor-tristis* Linn (Oleaceae)

Parijat, Night Jasmine

This is a small tree of 10m high with quadrangular strigose branches, rough, ovate acuminate leaves (scabrous above and densely pubescent below), small fragrant white flowers (having a bright orange tube), two stamens and sub-orbicular compressed chartaceous capsules separating into two flat 1-seeded mericarps. A native of India.

The flowers yield a jasmine like volatile oil. The color of the corolla tube is due to a diterpenoid nyctanthin (identical with crocetin) used for dyeing silk alone and in combination with safflower, turmeric, indigo or kath. For dyeing purposes the fabric is steeped in a hot or cold water decoction of the flowers. It provides with a beautiful, but fleeting golden color, the color can be stabilized by the addition of lime juice or alum.

The leaves contain tannin, methyl salicylate and resin. Bark yields an alkaloid and tannins. Seeds provide fatty oil.

The dye is extracted by steeping the flowers in plain water for one hour or with slight warming. Deeper colors are obtained when extracts of *Nyctanthes* flowers are mixed with *Butea,* turmeric and lime in acidulated water. Green color is obtained when indigo and *Butea* also are mixed. This dye is used for Tussar silk and cotton.

3.6. Porphyrins

Chlorophyll

Chlorophyll is a group of natural porphyrins present in all green plants. All of them are characterized by tetrapyrrole ring (porphin) containing magnesium. Though a number of chlorophylls exist in nature, higher plants contain chlorophyll a and chlorophyll b in the ratio 3:1. Chlorophyll *a*, a bluish black solid, is a diester of chlorophyllide *a* ($C_{32}H_{30}N_4O$ Mg $(COOH)_2$), methanol and phytol. Chlorophyll *b* is dark green in colour and contain Chlorophyllide *b* ($C_{32}H_{28}N_4O_2$ Mg $(COOH)_2$) in place of chlorophyllide *a*. Both these compounds are extracted by various solvents and are used as a non-toxic colouring material for foods, soaps and as a deodorant.

In presence of acids, the magnesium present in the porphyrin ring is replaced by two hydrogen atoms. The resultant product pheophytin is olive green to olive brown in colour. The traces of acids produced at higher temperature also cause this reaction and this explains the colour changes of canned as well as frozen vegetables. The enzyme chlorophyllase liberates phytol and propionic acid from chlorophyll, the latter compound causes the removal of magnesium subsequently. The bleaching of chlorophyll in dehydrated foods when exposed to sunlight for a long period of time is entirely a photo-oxidation process.

3.7. General

The plants which yield dyes of unknown chemistry are grouped here

1. *Dicoma schimperi* Bail. (Asteraceae)

In Arab countries this plant is a source of an yellow dye

2. *Rhynchosia* sp. (Fabaceae)

Pahariseem

The leaves are ground to a paste to which lime juice is added gradually till the extract turns red. The yarn is immersed for an hour. A red color is produced.

3. *Wendlandia tinctoria* DC (Rubiaceae)

A native of N. E. Himalayas, this small tree possesses elliptic-ovate leaves and panicles of white flowers.

The bark is employed in Assam and West Bengal as a mordant for dyeing.

3.8. Analytical Methods for Natural Dyes

The various natural dyes available belong to different classes of plant pigments. Though a majority of these are polyphenols, some are alkaloids (berberine, pelletierine), terpenoids (crocin, bixin, carotenoids), porphyrins (chlorophyll) or indoles (indigo). The extraction and analytical methods for all these groups are varied and diverse. In this section, only general methods of extraction and analysis of these compounds are provided. For further details the reader is referred to Harborne (1973) and Daniel (1991).

I. Phenolics

Phenolics are the largest group of coloring matters. They may be phenols (curcuminoids, litmus, etc), benzophenones (maclurin), quinones (alkannin, lawsone, ruberythric acid etc), flavonoids (quercetin, haematoxylin, anthocyanins etc) or tannins.

(a) Simple Phenols

Most of the simple phenols occur free in plant and are water insoluble. They are best extracted in methanol or ethanol. Chromatography on paper or TLC (silica gel) will separate the individual components. For example *curcuminoids* from turmeric are extracted by keeping the plant powder in methanol. The colored solution is concentrated and spotted / banded on TLC (silica gel) and chromatograms are developed using chloroform – ethanol – acetic acid (94:5:1). The visible yellow spots/bands separated will fluoresce bright yellow–white when seen in UV-365 lamp. The major component with the highest R_f is curcumin and the spot with lowest R_f is of bismethoxy curcumin whereas the middle spot is of monodemethoxy curcumin.

(b) Quinones

Quinones are aromatic diketones and are of three types: (1) possessing a single ring – benzoquinones, (2) two rings, naphthaquinones and (3) three rings – anthraquinones. Of these, the first two occur free while the anthraquinones occur free or as glycosides.

Benzoquinones

Embelin from fruits of *Embelia ribes* is the only benzoquinone used as a dye. All benzoquinones can be extracted in methanol/ether using a Soxhlet's extractor. For separation, the solvent system used in paper chromatography is BAW (n-Butanol: Acetic acid: water in the ratio 4:1:5 – two layers get separated. The upper organic layer is used for chromatography) and for TLC (silica gel) it is n-Hexane – Ethyl acetate (17:3). All benzoquinones will give absorption spectra in UV consisting of two peaks. The strong peak will be between 260-290 nm and a smaller peak, between 375-410 nm.

Naphthaquinones

Plumbagin, juglone and lawsone are the important members of this group. All of them can be extracted by using methanol/ethanol or ether. Solvent system recommended for silica gel TLC is petroleum ether (60-80°C) - ethyl acetate (7:3) or benzene - petroleum (2:1). All naphthaquinones have three to four spectral maxima, one or two below 300nm, one at about 330-340 and one above 400nm. Plumbagin has λ max 220, 266 and 410nm and juglone has 249, 345 & 422nm. The λ max of lawsone is 248, 270, 325 and 450nm.

Anthraquinones

Anthraquinones occur free or as glycosides. Therefore the extraction of these compounds is done by methanol or alkaline water. All the quinones turn red / purple with alkali. For separation of free compounds like emodin, chrysophanol, rhein etc, alkaline TLC plates (for e.g. 50 ml of 0.01M NaOH for 25 g silica gel) are used with benzene - ethyl acetate - acetic acid (75:24:1) as the solvent system. The anthraquinones which appear yellow will turn red/green etc on spraying with 10% methanolic KOH (or on standing for a day - color develops due to aerial oxidation). For glycosides like barbaloin the solvent system in TLC is ethyl acetate - methanol - water (100:16.5:13.5). All anthraquinones exhibit four or five absorption bands of which at least three lie between 215 and 300 nm and another above 430 nm for e.g. Emodin shows peaks at 223, 254, 267, 290 & 440. The λ max of chrysophanol is 225, 258, 279, 288 & 432.

(c) Flavonoids

This is a large group in which most of the compounds exhibit colors. The various subgroups used as dyes are flavones, flavonols, chalcones, aurones and proanthocyanidins.

Flavones

They are responsible for the white or cream color of flowers. These compounds are light yellow in color but give stable bright yellow colors on treatment with alkalies. Most of them occur as glycosides and both glycosides and aglycones (free compounds) can be extracted using methanol. This extract on concentration and addition of water will give an aqueous solution of flavones. This can be purified from the chlorophyll and waxes by filtering. The filtrate can be concentrated and the flavones are isolated by paper chromatography using BAW or 15% acetic acid. The developed papers are scanned under UV light (365 nm) when flavones appear as brown spots/ bands. On fuming the chromatogram with ammonia vapors or by spraying the paper with 10% Na_2CO_3, the flavones appear as bright yellow spots.

Since the identification of glycosides is cumbersome, the extract containing glycosides is hydrolyzed by 7% HCl in a water bath for 1h. The flavone aglycones precipitated can be extracted into ethyl acetate or solvent ether using a separating funnel. This extract on concentration and analysis using PC or TLC will isolate the flavone aglycones. In PC, solvent system used is 30% acetic acid or forestal (Con. HCl - Acetic Acid - Water, 3:30:10). The compounds can be seen using the same method as used for glycosides (UV or spraying with alkali). The chromatograms are marked in UV and the compounds are eluted in methanol and UV spectrum can be determined. In TLC the solvent system recommended is Toluene-formic acid - Ethyl formate (5:1:4).

Only two flavones, apigenin and luteolin, are widely distributed in plants. Apigenin is having a higher Rf, exhibits λ max 267, 336 while luteolin with low Rf has λ max 255, 350. The methylated derivatives of apigenin and luteolin show spectra similar to parent compounds.

Glycoflavones, the C-glycosides of flavones, do not get hydrolyzed with HCl and will remain in solution even after 10 hours of hydrolysis. The acidic solution containing them (after hydrolysis) is neutralized by slow addition of Na_2CO_3 and the solution is concentrated to dryness. To the dry powder, methanol is added and filtered. The methanol solution contains glycoflavones.

The glycoflavones are separated best by PC using water as the solvent. On fuming with ammonia or spraying with 10% Na_2CO_3, these compounds will give very stable bright yellow colors. The glycoflavones of apigenin (vitexin and isovitexin) will have the same spectrum as that of apigenin. Similarly orientin and isoorientin, the C-glycosides of luteolin, will have the same spectra of the parent compound. 6-Hydroxylated flavones (scutellarein & derivatives) are seen in advanced gamopetalous groups. They are visibly yellow (on chromatograms), brown in UV and turn yellow with alkali. They have λ max 280 - 290, 336.

Flavonols

The procedures for extracting flavonols are the same as that of flavones. These compounds are visibly yellow, yellow in UV and turn dark yellow on treatment with alkali. The color with alkali disappears on standing.

The three common flavonols, myricetin, quercetin and kaempferol are present in tea leaves. Myricetin has the lowest Rf in 30% acetic acid, turn green (color disappears soon) on spraying with 10% Na_2CO_3 and has a λ max 255, 372. Kaempferol has the highest Rf, turn yellow with 10% Na_2CO_3 (color persistent) and possesses peaks at 267, 367. Quercetin, the most common flavonol, is intermediate in Rf, turns yellow-brown with alkali (color fades on standing) and has λ max 255, 370. Reader is referred to section 1.3.2.4 for analysis of flavonols of tea. The methoxy derivatives of these compounds have spectra similar to the parent compounds.

Anthocyanins

These are red pigments, present in flowers, fruits and young/old leaves. Anthocyanins get decomposed on exposure to light. Only the salts of anthocyanins are stable and therefore these compounds are extracted by acidulated water (1% HCl) or acidic alcohols. The glycosides are separated in PC using 15% acetic acid. The aglycones (anthocyanidins), prepared by hydrolysis (7% HCl) and extracting in amyl alcohol, are separated well in 30% acetic acid or in Forestal. All of them are visibly red-magenta in color and turn blue/green on spraying with alkali and the color eventually disappears. The three common anthocyanidins, delphinidin, cyanidin and pelargonidin (corresponding to myricetin, quercetin and cyanidin), can be easily identified by Rf and λ max. Delphinidin has the lowest Rf and λ max at 546 nm. Pelargonidin has highest Rf and a peak at 520 nm while cyanidin shows medium Rf and a λ max 535 nm (Also see section 1.3.2.5).

Chalcones

Chalcones are dark yellow pigments found in a few plants like *Butea* and liquorice. The methods used for flavones are followed here. In UV they appear dark brown changing to a deep red color with ammonia or alkali. Chalcones give a broad peak between 365 and 390 nm. The spectral maxima of butein are 263, 382 and that of isoliquiritigenin are 235 & 372.

Aurones

Aurones also are rare in nature, found in advanced groups like Asteraceae. They also are yellow pigments which appear bright yellow in UV and turn orange - red with ammonia. These compounds possess an absorption peak in the visible region between 390-430 nm. The same solvent systems used for other flavonoids can be used here. The most common aurone, aureusidin exhibits spectral maxima at 254, 269 and 399nm.

Proanthocyanidins

The components of this group are essentially colorless, but on treatment with acid they liberate anthocyanidins and phlobaphenes, a group of complex polymers generally known as tannin reds. The different proanthocyanidins are distinguished by the anthocyanidins they liberate. Thus delphinidin, cyanidin and pelargonidin are produced by prodelphinidin, procyanidin and propelargonidin. To test for proanthocyanidins the plant material is hydrolyzed with 2N HCl for 30 min and the anthocyanidins are extracted from the hydrolysate with amyl/isoamyl alcohol. This extract is chromatographed and identified by using the same methods followed for anthocyanidins.

II. Betacyanins

Betacyanins are nitrogen containing indole derivatives, which are mutually exclusive to anthocyanins. All the plants of Caryophyllales (for e.g. Beet root, cocks comb, *Bougainvillea* etc) contain these red pigments while the rest of the plant world possess anthocyanins. Betacyanins also are water-soluble but are best extracted by methanol. They turn yellow on addition of alkali (anthocyanins turn blue) and thus are distinguished from anthocyanins. Betacyanins have high Rf in 1% aqueous HCl and low Rf in BAW. Similar to anthocyanins, betacyanins also have an absorption maximum in the range of 532-554 nm.

III. Alkaloids

Very few alkaloids like berberine or pelletierine are used as dyes. Methods for testing alkaloids are described in section 1.3.2.7.

For chromatography on paper, BAW is the solvent system preferred. The developed chromatograms are sprayed with Dragendorff's reagent when alkaloids appear as orange - red spots.

IV. Carotenoids

Carotenoids, being lipid soluble, are extracted in methanol or acetone. From fresh tissues also extraction is possible. The extract is diluted with water and reextracted with ether. The ether extract is concentrated at low temperature and chromatographed.

Since carotenoids are photosensitive, extraction and chromatography are carried out preferably in dimly lighted room. Chromatography is done by PC or TLC. For TLC, MgO [with petroleum (90-110°) - benzene (1:1)] or silica gel (Ethylene dichloride - ethyl acetate; 4:1) are used. For PC, petroleum ether- acetone (9:1) is used. All the carotenoids being colored can be detected easily.

All carotenoids show a characteristic spectrum between 400 and 500 nm, with a major peak around 450 nm and two minor peaks on either side.

Crocin from saffron is a glycoside and therefore it is water-soluble and moves in BAW solvent system in PC. The spectrum has peaks at 417, 437 and 458 nm.

V. Chlorophylls

Chlorophylls are lipids having a tetrapyrrole (porphyrin) nucleus with a chelated magnesium atom in the centre and a hydrocarbon (phytol) side chain. They also are labile in light and are to be protected from degradation. The enzyme chlorophyllase present in the plant removes the phytol chain and produce chlorophyllides. The central Mg atom also is lost during processing, producing protochlorophylls. Both these groups of decomposition products are dull grayish green in color and are relatively stable.

Chlorophylls are extracted by acetone or petroleum ether. The concentrated extract can be chromatographed in PC using petroleum ether (60-80°) - acetone (9:1). Chlorophyll a will be the blue green spot/band with higher R_f and chlorophyll b is yellow green with lower R_f. Chlorophyllides and protochlorophylls remain near the base line.

Chlorophylls give a spectrum having two peaks, the first peak in UV and another in visible region. The peak at visible region (between 600 - 700) is more distinctive. Chl.a gives a peak at 663 and Chl.b at 645 nm. The amount of different chlorophylls can be measured in these respective peaks.

3.9 Dyeing Techniques

The methods presented here are essentially those practiced in Maharashtra, Mysore, Tamil Nadu and Gujarat.

The clothes to be dyed are made into different types of pieces like sarees, robes, scarves etc. The clothes are subjected to a preliminary process of preparation which involves different steps such as washing, bleaching, dunging, galling, aluming, second washing and finishing.

1. *Washing:* Since the clothes contain materials like starch, paste, grease and other dirt during the process of spinning and weaving, these are removed by washing the clothes thoroughly and drying. However to remove grease, bleaching is resorted.
2. *Bleaching:* Bleaching is done by boiling the material in a solution containing chloride of lime or bleaching powder. This removes the grease and other oily material from the clothes.
3. *Dunging:* In this process the cloth is boiled with a solution containing cow dung. The dung contains phosphate of lime and aluminium salt and has a detergent effect on the fibers. Now dung is being replaced by phosphates of soda and lime, arsenites of soda or silicates of soda.

   The cloth is soaked in a solution of water and dung (of sheep, goat, camel or cow). This may be done in earthenware pots, utensils or in specially built pits. In Sind, potash is often added. The duration of dunging is one night. Next morning the cloth is taken out washed thoroughly (with thrashing) and dried in the sun with occasional sprinkling of water (this will help bleaching in the sun). The clothes are dried for the whole day in the sun.

   In some places the dunging and washing is repeated. In Sind, the clothes are then steamed for a night.
4. *Galling:* This is a process of impregnating the clothes with tannins so that the alum applied subsequently will be reduced to an insoluble state and can be washed off easily. The tannins are easily absorbed by most textiles like cotton, linen and silk, but not wool. The sources of tannins are myrobalans (in South India), pomegranate rind and galls of *Tamarix* (in North India). For dyeing silk, pistachio galls are preferred. Galling is done by soaking the cloth in solution of powdered myrobalans or behada (*Terminalia bellirica*) for 1-3 days. The clothes now assume a yellowish tint. The clothes are then washed and dried.

5. *Aluming:* This process is very important because it makes all the water soluble dyes into compounds which are insoluble in water so that they are not washed off from clothes after dyeing. This is accomplished by means of a mordant. The mordant has an affinity to the fibre and to the dye molecules. Normally a salt of a metallic oxide is used. In most cases it is alum. The term mordant is applied to certain substances which alter the shade of color or brighten the color. In some places gum or a paste of *Tamarind* seeds are added to make the solution sticky.

6. *Washing:* This is to remove the excess tannin or salts formed by the decomposition of alum.

7. *Dyeing:* This is done by immersing the cloth in a solution of the dye in cold or hot water, until the coloring matter has been absorbed and the desired color is obtained. In many places the cloth is boiled with dye solutions.

These are the steps /principles for dyeing practiced by most of the Dyers in India, but the exact methods followed, being founded on the varying experience of the ancestors or the variations based on the guess work with little or no knowledge of chemistry. The main variations are in the washing and dyeing processes and the time taken over for the different stages. It will take 2-3 weeks or even three months to dye a cloth.

Silk Dyeing

The process of silk dyeing is comparatively shorter than that of cotton. First silk threads are boiled in hot water, dried in the sun and fumigated with Sulphur. Fuming is done by burning stick sulphur in a stove in a stone chamber. Next the silk is steeped in a solution of lime and soda/potash in the ratio 2:1 and the solution is boiled. This alkaline dye bleaches the silk and provides it sheen and softness. Then the silk is taken out and washed.

Mordanting, the next step, is done by steeping silk in a 10% solution of alum in water for a night, the silk is then washed and dyed.

The wool is first washed with water and then soaked in solution of lime or potash for 12 h. It is them mordanted by alum, oxalic acid or sulphuric acid and dyed in hot solution of dye.

Compound colors

1. Purple, violet, lilac and lavender colors are produced by mixing red and blue.
2. Green and shades of green from blue and yellow.
3. Orange and shades of orange from mixing red and yellow.
4. Brown, grey, drab, maroon and brick colors are produced by mixing black with other colors.

In a work book "Colors from Nature silk dyeing using natural dyes" (Seri, 2000) a large number of plants have been listed along with the detailed methodology of dyeing. In this work about 150 plants have been studied for their dyeing properties against silk. The plant parts used are mostly leaves, though in a few cases flowers, fruits or bark are used. Same dye is used separately with different mordants. It is found that all these plants can be used as dyes for silk, but the colors produced is mostly light and dull. The plants included here are *Punica granatum* (fruit rind, leaves), *Eugenia jambolina* (bark), *Diospyros embryopeteris* (fruit), *Terminalia chebula* (Fruit), *Mesua ferreà* (Flower), *Butea frondosa* (Flower), *Eucalyptus globulus* (leaves), *Aerva lanata* (leaves), Areca nut (bark), *Cassia fistula* (fruit), *Exocaria agellocha* (bark), *Morus indica* (leaves), Sappanwood, *Maranta arundinacea* (Leaves), Bamboo (leaves), *Streblus asper* (leaves), *Cassia tora* (Leaves), *Ceriops roxburghiana* (bark), *Ruellia tuberosa* (flowers and leaves), *Cajanus cajan* (leaves), *Heliotropium indicum* (leaves), *Albizzia lebbek* (Gum), *Bassia latifolia* (leaves & bark), *Shorea robusta* (Bark), *Terminalia tomentosa* (bark), *Semecarpus anacardium* (bark), beet root, *Eclipta alba* (leaves), *Sweitenia mahogani* (tender leaves), Rose, *Cephalandra indica* (leaves), *Melia azedarach* (leaves), *Centella asiatica* (leaves), *Lantana camara* (leaves), *Blumea lacera* (leaves) *Bombax malabaricum* (flowers), *Bougainvillea* (flowers), *Mangifera* (leaves & bark), *Bixa orellana* (seeds), *Nyctanthes arbor- tristis* (leaves and flowers), *Ziziphus jujuba* (bark), Orange (fruit skin), *Thevetia neriifolia* (flowers), *Artocarpus integrifolia* (bark), *Mimusops elengi* (bark), *Nymphaea nouchalli* (leaves), *Prunus persica* (stem), *Polyalthia simiarum* (root), *Anogeissus latifolia*, *Calotropis gigantea* (flowers), *Ficus religiosa* (bark), *Terminalia bellerica* (fruit), *Spondias dulcis* (bark), *Piper longum* (leaves), *Vicus cunia* (leaves), *Tamarindus indica* (bark), *Moringa pterigosperma* (bark), cabbage, banana (corm), paddy (straw), *Carapa obovata* (wood), *Ammania baccifera* (leaves), *Celosia cristata* (flowers), *Vitex negundo* (leaves), cotton (leaves), *Ocimum basilicum* (leaves), *Lagenaria vulgaris* (leaves), *Emblica officinalis* (fruit), *Clerodendrum infortunatum* (leaves), *Marsilea quadrifoliata* (leaves), *Paedaria foetida* (leaves), *Pachyrrhizus erosus* (tuber), *Bacopa monnieri* (leaves), *Clitoria ternatea* (leaves), *Hibiscus rosa-sinensis* (flowers), *Amaranthus gangeticus* (leaves), *Erythrina indica* (flowers), *Melia azadirachta* (bark), *Acacia moniliformis* (bark), *Quisqualis indica* (leaves), *Adhatoda vasica*, potato (leaves), *Amaranthus gangeticus* (leaves and flowers), *Datura fastuosa* (flowers), *Eicchornia crassipes* (leaves), *Jatropha gossypifolia* (leaves), sandalwood (leaves), *Caesalpinia pulcherrima* (leaves), *Vinca rosea* (leaves), *Bryophyllum* (leaves), *Cynodon dactylon* (leaves), Fenugreek (leaves), *Dolichos* (leaves), *Abutilon indicum* (leaves), *Ipomoea aquatica* (leaves), *Delonix regia* (leaves), *Cleome viscosa* (leaves), *Andrographis paniculata* (leaves), *Impatiens balsamina* (Flowers), *Phoenix sylevestris* (leaves), turmeric, *Borassus flabellifer* (fruit), *Anacardium occidentale* (nut), *Aloe* (leaves), *Ixora* (Flowers), *Corchorus olitorius* (leaves), *Pongamia glabra* (leaves), *Tinospora cordifolia* (leaves), *Leucas linifolia* (leaves), *Acalypha indica* (leaves), *Bauhinia* (flowers), *Scoparia dulcis* (leaves), *Justicia gendarussa* (flowers), *Malachra capitata* (leaves), *Anisomeles ovata* (leaves), *Tridax procumbens* (leaves), *Polygonum barbatum* (leaves), *Atylosia scarabaeoides* (leaves), *Sapindus mukorossi* (leaves), *Aerva aspera* (leaves), *Saraca indica* (flowers), *Annona squamosa* (leaves), *Lippia geminata* (leaves), *Eupatorium odorum* (leaves), *Cosmos bidentata* (flowers), *Hibiscus sabdariffa* (leaves), *Hibiscus mutabilis* (flowers), *Ageratum* (leaves), *Phaseolus mungo* (leaves), *Cleome viscosa* (leaves), *Gloriosa superba* (petals), *Avicennia officinalis* (fruit) and *Sterculia foetida* (fruit).

□□□

4

Biopesticides

Synthetic pesticides have become a menace in the present world. Due to their non-degradable and bioactive nature they pollute the environment and cause irreparable damage to man and the beasts. 25 Million cases of acute occupational pesticide poisoning occur in developing countries each year. 14% Of all known occupational injuries and 10% of all total injuries are caused by pesticides. Obsolete pesticides stored in developing countries amount to 20000 tonnes in Africa alone.

About 80,000 tons of pesticides are used in India annually, mostly on cotton and rice. While cotton is planted on about 5% of the total cultivable area (on about 8 million hectares out of 170 m.), it accounts for 45% of pesticide application. Rice accounts for another 23%, vegetables and fruits account for 7%, jowar 8.9% and wheat 6.4%.

Pesticide resistance to the two most common pesticides (DDT and HCH) and the case in which bollworm developing resistance to almost all pesticides in a number of regions (other pests of cotton like white fly and jassid also developing resistance in some area) means that a large proportion of produce is lost to pests. This is mainly due to excessive and indiscriminate case of pesticides as well as due to the pesticides of spurious quality sold in small towns and villages. In Bidar (Karnataka) where there occurred an acute problem of pest resistance, more than 50 brands of pesticides were found to be substandard in 1998-99. In another incident the licenses of more than 115 pesticide producers were cancelled for selling substandard pesticides. In case of sub-standard pesticides the pests are exposed to the low concentration of pesticides for a longer and repeated period of time and this contributes to the building up of resistance in pests.

The residual pesticides remaining in plants for longer times enter the foods such as dairy products and meat. The incidence of pesticide residue is much higher in India than in developed countries. In one study more than 80% of milk samples tested in India was found to contain residues of DDT and HCH. In another study, breast milk samples collected from mothers in Punjab, were found to contain DDT and benzene hexachloride, two suspected carcinogens, in about 21 times more than the acceptable limits.

Due to the very many reasons cited above, the whole world now turns to the biological ways of pest control. The biopesticides include all the products derived from living organisms. They are biodegradable, mostly harmless to humans and other animals and in many cases improve the performance of the crops. Though, at present the biological agents cannot replace the synthetic pesticides completely, they can, to some extent, reduce the deleterious influence of the latter in the environment and improve the quality of life.

Biopesticides are being promoted in India by the Ministry of Agriculture and Department of Biotechnology, via the Central IPM center at Faridabad, The National Centre for IPM (NIPM under ICAR) and the Directorate of Biological control. NIPM was successful in destroying stem borer and leaf folder of rice using *Trichogramma* in some parts of west UP.

Biological Pest Control Agents include 1) microorganisms that control pests (microbial pesticides), 2) naturally occurring compounds that control pests (biochemical pesticides and or botanical pesticides), 3) pesticidal substances produced by plants containing added genetic material (Plant incorporated protectants) or PIPs, 4) parasitoids and predators, 5) pheromones and other semiochemicals, 6) ecdysteroid agonists, 7) nematodes and 8) other substances. By the end of 2001, there were approx. 195 registered biopesticide compounds and 780 products. The known biopesticides are the following.

4.1 Microbial Biopesticides

Since the insects and other pests are hosts to a number of parasitic micro-organisms, a large number of these microbes are found to destroy many harmful insects. They include bacteria, viruses, fungi and protozoa. Bacteria which are used thus are *Bacillus thuringiensis, Bt.* var. *tenebrionis, B. sphaericus, B. popilliae, Serratia entomophila, Agrobacterium radiobactor* and *Ampelomyces*. Nuclear polyhedrosis viruses, granulosis viruses, non-occluded baculoviruses are the parasitic viruses used. Among the fungi, *Beauveria, Metarhizium, Entomophaga, Zoophthora, Paecilomyces, Normurarea, Pseudomonas, Candida* and *Geocladium* are used in biological control. The protozoa used thus are *Nosema, Thelohania* and *Vairimorpha*. Entomopathogenic nematodes such as *Steinernema, Heterorhabditis* and *Romanomermis* also are used to control insect pests.

4.1.1. Bacteria

1. *Agrobacterium radiobacter* : Common worldwide near plant roots. It controls *Agrobacterium tumefaciens* which causes crown gall infection in certain plants by competing directly for food and space. *A. radiobacter* normally releases a protein which inhibits *A. tumefaciens.* Found in USA and Australia.
2. *Bacillus popilliae*: This gram-negative spore forming bacterium is grown in a rich medium and the spores formed are sprayed in soil. These spores are ingested by grubs in the soil and multiply in the hemocoel. The infected larvae do not moult and then die. Used to control larvae of Japanese beetles, chafers, some May & June beetles in USA.
3. *Bacillus sphaericus:* This bacterium strain 2362 is used to control specific types of mosquitoes esp. *Culex* (including some that transmit disease such as encephalitis), *Psorophora* and *Anopheles* spp. Once ingested by larva of insects, these bacteria produce lethal doses of an endotoxin killing the mosquito larvae. This bacterium is sprayed on the ground / water. Used in USA.
4. *Bacillus subtilis*: This is a gram - positive harmless bacterium growing in the roots of host plants and thus prevents fungi like *Rhizoctonia, Fusarium, Alternaria* and *Aspergillus* which cause root rots and seedling diseases. This is applied directly on seed, also used for production of enzymes and speciality chemicals useful for cotton, peanut, soybean and vegetables in USA.
5. *Bacillus thuringiensis* subsp. *kmurstaki* (Btk).

   This bacterium produces parasporal proteinaceous, crystal inclusion bodies during sporulation. Upon ingestion, these are insecticidal to larvae of the order Lepidoptera and to both larvae and adults of a few Coleoptera. Once taken inside the insect, the crystal proteins are solubilised and the insect gut proteases convert the original pro-toxins into a combination of up to four smaller toxins. These hydrolysed toxins bind to the insect's midgut cells at high affinity, to the specific receptor binding sites, where they interfere with the potassium-ion dependant active amino acid support mechanism. This disruption causes the formation of large pores that increase the water permeability of the membrane. A large uptake of water causes cell swelling and eventual rupture, disintegrating the midgut lining. Different toxins bind to different receptors in different insect species and with varying intensities and this explains the species specificities.

The crystal inclusions derived from Btk are generally Lepidopteran specific. Because they have to be ingested and processed within the gut of insects, they are often slow in action i.e. will take two to forty-eight hours in comparison to the conventional chemicals. The toxin results in starvation leading to death. This will take several days. Different strains and serotypes have been developed by different companies. In addition to producing endotoxins, many strains of Bt are potent insect pathogens.

The target pests of Btk are *Helicoverpa armigera* (American boll worm), *Pectinophera* (pink boll worm), *Erias insulana* (Spotted boll worm), *Plutella xylostella* (diamond black moth), *Leptinotarsa decemlineata* (Colarado potato beetle), *Crocidolomia binotalis & Pieris brassicae* of Crucifers, *Helothis armigera* of tomato and sunflower, *Spodoptera litura* of soybean, *Athlia lugens proxima* of raddish, *Papilio demoiecus* of citrus, *Achaea janata* of castor, *Acigona steniellus* of sugar cane, *Earias vitella* of Okra, *Spoiadea recurvalis* of Amaranthus and *Phthorimaea operculella* of potato and many forest insects.

Bt is available as a combination of endotoxin crystals and living bacterial spores, in the form of suspension, granules, oil-miscible suspension, dispersible powder or a wettable powder. It is sprayed while the larvae are small and the process is repeated every 5-7 days when infestation is high.

6. *Bacillus thuringiensis* var. *tenebrionis*: This occurs naturally in soil and plants. It produces a crystalline protein which is specific to certain insects. The bacterium can be sprayed. The bacteria should be eaten by the insects to be effective. It is active against larvae of Colarado beetle. Used in USA.

   The endotoxins of Bt are also produced by transferring the genetic material that codes for two protein toxins to *Pseudomonas fluorescens*. The advantage here is that *P. fluorescens* produces higher concentration of toxins Cry1Ac & Cry1C, the toxins remain in *P. fluorescens* cells (encapsulated) making collecting the toxin relatively easy, and *P. fluorescens* does not produce the large number of additional toxins made by Bt, which may harm the non-target species. These bacteria are applied by spraying.

Other spp. of Bt used are (1) *Bt* subsp. *israelensis* (used against black flies, mosquitoes, fungus gnats etc) (2) *Bt.* var. *sandiego* (used against larvae of Colarado beetle and adults of elm leaf beetle) (3) *Bt.* var. *aizawai* (useful on Lepidoptera of vegetables and corn) and (4) *Bt.* var. *kursataki* (as above)

7. *Pseudomonas chlororaphis*: This bacterium also is useful in controlling *Pythium, Rhizoctonia* and *Fusarium* by (1) producing cytokinins, which help the plant to limit fungal damage, (2) producing antibiotics or (3) by producing a protein which binds with iron that the fungi might need for growth and reproduction. USA.
8. *Pseudomonas syringae*: This is a harmless strain used to protect post harvest pathogens on apples, pear, citrus etc. They out compete the pathogenic fungi for space and nutrients on the fruit, thereby preventing the fruits from rotting.
9. *Serratia entomophila*: This bacterium is a non-spore forming, gram-negative, anaerobic, rod-shaped member of the family Enterobacteriaceae. It adheres to the foregut intima and are numerous around the insect heart, the insect ceases feeding and clear the gut depleting fat bodies. Effective against grass grub (*Costelytra zealandica*) and the root knot nematode *Meloidogyne incognita* which attacks the roots of tomato plants. Used in New Zealand.
10. *Streptomyces griseoviridis & S. lydicus*: These organisms colonize plant roots early and produce toxins to attack harmful fungi. Used against *Pythium, Fusarium, Alternaria, Rhizoctonia, Phomopsis, Phytophthora* etc. which cause wilt, seed, root and stem rots. Common in USA
11. *Pseudomonas cepacia* - USA: This bacterium controls soil pathogens like *Rhizoctonia, Fusarium* and *Pythium.*

4.1.2. Nuclear polyhedrosis viruses

More than 250 viruses are there which are pathogenic to insects. Some of them are occluded (the nuclear material embedded in protein matrix); others are non-occluded (without protein matrix). A major group of insect pathogens are *polyhedroses*, forming polyhedral inclusions (seen in a light microscope) in the infected cells. These inclusions consist of a matrix of protein molecules that occur in lattice arrangement and numerous virions (viruses) which are embedded within the protein matrix. In *nuclear polyhedroses*, the polyhedra occur within the nucleus of infect host cell. The virions are rod-shaped and consist of a double-stranded DNA surrounded by a protein coat and enclosed within an envelope. The polyhedra enlarge and cause the rupture of the nucleus and the subsequent disintegration of the cell. The nuclear polyhedroses are found in the larvae of *Diptera*, *Hymenoptera* and *Lepidoptera*. In *cytoplasmic polyhedroses*, the polyhedra occur in cytoplasm, and are icosahedral and contain double-stranded RNA. They occur in larvae of Lepidoptera.

Some nuclear polyhedrosis viruses have been used in the control of boll worm in cotton (in California), gypsy moth (in North-eastern US) and spruce budworm (Canada). *Granulosis viruses* produce granuloses – small oval inclusions each containing a protein matrix in which two rod shaped DNA containing viruses are embedded in the nucleus of host cells.

Baculoviruses are a group of nuclear polyhedroviruses used against codling moth (Apple, pear, Walnut and Plum), cabbage moth, American bollworm, potato tuber moth, *Spodoptera littoralis,* (Cotton, corn, tomatoes), *Helicoverpa zea* (Tobacco), *Heliothis virescens* (Cotton), *Spodoptera exigua* (Beet, vegetables), *Anagraphis falcifera* (Celery looper), *Autographa californica* (Alfalfa) and *Lymantria dispar* (Gypsy moth – on wood).

The infectious virus particles (virions) on ingestion by insect larvae, reproduce in midgut cells and other tissues and infect the epidermis, fat bodies and blood cells. The larvae usually die from 5-12 days. Just before dying, the larvae often crawl to the tops of plants where they die and decompose. Millions of polyhedra, contained in the fluid mass of disintegrating larvae, fall on leaves or leaf litter, which are then ingested by other larvae.

Nonoccluded baculoviruses (nudiviruses) are not occluded in polyhedra. They are less effective in peroral inoculation, but highly injected to insect hemocoel.

1. *Anagrapha falcifera* : This is a baculovirus (large rod-shaped DNA containing viruses) isolated from Lepidoptera, Hymenoptera and

Diptera. This virus on oral ingestion by moth, enters midgut cells, replicate and pass on to other tissues. It is used to control celery looper in USA.

2. *Autographa californica* (NPV): This baculovirus, similar to *A. falcifera,* is used to control alfalfa looper in USA. They are introduced to the larva by food.

3. *Helicoverpa zea,* Effective against *Heliothis virescens* and other Lepidopterans in USA.

4.1.3. Fungi

A large number of fungi are known to infect insects. Most of the entomopathogenic fungi are seen in the orders Entomophthorales and the Moniliales. All these are mitosporitic fungi except for *Lagenidium.*

1. *Beauveria bassiana:* A Deuteromycetous fungus reproducing by conidia, this fungus is used to control adults and larvae of many kinds of soft-bodied insects, eggs of lepidopteran pests like moths, mole cricket, ants, boll weevil, thrips etc. The fungus grows on the insects, secreting enzyme that weaken the outer coat of insects, getting inside and growing eventually killing the pest.

2. *Beauveria brongniartii,* specifically useful against *Hoplochelis marginalis* and *Melolontha melolontha.*

   It was also found to be active against fourth instar nymphs of sweet potato whitefly, *Bemisia tabaci,* and greenhouse whitefly, *Trialeurodes vaporariorum,* at a concentration of 1×10^7 conidia/ml. All isolates were pathogenic for both whitefly species, whereas mortality rates varied from 3 to 85% (Quesada-Moraga *et al.,* 2006). Used in countries of USA & Europe.

3. *Burkholderia cepacia* (Syn. *Pseudomonas cepacia)*: Controls soil pathogens like *Fusarium, Pythium* and some nematodes. Used in USA.

4. *Candida oleophyla* : This yeast is used to control post harvest pathogens like *Botrytis* and *Penicillium.* It, on spraying, colonizes the fruit surfaces especially the wounded tissues and thus prevent other microbial contamination. Used in USA.

5. *Clitocybe inversa* (Mushroom) : Chemical investigation of pest insect control agents produced by this mushroom yielded a nucleoside clitocine, which showed strong insecticidal activity against the pink bollworm *Pectinophora gossypiella* (Kubo *et al.*, 1986).
6. *Entomophaga grylli* : Two phenotypes of this zygomycetous fungus i.e. *E. macleodii* and *E. calopteni* attack all grasshoppers which are major pests of crops.
7. *Entomophthora muscae* : This fungus attacks housefly and destroys the host.
8. *Gleocladium* spp. esp. *G. virens* : These fungi found in soil, control *Rhizoctonia solani* and *Pythium* spp. which cause damping off and root rot. Used in areas of USA.
9. *Lagenidium giganteum* : This fungus is specific to larvae of most of mosquitoes and is effective only in fresh water fields or ponds. Seen in USA.
10. *Gloeophyllum odoratum* : The brown-rot fungus *Gloeophyllum odoratum*, when elicitors chitin, chitosan and D- (+)-glucosamine were used, produced linalool, citronellol, geraniol and methyl *p*-methoxyphenylacetate or drimenol depending on the culture type and elicitor. The composition of volatiles in the natural fungus was different from that of the cultivated fungus since the major compound was methyl *p*-methoxyphenylacetate. The volatile oils were toxic to larvae of the brine shrimp, *Artemia salina* (Kahlos *et al.*, 1994).
11. *Metarhizium anisopliae* var. *acridium* : This deuteromycetous fungus enters the locusts and grasshopper hosts by spores. Within the host it multiplies rapidly and the extensive mycelial colonization causes asphyxiation or starvation or by toxins released. Commonly used in USA.
12. *Metarhizium flavoviride* : This is especially effective against grasshoppers and locusts
13. *Metarhizium anisopliae* : Specific to soil inhabiting beetle and termites, used in USA.
14. *Paecilomyces fumosoroseus:* Useful against whitefly, thrips, aphids and spider mites.

15. *Neotyphodium uncinatum:* This common endophyte of meadow fescue (*Lolium pratense*, Syn. *Festuca pratensis*) produced loline, *N*-acetylnorloline, and *N*-formylloline when grown in the defined minimal media at pH 5.0–7.5, with both organic and inorganic nitrogen sources and sugars as carbon sources (Blankenship *et al.*, 2001). Lolines (saturated 1-aminopyrroli-zidines with an oxygen bridge) are insecticidal alkaloids found produced in symbioses of certain *Epichloë* (anamorph-*Neotyphodium*) species (fungal endophytes) with grasses, particularly of the genera *Lolium* and *Festuca*.

16. *Oospora destructor:* This entomopathogenic fungus yielded destruxins, which are cyclic hexadepsipeptides composed of a α-hydroxy acid and five amino acid residues. The name "destruxin" is derived from the species name "destructor". Individual destruxins differ on the hydroxy acid, N-methylation, and R group of the amino acid residues; where established, the configurations of the amino acid residues are *S*, and those of the hydroxy acids are *R*. Destruxins exhibit a wide variety of biological activities, but are best known for their insecticidal and phytotoxic activities. The great interest in destruxins derives from their potential role as virulence factors in fungi, whether such microorganisms are useful insect biocontrol agents or detrimental, causing great plant disease epidemics (Soledade *et al.*, 2003)

17. *Phlebiopsis gigantea (Peniophora gigantea)*: This is a weakly parasitic fungus but poses no danger to healthy trees. But if it is applied on coniferous trees, it will prevent *Heterobasidion annosum,* the most damaging root pathogen. The hyphae of *P. gigantea* disrupt the hyphae of pathogen on contact causing disorganization and loss of membrane integrity. Commonly used in European countries.

18. *Pythium oligandrum* : Antagonizes *P. ultimatum* in sugar beets. Used in Slovak Republic

19. *Tolypocladium niveum* : *Tolypocladium niveum* is found to contain peptides such as efrapeptins, which occur as a crude mixture of five closely related compounds which were found to be active against two spotted spider mites, Colorado potato beetle, southern armyworm, and tobacco budworm in a foliar spray assay. Efrapeptin F (LC_{50} = 8.4 ppm) was about twice as potent as

efrapeptin D (LC_{50} = 18.9 ppm). ATPase inhibitory activity of the efrapeptins was demonstrated using mitochondrial preparations from two fungal species, *Metarhizium anisopliae* and *T. niveum*, and from an insect, house fly. Against house fly mitochondria, efrapeptin D was the most potent ATPase inhibitor among the five efrapeptins tested (Krasnoff, 1991).

20. *Trichoderma harzianum, T. polysporum and T. viride* : These are mycoparasites and are effective against tree wound pathogens and soil pathogens Used in Europe.
21. *Verticillium lecanii* : These are insect eating fungi used against greenhouse pests like whitefly, aphid etc.
22. *Ampelomyces quisqualis* : A fungus first isolated from a powdery mildew of certain plants, this can be grown in a liquid nutrient medium, spores are collected and they can be applied to crops to control powdery mildew. Used in USA.
23. Mushroom and toadstool carpophores: Powdered fruit bodies of 79 edible and poisonous mushrooms are found to inhibit development of *Drosophila melanogaster* (Mier *et al.*, 1996).

4.1.4 *Algae*

(1) *Laurencia pinnata* Yamada (marine red alga)

This marine red alga is found to contain two acetylenic cyclic ethers laurepinnacin and isolaurepinnacin which are insecticidal in nature (Fukuzawa and Masamune, 1981).

(2) *Plocamium cartilagineum* (Rhodophyceae)

Four polyhalogenated monoterpenes; based on the 1-(2-chlorovinyl)-2, 4,5-trichloro-1,5-dimethylcyclohexane skeleton were isolated from this alga of the Chilean coast, together with four known cyclic monoterpenes. These compounds are found to be insecticidal/ ascaricidal and fungicidal in nature (San-Martin *et al.*, 1991).

(3) *Plocamium telfairiae*

The red alga *Plocamium telfairiae* yielded two insecticidal polyhalogenated monoterpenes; aplysiaterpenoid A and telfairine. Both compounds showed strong insecticidal activities against the German cockroach (*Blatella germanica*) and mosquito larvae (*Anopheles gambiae*). The dieldrin-resistant strain of mosquito larvae (*A. gambiae*) which exhibited the significant cross-resistance to 1,2,3,4,5,6-

hexachlorohexane (γ-BHC) also revealed weak but definite cross-resistance to aplysiaterpenoid A and telfairine. The electrophysiological symptom that telfairine increases the frequency of spontaneous discharge and prolongs the synaptic after-discharge of the central nervous system of the German cockroach (*B. germanica*) in the same manner as γ-BHC indicated that telfairine also has some excitatory effect on the central nervous system. These monoterpenes from marine alga seem to interact with the picrotoxinin receptor, indicating that their mode of action is cyclodiene-type (Watanabe, *et al.*, 1990).

4.1.5. Protozoa

(1) *Nosema locustae*: This protozoan infects and kills grass hoppers and mormon crickets in USA.

4.2. Botanical Pesticides

More than 2000 plants are known to have insecticidal properties (Anonymous, 1958, 1975). Of these, the plants commonly used or about which experimental evidences are available are explained below. These plants are arranged on the basis of the active components they contain. The active components belong to the classes like volatile oils, sesquiterpenes, diterpenes, triterpenes, limonoids & quassinoids, acetogenins, alkaloids, phenolics, rotenones, tannins, glycolipids etc.

4.2.1. Volatile Oils

Since many volatile oils contain insect pheromones, these oils repel or deter other insects. They are found very useful in controlling pestiferous insects in confined environments such as greenhouses and granaries and are found effective as fumigants for stored-product pests as well as for parasitic mites of honey bees (Rice and Coats, 1994; Calderone and Spivak, 1995).The essential oils containing linalool, terpen-4-ol, 1,8-cineole and pulegone, isolated from four Lamiaceae members repelled and strongly reduced the fecundity of the carmine spider mite *Tetranychus cinnabarinus* (Mansour *et al.*, 1986).But for greenhouse crops some of the essential oils are found to be phytotoxic due the closed environment.

A total of 53 plant essential oils were tested for their insecticidal activities against eggs, nymphs and adults of *Trialeurodes vaporariorum* using an impregnated filter paper bioassay without allowing direct contact. Bay, caraway seed, clove (leaf), lemon, eucalyptus, lime, pennyroyal, peppermint, rosewood, spearmint and tea tree oils were highly effective in this vapour phase (Choi *et al.*, 2003). Another such study was conducted by Kim *et al.* (2003) on *Sitophilus oryzae* and *Callosobruchus chinensis.*

Essential oils of cedar, lavender, Eucalyptus, pennyroyal and Citronella are used as insect repellants on pets and humans, primarily to discourage fleas and mosquitoes. These oils are believed to interfere with insect perceptions of chemical cues (odours) that help them to locate their mammalian hosts.

Monoterpenes

The individual monoterpenes which exhibits insect repellant or insecticidal activity are the following.

1. Limonene and linalool

Citrus extracts and two components of these extracts are added recently to the list of commercial botanical insecticides (Weinzierl, 1998). Limonene (specifically *d*- limonene) and / or linalool are reported to act as neurotoxins, insect growth regulators and repellents and fumigants (Coats *et al*, 1991). Limonene affects the sensory nerves of the peripheral nervous system, but it does not inhibit acetyl cholinesterase. In insects it causes spontaneous stimulation of sensory nerves and subsequent signaling to motor nerves that results in muscle twitching, convulsions followed by paralysis. Both these compounds are very safe to humans and other mammals. The activity of linalool is not known exactly, but it is synergized by piperonyl butoxide.

Antioxidant compounds like piperonyl butoxide (PBO) and N-acetyl bicycloheptene dicarboximide (MGK 264) increase the shelf life of insecticides like pyrethrum, linalool etc. by delaying the oxidative damage of these compounds. They are called synergists. They are normally added to the insecticides in the ratio 2:1 or 10:1 (synergist: insecticide).

Since limonene and linalool are volatile, they evaporate rapidly from treated surfaces and therefore they are used in formulation with synergists like PBO or MGK. They are used in flea dips and shampoos for cats and dogs and in insecticidal soaps. Limonene is useful for all external pests of pets including fleas, lice, mites and ticks. Linalool repels mosquitoes, fleas, mites, ticks and spiders and best for dog and cat flea sprays and for carpets.

2. **Cineole** from *Eucalyptus* spp. and **limonene** from *Citrus* spp. were bioassayed to determine possible fumigant, contact, and ingestion activity against *Rhyzopertha dominica* (F.), lesser grain borer, and *Tribolium castaneum* (Herbst), red flour beetle, which are important pests of stored grain. It was seen that both these compounds are insecticidal against the two insects, mainly in the contact and/or ingestion action (Prates *et al.*, 1998).

3. **Cinnamaldehyde:** A methylene chloride extract of the spice, cinnamon, *Cinnamomum aromaticum* Nees, was shown to be insecticidal to *Tribolium castaneum* (Herbst) and *Sitophilus zeamais* Motsch. The contact fumigant and antifeedant effects of cinnamaldehyde were tested against *T. castaneum* adults and larvae and *S. zeamais* adults. *T. castaneum* and *S. zeamais* adults showed similar susceptibility to the contact toxicity of cinnamaldehyde, both having an LC_{50} of 0.7 mg cm^{-2} and an LC_{95} of 0.9 mg cm^{-2}. However, cinnamaldehyde had a higher level of fumigant toxicity to *T. castaneum* than to *S. zeamais*, with LC_{50} values of 0.28 and 0.54 mg cm^{-2}, respectively, and LC_{95} values of 0.32 and 1.78 mg cm^{-2}, respectively. *T. castaneum* adults were more susceptible than larvae to the contact and fumigant actions of cinnamaldehyde. It significantly ($P < 0.05$) reduced food consumption, growth and dietary utilisation in *T. castaneum* larvae, and had obvious antifeedant action against the larvae at concentrations of 27.2 and 54.4 mg g^{-1} food. Antifeedant action was observed to increase with increasing cinnamaldehyde concentrations. The combined contact, fumigant and antifeedant properties of cinnamaldehyde make it a potentially useful grain protectant (Huang and Ho, 1998). Cinnamaldehyde is also used on ornamentals and many crops to control various insects and to attract corn root worms and the related beetles as also to control fungi and many insects. This is preferred for many food crops, cotton, ornamentals and processed foods.

4. **Geraniol** attracts Japanese beetles ant repels cats and dogs

5. **Eugenol, isoeugenol and methyleugenol**

Eugenol, isoeugenol and methyleugenol showed similar contact toxicity to *Sitophilus zeamais* (LD_{50} values approximately 30 µg/mg insect). For *Tribolium castaneum*, the order of potency of these chemicals compared by the LD_{50} level was isoeugenol>eugenol>methyleugenol. Furthermore, *S. zeamais* was generally more susceptible than *T. castaneum* to these three compounds except for isoeugenol based on the LD_{50} level. When sublethal doses (doses that produce no appreciable mortality) of eugenol, isoeugenol and methyleugenol were topically applied to adults of either species of beetles pre-treated with deltamethrin, the toxicity of deltamethrin was enhanced. A flour disc bioassay using no-choice tests was employed to assess, in terms of food consumption and growth rate, the susceptibility of *S. zeamais* and *T. castaneum* to media treated with eugenol, isoeugenol, and methyleugenol. Only eugenol significantly ($P<0.05$) reduced food consumption (RCR) in the adults of *S. zeamais* at a concentration of 13.2 mg/g food. However, it reduced the growth rate (RGR), food consumption and food utilisation (ECI) in adults and larvae of *T. castaneum* at concentrations of 35 and 99 mg/g food, respectively. Isoeugenol reduced the RGR and RCR in *S. zeamais* adults and *T. castaneum* adults and larvae, in a concentration-dependent manner, but only reduced ECI in the adults of *T. castaneum* at the concentration of 76 mg/g food. In comparison, methyleugenol reduced the RGR and RCR in the three groups of insects, and the ECI of *S. zeamais* adults and *T. castaneum* larvae (Huang, *et al.* 2002).

Eugenol is found to attract Japanese beetles and kills insects, recommended for many food crops, ornamentals, buildings (inside and outside) and for pets. Eugenol is used on ornamentals and many crops to control various insects.

6. **α-Ionone** attracts adult rose chafers; used only outdoors like lawns.

7. **1-Octen-3-ol**, present in clover and alfalfa attracts mosquitoes and dragon flies to electronic insect killers; used in non-agricultural sites only.

8. **2-Phenylethyl propionate** present in peanuts attracts Japanese beetles and kills insects, mites, ticks, spiders etc. Useful in food and feed crops and ornamentals both indoor and outdoor plants.

9. **1,2,4- Trimethoxybenzene** : present in squash flowers- attracts corn rootworms and cucumber beetles and useful for fruits, vegetables and feed crops.

10. **Indole** is another compound (not a monoterpene) which attracts corn rootworms and the corresponding beetles useful for fruits and vegetables.

The insect-repellant/insecticidal plants containing volatile oils are the following.

1. *Abelmoschus moschatus* Medic. (Malvaceae)

Syn. *Hibiscus abelmoschus* Linn.

Kasturi Bhindi

An erect annual/biennial hirsute herb up to 2m in height possessing irritating hairs, polymorphous leaves with lower ovate acute and upper palmately 3-7 lobed, large bright yellow flowers solitary or in few flowered racemes and 6-8cm long ovate acute hispid capsules containing many reniform musk-scented black seeds. Common in foothills of Deccan, Western Ghats and Himalayas.

Seeds, when crushed and steam distilled, yield up to 0.6% volatile musk seed oil / **ambrette seed oil.** The oil has a characteristic musk odour due to a ketone ambrettolide (a lactone of ambrettolic acid). Farnesol forms the major constituent of this oil, present up to 60%. The vol. oil also contains a large amount of higher fatty acids mainly palmitic acid. Seeds yield a fatty oil (15%, rich in linoleic acid and contain cephaelin and phosphatidyl serine), starch (14%) protein (2%) and β-sitosterol.

Seeds are used to protect woollen garments against moth. They are also used as a tonic, stimulant and carminative.

2. *Acorus calamus* Linn. (Araceae)

Vacha, Sweet flag, Calamus

Calamus is the sweet and aromatic rhizome of sweet flag, growing in marshy places of Europe, Asia and America. The rhizomes are furrowed longitudinally on the upper surface, jointed and spongy. Leaves are long and linear. Flowers are small yellow green, born in a spadix enclosed by a spathe. Berries green; 1-3 seeded; seeds are oblong.

All parts of the plant contain an essential oil having asarone (80%) asaraldehyde, acoradin, phenylindane, a phenyl propene [1-(*p*-hydroxy phenol)-1-(O-acetyl) prop-2-ene], calamenol, calamene and sesquiterpene ketones like acorone, calarene and calacone and monoterpenes 1,8-cineole, linalool, α-terpeneol and eugenol methyl ether. The oil yield in dried rhizomes is 3.5%. The roots yield choline and acoric acid. Acorin, a bitter principle, luteolin glycoside, tannins, mucilage, β-sitosterol, and calcium oxalates are the other constituents reported from the plant.

The rhizome powder is a well-known **insecticide** against fleas, bedbugs and used to prevent insect pests in rice and other grains and silverfish in libraries. ß-Asarone, a derivative of asarone is suspected to be carcinogenic. The extract of rhizome is found to be active against *Dysdercus cingulatus.* Aqueous extract is used against aphids, armyworm, cotton stainer, housefly, large cabbage worm, mosquito, mustard sawfly, wooly bear and stored grain pests.

The powder and the volatile oil are found nematicidal against *Meloidogyne incognita.*

3. *Acorus gramineus* Sol. (Araceae)

Japanese Rush Plant

Acorus gramineus rhizome containing (Z)-asarone and (*E*)-asarones was found to be insecticidal to adults of *Sitophilus oryzae* (L.), *Callosobruchus chinensis* (L.) and *Lasioderma serricorne* (F.) in direct contact application and fumigation methods (Park *et al.,* 2003) It was also found effective against *Nilaparvata lugens* and *Plutella xylostella* by direct contact application method (Lee *et al.,* 2002).

4. *Anethum sowa* Roxb.

Syn. *A. graveolens* DC,
Peucedanum sowa Roxb.

Sowa, Indian Dill

A. sowa is an aromatic annual/biennial, up to 1m in height with glabrous branching stems bearing tripinnatipartite finely dissected leaves. The yellow flowers are borne in compound umbels and the mericarps are subelliptical with three longitudinal ridges and vittae having marginal walls appearing irregular. A native of India.

The fruits yield a volatile oil (1.5 to 3.5%), proteins (20%), carbohydrates (31%), ß-sitosterol, piperine and flavonoids (kaempferol and quercetin). The major constituent of oil is carvone (20-70%), with dihydrocarvone, *d*-limonene, *d*-phellandrene, myristicin, apiole and dill-apiole. The tender plant also yields a volatile oil rich in *d*-phellandrene.

The volatile oil of *Anethum sowa* also is found nematicidal against the nematode, *Meloidogyne incognita.* Sowa is otherwise used as a carminative, antipyretic and anthelmintic and is a component of gripe water. It is also used as a condiment and in confectionery. The seed oil is useful in preventing flatulence in children.

5. *Artemisia absinthium* Linn. (Asteraceae)

Damar, Absinthe

This is an aromatic and bitter shrub with pinnatifid leaves and yellow heterogamous heads and elliptic-oblong achenes,. This species is common in Kashmir. *A. absinthium* yields a volatile oil called wormwood oil or absinthe, from the leaves. The vol. oil contains thujyl alcohol, phellandrene, pinene, *s*-guaiazulene and chamzulene. Also present in the plant 24-zeta-ethyl cholesta-7, 22-dien-ß-ol, flavonoid artemisetin, homoditerpene peroxides 1-(E)-8-isopropyl-1,5-dimethyl-nona-4,8-dienyl-4-methyl-2,3-dioxa-bicyclooct-5-ene and its diasteriomeric absinthin, anabsinthin, anabasine and thujone. Methanol extract contained α-santonin and ketopelenolid A.

A. absinthium and related *A. annua* (Annual wormwood), containing similar principles are effectively used against the nematode, *Meloidogyne incognita*

This is an important drug in India. The powdered plant material, which forms the drug 'Afsanteen', is used in India in chronic fever, swellings, inflammation of liver and as a tonic and stimulant. The wormwood oil is applied externally to swellings and is antimicrobial in nature. Fresh wormwood is a good source of azulenes. Artemisetin exhibited marked antitumour activity against melanoma-B 16. Homoditerpenes are antimalarial substances (Rucker *et al.*, 1991).

6. *Artemisia monosperma* Delile

The essential oil of this plant containing 3-methyl, 3-phenyl-1,4 pentidyne was shown insecticidal against house fly, cotton leaf worm and the rice weevil (Saleh *et al.*,1984)

7. *Artemisia tridentata* ssp. *vaseyana* Rydb.

Big sagebrush

Vapors released from foliage of mountain big sagebrush, *Artemisia tridentata* Nutt. ssp. *rasayana* (Rydb.) containing 1,8-cineole and camphor delayed adult emergence of the lesser grain borer, *Rhyzopertha dominica* (F.), suppressed oviposition of *Plodia interpunctella* (Hübner), and was insecticidal to adult red flour beetle, *Tribolium castaneum* (Herbst). These can be an alternative to methyl bromide for protection of stored grain, commodity, and space fumigations (Dunkel and Sears, 1998).

8. ***Bidens pilosa*** **Linn. (Asteraceae)**

Spanish needle, Hairy Beggarticks

This is a weak annual with spreading branches, hairy serrate leaves in opposite pairs, yellow heterogamous heads (ray florets white, disc florets yellow) and quadrangular achenes. Native to American tropics, now a weed everywhere.

Heads contain polyacetylenes, chalcone and aurone glycosides, volatile oil, sterols, acylated okanin glycoside, gallic acid and phenols.

Seeds and the whole plant are insecticidal. They are used for aphids, cut worm and termites.

9. ***Blumea eriantha*** **DC. (Asteraceae)**

Nimurdi

A pubescent aromatic annual with dichotomously branched stem, lower petiolate elliptic oblong, silky – villous and upper sessile leaves and irregularly toothed margins, yellow silky homogamous heads and minute angled achenes. A common weed throughout.

Leaves yield an essential oil rich in ketones such as carvotanacetone and *l*-tetrahydrocarvone. Seeds also yield a volatile oil with a camphor-like smell.

The essential oil is found larvicidal against *Culex pipiens* and *C. fatigans*. It is generally used as an insect repellant.

10. ***Caesulia axillaris*** **Roxb. (Asteraceae)**

Bangra, Caesulia

C. axillaris is a glabrous decumbent / semierect herb with sessile, semi-amplexicaul, oblong-lanceolate leaves, white/pale blue disc florets clustered in the axils of leaves and dark brown ribbed achenes. A common weed in rice fields throughout India.

The whole plant yielded a volatile oil consisting of γ-asarone (48%), limonene (34%), cineole (4%) and pinene (3%). Another sample from North-West Himalayas is found to contain volatile oil which consists of dilapiole (50%) and methyl heptanol (42%). Catechol, 2,4,5-trimethoxy cinnamaldehyde, 2,4,5-trimethoxy benzaldehyde and asarone are reported from aerial parts. Seeds are found to yield fat (15%) consisting of stearic (32%), linoleic (46%), palmitic (15%) and oleic (6%) acids and protein (5%).

The volatile oil shows anti-feedant activity against insects *Spodoptera litura* and *Achoea janata,* due to γ-asarone.

11. *Carum carvi* Linn. (Apiaceae)

Kala zira, Caraway

An aromatic glabrous annual/biennial herb reaching a height of 1m, with thick fusiform roots, grooved hollow branches, pinnately dissected leaves having laciniately cleft leaflets, white flowers in compound umbels and oblong-oval yellowish fruits. Mericarp curved, brownish with five prominent ribs. Found wild in North Indian hills, cultivated elsewhere for aromatic seeds.

The mericarps ('seeds') contain protein (40%), fixed oil & resin (10%) and a volatile oil (3-8%). The principal constituents of vol. oil are carvone (55-60%) and limonene. Other monoterpenes present are α-pinene, β-pinene, and *p*-cymene. The fixed oil is rich in oleic (60%), linoleic (20%) and petroselinic (17%) acids. The phenolics reported from seeds are glycosides of kaempferol, quercetin and isorhamnetin and coumarins such as 5- and 8-methoxy psoralens, umbelliferone, herniarin, and scopoletin. The whole plant also yields a volatile oil consisting of cadinene (37%), carvone (31%) and dihydrocarveol (10%). The roots contain polyenes, falcarindione, falcarinolone, phytofluene, β-sitosterol, umbelliferone and scopoletin. Leaves contain glycosides of kaempferol and quercetin.

The hexane extract of the fruit was found to have excellent larvicidal activity against the mosquito *Culex pipiens faligens* Wiedm. (Deshmukh *et. al.,* 1982). Caraway is a popular spice and condiment. They are prescribed in flatulent colic and stomach problems. Alcoholic extract is antispasmodic. Caraway roots are edible and leaves are eaten as salads.

12. *Chamaecyparis obtusa* Siebold et Zuccarni (Cupressaceae)

The vol. oil of *Chamaecyparis obtusa* contains bornyl acetate, (+)-limonene, myrcene, α-phellandrene, α-pinene, sabinene and terpinolene. This oil is found causing mortality to adults of *Callosobruchus chinensis* (L.) and *Sitophilus oryzae* (L.). Bornyl acetate, α-phellandrene and terpinolene caused 97%, 97% and 87% mortality, respectively, against *C. chinensis* adults, 1 day after treatment. Against

S. oryzae adults, at 0.26 mg/cm^2, terpinolene caused 93% mortality, 2 days after treatment, whereas 80% mortality, 4 days after treatment was observed with bornyl acetate and α-phellandrene. Bornyl acetate and terpinolene were more effective with *S. oryzae* adults in fumigation treatments (Park *et al.*, 2003).

13. *Chenopodium ambrosioides* Linn. (Chenopodiaceae)

Kshetravastuka, Indian wormseed

A much branched woody herb with a smell of camphor, having angular-ridged purple-tinged fleshy stems, oblong-ovate leaves, green flowers in axillary / terminal spikes and small reddish brown seeds. A native of tropical America now found as a weed in waste places.

The flowering and fruiting shoots yield a volatile oil (0.25-0.45%) containing ascaridole (50%), xylene, *p*-cymene, tetradecanol, caproic and caprylic acids. The leaves contain kaempferol and organic acids. Fruits yield glycosides of kaempferol as well as isorhamnetin, quercetin and 4'-O-demethylarbectorin (Jain *et al.*,1990).

Leaves are repellants and aqueous extract is a wide -spectrum pesticide. The powdered seed or its decoction also is used as an insecticide.

14. *Chromoleana odorata* King &Robinson

Syn. *Eupatorium odorata Linn.*

Siam weed

This plant is a shrubby weed of 3m having small 3-nerved obovate (almost truncate) yellowish green leaves, pinkish white homogamous heads and linear straw-colored fruits. A native of Tropical America and now a common weed in India.

The volatile oil consists of trans-caryophyllene, δ-cadinene and α-copaene etc. Aerial parts contain flavones scutellarein, tamarixetin, chalcone (odoratin) and sterols,

This plant is found to have insecticidal activity and is used as stored grain protectants. The volatile oil is found to have significant oviposition deterrent effect on *Phyllotreta striolata* and *Plutella xylostella*.

15. *Cinnamomum aromaticum* Nees. (Lauraceae)

Syn. *C. cassia* Blue

Chinese Cassia

C. aromaticum is an aromatic evergreen tree with smooth grey bark oblong, elliptic large leaves (3-nerved from base), small flowers in terminal or axillary panicles and black aromatic fruits containing single seeds. A native of China.

Bark is rich in diterpenes such as cinnzeylanin, anhydrocinnzeylanin. cinnzeylanol, anhydrocinnzeylanol, cinnacassiols A, B, C, D_1, D_4 and E and their glycosides. Other compounds in the bark are cassioside, cinnamoside, lyoniresinol glucoside, syiringaresinol, cinnamic aldehyde, epicatechin, cinnamtannins A_1-A_4, procyanidins and a volatile oil (0.3%). The constituents of vol. oil are cinnamaldehyde (82%) eugenol, cinnamic acid, cinnamyl acetate etc. The leaves yield an essential oil (0.4%) containing 71% cinnamaldehyde and 10% eugenol. The flowers and unripe fruits also yield similar volatile oils.

Cinnamaldehyde is found to attract corn rootworms and the corresponding beetles. It controls insects and fungi also.

Cinnamon, *Cinnamomum zeylanicum* Breyn., containing cinnamaldehyde as the principal component of the volatile oil from the bark also exhibits the same properties.

16. *Cinnamomum camphora* Presl.

Kapoor, Camphor

All the parts of this tree contain a volatile oil rich in camphor. Camphor separates out on cooling the steam distilled oil. Other compounds in mother liquor are safrole, cineole, borneol, camphene, dipentene and terpeneol. A number of varieties of *C. camphora,* produce volatile oil rich in compounds like safrole, linalool, cineole, etc.

Camphor is a well-known insect-repellant and is used as a remedy against the plague in Persia, now Iran. It was also used in embalming. The therapeutic properties of camphor oil include the following: anti-inflammatory, antiseptic, carminative, diuretic, febrifuge, laxative, rubefacient, and stimulant action.

17. *Citrus* spp. (Rutaceae)

Citrus is a genus of evergreen aromatic shrubs/small trees with axillary thorns, winged petiole, gland-dotted leaves, fragrant white flowers and fruit a hesperidium. The various species useful are *C. limon* Burm.f (Lemon), *C. medica* Linn.(citron), *C. paradisi* Macfad. (grape fruit), *C. reticulata* Blanco(orange) and *C. sinensis* Osbeck.(mosambi).

Lime oil obtained from the peels of lime and orange oil from the peels of orange contain limonene as the principal component. All the other species such as *C. aurantifolia* (acid lime), *C. aurantium* (sour orange), *C. macrocarpa*(China orange), *C. grandis* (pummelo), *C. medica*, *C. paradisi and C. limonia*(lemandarin)also contain limonene as the principal component in the vol. oils of fruits. Petit grain oils (oils from the leaves and shoots of the *Citrus* plants) of *C. aurantifolia*, and *C. limonia* also contain limonene as the main constituent. The oils of *C. aurantium* and *C. paradisi*(acid lime) contain linalool whereas the oil of *C. microcarpa* contains citral. Bergamot oil is the volatile oil obtained from the fresh peels of sour orange. This contains linalyl acetate (major component), limonene, pinene and terpeneol. Limonoids are the compounds providing bitterness to many *Citrus* fruits. Limonin, the most common limonoid, is the principal bitter principle. Limonoids tend to disappear when the fruit ripens, being converted to non-bitter derivatives. Seeds contain glucosides of limonin, obacunone, deacetyl nomilin, nomilin, ichangin, isolimonic acid etc.

Limonene is useful for all external pests of pets including fleas, lice, mites and ticks. Linalool also repels insects and mosquitoes. Bergamot oil is preferred for ornamental plants and lawns. It is also found to be adulticidal and larvicidal for *Culex pipiens* and *Culex quinquefasciatus.*

18. *Combretum sundaicum* Miq. (Conbretaceae)

Dukhar-rau, jungle weed, opium anti dote

This is a scandent shrub/liana growing up to 30 m in length, with elliptic leaves, greenish white flowers and sub-orbicular fruits. Seen in Assam Hills.

Leaves possess a tea-like flavour. Decoction of the roasted leaves is given as anti-opium agent. The leaves of this plant and of *C. wallichi* are considered pest repellant.

19. *Conyza newii* Oliv. and Hiern. (Asteraceae)

The oil of *Conyza newii* contains perillaldehyde and perillyl alcohol and exhibited higher fumigant toxicity to *Anopheles gambiae* (Omolo *et al.*,2003).

20. *Coriandrum sativum* Linn. (Apiaceae)

Dhanya/Kottmir, Coriander

This is a glabrous aromatic annual with a slender branching stem. Basal leaves entire with crenate margin whereas the leaves from stem are pinnate (leaflets branched dichotomously to linear segments but outline of leaflet remains ovate), leaf base almost sheathing. White flowers are borne in umbels in which the outer flowers have enlarged ray like petals. The spherical cremocarp contains two hemispherical mericarps united by their margins. Each mericarp has five wavy inconspicuous primary ridges alternating with four prominent secondary ridges. Coriander is a native of Mediterranean region widely cultivated everywhere.

The leaves yield isocoumarins like coriandrones A & B, coriandrin and dihydrocoriandrin and are good sources of vit. C and carotenoids. Fruits contain fixed oil (15%), volatile oil (up to 1%), tannins and calcium oxalate. 65-70% of volatile oil is linalool. Other components are α- and β-pinenes, limonene, γ-terpenene, *p*-cymene, borneol, citronellol, camphor, geraniol and geranyl acetate. Heterocyclic components such as pyrazine, pyridine, thiazole, furan and tetrahydrofuran derivatives., isocoumarins *viz.* coriandrin, dihydrocoriandrin and coriandrones A-E, flavonoids like 6-methoxyquercetin, sterols, phthalides such as neochidilide and 2-digustilide are the other constituents reported from coriander. (Gijibels *et al.*, 1982; Tanaguchi *et al.*, 1996; Lamparsky and Linies, 1988),.

Coriander is used against spider mites and fungal diseases. Linalool repels insects and mosquitoes. Coriander is a well-known spice. It has got stomachic, spasmolytic and carminative properties. The essential oil is antifungal in nature.

21. *Cuminum cyminum* Linn. (Apiaceae)

Jira, Cumin

Cumin is a slender glabrous (except for fruits) herb, indigenous to Mediterranean region, now widely cultivated for its spicy fruits. The leaves are dissected and the small pinkish flowers are borne on compound umbels. Fruits resemble caraway, but are straighter, longer and covered with dense short compressed hairs.

The fruits yield up to 4% volatile oil containing cuminaldehyde (30%), pinenes and α-terpeneol. Also present are fatty oil (10%), protein and flavones such as apigenin and 7-OMe apigenin and phenolic acids (Umadevi and Daniel, 1990).

Vapours of essential oils are larvicidal and are found to be toxic to two greenhouse pests, viz. the carmine spider mite, *Tetranychus cinnabarinus* and cotton aphid, *Aphis gossypii*. A minimum dose of 0.5 µl/l air and 2–3 days of exposure was required for 99% mortality (Tunc and Sahinkaya, 1998).

22. *Cymbopogon citratus* Stapf. (Poaceae)

Lilichai, Lemon grass

This is a tall aromatic grass with linear, distichous leaves, elongate panicles and many sessile spikelets. Found wild or cultivated.

Leaves yield a volatile oil rich (78-82%) in citral, waxes, flavonoids (luteolin, isoorientin and orientin), phenolic acids (chlorogenic, caffeic and *p*-coumaric acids), triterpenoids such as cymbopogonol and cymbopogone. Other constituents of volatile oil are d-limonene, citronellol, geraniol and myrcene.

Citral is an insect repellant. Lemon grass is used in elephantiasis, as CNS depressant, analgesic, antirheumatic, antipyretic and expectorant. Lemongrass oil is anticarcinogenic in that *d*-limonene and geraniol induce activity of glutathione transferase an enzyme that inhibits cancer development. The oil is also used as a remedy for ringworm.

23. *Cymbopogon nardus* Linn. (Poaceae)

Contains a volatile oil consisting of 78-82% citral, cymbopogonol, cymbopogone, β-sitosterol, hexacosanol, triacontanol and saponin.

The oil shows larvicidal activity at a dose of 63.2 ppm and caused 50 percent mortality of *Anopheles stephensi* larvae.

The water extracts of *C. marginatus, C. nardus* and *C. citratus* are used against leaf blight, bacteria and rice pests. The volatile oil is found nematicidal for *Meloidogyne incognita, Heterodera avenae, H. cajani,* and *H. zeae*

All species of ***Cymbopogon act as*** insect repellants (as topical application). The volatile oil of ***Cymbopogon flexuosus*** is found active against *Meloidogyne incognita, Heterodera avenae, H. cajani,* and *H. zeae.* The oil of ***C. martini*** is used for *M. incognita* while that of ***C. winterianus*** is effective against *Meloidogyne incognita, Heterodera avenae, H. cajani,* and *H. zeae.* Citronella oil repels insects and ticks-used for protecting clothes and for both indoor and out door plants.

24. *Eucalyptus globulus* Labill. (Myrtaceae)

Nilgiri, Eucalyptus

E. globulus is a tall aromatic tree with greyish white smooth bark exfoliating occasionally and elliptic lanceolate leaves and greenish yellow flowers. Cultivated in many parts as a social forestry tree

The leaves yield a volatile oil (up to 1.5%) consisting of cineole (65-75%), caryophyllene, α-pinene, limonene etc. The bark and wood contain polyphenols. Euglobals (acyl-phloroglucinol-monoterpene/ sesquiterpene, 11 in number, such as euglobal 1a1, 1a2, 1b, 1c, lla, llb, llc, lll-lX) form another group of compounds isolated from this plant. Also present in the leaves are acetogenin mevalonates, eucalyptin, quercetin, rutin, chrysin, phloroglucinol, oleanolic acid, ursolic acid and phenolic acids such as caffeic, ferulic, gallic, gentisic and protocatechuic acids (Dayal, 1988).

The oil has insecticidal properties against bedbugs, black beetles, flies, lice, and mosquitoes. *Eucalyptus maculata* essential oil is repellent to *Anopheles gambiae and A. funestus* and other fleas. Klingauf *et al.*, (1983) obtained 100% mortality against the rose-grain aphid, *Metopolophium dirhodum* with *Eucalyptus* oil at doses of 31.4 µl/1 air for 3 hours. The vapours of this oil strongly reduced fecundity, decreased egg hatchability and increased neonate larval mortality in the bean bruchid, *Acanthoscelides obtectus* (Stamopoulos,1991). The oils from *E. citriodora* and *Eucalyptus* hybrid are found effective against the nematode, *M. Incognita.*

25. *Eucalyptus resinifera* Sm.

Red mahogany

This is a large tree with a tall clean bole grown in Nilgiri and Shimla Hills. The methanolic extract containing monoterpenes and flavonol glycosides resinosides A & B, exhibited significant insect repellant activity against Blue Mussel *Mytilus edulis*.

26. *Eucalyptus umbellata* Domin. (*E. tereticornis* Sm.)

The leaf powder mixed with paddy (1% by wt.) is effective against the development of Angoumois grain moth, *Sitotroga cerealella* and prevents cross infestation by the lesser grain borer, *Rhizopertha dominica.*

27. *Eupatorium triplinerve* Vahl. (*E. ayapana* Vent.) (Asteraceae)

The crude extract of the aerial parts is found to deter (up to 90%) feeding of the caterpillar of a polyphagus pest, *Diacrisia obliqua.*

The aqueous extract of leaves of some other species of *Eupatorium* exhibits insecticidal and repellant properties against sweet potato weevil, aphids, diamond black moth and snails.

28. *Feronia limonia* Swingle (Rutaceae)

Syn. *F. elephantum* Corr.

Kota, wood apple

Since this plant provides a dyestuff also, the description and chemistry are given in in the previous section on Natural Dyes

The leaves yield an essential oil (0.73%), containing estragol (90%) as the main constituent, a flavonol, 3'OMe-quercetin and phenolic acids like *p*-hydroxy benzoic, vanillic, syringic, *p*-coumaric and ferulic acids.

Methanol extract of leaves are found repellant to *Aedes aegypti.* The fruit is a tonic and used as a substitute for *Aegle marmelos*, for diarrhoea and dysentery. The gum is used as a substitute for gum acacia.

29. *Foeniculum vulgare* Mill. (Apiaceae)

Madhurika/Saunf, Fennel

Fennel is an aromatic, glabrous erect perennial reaching a height of 1m with finely dissected leaves and yellow flowers in large umbels. The glabrous fruits bear a stylopod at the apex and each mericarp is linear oblong, slightly curved and possesses five prominent primary ridges. The plant is a native of Mediterranean region, now cultivated all over the world.

The fruits contain fixed oil (12%), proteins, a volatile oil (1-4%) and phenolics. The principal component of volatile oil is a phenolic ether, anethole (60%) followed by the ketone, fenchone (10-30%). Other components of oil are methyl chavicol, limonene, α-pinene etc. The phenolics include furanocoumarins (imperatorin, bergaptene, and xanthotoxol), stilbene trimers (miyabenol C and *cis*-miyabenol C and the glycosides of the latter such as foeniculosides 1-4) and flavonoids. Also present are monoterpene glycosides viz. foeniculosides V-IX, and other compounds such as zizybeoside I, icaviside A, syringin, sinapyl alcohol, and adenosine and anethole glycol.

The essential oil exhibited insect repellent property against the storage pest *Tribolium castaneum* and strong fungitoxicity against storage fungi *Aspergillus niger* and *A. flavus*. It is effective against larvae of root knot nematode *Meloidogyne incognita*.

Fennel is generally used as a spice and carminative. The fruits are anabolic, estrogenic, analgesic, antipyretic and antimicrobial agents. It reduces flatulence in infants.

30. *Heliotropium peruvianum* Linn. (Boraginaceae)

Syn. *H. corymbosum* Linn.,

H. arborescens Linn.

Vanilla heliotrope

A highly aromatic (vanilla fragrance) annual of ½ to 1 m ht. with glossy ovate leaves having a corrugated appearance and blue/white flowers in apical umbellate clusters of scorpioid cymes. Cultivated for the fragrant flowers. A native of Peru.

The volatile oil is found to contain methyldihydrojasmonate. The leaves contain an alkaloid lasiocarpine. Roots contain cyanoglossin. The whole plant is insecticidal.

31. *Hyptis verticillata* Jacq. (Lamiaceae)

This plant contains squamulosone (aromadendr-1(10)-en-9-one), which when incubated with the fungus *Curvularia lunata*, six metabolites were produced. All these aromadendranes showed insecticidal activity against the sweet potato weevil, *Cylas formicarius elegantulus*. (Collins *et al.*, 2001).

32. *Hyptis suaveolens* Poit. (Lamiaceae)

Ganugatulsi, Hyptis

This is a rigid erect aromatic annual with petiolate ovate (lower cordate) serrate leaves and blue flowers in axillary cymes and a carcerule containing 4 elongate black nuts. A native of Tropical America grown in gardens or found as an escape.

Leaves, twigs and flowers contain a volatile oil consisting of sabinene, β-pinene, limonene, β-caryophyllene, 1,8-cineole, spathulenol, caryophyllene and bicyclogermacrene. Aerial parts contain a triterpenoid, hyptadienic acid. The ethyl acetate extract containing 5-keto-pent-3, 4-enyl-2′-phenol and 5-pentyl methylene oxy-4, 4-dimethyl- cyclohexanol was found to exert antifeedant and ovicidal activity against *Helicoverpa armigera* and *Spodoptera litura* (Raja *et al.*, 2005).

This plant is a stimulant, carminative, antiseptic, sudorific and galactogogue. Infusion of the plant is used in catarrhal conditions, affections of the uterus, diabetes and cancer.

The volatile oil from *H. martiusii* also is found to be insecticidal.

33. *Illicium verum* Hook.f. (Magnoliaceae)

Star anise

Star anise is a star shaped fruit of a small evergreen tree of China. It consists of 8 partly dehisced one-seeded follicles. The pericarp of the fruit is reddish brown and each seed has a brittle shining testa and an oily kernel.

The fruits as a whole contain 2.5 - 5% volatile oil (5 - 6% in the pericarp and 1.8 – 2% in seeds) of which 80 – 90 % is anethole. Chavicol methyl ether, pinene and safrole form the minor components.

The volatile oil is found to be effective in controlling *Blatella germanica* by direct contact application and fumigation methods.

34. *Juniperus virginiana* Linn. (Pinaceae)

Eastern Red Cedar

This is a tall conical tree with opposite closely overlapping scale-like leaves on older branches and pointed awl-shaped leaves in young branches and small fleshy berry like cones containing 1-4 seeds.

The wood contains 1-3% of a pale yellow oil containing tricyclic sesquiterpenes cedrene (up to 80%), cedrol and cedranol.

Cedar wood oil is recommended for mothproofing and repels larvae of moths attacking clothes. Used as a natural pest control oil for other insects also.

35. *Kaempferia galanga* Linn. (Zingiberaceae)

A rhizomatous herb with aromatic tuberous root stock and fleshy cylindric root fibers, rotund ovate leaves, spreading flat upon the ground, central radical spikes containing 6-12 white fragrant flowers (single stamen, deeply 2- 10 lobed lip) and an oblong capsule containing many arillate seeds. A native of India and Malaya.

The root stock yields a volatile oil containing ethyl *p*-methoxy cinnamate, ethyl cinnamate, Δ^3 carene, camphene and borneol.

The rhizomes are used to protect clothes from insects.

36. ***Lavandula angustifolia*** **Mill. (Lamiaceae)**

Lavender

This is an evergreen aromatic shrub of 1m. with linear-lanceolate leaves and blue flowers in terminal spikes. This is indigenous to S.Europe.

Fresh flower spikes and leaves yield about 2% of a volatile oil comprising of linalyl acetate, linalool, limonene, and cineole.

Lavender oil repels moths attacking clothes and recommended for home, clothes, drawers and storage containers.

37. ***Lippia stoechadifolia*** **HBK.(Verbenaceae)**

Poleo

This plant contains a monoterpene, pulegone-1,2-epoxide which is insecticidal (Grundy and Still, 1985).

38. ***Melaleuca leucadendron*** **Linn. (Myrtaceae)**

Syn. *Myrtus leucadendron* Linn.

Cajuput oil tree

A large sized aromatic tree with thick spongy bark peeling off in layers, pendulous branches, alternate lanceolate/elliptic leaves, small sessile flowers in spikes (stamens many in five bundles, anthers versatile) and 3-valved capsules containing numerous obovoid/cuneate seeds. A native of Australia cultivated in India.

The volatile oil obtained from leaves of terminal branches. The oil contains α-terpeneol, l-apinene, l-limonene, sesquiterpenes, azulene, sesquiterpene alcohols, valeraldehyde and benzaldehyde.

Fresh leaves on steam distillation yield cajuput oil, which is valued as a mosquito repellent and has the advantage over citronella oil in that it is less volatile. It is also effective against lice and fleas.

39. ***Mentha piperita*** **Linn. (Lamiaceae)**

Pudina, Peppermint

Peppermint is a small perennial aromatic herb (with an aroma of pepper) having a running rootstock, opposite oblong lanceolate serrate glabrous leaves and purple flowers in thick terminal spikes. It is a native of Europe now cultivated everywhere.

The whole plant yields a volatile oil (1-3%), flavonoids, phenolics and steroids. The main constituents of vol. oil are menthol (30-35%) and menthone (15-30%). Other components are menthyl acetate, menthofuran, cineole, limonene, neomenthol, isomenthol and a sesquiterpene (viridofloral). The flavonoids present are apigenin and luteolin along with highly oxygenated flavones. Caffeic, chlorogenic and rosemarinic acids are the phenolic acids present whereas the triterpenes located are α-amyrin, ursolic acid and sitosterol.

The essential oil is found larvicidal, repellant against *Culex quinquefasciatus, Anopheles stephensi* and *Aedes aegypti.* Mint oil kills aphids on plants when used with thyme- safe for water plants with or without fish and ornamentals.

Aqueous extract of leaves is used against bacterial leaf spot and other bacterial diseases. The oil is found effective against *Meloidogyne incognita, Heterodera avenae, H. cajani,* and *H. zeae.*

Peppermint is a carminative and flavouring agent. It is a spasmolytic, smooth muscle relaxant and anti-inflammatory. It helps in production of bile and is a diuretic and mild sedative.

The yield and composition of volatile oil are influenced by environmental factors, harvesting time and planting time.

Japanese peppermint *Mentha arvensis* L. yields a similar volatile oil rich in menthol and is used like wise. This oil is found effective in controlling *M. incognita.* Similarly the oil from *M. spicata* also is nematicidal to *M. incognita.*

40. *Minthostachys setosa* Brig.(Lamiaceae)

Muna-Muna

This native herb of Peru is a well-known medicinal plant of that region used for cough and other respiratory problems. The dichloromethane extract of *M. setosa* was quite potent against *A. aegypti* larvae showing an LC_{50}=9.2 μg/ml (LC_{100}=25.2 μg/ml) (Ciccia *et al.,* 2000).

41. *Minthostachys tomentosa* Epling. (Lamiaceae)

This plant, also a native of Peru, contains (-)-(1*S*,2*R*,3*R*,4*S*)-1,2-epoxy-1-methyl-4-(1-methylethyl)-cyclohex-3-yl acetate which is found to be insecticidal to *Oncopeltus fasciatus* (Cantin *et al.,*2001).

42. ***Myristica fragrans*** **Houtt. (Myristicaceae)**

Jatiphala, Nutmeg

Nutmeg is the seed of an evergreen dioecious tree, a native of Molucca islands. Leaves are simple, ovate. The yellow flowers are fleshy and aromatic. The golden yellow fruits contain a single seed, covered by a brilliant red branching aril, the 'mace' (*Javantri*) of commerce. The seed has an outer dark brown perisperm and an inner light brown endosperm. The perisperm is infolded and penetrates the endosperm giving it a marbled appearance.

Nutmeg contains fat (24-40%), occurring as prismatic crystals-nutmeg butter; volatile oil (5-15%), phytostearin, large amounts of starch, protein and amylodextrin. The volatile oil consists of 80% terpenes like pinene, camphene and sabinene, 8% dipentene, 4%myristicin, 2% elemicin and isoelemicin; up to 1% safrole and 1% other constituents like eugenol, methyl eugenol, isoeugenol etc.

The leaves (herbage) of this plant are found effective in controlling *M. incognita.*

43. ***Nigella sativa*** **Linn. (Ranunculaceae)**

Kalajira, Black cumin

A small annual herb of 60 cm with compound 2-3 pinnatisect leaves having linear-lanceolate segments, pale blue solitary flowers on long peduncles and black trigonous seeds. Cultivated in North India.

Seed contains volatile oil, which contains carvone (45-60%), d-limonene and cymene, a carbonyl compound, nigellone ($C_{18}H_{22}O_4$), bitter nigellin and a saponin melanthin (sapogenin-melanthigenin).

Seeds are kept scattered between folds of linen or woolen clothes to preserve them against insect attack.

44. ***Ocimum americanum*** **Linn. (Lamiaceae)**

Syn. *O. basilicum* Linn.

Sweet Basil

A smooth aromatic herb with green stems, ovate-lanceolate leaves, and white slightly larger (2cm long) flowers in much branched thyrsoid racemes. The two middle teeth of the lower lip of calyx is awned and longer than the rounded upper lip and the nutlets are pitted. A widely cultivated herb.

The leaves yield a volatile oil containing methyl cinnamate (major component), camphor and citral

The essential oils of the leaves and seeds are repellant, toxic or growth inhibitory to many insects (Grainge and Ahmed, 1988). In field and laboratory experiments, treatments with seed extracts increased mortality between 50-90% against bugs in turnip, *Bagrada cruciferarum*, the American bollworm *Helicoverpa armigera* and aphis, *Aphis gossypii*.

O. sauve and *O. canum* also are traditionally used against pests. Branches of *O. sauve* are often placed around windows and doors to keep mosquitoes away due to eugenol content in its vol. oil. Linalool in *O. canum* was found responsible to the toxic effects of dried leaves to the bruchid *Zabrotes subfasciatus* and other storage pests. *O. sauve* is found toxic to maize weevil *Sitophilus zeamais*. Eugenol and other components have a strong inhibitory effect on many fungi.

45. *Ocimum gratissimum* Linn. (Lamiaceae)

Ram Tulsi, Shrubby Basil

This is a shrubby perennial with a much branched woody quadrangular stem, pubescent gland dotted, crenate, serrate, elliptic lanceolate leaves, greenish yellow pubescent flowers in short racemes and subglobose nutlets.

The wild species growing in Asia and Africa are said to have at least six chemotypes each containing eugenol, ethyl cinnamate, thymol, citral, geraniol or linalool. The most common variety is eugenol type; but the pungent types have thymol as the principal ingredient.

This plant is used as a mosquito repellant and growing *O. gratissimum* is recommended as a means of controlling this pest.

Both *Ocimum gratissimum* and clove are found best in preventing mould growth in stored grains in Ghana.

46. *Ocimum tenuiflorum* Linn (Lamiaceae)

Syn. *O. sanctum* Linn.

Tulsi, Basil

This is a perennial herb with brownish pink stem, elliptic-oblong pubescent (minutely gland-dotted) leaves on a hairy slender petiole, whitish pink flowers on terminal racemes and a carcerule fruit enclosing 4 nutlets. A native of India cultivated as a sacred plant.

The leaves yield a volatile oil (up to 0.8%), triterpenes, flavonoids and fatty acid esters. Eugenol (content of which is maximum in spring, up to 20%) and caryophyllene (38%) are the principal components of vol. oil. Other constituents are bornyl acetate, β-elemene, methyleugenol, neral, β-pinene, camphene and α-pinene. Ursolic acid, campesterol, cholesterol and stigmasterol are the triterpenoids present.

The volatile oil is found larvicidal for *Culex quinquefasciatus, Anopheles stephensi* and *Aedes aegypti.* Basil is used effectively against caterpillars, fruit flies, red spider mites, leaf beetles, fungal diseases and nematodes. The crude leaf extract is found to be very effective against the nematode, *M. incognita.* The ethanolic as well as acetone extract were toxic to *M. incognita, M. javanica, Heterodera avenae, H. cajani and H. zeae.* The toxicity of the aqueous extract against the eggs of cyst nematode *Heterodera avenae* was also proved.

The herbage of ***O. basilicum*** is found to be effective in controlling *M. incognita, H. avenae, H, cajani* and *H. zeae,* whereas the leaves of ***O. canum*** are useful against *M. incognita* only.

47. *Origanum majorana* Linn. (Lamiaceae)

Syn. *O. vulgare* Linn.

Majorana hortensis Moench

Sweet Marjorum

Sweet marjorum is an aromatic bushy perennial with much branched stems which are often descending, small fragrant finely pubescent oval leaves, and small yellow flowers.

The leaves contain a volatile oil the constituents of which vary with the chemotype. Most of the varieties contain terpen-4-ol, *p*-cymene and sabinene as major constituents.

The oil is found to interfere with feeding activity of onion thrips females and reduced oviposition.

The oil of *O. vulgare* containing thymol and carvacrol is used for killing lice.

48. *Pelargonium graveolens* L'Herit. (Geraniaceae)

Geranium

This is a fragrant bush (up to 1m. high) with broadly cordate-ovate to nearly circular deeply 5-7 lobed leaves (segments again lobed, toothed) and small rose zygomorphic and spurred flowers in dense umbels. A native of South Africa, now grown as an ornamental.

The leaves yield a volatile oil (oil of *Geranium*) containing geraniol and citronellol as principal components occurring usually as tiglates or formates. Also present in the oil are isomethone, isoamylalcohol, β-phellandrene, linalool etc. In plants oil occurs free or as glycosides.

Geranium oil is used as an all-purpose insecticide. The herbage of this plant is found to inhibit *Meloidogyne incognita* (Pandey *et al.*, 2000).

49. *Pimenta racemosa* Moore (Myrtaceae)

Bay oil

P. racemosa, a native of West Indies, is a small tree with fragrant obovate, white flowers in corymbs and black aromatic berries.

The leaves yield a clove-like volatile oil (1.5%), containing eugenol (60-70%), methyl eugenol, chavicol and pinene.

Bay oil is an all purpose pesticide.

50. *Pimpinella anisum* Linn. (Apiaceae)

Saunf, Anise

Anise is an aromatic annual herb native to the Mediterranean region, cultivated extensively in Europe, India and Mexico. The lower leaves are simple and upper ones 1-2 pinnate. Flowers occur in umbels and the fruits are greyish brown pear-shaped compressed and 3-ribbed with pubescent mericarps.

Anise contains fixed oil (25-30%), proteins, sugars and volatile oil (2-3%) consisting of anethole (80-90%), methyl chavicol, limonene and *p*-methoxy phenyl acetone.

Oil is used externally as insecticidal against small insects such as head lice, mites, and vermin. Vapours of this oil are larvicidal and are found to be toxic to two greenhouse pests, viz. the carmine spider mite, *Tetranychus cinnabarinus* and cotton aphid, *Aphis gossypii*. A minimum dose of 0.5 μl/l air and 2–3 days of exposure was required for 99% mortality (Tunc and Sahinkaya, 1998). But for confused flour beetle, *Tribolium confusam*, a higher dose (81μl/l air) for 24h. was needed to achieve 95% mortality (Sarac and Tunc, 1995).Anise is preferred for ornamental plants and lawns.

51. ***Plectranthus marrubioides*** **Hochst.(Lamiaceae)**

The total volatile oil of this plant is found to be potent against *Anopheles gambiae* (Omolo, *et al.*,2003).

52. ***Pogostemon cablin*** **Benth. (Lamiaceae)**

Syn. *P .patchouli* var. *sauvis* Hook.f.

Patchouli

An erect much branched aromatic herb (up to 1.2m high) with ovate crenate-serrate tomentose leaves, white flowers with purple streaks in whorls on terminal and/or axillary spikes. A native of Philippines.

Leaves yield the volatile oil, patchouli oil, (up to 5.8% dry wt. basis) and a resin. The oil consists of sesquiterpenes, patchouli alcohol (major compound), β-patchoulene, α-guaiene, α-bulnesene and benzaldehyde.

Oil is an all purpose insecticide. Leaves and tops are antirheumatic and said to be good for menstrual problems.

53. ***Ruta graveolens*** **Linn (Rutaceae)**

Garden Rue

A strongly odoriferous glabrous herb reaching a height of about 1m. with 2-3 pinnate leaves having oblong/spathulate strongly aromatic leaflets covered with a bloom, small yellow flowers in corymbs and small capsules. A native of Mediterranean region, cultivated widely.

The herb contains a volatile oil, alkaloids, flavonoids and coumarins. The volatile oil, rue oil, is strongly odoriferous and pungent and contain methylnonyl ketone (80-90%), methyl heptyl ketone, α-pinene, limonene, cineole etc. Rutin amounts to 2% of the plant. Imperatorin, iso- imperatorin, xanthotoxin, bergaptene and psoralen are the coumarins present. The alkaloids obtained from stem and leaves are skimmianine, graveolinine and kokusaginine, whereas the roots yield dictamine and γ-fragarine.

Rue oil is customarily used to drive away the insect pests indoors and granaries. The oil is anthelmintic, antispasmodic, antiepileptic, rubefacient and emmanagogue.

54. *Salvia moorcroftiana* Wall.(Lamiaceae)

The crude acetone extract of aerial parts of *Salvia moorcraftiana* Wall., a native of S. E. Asia, was found to possess strong phytotoxic activity against *Lemna aequinoctials* Welve. and moderate antifungal activity against animal and plant pathogens. Two constituents were found to be effective inhibitors of α-glucosidase (Khan *et al.*, 2002).

55. *Santalum album* Linn. (Santalaceae)

Safed Chandan, Sandalwood

Sandalwood is the heartwood of a small tree native to peninsular India. This is a root parasite with glabrous elliptic ovate leaves, small purple/violet flowers in terminal or axillary paniculate cymes and purple black globose drupes.

The oil, amounting to 4-5% of the wood, is extracted by water distillation, which is further purified by a subsequent distillation. The roots are richer in oils, yielding up to 10% dry weight. The refined oil is a pale yellow clear liquid consisting principally a sesquiterpene alcohol, santalol (90%). Santalol exists both in α- and β-forms, the former being predominant. Also present in the oil are santene (C_9H_{14}), α- and β-santalenes and santalone.

Sandalwood oil and the wood possess insect repellant properties. Boxes made up of this wood are ideal to keep clothes. The oil is one of the very important raw materials in perfumery, soaps, face creams, powders etc.

56. *Saussurea lappa* C.B.Clarke (Asteraceae)

Kushta, Costus

This is a woody erect perennial reaching a height of 2m, with stout robust pungent roots (up to 60cm long), fibrous thick stem, large (up to 1m) radical leaves having long lobately winged stalks, large flower heads (3-5cm in diam.) containing dark blue-purple or black flowers and small curved compressed achenes. Endemic to Kashmir valley at altitudes of 2500-3000m.

Roots, yield a resin (6%), volatile oil (1.5%), inulin (18%) and alkaloids. The oil consists of costunolide (a sesquiterpene lactone-principal component), dehydrocostuslactone, dihydrocostuslactone, 12-methoxy dihydrocostunolide, costol and β-ionone. β-Sitosterol, stigmasterol and betulin are the components of resin. The alkaloids reported are saussurine and kushtin.

Kushta is a well-known incense, insecticide and used in asthma and chronic bronchitis.

57. *Schinus molle* Linn. (Anacardiaceae)

Brazilian pepper tree

This is a small aromatic tree (up to 10m) with narrow spiny leaves, small white flowers in panicles and many small fruits. A native of Brazil & Peru.

All parts of the tree are rich in an essential oil containing limonene, Δ^3 carene, phellandrene, thujene, pinene, sabinine etc.

Leaves & fruits are repellants to aphids

S. terebinthifolius, also known as pepper tree, is used similarly.

58. *Scleria pergracilis* Kunth (Cyperaceae)

A native of tropical Africa, this glabrous aromatic perennial herb has an erect stem (up to 50 cm), narrow linear leaves and spikelets clustered on a linear interrupted spike.

The lemon-scented leaves are used against mosquitoes.

59. *Syzygium aromaticum* Merr. & Perry (Myrtaceae)

Syn. *Eugenia caryophyllata* Thaub.

E. aromatica Bail

Laving, Clove

Cloves are the unopened flowers of a small, evergreen tree, native of the Molucca islands. Cultivated in many tropical countries; the trees are with oblong-ovate leaves. In wild state they produce purplish or crimson flowers in terminal branching racemes, but never reach the flowering state when cultivated. The flower buds are greenish or reddish when fresh and have slightly cylindrical receptacles surrounded at the top by 4 calyx teeth enclosing the plump globular unopened corolla. When dried the buds are brown and brittle.

Cloves contain 14-20% volatile oil, 10-13% tannins, oleanolic acid, vanillin, and a chromene-eugenin. 70-90% of volatile oil is eugenol. Also present in the oil are minor quantities of eugenol acetate, caryophyllene, methyl-n-amyl ketone and some esters.

Clove is kept in granaries, bottles, boxes etc. which keeps the pests away. Eugenol attracts Japanese beetles and repels cats and dogs.

60. *Tagetes erecta* Linn. (Asteraceae)

Galgoto, Marigold

An erect aromatic herb of 1m height with opposite odd-pinnately compound leaves (leaflets oblong dentate) and large yellow heads.

The flower heads yield a volatile oil consisting of tagetone, limonene, valeric acid, ocimene etc.

The Acetone extract as well as the steam distillate are found larvicidal, growth regulator and adulticidal against *Culex quinquefasciatus, Anopheles stephensi,* and *Aedes aegypti*. CH_2Cl_2 and MeOH extracts of this plant showed a significant activity against *Sitophilus oryzae* (Broussalis, *et. al.*, 1999). This plant is used as intercrop with *Abelmoschus esculentus* Moench (*Hibiscus esculentus* Linn.) for controlling root infesting nematodes.

Almost all species of *Tagetes* (Flowers, leaves, roots) are found insecticidal, repellant, fungicidal and nematicidal. Aqueous extract are used for ants, aphids and grass hopper whereas fermented extract is found better for rice blast and tomato blight.

61. *Tagetes minuta* Linn. (Asteraceae)

Gaindri

This is a highly aromatic annual herb of 1-2m in height having pinnatisect leaves (lobes linear/lanceolate), pale yellow homogamous heads of disc florets in corymbose clusters and black achenes. A native of S. America now occurs as a weed in North India.

The plant yields a volatile oil consisting of tagetone, phenyl ethyl alcohol, salicylaldehyde, eudesmol, linalyl acetate, limonene and linalool.

The essential oil proved adulticidal and larvicidal for *Anopheles stephensi,* and adulticidal, larvicidal for *Aedes aegypti* (Green *et al.*, 1991). The roots of this plant show nematicidal activity and therefore grown in tobacco fields. It exhibits strong larvicidal effect and is a slow repellent.

Various species of *Tagetes* were found effective against *M. incognita, M. hepla, M. javanica* and *Rotylenchus penetrans.*

Sesquiterpenes

1. ***Anamirta cocculus*** **W & A (Menispermaceae)**

Kakamari, Fish berries

A. cocculus, a native shrub of India, yields the 'fish berries'. The plant is a climber with glabrous corky shoots, ovate leaves, unisexual flowers in panicles and a drupe containing globose seeds.

Seeds contain approximately 1.5% of a bitter principle picrotoxin (cocculin) consisting of equimolecular proportions of highly oxygenated sesquiterpenes picrotoxinin ($C_{15}H_{16}O_6$), picrotin ($C_{15}H_{16}O_6$) and about 50% fat.

Seeds are highly toxic to fish. Picrotoxin is insecticidal.

2. Celastraceae

Forty one sesquiterpenes with a dihydro-β-agarofuran skeleton are assayed for antifeedant and insecticidal activities against the Egyptian cotton leafworm *Spodoptera littoralis* using the leaf disk method. Thirty eight of the sesquiterpenes assayed show antifeedant activity, including 38 more active than triphenyl tin acetate used as a positive control in this assay. Three sesquiterpenes demonstrate insectidical activity. (Gonzalez *et al.*, 1997).

3. ***Celastrus angulatus*** **Maxim. (Celastraceae)**

Three sesquiterpene polyol esters based on β-dihydroagarofuran skeleton such as 1α,2α-diacetoxy-8β-isobutanoyloxy-9α-benzoyloxy-13-(α-methyl)butanoyloxy-4β- 6β-dihydroxy- β-dihydroagarofuran and related compounds are isolated and are found to be insecticidal (Wenjun *et al.*, 2001).

4. ***Nardostachys jatamansi*** **DC. (Valerianaceae)**

Jatamansi

A native of Alpine Himalayas, jatamansi is an erect, perennial herb with a long woody stout rootstock, often covered by a dense mat of reddish brown fibres formed from the petiole of withered leaves. The stem is up to 50 cm long and bears radical leaves, red flowers in cymose heads and one-seeded white tomentose indehiscent fruits.

A sesquiterpene, jatamansone (valeranone), forms the major chemical component (up to 0.1%) of the rhizome. Other sesquiterpenoids present are spirojatamol, patchouli alcohol, norseychelanone and α-, & β-patchoulenenes. Also present are jatamol, jatamanshic acid, terpenoid coumarins oroselol and jatamansin and neolignans such as l-hydroxy pinoresinol and virolin. (Bagchi, 1991a, 1991b; Rucker *et al.*, 1993).

The volatile oil of this plant is found useful in controlling *M. incognita.* Jatamansi is used as a bitter tonic, antiseptic and for treatment of epilepsy and hysteria. It is also used for insomnia and exhibits negative inotropic and chronotropic effects.

5. *Pseudowintera colorata* Dandy (Winteraceae)

This native of New Zealand, is an evergreen shrub or small tree whose foliage is conspicuously blotched with red and yellow and having small black fruits.

The leaves of this plant contain sesquiterpene dialdehydes polygodial and 9-deoxymuzigadial which are insecticidal in nature.

6. *Tithonia diversifolia* A.Gray. (Asteraceae)

Mexican sunflower. Wild sunflower

This is a prolific shrub with dark green palmately divided leaves and yellow heterogamous heads. A native of Mexico.

The plant yields a vol. oil containing a number of sesquiterpene lactones such as tagitinin A & C, tirotunidin, tithofolinolide etc.

The extracts were found to cause high mortality rate of the second stage juveniles of *Meloidogyne* spp. (Ali *et al.*, 1985).

7. *Vetiveria zizanoides* Nash. (Poaceae)

Vetiver, Khus

This plant is a native of India, the roots and rhizomes of which yield a volatile oil rich in two sesquiterpene ketones α- and β-vetivones and vetiverols. One component of vetiver oil kills formosan termite. Farmers use a mulch of dried chopped roots and tops as insecticides in Thailand.

8. *Xanthium strumerium* Linn. (Asteraceae)

Chota gokhru, Ditch Bur

This is a tall, coarse erect weed with triangular, often three-lobed irregularly toothed scabrid/hispid leaves, white homogamous heads in terminal/axillary racemes and obovoid, thick achenes enclosed in hardened involucral bracts armed with strong hooked spines. A common weed throughout.

The aerial parts contain a volatile oil rich in sesquiterpene lactones, xanthinin and xanthatin and alkaloids. The seeds contain a fixed oil (35%). The leaves exhibit nematicidal activity against various *Meliodogyne* sp.

4.2.2. Iridoids

1. *Duranta plumeri* Jacq. (Verbenaceae)

A native of S. America, this erect bushy shrub possesses spinous branches, oblong- lanceolate acuminate leaves and blue flowers borne in great profusion in racemes at the end of branches and orange colored berries.

The whole plant is found to contain iridoids such as lamiide, durantosides and their acetyl derivatives. The leaves yield pectolinaringenin and scutellarien. Fruits also contain durantoside and triterpenoids.

This plant is considered insecticidal.

2. *Lantana camara* Linn. (Verbenaceae)

Lantana

L. camara is a straggling herb having numerous recurved prickles on the branches, simple opposite ovate crenate leaves, variously colored flowers in pedunculate capitate spikes and a cluster of black drupes containing two seeds. A native of tropical America, now a weed everywhere.

The leaves contain a volatile oil consisting of caryophyllene (80%) and l-α-phellandrene, triterpenoids lantadene A (angeloyloxy oleanolic acid) and lantadene B (dimethylacryloyloxy oleanolic acid), lantoic acid (3, 25-epoxy-3α-22β-dihydroxy ursa-12-ene-28-oic acid) and iridoids.

Methanol extract of flowers is found repellent to *Aedes albopictus* and *A. aegypti*. Live plants of *Lantana camara* and *Lippia uckambensis* repel *Anopheles gambiae* mosquitoes.

3. *Plumeria rubra* Linn. (Abocynaceae)

Frangipani

This native of Mexico is a soft-stemmed laticiferous tree with obovate leaves having an acute base and acuminate apex and large red/white fragrant flowers borne in apical cymes. Widely cultivated near temples.

The bark contains iridoids such as plumericin 1, isoplumericin 2 and plumieride, acetophenone etc.

The extracts of the leaves and bark of this plant were found to cause 100% mortality to *Meloidogyne javanica* within 12 hours (Ahmed *et al.*, 1990). Plumericin and isoplumericin are found molluscidal and antibacterial. Plumericin and plumieride are algicidal too.

4. *Vitex negundo* Linn. (Verbenaceae)

Nirgundi, Vitex

Nirgundi is a small tree with grey pubescent branches, 3-5 foliolate leaves (leaflets lanceolate , white tomentose beneath, petioluled, lateral leaflets small, lanceolate and terminal leaflets larger), small blue bilabiate flowers (stamens 4-didynamous) and ovoid black drupes containing 4 obovate seeds. Common in India, Afghanistan, Philippines and Ceylon.

The leaves contain aucubin, agnuside; alkaloids nishindine ($C_{15}H_{21}ON$) and hydrocotylene($C_{22}H_{33}O_8N$), glycoflavones orientin and isoorientin and 5-hydroxy, 3,6,7,3',4'-penta methoxy flavone.

The leaves are used to repel insects in grain stores. Extracts of the leaves have insecticidal activity and used against armyworm, diamondback moth, hairy caterpillar, rice leaf folder, rice stem borer and semi-looper. The fresh leaves are burnt with grass as a fumigant against mosquitoes. The leaves are laid over stored grains to ward of insects. It was found that either plant extracts in aqueous suspensions or cold alcoholic extracts possess insecticidal properties against insects *Spodoptera litura* F and *Lipaphis erysimi* K. in the laboratory. The petroleum ether extract is found to have promising insecticidal property with more than 50 percent mortality. The plant is an antifeedant and repellant also.

5. *Vitex trifolia* Linn. (Verbenaceae)

Simple leaf Chaste tree

The leaves of this ornamental tree possess insecticidal and antibacterial properties.

4.2.3. *Mixed Terpenoids*

These compounds contain a terpenoid component, normally a monoterpene skeleton, combined with a non-terpenoidal component.

Pyrethrins and cannabinoids are the principal compounds in this group.

Pyrethrins – are a group of six closely related esters present in pyrethrum flowers, which were used as insecticides since the 1800's. The advent of synthetic analogs like allethrins reduced the utility of these compounds in large scale. The demand worldwide is about 25,000 tons of flowers.

1. *Chrysanthemum cinerariifolium* Vis. (Asteraceae)

Syn. *Tanacetum cinerariifolium* Schultz-Bip.

Kriminaasahoova, Pyrethrum

Pyrethrum is a tufted perennial grey pubescent herb reaching a height of 60 cm with alternate pinnate (up to 30 cm long) leaves, large heads (3-4 cm in diam) bearing several yellow disc florets in the centre and a ring of white / cream colored ray florets and 5-ribbbed tapering achenes. It is a native of Yugoslavia now cultivated in colder areas of Kenya, Japan, USSR, USA, Brazil and India.

The dried flowers are commonly known as pyrethrum flowers in commerce. Pyrethrins are contained mostly in oil glands which are concentrated in ovary and developing achenes. Unfertilized achenes contain less pyrethrins. Disc florets and ray florets are very poor in pyrethrin content. Pyrethrins are mixed terpenoids in which a keto alcohol is esterified with chrysanthemic acid (a cyclopropane terpenoid carboxylic acid). The different pyrethrins are pyrethrins I and II, cinerins I and II and jasmolins I and II. Also present in the flower are 1-2% of a volatile oil, carotenoids, pyrethrosin, pyrethrol, rotatridins, tatridins, steroids and flavonoids.

Pyrethrin I is the pyrethrolone ester of chrysanthemic acid; Pyrethrin II the pyrethrolone ester of pyrethric acid; cinerin I, the cinerolone ester of chrysanthemic acid; cinerin II, the cinerolone ester of pyrethric acid, Jasmolin I, the jasmolone ester of chrysanthemic acid and Jasmolin II, the jasmolone ester of pyrethric acid. Pyrethrin I & II are major compounds and Jasmolin I & II are in lowest concentrations.

Pyrethroids are synthetic insecticides with a pyrethrin skeleton to which radicals like chlorine, bromine or fluorine are added. These derivatives are more stable and persist in the environment for more time, but are more toxic to humans than natural pyrethrins. Pyrethrins are best extracted in petroleum ether, acetone, glacial acetic acid, ethylene dichloride or methanol. Highly concentrated extracts (90%) are more unstable and therefore solutions having a concentration of 25-50% pyrethrins are generally used. Pyrethrins are extremely unstable in sunlight.

Pyrethrins interfere with nerve transmission of insects by slowing or preventing the shutting of sodium channels in nerve axons. This results in hyperactivity and convulsions. Pyrethrins are not very toxic to mammals.

Pyrethrins are generally used mixed with other stable botanical pesticides. They are best used for ectoparasites for humans & pets (lice, fleas etc) and also in greenhouses, grain storages and food processing plants. Due to their effectiveness against a broad range of insects, low toxicity to mammals and other warm-blooded animals they are an insecticides highly effective against flies, fleas, cockroaches, aphids, lice and yellow fever and malarial mosquitoes. Pyrethrum is used as insecticide in a number of ways in form of powder, sprays, aerosol, coils, creams and ointment. Powder is used against cockroaches, crickets, silverfish, ant, lice and insects found in kitchen, food stores and refuge damps. Liquid sprays are effective in control of parasitic insects like ticks, mites, flies, midges in animal houses and barns. Coils are employed to control mosquitoes in humid tropics, (Japan, Hong Kong). Pyrethrum extract incorporated in varnishing creams has found use as a mosquito repellent in Asian countries. Pyrethrum is used for protection of dried fish and fishmeal against flies. It is also used against pest of common mushroom. Pyrethrum is employed on certain crops in the final stages immediately prior to harvesting in place of higher residual activity products. Pyrethrum is used in ultra low volume preparations for agricultural use. It is also effective for herbarium cases for long-term protection against insects. It is effective against pests of fruits and vegetables especially *Myzus persicae* Sulz. and *Aphis fabae* Scop. The pests of stored flood grains such as *Calandra granaria, C. oryzae* and *Tribolium castaneum* as well as *Trogoderma granarium* (khapra grub of wheat), are effectively controlled by pyrethrins. The attack of *Dermestes maculatus, Calliphora vicina* and *C. wealensis* on stored fish and meat also are repelled by pyrethrins. The pyrethrum marc, the grist from flowers left after extraction of pyrethrins, is used for preparing mosquito coils and is useful against beetles, *Aulacophora foveicollis, A. atripennis* and the honey bee *Apis cerana indica.*

Pyrethrum is an effective fungicide and nematicide also. Alcoholic extract is used for aphids, cabbage loopers, codling moths, Mexican bean beetles, spider mites, stink bugs, thrips, tomato pinworms and white flies. Water extract is used with soap or the powder is used mixed with water.

Addition of decomposed leaves of *Chrysanthemum coronarium, Azadirachta indica, Tagetes patula, Datura fastuosa* and *Nerium indicum* reduced the incidence of *M. javanica* and increased plant growth (Hameed, 1968).

2. *Cannabis sativa* Linn. (Cannabinaceae)

Indian hemp, Marijuana, Ganja

C. sativa is a dioecious strong-smelling rough puberulent herb, up to 4 m in height, with alternate, long-petioled palmately divided leaves having long lanceolate serrate lobes, staminate flowers in long panicled cymes (in male plants) and pistillate flowers in short leafy spikes in female plants and seeds contained in achenes. A native of Central and W. Asia.

The minute stalked glands of the flowering tops of female plants yield a sticky yellow resin. The resin contains a group of isoprenoid resorcinols collectively known as cannabinoids. Δ^9-Tetrahydrocannabinol, a dibenzopyran derivative, is the principal component with lesser quantities of cannabinol, cannabidiol, cannabidiol-carboxylic acid and cannabichromene. A volatile oil consisting of about 30 monoterpenes, choline and alkaloids (cannabisativine and trigonelline) are the other constituents of the resin.

It has mild **insecticidal** properties and thus is a good insect **repellent.** Its alcoholic extract has a limited lethal effect on flies and mosquitoes but has a marked repellent action against both.

4.2.4. Glucosinolates and Sulphides

1. *Allium sativum* Linn. (Liliaceae)

Lashuna, Garlic

A. sativum is a hardy perennial with an underground bulb consisting of a disc shaped stem, swollen axillary buds (cloves) encircled by 2-3 layers of tunica and long flat leaves. Flowers are small, white, borne on scapes. A native of Central Asia now cultivated everywhere.

Garlic bulbs contain carbohydrates (30%), protein (6%) and essential oil (0.1%) besides minerals and vitamins. Also present are peptides such as γ-glutamyl phenylalanine, γ-glutamyl-S-methyl cysteine, S-allyl mercapto-L-cysteine and γ-glutamyl-S-allyl mercapto-L-cysteine. In addition, garlic contains sulphur containing amino acids such as S-propyl-L-cysteine, S-propenyl-S-cysteine and S-allyl cysteine. Other compounds reported from garlic are a thioglucoside, scordine, scordinines A, A_2 and B and cyanidin. The volatile oil of garlic consists of allyl alcohol, allyl propyl disulphide, methylallyl disulphide, diallyl disulphide, dimethyl trisulphide, methyl allyl trisulphide and diallyl trisulphide.

Garlic is used against aphids, army worms, colarado beetle, kharpa beetle, codling moth, cabbage worm and wire worms. Garlic in kerosene / mineral oil / soap is used against American bollworm, onion thrips, root knot nematode, potato tuber moth, sugar cane shoot borer, bacterial diseases, anthracnose, downy mildew and rice blast. Garlic extracts are found effective against a number of plant parasitic nematodes such as *Meloidogyne incognita, M. javanica, Radopholus similis, Aphelenchoides sacchari* and *Tylenchulus semipenetrans.* Almost all the compounds from garlic exhibit anti-microbial activity.

Onion (*Allium cepa*) bulb aqueous extract is found useful against white flies, ants, scales, spider, mites and thrips. All the tested species of *Allium* including *A. grayi* and *A. fistulosum* are found to be nematicidal (Pandey *et al.,* 2000).

2. *Brassica nigra* Koch. & *B. alba* Hook.f. (Brassicaceae)

Mustard

The volatile oil obtained from the seeds of various spp. of *Brassica* is the mustard oil. The pungent volatile oil of *B. nigra* contains allyl isothiocyanates while the oil of the latter contains acrinyl isothiocyanate.

Mustard oil repels and kills insects, spiders and centipedes-used for indoor plants, ornamentals and garbage cans.

3. *Cleome gynandra* Linn. (Capparaceae)

Syn. *Gynandropsis pentaphylla* DC.

G. gynandra Briq.

Pandri-tilwan

This plant is a glandular pubescent annual herb with palmate 3-7 foliolate leaves (leaflets elliptic-ovate), pale pink or white flowers (sepals 4; petals 4, clavate, stamens 6 on a gynophore) in dense bracteate racemes, and viscid-pubescent capsules containing dark brown orbicular, muricate seeds. A common weed.

The whole plant yields an essential oil containing isothiocyanates. Seeds contain a fixed oil, glucocapparin and steroids. The volatile oil shows insecticidal properties.

4. *Eruca sativa* Mill (Brassicaceae)

Syn. *E. vesicaria* Cav.

Taramira, Rocket Salad

The glucosinolate extract (2%) from the leaves is reported to adversely affect the survival and development of nymphs, adult longevity, reproductive period and fecundity of mustard aphid, *Lipaphis erysimi*.

5. *Petiveria alliacea* Linn. (Phytolaccaceae)

Anamu

A native of Central and South America and Africa, this plant is an odoriferous herbaceous perennial with ovate leaves, small white flowers in terminal /axillary spikes and apically two-lobed spiny (spines 1-3, reflexed) achenes containing a single seed.

Roots contain an essential oil consisting of benzaldehyde, dibenzyl disulphide, dibenzyl trisulphide, coumarin trithiolniacine and stilbene. The leaves contain benzyl-2-hydroxy ethyl trisulphide, tannins and other polyphenols.

Dibenzyl trisulphide is found to be effective against insect pests such as *Cylas formicarius elegantulus* and *Hypothenemus hampei* and ascaricidal (*Boophilus microplus*). Root extract is antifeedant and used against fall armyworm, leaf-cutting ants and whiteflies. When applied to the soil, populations of *Melidogyne incognita* were found decreased.

The plant is medicinally important as antileukemic, antitumourous and cytotoxic. It acts as an immunostimulant, antiarthritic and antimicrobial.

4.2.5. Diterpenes

1. *Aleurites fordii* Hemsl. (Euphorbiaceae)

Tung Oil Tree

A small much-branched milky tree, *A. fordii*, possesses simple trifid leaves, unisexual flowers in clusters in which one or more pistillate flowers are surrounded by a number of staminate flowers, spheroid/ top-shaped fruits containing 3-5 one seeded segments. It is a native of China cultivated widely in India.

Leaves yield three ellagitannins, aleurinins A, B & C along with corilagin, geraniin and chebulagic acid. Wood is found to contain a coumarinolignoid and 5, 6, 7-trimethoxy coumarin. Seeds yield 30-35% oil containing α-elaeostearic (75-85%) and linoleic acids (10%) as the principal fatty acids, about 10% proteins, starch and 2 diterpene esters 13-O-acetyl-16-hydroxy phorbol and 12-O-palmityl-13-O-acetyl-16-hydroxy phorbol and a non-protein amino acid, L-3-carboxyl-1, 2, 3, 4-tetrahydro-β-carboline. The bark and fruit husk contain 12 and 4.5% tannins respectively (Fozdar *et al.*, 1989).

Phorbol derivatives are insecticidal. Seed oil is used for treating boils, ulcers, swellings and burns.

2. *Andrographis paniculata* Wall. (Acanthaceae)

Kalmegh, Creat

A. paniculata is a small erect annual with winged quadrangular stem, glabrous lanceolate leaves, small white purple flowers borne on horizontal axillary racemes and linear-oblong capsules containing numerous subquadrate brown seeds. A native of tropical Asia.

The whole plant is intensely bitter due to the presence of diterpenoid lactones such as andrographolide (0.6%) 1, 4-deoxy-11-oxoandrographolide, 14-deoxy-11,12-didehydroandrographolide, 14-deoxy andrographolide and a non-bitter neoandrographolide. Also present are flavones such as 5-OH,7,2′6′-triOMe flavone, 5-OH,2′,7,8 tri OMe flavone, 2, 5-diOH-7,8-diOMe flavone, mono-O-methyl wightin, 5-OH-7,8-diOMe flavone and 5-OH-3,7,8,2′-tetraOMe flavone; a C_{23} terpene, 14-deoxy-15-isopropylidine-11,12-didehydro-andrographolide and an iridoid glucoside - procumboside. The leaves contain up to 1% andrographolide. The roots, in addition to andrographolide, contain flavones such as 7,4′-diOMe apigenin, echioidin, andrographin and panicolin, as also α-sitosterol (Abeyasekara *et al.*, 1990; Badar, *et al.*, 2003).

Whole plant extract is useful for aphids, melon worms, thrips and white flies (Stoll, 2000). This herb is extensively used as a bitter tonic and febrifuge. It is also used in dysentery, cholera, diabetes, influenza, piles and gonorrhea. It is also considered a tonic, blood purifier and a cure for liver dysfunction.

3. *Croton tiglium* Linn. (Euphorbiaceae)

This is a small evergreen tree with stellately hairy shoots, ovate acuminate serrate leaves, male flowers with 15-20 stamens and white turbinately ovoid, trigonous capsules. A native of China, now naturalized in Konkan regions.

Seeds yield oil and a number of phorbol derivatives, β-sitosterol and lectin. Phorbol derivatives are carcinogenic.

The plant is used in the preparation of insecticides to control aphids, cabbage caterpillar and bollworm.

4. *Euphorbia milii* Des Moulins (Euphorbiaceae)

Crown of Thorns Plant

This is a woody, spiny, laticiferous climbing succulent shrub with shoots reaching a height of 6 feet. Leaves are ovate and the inflorescence (cyathium) is characterised by two cordate colored (pink/yellow) petaloid involucral limbs. Cultivated for the colored inflorescences.

The latex contains triterpenes, ingenol, phorbol esters, flavonoids, a macrolide and ingenol peptides such as miliamines A-L. It inhibits significant molluscicidal activity against Schistosomiasis-transmitting snails like *Biomphalaria glabrata* and *B. tenagophila* due to miliamine-L.

5. *Euphorbia thymifolia* Linn.

Syn. *E. prostrata* Ait.

The whole plant yields deoxyphorbol acetate derivatives, epitaraxerol, hexacosanol, euphorbol, 24-methylene cycloartenol, taraxasterol, tirucallol and quercetin.

This also is considered as an insecticide.

6. *Euphorbia tirucalli* Linn. (Euphorbiaceae)

Milk bush

Milk bush, a native of Africa, is a small laticiferous succulent tree with cylindric green branches in pseudowhorls, small deciduous alternate oblanceolate leaves and yellow flowers (cyathium) at the end of twigs.

The latex and stem bark contains triterpenoids such as ingol, its acetyl and isovaleryl ester, euphorbinol, cycloeuphorbinol, euphorcinol, euphorginol, glutenol, cycloart-23-ene-3β,25-diol, α-taraxerol, cycloartenol, α-amyrin, isoeuphorbol, euphorbol, euphol, its acetate, taraxerol and taraxerone. Also present are diterpenoids like ingenol and phorbol esters and ellagic acid. Rubber and n-hentriacontane are the other components of stem.

The plant is not eaten by cattle due to the milky latex. Latex is toxic to fish and rats. Africans regard the tree as a mosquito repellant. Stem, branches and buds have insecticidal and repellant properties against aphids, grass hoppers and mosquitoes. The plant sap is used as a seed dressing against the plant parasitic nematodes *Tylenchorhynchus brassicae* and *Rotylenchus reniformis*. Leaves, roots and seeds are soaked in water and this solution is sprayed to protect vegetables from caterpillars and seedlings from termites. The latex is toxic and cause irritation to eyes, must be handled with care. The plant is planted near coconut trees to prevent attack from Rhinoceros beetle *Oryctes* sp. in Kenya. *E. tirucalli* latex and stem bark are found effective against snails which are vectors of liver fluke. This plant is used against citrus aphids, cutworm, termites.

7. *Kingiodendron pinnatum* Harms. (Caesalpiniaceae)

Syn. *Hardwickia pinnata* Roxb.

Anjam

This is large, handsome evergreen tree of 30-40 height, with imparipinnate leaves (only 2 sessile oblong - ovate oblique acute leaflets, similar to that of *Bauhinia*), small white flowers (petals absent) in dense racemes arranged in panicles and an ovoid - oblong pod containing a single seed. A native of Tropical Asia common in Western Ghats.

The tree yields an oleoresin containing barbinatic acid, β-caryophyllene and cadinene.

The plant is reported to have insecticidal properties due to its volatile oil.

8. *Jatropha curcas* Linn. (Euphorbiaceae)

This native of Tropical America is a soft-wooded shrub with angular 3-5 lobed leaves, yellowish-green male and female flowers and a 3-lobed capsule containing three oblong seeds.

Leaves are reported to have insecticidal properties.

9. *Pedilanthus tithymaloides* Poit. (Euphorbiaceae)

A native of S. America, this plant is a laticiferous succulent herb of 1m with fleshy ovate, often variegated leaves in two rows and scarlet slipper-shaped involucre.

This plant and *Pedilanthus cucullata* are planted near coconuts to prevent Rhinoceros beetle and to protect seedlings from termites.

10. *Persea indica* Spreng.(Lauraceae)

This ornamental tree has alternate, elliptic to oblong-lanceolate or lanceolate leaves, green to yellowish flowers in terminal panicles and elliptic, long, black fruits. A native of India

It is a natural source of the insecticidal diterpenes ryanodol and cinnceylanol (more in stems) which are found toxic against *Macaronesia fortunata* and *Heliothis armigera* (Gonazalez-Coloma *et al.*, 1993).

11. *Polyalthia longifolia* B&H (Annonaceae)

Asupalav

This is a tall handsome tree with narrowly lanceolate leaves tapering to a fine point, yellowish green flowers in fascicles and an etaerio of l-seeded drupes. A native of South India.

The seeds are found to contain two clerodane diterpenes kolavenic acid and 16-oxo-cleroda-3, 13(14) E-diene-15-oic acid. Roots yield isoquinoline alkaloids pendulamine A & B and penduline along with stigmasterol, allantoin, kolavenic acid and an azofluorine alkaloid, isoursuline.

Leaf extract is reported larvicidal to *Culex quinquefasciatus.*

12. *Rhododendron molle* G. Don.(Ericaceae)

Dried flowers of this Chinese insecticidal plant contain three grayanoid diterpenes rhodojaponin III, grayanotoxin III, and kalmanol which are the most active constituents. Rhodojaponin III, the major compound, exhibited antifeedant, growth inhibitory and insecticidal activities against larvae of *Leptinotarsa decemlineata* and *Spodoptera frugiperda*. Grayanotoxin III and kalmanol occurred in much lower amounts and were less active as antifeedants. The 2ß,3ß-epoxy group of rhodojaponin III was found to be important for the antifeedant activity, while the 14-acetate of grayanotoxin I and the rearranged B and C rings of kalmanol were not (Klocke *et al.*, 1991).

13. *Sideritis trojana* Bomm. (Lamiaceae)

This plant, endemic to Turkey contains volatile oils and diterpenoids. The *ent*-kaurene diterpenoids 7-epicandicandiol and 18-acetylsideroxol of this plant showed toxicity against *Acanthoscelides obtectus* and *Sitophilus granarius* while 7-epicandicandiol diacetate showed toxicity against *Ephestia kuehniella* (Aslan *et al.*, 2006).

4.2.6. Triterpenes

1. *Cucumis sativus* Linn.

Trapusah, Cucumber

This is a hispid climbing/ trailing annual herb with deeply cordate 3-5 lobed leaves, yellow flowers (male flowers clustered, female flowers solitary) and a cylindrical pepo of differing forms and sizes containing white flattened seeds. A native of S. India.

Seeds and leaves contain cucurbitasides B & C and sterols such as stigmasterol and spinasterol.

The allelopathic potential of cucumber is used in weed control.

2. *Cucurbita pepo* Linn. (Cucurbitaceae)

Kumra, Pumpkin

C. pepo is a stout hispid annual wine (hairs are pungent) with palmately lobed leaves and pear-shaped large fruits. A native of Hungary, now cultivated in India.

Seeds contain fixed oil, cucurbitacins and phytosterols. The seeds are used as taenicides. Plant organs are found to contain light dependant acetyl choline.

3. *Diploclisia glaucescens* Diels. (Menispermaceae)

The stems yield ecdysterone (over 3%) and triterpenoids such as serjanic acid, phytolaccagenic acid and its 3- glucoside. The seeds contain five ecdysteroids.

Ecdysterone is hemolytic, insecticidal and spermicidal. The seed ecdysteroids show activity against larvae of the European corn borer (*Ostrinia nubialis*).

4. *Junellia aspera* Moldenke (Verbenaceae)

Triterpenes such as maslinic acid , daucosterol , and 3β-hydroxy-12α-bromine-(28→13)-oxide-oleanane from this plant showed the highest toxic effects against adults of *Sitophilus oryzae* (L.) when ingested, while oleanolic acid and oleanonic acid showed less toxicity. No toxicity was observed when these compounds were topically applied (Pungitore *et al.,* 2005). This plant is a native of S. America.

5. *Myrtillocactus geometrizans* Console (Cactaceae)

Blue Candle, Whortleberry Cactus

A methanol extract from the roots and aerial parts of this plant containing sterols such as peniocerol, macdougallin and chichipegenin showed insecticidal and insect growth regulatory activity against fall armyworm (*Spodoptera frugiperda,* an important insect pest of corn) and *Tenebrio molitor* (Coleoptera), a pest of stored grains in Mexico. These compounds appear to have ecdysone mimicking action in exerting selective effects on the pre-emergence metabolism of Coleoptera, because in all treatments of the larvae of *T. molitor,* pupation were shortened and this process shows precociousness in relation to controls. In contrast to *S. frugiperda* larvae, onset of pupation was noticeably delayed. Emergence in both cases was drastically diminished. In both pupae and in the few adults that were able to emerge, many deformations were observed. The results of these assays indicated that the compounds were more active than other known natural insect growth inhibitors such as gedunin and methanol extracts of *Cedrela salvadorensis* and *Yucca periculosa*(Cespedes *et al.,* 2005).

6. *Nothofagus dombeyi* Oerst. (Fagaceae)

Ethyl acetate extracts of dried leaves of *N. dombeyi* contain several triterpenes and flavonoids. 12-Hydroxy oleanolic lactone and pectolinarigenin were found to have marked anti-feeding activity against 5th instar larvae of *Ctenopsteustis obliquana,* a leaf roller. Dihydrooroxylin isolated from *N. pumilio* Oerst. also showed similar antifeedant activity, while in *N. alessandri* Espinosa and *N. fusca* Oerst. the same activity was due to the combined presence of stilbene pinosylvin and flavone galangin (Thoison *et. al.* 2004).

7. *Schleichera oleosa* Oken. (Sapindaceae)

Kusum, Lac tree

A large tree of 20m with pinnate pubescent leaves (leaflets 2-4 pairs, lowest pair largest), white flowers in branched racemes and echinate fruits containing 1-2 seeds enclosed by a fleshy aril. Native of S.E. Asia.

Bark and stem yield sterols such as schleicherastatins 1-7 and schleicheols 1 and 2. The plant is reported to have insecticidal properties.

4.2.7. Limonoids and Quassinoids

1. *Ailanthus excelsa* Roxb. (Simaroubaceae)

Ardusa, Tree of heaven

A. excelsa is a large majestic, deciduous tree with a tall cylindrical bole, pinnate leaves (up to 90cm long having 8-14 pairs of unequal sided leaflets), small yellowish flowers in panicles and one seeded samaroid fruits. This tree is very fast growing and is introduced all over India.

The stem bark contains quassinoids such as glaucarubin and excelsin, ailanthic acid, 2,6-dimethoxybenzoquinone, β-sitosterol, malanthin, triacontane and hexatriacontane. The root bark contains quassinoids such as ailantheone and glaucarubinone.

The quassinoids possess insecticidal properties. They are also found to be anti-tumour and cytotoxic against leukaemia cells.

2. *Azadirachta indica* A. Juss. (Meliaceae)

Nimba, Neem

Since the bark of this plant yields a dye the description of this tree is given in the previous section.

All the parts of the plant contain limonoid bitter principles. The stem bark yields nimbin, nimbidin, β-sitosterol, tannins (12%) and a gum. Root bark also contains nimbin and nimbidin. Leaves contain nimbin, nimbinene, 6-desacetylnimbinene, nimbandiol, nimbolide, quercetin and β-sitosterol. The flowers yield β-sitosterol (along with its glucoside), quercetin, kaempferol, thioamylalcohol and benzyl alcohol and a volatile oil containing sesquiterpenes, margosene and azadirachtene. The fruits are found to contain gedunin, azadirachtin, 7-deacetoxy-7α-hydroxy gedunin, azadiradione, azadiradone, 17-β-hydroxy azadiradione, 17-*epi* azadiradione and nimbiol. Seeds yield oil and a number of bitter principles both in oil and outside. The oil consists of oleic (56%), palmitic (16%), stearic (14%) and linoleic (9%) acids and 0.8% unsaponifiable matter containing nimbidin, nimbin, nimbinin, meliantriol as major components along with gedunin, meldenin, desacetylgedunin, salanin, azadirone, epoxyazadirodione, vepinin, nimbinene, 6-O-acetyl nimbandiol etc.

Almost all limonoids of this plant show insect repellence or suppression; seeds or seed kernels provide the largest amounts of insecticidal compounds. Azadirachtin is extracted in methanol (from seeds).

The effects of neem include repellence, feeding deterrence, oviposition deterrence, reduced growth and development and interference with reproduction, but the biochemical or molecular mode of action for such effects remain unknown (Schmutterer,1990). The compounds of neem show extremely low toxicity to mammals. Neem is most effective on soft bodied immature stages of pests including whiteflies, thrips, mealy bugs and various caterpillars.

The leaves are used as insect repellents and are kept among clothes and books. The dried leaves, when mixed with the grains of wheat and sorghum, are used to provide protection against moth (*Ephestia cantella* Walker), rice weevil (*Calandra oryzae* Linn), red flour beetle (*Trifolium castaneum* Herbst.), grain moth *(Sitotroga cerealella* Olivier) and grain borer (*Rhizopertha dominica*). Larvae of Lucerne weevil and locust (*Schistocerca gregaria* Forsk.) are the other organisms repelled or destroyed by the leaves. The pulp extract of fruit also is used against locusts. The powdered fruit/kernel mixed with wheat or legumes (green gram, cow pea or pea) are used against rice weevil (*Sitophilus oryzae* Linn.), lesser grain borer (*Rhizopertha dominica*), khapra beetle (*Trogoderma granarium* Everts), bruchid (*Callosobruchus maculatus* and flour beetle. The anti-feeding property is attributed to azadirachtin and salanin. But the crude preparations of kernel are found to be more effective than refined ones. Other insects on which the kernel exhibited a toxic/deterrent activity are *Locusta migratoria, Schistocerca gregaria, Spodoptera litura* and *Boarmia (Ascotis) selenaria*. The essential oil from seeds is found to be effective against pulse beetle and *Callosobruchus* sp. The oil cake also is used as an insecticide and nematicide

In general, the leaf extract is used against aphids, Colorado beetles, grass hoppers, grubs, Japanese beetles, leaf hoppers, locusts, scales and snails. Seed powder (aqueous extract.) is preferred for cotton boll worm, aphids, colarado beetles, cotton leaf roller, cutworm, diamond black moth and fall armyworm. Seed oil suspension (with soap) is applied for aphids, plant hopper, flea beetle, leaf hopper, potato tuber moth, psyllid, scale insects, white flies and whorl maggot.

When the natural neem oil is removed from the seeds and treated with alcohol, virtually all of the azadirachtin and related substances separate from the oil itself. The remaining oil - without the azadirachtin is called Clarified Hydrophobic Extract of Neem Oil. Unlike azadirachtin, this active ingredient is also active against fungal diseases such as mildews and rusts.

Azadirachtin and clarified hydrophobic extract of neem oil are used for many food and non food crops. Adverse effects to humans are to be avoided and this insecticide should not be applied when honeybees are actively visiting the flowers. Safe for indoor plants.

Azadirachtin interferes with the normal life cycle of insects including feeding, moulting, mating and larvae and thus deter certain insects like locusts from feeding

Neem leaves, kernel and oil cake are found to be effective against nematodes such as *Hoplolaimus indicus, Rolylenchulus reniformis, Tylenchorhynchus brassicae, Meloidogyne incognita* and *M. halpa.* Treatment of the seeds of tomato with crude extracts of the seeds of *Azadirachta indica,* inhibited the penetration of *Meloidogyne javanica* into tomato roots (Prot and Kornprobst, 1983).

3. *Cedrela sinensis* Juss.(Meliaceae)

A tall tree of China and Korea. The leaves contain limonoids like 11β- hydroxy-7α- obacunyl acetate, 11-oxo-7α- obacunyl acetate, 11-oxo-7α-obacunol,11β-hydroxy cneorin G and 11-oxocneorin G (Mitsui *et al.,* 2004).

These limonoids exhibit marked insecticidal properties.

4. *Guarea grandiflora* A.D.C. (Meliaceae)

Three protolimonoids, namely 21-α-acetylmelianone, melianone and melianodiol, and three limonoids, gedunin, 7-deacetoxy-7-oxogedunin and 6-α-acetoxygedunin, are present in the seeds of this plant. These compounds are found insecticidal to the European corn borer, *Ostrinia nubialis* (Jimenez *et al.,* 1998).

5. *Hannoa undulata* Planch.(Simaroubaceae)

Syn. *Quassia undulata* Gaertn

This is a perennial shrub native to Tropical America. The seeds and roots of this plant are found to contain quassinoids chaparinone, klarneanone and glaucarubolone. The seed extract when admixed in the soil before sowing any crop, show nematicidal activity against

Meloidogyne javanica. Root extract is found to inhibit penetration and reproduction of *M. javanica* in tomato.

The roots of *H. klaineana,* a related plant containing quassinoids, also are found to have similar nematicidal activity. Treatment of the seeds of tomato with crude extracts of the seeds of *Hannoa undulata* and *H. klaineana,* inhibited the penetration of *Meloidogyne javanica* into tomato roots (Prot and Kornprobst, 1983). Seed amendment with seed powder of *H. undulata* inhibited the reproduction of *M. javanica* fully in tomato roots.

6. *Lansium domesticum* Corr. (Meliaceae)

Lansone

A native of China & Indonesia, this is a large tree with large pinnately compound leaves (leaflets 5-7, elliptic acuminate), small white flowers in cauliflorous spikes and yellow ovoid fruits in clusters of 4-40 containing 1-3 greenish seeds in an edible pulp.

The fruit skin is found to contain limonoids lansic acid and lansiosides A, B & C. Seeds yield dukunolides A, B and C.

Fruit peel is dried and burned to repel mosquitoes. Seeds also are insecticidal and used against armyworm and leaf eating caterpillars. Bark is used to treat malaria and seeds for fever.

7. *Melia azedarach* Linn. (Meliaceae)

Bakum-nimb, Persian Lilac

This is a small tree of 15m with bi- or tripinnate leaves (leaflets oval, serrate, unequal-sided), small white fragrant flowers having a purple staminal tube in long-peduncled panicles and ellipsoid-globose 4-seeded drupes. A native of Persia now cultivated everywhere as a social forestry tree.

All the parts of plants contain limonoids. Besides azadirachtin and nimbolin A which are common with neem, the other limonoids reported from this plant are very similar in structure to the neem limonoids. They are sedanolactone, methyl kulonate, 3-deoxy melianone, melianin A, melianin B, sedanin, desacetyl sedanin, sedanal, ochinal, ochinin acetate, nimbolinin B, nimbolidin A &B, ochinolide, 7-deacetyl azadirone and 7-deacetyl proceranone. But it is the three meliacarpin derivatives (C-seco limonoids) namely 1,3-dicinnamoyl-11-hydroxymeliacarpin, 1-cinnamoyl-3-methacrylyl-11-hydroxymeliacarpin and 1-cinnamoyl-3-acetyl-11-hydroxymeliacarpin which are found as the principal insecticidal and growth disrupting constituents of *Melia azedarach* leaves (Bohnenstengel *et al.*, 1999). Toosendanin is a commercial insecticide derived from this plant.

The insecticidal properties of this plant are the same as that of neem and often this is used as a substitute for neem.

8. *Phellodendron amurense* Rupr. (Rutaceae)

Amur cork

This is a broad spreading tree with a short trunk, opposite pinnate leaves having 5-11 leaflets and blue black fruits in female trees. A native of China.

The leaves and wood/bark are found to contain alkaloids such as candicine, magniflorine, berberine, jatrorrhizine and palmatine, limonoids like obacunone and limonin and 8-isoprenyl dihydro-flavonols (phellamurin) and flavonols (amurensin, phellamurin and phelloside)

A natural insecticide is prepared from this plant.

9. *Picrasma excelsa* Planch. (Simaroubaceae)

Quassia

A native of West India, this plant is a tall polygamous shrub with very large compound leaves having numerous ovate unequally sided leaflets.

The fine shavings of the wood, the Quassia of commerce, contain a mixture of bitter terpenoids like the lactone, quassin (which form the chief constituent), neo-quassin (the hemiacetal of quassin) and picrasmin (*iso*-quassin).

Quassia is an insecticide especially for lice. It is also used as anthelmintic and antitumour agent.

10. *Quassia amara* Linn.(Simaroubaceae)

Quassia

This is a shrub of about 4m bearing compound leaves with a winged leaf rachis and attractive bright red flowers on terminal racemes. It is a native of South America, but now extensively cultivated worldwide as an ornamental tree.

The wood is extremely bitter and contains quassinoids such as quassin, neoquassin and 18-hydroxy quassin and indole alkaloids including various β-carbolines and several canthinone type (canthin-6-one,4-methoxy-5-hydroxy-canthin-6-one) alkaloids.

This plant is a contact and stomach poison, insecticide, larvicide and nematicide and acts against aphids, caterpillars, colarado beetle, leaf miners, melon worm, mites, silkworm and carpet beetle.

11. *Swietenia humilis* Zucc. (Meliaceae).

The seeds of this plant contain limonoids such as humilinolides A- F, humilin B, swietenin C, methyl-2-hydroxy-3-α-isobutyroxy-1-oxomeliac-8 (30)-enate, methyl-2-hydroxy-3-β-tigloyloxy-1-oxomeliac-8 (30)-enate and swietemahonin C. All these compounds are found insecticidal (Jimenez *et al.*, 1998).

12. *Toona ciliata* Roem. (Meliaceae)

Syn. *Cedrela toona* Roxb.

Indian Mahogany

T. ciliata is a tall tree with a shining fine brick red colored aromatic wood, long (up to 1m) pinnate leaves having 8-30 obliquely ovate/ lanceolate shining leaflets, white honey-scented flowers (petals ciliate) in large terminal pubescent panicles and oblong capsules containing many winged seeds. A native of South India.

The stem bark contains limonoids such as cedrelone, cedrelonelide, epoxyazadirone 12-deacetoxy toonacilin, norisotoonafolin, pregnanes and a coumarin, siderin. The timber yields limonoids cedrelone and 1,2-dihydrocedrelone and procyanidins. Roots contain 9,10-dihydroxy phenanthrenes. Leaves are found to contain toonafolin and surenolactone and coumarins. Seeds yield bergaptene, a coumarin.

Leaf extract is used for treatments against pod borer of groundnuts, termites, wireworms and white grubs.

13. *Trichilia emetica* Vahl (Meliaceae)

Natal Mahogany

T. emetica is a native tree of Tanzania with alternate pinnate leaves (leaflets obovate), creamy green flowers in bunches and small 3-4 lobed crimson capsules split open to expose 6-8 bean like black seeds surrounded by a red aril.

Contain limonoids which are insecticidal.

14. *Walsura trifoliolata* Harms (Meliaceae)

Syn. *W. piscida* Roxb.

Walasura

A native of S.E. Asia, this tree reaches a height of 30 m with 3-foliolate leaves (leaflets oblong - elliptic, sub-coriaceous), yellowish white flowers in corymbosely branched panicles and orange - yellow berries containing a single seed surrounded by a white juicy aril.

The bark is used as a fish-person. The pulp of fruit also is a fish-poison but used to kill vermin in hair.

4.2.8. Saponins

1. *Acacia auriculiformis* A.Cunn. (Mimosaceae)

This is a tree native to Australia now cultivated extensively as a social forestry tree. The plant is almost leafless and the leaf-rachis is modified to leaf-like lanceolate phyllodes. The flowers are small , yellow born on catkins and the fruits are spiral legumes.

The fruits yield two acylated triterpenoid bisglycosides acaciasides A & B. Heartwood and bark contains flavones.

The fruit extract and acaciasides are found effective against the nematode, *Meliodogyne incognita.*

2. *Adina cordifolia* Hook. f. (Rubiaceae)

Haldarvo, Haldu

Haldu is a large deciduous tree having a straight fluted bole, light grey bark exfoliating in patches (bark light red inside with white streaks), broadly ovate abruptly acuminate long petioled leaves and intrapetiolar foliaceous stipules generally enclosing the terminal bud. Flowers are small yellow borne in globose pedunculate heads and the fruits are capsules splitting to two dehiscent cocci containing many minute seeds.

The heartwood yields adifoline, cordifoline, 10-deoxyadifoline, 10-deoxycordifoline, benzoic acid, umbelliferone and a yellow pigment adinin. An orange-coloured oleo resin is induced by incision of the trunk or extracted from saw dust.

The wood is useful as a plywood timber. Bark is used in biliousness and as a febrifuge and antiseptic. Adinin is fish poison. Wood is untouched by white ants or insects and is used extensively for structural work.

3. ***Aesculus hippocastanum*** **Linn. (Hippocastanaceae)**

Horse chestnut

This is a tall deciduous tree with a palmately compound leaf bearing 5-7 oval to spindle shaped leaflets, white/pink flowers borne on large terminal racemes and a spiny green fruit containing 3 rounded shiny brown seeds. This is a native of Western Asia now cultivated around the world as an ornamental.

Seeds yield saponins, coumarins, flavonoids and lipids. There are over 30 pentacyclic saponins reported from this plant of which "aescin" was considered a principal saponin. But aescin is a group of saponins, based on the sapogenins proaescigenin and barringtenol, combined with angelic, tiglic or acetic acids.

The coumarin, aesculin, present as also the saponins contribute to the insecticidal and piscicidal activities of this plant. Seeds are useful in varicose veins and to prevent thrombosis, thrombophlebitis, calf cramping, edema and hemorrhoids.

4. ***Aesculus assamica*** **Griff. (Hippocastanaceae)**

Sarlok-asing

This is a deciduous tree, up to 15 m in height, with long petioled palmately compound leaves, white yellow flowers in panicles and loculicidal capsules containing dark brown seeds. A native of India

The powdered bark is used as a fish poison.

Aesculus californica also is used as an insecticide.

5. ***Agave americana*** **Linn. (Agavaceae)**

Kantala, American Aloe

A. americana is a short-stemmed plant bearing a rosette of long erect pointed fleshy leaves with marginal spines and longitudinal white yellow streaks or bands. The plant flowers only once during its life and the flowering stem arises from the center of the plant as a thick pole. The flowers are funnel shaped and capsules oblong clavate and beaked. A native of America, it is introduced in India for the leaf fibre.

The leaves yield saponins such as agavasaponins A-I (10 in number) based on hecogenin (aglycone), chlorogenin, rockegenin, tigogenin, dehydrohecogenin and piscidic acid. Flowers contain chlorogenin and kaempferol and the seeds neotigogenin, hecogenin and kammagenin.

The saponins are piscicidal and insecticidal.

6. *Albizzia lebbek* Benth. (Mimosaceae)

Siris

Siris is a tall deciduous tree with abruptly bi-pinnate leaves (lateral leaflets elliptic oblong, terminal leaflets obovate oblong, glands between leaflets, leaflets unequal sided), white fragrant flowers in globose umbellate heads (corolla united) and long linear oblong reticulately veined pods containing 4-12 ellipsoid compressed seeds. A common avenue tree.

Pod yields a saponin, lebbekanin C. Seeds contain lebbekanin A (major) & B. Heartwood yield lebbekanin E, melanoxetin, okanin, melacacidin, 3-OMe melacacidin and procyanins. Bark contains tannins, friedelan-3-one and γ-sitosterol. Leaves are found to have echinocystic acid, β-sitosterol and a flavone, vicenin II. Flowers possess lebbekanins D, F, G & H.

The saponins are found pesticidal.

7. *Asparagus racemosus* Willd (Liliaceae)

Shataveri, Asparagus

A. racemosus is a scandant much-branched spiny undershrub with tuberous short root stock bearing a number of tuberous roots, stem armed with numerous recurved spines, sickle-shaped cladodes, leaves reduced to scale leaves, small whilé flowers in racemes and globose berries. Occurs wild and cultivated throughout India.

The tubers contain saponins, named shatavarins I, II, III, IV, glycoside AR-4 ($C_{45}H_{74}O_{16}$), tridecaacetyl shatavarin-I ($C_{77}H_{112}O_{36}$) and tetradeca-O-methyl shatavarin-I ($C_{65}H_{114}O_{23}$). Shatavarin-IV is based on sarasapogenin. Also present are sarasapogenin, β-sitosterol and 4, 6-dihydroxy-2-O-(2′-hydroxy isobutyl) benzaldehyde. Carbohydrates (53%) inclusive of mucilage, polyfructosans, free sugars and insoluble polysaccharides as well as proteins (3%) form the other components of the tubers. Asperagusic acid is reported from the roots. A polycyclic alkaloid, asparagine-A is reported from plants of Thailand. The cladodes are said to contain diosgenin and quercetin. Flowers yield a volatile oil, rutin, hyperoside and quercetin. Fruits also yield quercetin, rutin and hyperoside along with cyanidin glycosides (Joshi and Dev, 1988).

The asparagusic acid, which occurs up to 35 mg in roots, is found to be nematicidal for *Heterodora rostochinensis, Meloidogyne halpa, Pratylenchus penetrans* and *P. curvitatus.* Intercropping with tomato is found to decrease the nematodes like *Paratrichodorus minor* from the soil.

8. *Carthamus tinctorius* Linn. (Asteraceae)

The seeds contain a fatty oil, matairesinol (a monoglucosidic bitter principle), steroid diglucosides, aellobioside and tracheloside, lignan glycoside, 2-OH-arctiin, polyacetylenes and conjugated serotonins.

The seeds are found nematicidal against *Meloidogyne incognita*

9. *Catunaregam spinosa* Tirveng. (Rubiaceae)

Syn. *Randia spinosa* Poir.

R. dumetorum Poir.

Xeromphis spinosa Keay.

Madana, Emetic nut

For the description of this plant see the previous section (Natural Dyes)

The bark contains saponins based on randialic acid A & B, mannitol (up to 6%), 10-methyl lixoside (an iridoid) and scopoletin. Stems yield iridoids such as randinoside, galioside, deacetylasperulosidic acid methyl ester, scandoside methyl ester, geniposide and gardenoside. Roots also contain mannitol, scopoletin and a yellow dye. The leaves yield flavonoids, kaempferol, 4-OMe kaempferol and quercetin and phenolic acids *p*-hydroxy benzoic, gentisic, melilotic, vanillic and syringic acids (Thomas, 1989; Hamerski *et al.*, 2003).

The fresh fruits contain 2-3% saponins (10% dry wt.) which mostly are concentrated in pulp. They are dumetorins A, B, C, D, E and F (based on oleanolic acid), randioside A (on oleanolic acid), ursosaponin (on ursolic acid), two triterpenoid glycosides based on oleanolic acid and 10-methyl lixoside. Seeds contain protein (14%) and saponins based on oleanolic acid.

Roots possess marked insecticidal and insect repellant properties. 10% aqueous extract of roots is used against green-scale (*Coccus viridis*) of coffee. Roots are also used as fish poison. The bark is used in diarrhoea, dysentery and is abortifacient, anthelminitic, antipyretic and emetic.

10. *Dioscorea hispida* Dennst. (Dioscoreaceae)

Syn. *D. daemona* Roxb.

D. hispida is a slender biennial with a tuberous root, prickly stem, reticulate veined 3-foliolate leaves (petiole prickly, leaflets petiolulate, broadly cuneate-obovate, villous), fleshy unisexual flowers in dense long cylindric spikes and quadrately oblong capsules containing 6 winged seeds. A native of S. E. Asia.

Tubers contain alkaloids, dioscorine and 4-epidioscorine and saponins like diosgenin.

Tubers are used as fish poisons and exhibit molluscicidal activity. Aqueous extract used as pesticide in Philippines.

11. ***Ficus benghalensis*** **Linn. (Moraceae)**

Nyagrodhah / Vad, Baniyan Tree

Please see Natural dyes for the description and chemistry of this plant

Leaf juice of this tree is used as a vermicide.

12. ***Hedera nepalensis*** **Koch . (Araliaceae)**

Syn. *H. helix* Linn.

H. rhombea S. & Z.

Ivy

A hardy climber, climbing adhesively by the rootlets. Leaves variable, simple linear-lanceolate or cordate-ovate or palmately lobed. Flowers small in panicled umbels. Fruit, a globose black berry. Common from W. Europe to Japan.

The whole plant yields emetine. Leaves contain hederasaponins such as α- and β-hederins and hederasaponins B&C. Sesquiterpenes like β-elemene, elixene and germacrene B as also β-lectins are other components of leaves. Inflorescence contains β-amyrin, β-sitosterol and its glucoside, oleanolic acid and triterpene glycosides, nepalin-1 (hederagenin arabinoside), napalin-2 (hederagenin) and napalin-3

The plant is used as an insecticide. It in various combinations, is used in cancer and indurations of different kinds. Extract is found to be anti-tumour in nature. Leaves are antirheumatic and antineuralgic. Hederin is found to be acting as vasoconstrictor and hypotensive. Napalins 1,2 and 3, at concentrations of 0.5%. 0,25% and 0.125% respectively, immobilize human sperm completely.

13. ***Madhuca indica*** **Gmel. (Sapotaceae)**

Syn. Bassia longifolia Koenig

B. latifolia Roxb.

Mawra

Please see Natural dyes for the description and chemistry of this plant.

Mahuda cake possesses insecticidal and piscicidal properties. It is applied to lawns and golf greens, as the saponin present in it has a specific action against earthworms.

14. ***Mitragyna inermis*** **Korth. (Rubiaceae)**

The bark contains triterpenoid saponins, quinovic acid glycosides (3 in number), two 27-nor-triterpenoid glycosides named inermiside I and II and oxindole alkaloids.

The bark extract is used as an insecticide.

15. ***Nicandra physaloides*** **Gaertn. (Solanaceae)**

Ranpopti

The fresh herb contains 0.65% of a glycosidal bitter principle designated as nicandrin ($C_{27}H_{37}O_7$). The seed yields oils which are useful to produce varnishes.

The plant is reported to have insecticidal, diuretic and antihelmintic properties.

16. ***Sapindus mukorossi*** **Gaertn. (Sapindaceae)**

Aritha, Soap nut

Please see Natural dyes for the description and chemistry of this plant. Fruits are used as fish poison. Powdered seeds are said to possess insecticidal properties.

17. ***Sapindus trifoliatus*** **Linn.**

The saponin present in the fruits, emarginatoside (a hederagenin glycoside), is used in the preparation of certain fungicides and insecticides. Fruit and root bark are used as fish poison.

18. ***Swartzia madagascariensis*** **Desv. (Fabaceae)**

Syn. *Bobgunnia madagascariensis* K. & W.

Snake bean

A native of Africa, this is a small tree with imparipinnate leaves, flowers having single petal and numerous stamens and woody indehiscent, cylindric pods.

Seeds contain diterpenes and triterpenoid saponins.

The seeds are found to exhibit molluscicidal activity and toxicity to mosquito larvae.

19. *Tribulus terrestris* Linn. (Zygophyllaceae)

Goksura/Gokhru, Tribulus

T. terrestris is a prostrate wooly perennial with paripinnate leaves (leaflets ovate/elliptic), yellow flowers, and a pentagonal spiny fruit, A common weed throughout India.

The fruits, which form the drug, contain steroidal saponins, alkaloids and steroids. The saponins, terrestrosins A-E, desgalactotigonin, F-gitonin, desglucolanatigonin and gitonin are based on sapogenins, diosgenin, hecogenin and neotigogenin. β-Sitosterol, stigmasterol, terrestiamide (a cinnamic acid amide) and 7-methyl hydroindanone are the other compounds reported. Leaves contain kaempferol, isorhamnetin, vanillic acid, syringic acid, melilotic acid and *p*-coumaric acid. (Ren *et al.*, 1995; Yan *et al.*, 1996).

The seeds are found to inhibit *Meloidogyne incognita* (Singh *et al.*, 1991). Gokhru is a well known diuretic and antiurolithiatic. It is also nephroprotective and cardiac stimulant.

20. *Trigonella foenum-graecum* Linn. (Fabaceae)

Methika, Fenugreek

Fenugreek is an herbaceous annual with pinnately trifoliolate leaves having long oblanceolate dentate leaflets, white/yellow flowers singly or in pairs from axils and long (up to 15cm) cylindrical pods containing 10-20 yellow grooved trapezoid seeds. A native of Mediterranean grown for seeds and as pot-herb.

Seeds contain protein (26%), fat (6%), carbohydrates (44%), flavonoids such as quercetin and luteolin (free and as glycosides) saponins based on diosgenin, tigogenin and gitogenin, trigonelline and minerals. A galactomannan is the main component of carbohydrates. Leaves yield proteins (41%), carbohydrates (6%) and free amino acids. Also present in leaves are saponins (similar to those of seeds), flavonoids such as kaempferol and 7, 3′-diOMe quercetin (Daniel, 1989a).

Topical applications of extracts of seeds and leaves of this plant produced a high degree of mortality in both stored product pests *Tribolium castaneum* and *Acanthoscelides obtectus* (at 6 and 30 μg/insect). Powdered fenugreek seeds or extracts applied to *Phaseolus vulgaris* beans produced mortality and inhibited opposition and larval penetration by *A. obtectus*. The presence of the plant material decreased *A. obtectus* fecundity and reduced its longevity. Fenugreek seed appeared moderately toxic to young larvae of *T. castaneum* (LD_{25} = 18% in diet), and surviving adults showed progressively decreased fecundity as the dose of seeds, especially ripe seed, was increased. The seed affected the fertility of both sexes (Pemonge *et al.*, 1997)

Its strong smell somewhat discourages livestock from eating it. Farmers keep dried plants with stored grains to protect the latter form insects in the rainy season.

21. *Tripterygium wilfordii* Hook.f. (Celastraceae)

Lei gongteng, Thunder god wine

This is a deciduous climbing wine growing up to 12 m with brown angular downy twigs, light green ovate leaves (hairy underneath), small scented white flowers and 3-winged brownish red fruit. Native of China, Taiwan & Burma.

All parts of the plant contains saponins such as wilforgine, wilfordine, wilfornine, celacinnine, triptofordin, triptofordinine etc, diterpenes such as triptolide, tridiolide, triptonoterpene, triptolidenol etc, sesquiterpene alkaloids wilforine and wilfordsuine, triterpenoid lactones, anthraquinone syringaresinol, dulcitol and tannin. This plant is a well-known insecticide of China.

Roots (aqueous extract) are used against european corn borer.

22. *Viguiera decurrens* Gray(Asteraceae)

A bisdesmoside oleanolic acid saponin, 3-*O*-(methyl-β-D-glucuronopyranosiduronoate)-28-*O*-β-D-glucopyranosyl-oleanolate, along with two monodesmoside oleanolic acid (chikusetsusaponin methyl ester)saponins, two diterpenic acids, one chromene, three triterpenes and one steroidal glycoside were the components of this plant. A mixture of monodesmoside saponins displayed cytotoxic activity against P388 and colon cell lines (ED_{50}=2.3 and 3.6 μg/ml, respectively). The extract containing saponins showed insecticidal activity against the Mexican bean beetle larvae (*Epilachna varivestis*) (Marquina *et al.,* 2001). This plant is a native of South America.

4.2.9. Cardiac glycosides

1. *Calotropis gigantea* Ait.f (Asclepiadaceae)

Arkah, Milkweed

An erect tomentose laticiferous shrub reaching a height up to 2m with large ovate sub sessile leaves, white/purple-tainted flowers in umbellate cymes and a pair of subglobose inflated follicles enclosing broadly ovate comose seeds. This is a very common weed occurring in the dry regions of India.

The root bark yields cardiac glycosides such as gigantin, giganteol, isogiganteol, calotroposides A-G, steroids like α- and ß-amyrins, taraxasterol and its derivatives, ß-sitosterol and a wax. The latex, found in all parts of the plant, contains caoutchouc (rubber), resin, cardiac glycosides gigantin, calotropin, uscharin, calotoxin, calactin and uscharin (all based on the genin calotropagenin), proteases like calotropain, calotropin D_1 and D_2 , calotropain F_1 and F_2 and enzymes like invertase. Flowers yield a resin containing resinols α- and ß-calotropeol, giganteol, calotropin, ß-amyrin, flavonoids and fatty acids. Seeds contain a fatty oil (30%; rich in oleic acid and linoleic acid), stigmasterol, melissyl alcohol and laurane. Leaves contain free sapogenins, ß-amyrin, ß-sitosterol, taraxasterol, ø-taraxasterol, tannins and resin. The flowering tops contain holarrhetine and cyanidin glycoside.

Arkah is used as an insecticide by many a rural folk in Gujarat.

2. *C. procera* Ait.f

Alkarka, Dead Sea Apple

C. procera is similar to C. *gigantea* in overall appearance but differs in having a broad, but short, corona which is equaling or longer than the staminal column and not having the two auricles below the apex of corona. The height of entire staminal column is about 0.5 to 0.8 *i.e.* less than a centimeter and the stigma is almost sunken in the corona lobes. [In C. *gigantea,* the corona is narrower and longer, much curved back to the staminal column towards halfway down, more than 1cm in height, but always lower than the gynostemium (stigma always elevated up) and possesses two obtuse structures (auricles) at the apex of each lobe].

The root bark contains α- and β- amyrins, taraxasterol and its isomers, isovalerates and acetates, giganteol, β-sitosterol and a wax. The latex, present in all the parts, consists of caoutchouc, resin and cardiac glycosides, proteases and steroids. The cardiac glycosides are calotropin, uscharin, calotoxin, calactin, uscharidin (all based on the genin calotropagenin), voruscharin, proceroside and two genins, uzarigenin and syriogenin. The proteases encountered are calotropain F_1, F2, calotropin D_1 and D_2. α-Amyrin, β-amyrin and β-sitosterol are the steroids present. Flowers contain α-, β-amyrins, stigmasterol, β-sitosterol, multiflorenol, cyclosadol, procerasterol and calotropenyl acetate (Khan & Malik, 1989).

The larvae of *Anopheles stephensi* List., *Aedes aegypti* Linn and *Culex fatigans* Wiedm died within three hours of putting in 500 ppm latex water. It is also an effective fish poison. Aqueous extract of leaves inhibited the larval hatching of the root knot nematode *Meloidogyne incognita.* The alkaloidal fraction of leaves serves as an antifeedant to desert-locust (*Schistocerca gregaria*). A decoction of leaves, along with soap, is an effective remedy against aphids and white ants. Leaf powder is used to control the pulse beetle *Callosobruchus chinensis* Linn, a serious pest of stored cowpeas. Flower tops showed antifeedant activity against *Rhyzoptera dominica* Fabr., a pest of stored grains.

3. *Cryptostegia grandiflora R.Br. (Asclepiadaceae)*

This is a lofty laticiferous climber with large glossy elliptic leaves, long large rose-colored showy flowers and 3-winged follicles. A native of Tropical Africa, this plant is extensively cultivated in gardens.

Aerial parts contain proteolytic enzymes (in latex), cardenolides such as oleanodrigenin, rhodexin B, 16-propyonyl- and 16-anhydrogitoxygenin and gitoxygenin.

The leaves exhibit molluscicidal activity..

4. *Kalanchoe daigremontiana × tubiflora* (Crassulaceae)

The leaves of this hybrid plant yield methyl daigremonate, an insecticidal bufadienolide, which was active against the third instar larvae of silkworm (*Bombyx mori*) (Supratman *et al.*, 2001).

4.2.10. Glycolipids

1. *Calonyction muricatum* G. Don. (Convolvulaceae)

Syn. *Ipomoea muricata* Jacq.

Gariya

C. muricatum is a large twiner with muricate stem, broadly ovate (base deeply cordate) long petioled leaves, large rose purple flowers and capsules containing black smooth seeds. Pedicel is usually much thickened in fruits and is edible. A native of S. E. Asia.

Seeds contain glycosides muricatin A & B and an alkaloid ipomine. Muricatin B is the glycoside of 11-hydroxy palmitic acid and muricatin A is the glycoside of 4-hydroxy ethyl stearate.

The whole plant is used as an insecticide.

2. *Ipomoea fistulosa* Mart. (Convolvulaceae)

Syn. *I. carnea* Jacq.

Naffattia

This native of S. America is a suberect shrub having ovate cordate petiolate leaves, large pale rose colored showy flowers and globose capsules. Common weed on edges of water bodies or planted as a fence.

The leaves yield glycolipids such as glycosides of decanoic acids, ergoline alkaloids like agroclavine and α-dihydrolysergol and a saponin, ipamotocin

The leaf extract(200 and 300 ppm) caused larval (60%) and pupal (79%) mortality of filaria vector, *Culex quinquefasciatus*. Chopped shoots are used as soil amendments in effectively reducing the population of phyto-nematodes in tomato and brinjal crops. The leaves mixed with green gram proved a potent oviposition inhibitor and a safe protectant against the pulse beetle (*Callosobruchus chinensis*) infestation. The aqueous extract shows marked parasiticidal activity against mites (*Psoroptic* spp.) and lice (*Linognathus* spp.) attacking buffaloes. Extracts of this plant is found to cause 100% mortality on *Meloidogyne javanica* within 12 h.

4.2.11. Annonaceous acetogenins

1. *Annona reticulata* Linn. (Annonaceae)

Ramphal, West Indian Custard Apple

This is a small deciduous tree with a rough longitudinally fissured bark, oblong-lanceolate leaves having an unpleasant odour, solitary, leaf-opposed greenish white flowers and a yellow/brownish red heart-shaped fruit having pentagonal areoles enclosing many black smooth seeds. A native of Tropical America now naturalized in many parts of India.

The root bark yield alkaloids such as annonaine, liriodendrine, norushinshunine and reticuline. Leaves also contain alkaloids. Shoots yield a blue/black dye. The seeds of *Annona reticulata* contain one γ-lactone acetogenin (*cis/trans* - isomurisolenin), annoreticuin, annoreticuin-9-one, bullatacin, squamosin, *cis/trans* bullatacinone, *cis/trans* murisolinone, squamone, solamin, annomonicin and rolliniastatin.

The seeds, leaves, stem and roots possess insecticidal properties. The alcoholic extract of the aerial parts of the plant exhibited insecticidal activity against *Tribolium castaneum. Annona* extractives are both contact and stomach poisons. Contact action is equivalent to that of rotenone or nicotine. Stomach poisoning is variable and there is little or no ovicidal activity. Most effective applications are against various aphids and human body pests such as lice.

2. Annona squamosa *Linn (Annonaceae)*

Sitaphal, Custard apple

This plant is similar to *A. reticulata*, but differs in having a tubercled fruit. A native of India.

Bark yields acetogenins squamotacin, molvizarin, bullatacin, bullatacinone and squamone - all are cytotoxic (Cortes *et al.*, 1993, Hopp *et al.*, 1996, Li *et al.*, 1990). Other acetogenins are 4-deoxyannoreticuin, *cis*-4 deoxy annoreticuin and squamoxinone. Leaves and stem contain annonaine, roemerine, norcorydine, corydine, norlaucreline, isocorydine, norisocorydine, glaucine and nitroheterocide. Roots contain alkaloids annonaine, michelalbine, exoushinsuine, reticulatine, analobine, diazepine and squamolone. Root bark yields β-sitosterol, camphor, and borneol. Fruits contain polyphenols.

Seeds yield annonin, annonacin, neodesacetyl uvaricin, neo-annonin B, neo-reticulatin A, annotemoyin-1 and annonin-I.

Seeds possess insecticidal and piscicidal properties. The powdered seeds are used to destroy worms. It yields an acrid principle which is a contact poison to the flies, *Musca nebulo, M. domestica* and to the larvae of several beetles and aphids. The seed oil of *A. squamosa* was used in Vietnam for protecting rice against leaf hoppers and plant hoppers. The seed extract was found toxic to fruit flies, azuki bean weevils and diamond black moth (*Plutella xylostella*). Annonin-I was found more toxic than pyrethrum against *Phaedon cochlearial* and *P. xytostella*. Annotemoyin-I is found to be effective in controlling both larvae and adults of *Tribolium castaneum*. The leaf poultice also kills lice and prevents bed bugs and maggots. Roots, leaves, & fruit also are insecticidal, antifeedant & repellant. The whole plant extract was found larvicidal, growth regulator & chemosterilant against *Anopheles stephensi*.

The seeds and leaves of *A. muricata* Linn and *A. montana* Macf. also are said to have insecticidal properties. Seeds of *A. muricata* are found to contain *cis*-annonacin-9-one, *cis*-goniothalamicin, arianacin, javoricin, and annomuricin A & B. From the roots cohibins A and B are reported.

4.2.12. Alkaloids

Nicotine and related nornicotine were used as insecticides since 1600's. But the use is reduced due to high cost of production, disagreeable odour, extreme toxicity to mammals and limited toxicity to insects. Structurally related anabasine is the choice in Russia. Other alkaloids used such are cevadine, veratrine, sabacilline and ryanodine. Physostigmine, another alkaloid from *Physostigma venenosum*, served as a model compound for the development of carbamate insecticides.

1. *Aconitum napellus* Linn. (Ranunculaceae)

Ativish/Bachnag, Aconite

The roots of *A. napellus*, a native of Alps, and other mountainous regions of Europe and Asia, provide the drug Aconite. It is an erect, showy, herbaceous perennial with turnip-like tuberous roots, large lobed leaves, blue-violet flowers in spike-like terminal racemes and follicular fruits.

Total alkaloid yield of the roots varies from 0.3 to 1.5%. Two types of alkaloids, aconitines and atisines have been recognised in the drug. Aconitines are diacylesters of polyhydric amino alcohols and are highly poisonous. Atisines are simple amino alcohols and possess less toxicity. Aconitine-$C_{34}H_{47}O_{11}N$- is the major alkaloid amounting to 30% of the total alkaloid fraction and yield aconine ($C_{25}H_{41}O_9N$), acetic acid and benzoic acid on hydrolysis. Hypaconitine, neopelline and neoline are the other alkaloids present in appreciable quantities. Also present are trace amounts of ephedrine, sparteine, aconitic acid and a large amount of starch.

Aconite is a highly potent and quick acting poison and therefore it is used as an insecticide. It causes drastic cardiac effects in humans. The drug is now used as a local analgesic in liniments. The other species of *Aconitum* used as insecticides and in medicine are:

2. *Aconitum ferox* Wall.

Bacchanag, Indian aconite

This is a herbaceous perennial with tuberous dark brown ovoid/ ellipsoid roots, palmately 5-fid cordate/reniform leaves, pale blue flowers borne on a terminal raceme and oblong follicles enclosing obovoid winged seeds. Common in alpine regions of Kashmir and Nepal.

The alkaloid content of tubers reaches up to 5%. The major alkaloids are pseudaconitine, chasmaconitine, indaconitine, bikhaconitine, veratryl pseudaconitine, diacetyl pseudaconitine, picro-aconine, aconine, veratryl -ψ-aconine and di-Ac-ψ aconitine. The drug is highly toxic. The tubers are used in medicine only after keeping them in milk for 2-3 days to reduce their depressant action on the heart and instead become mild cardiotonics. They are then used as diaphoretic, diuretic, anodyne and antidiabetic.

3. *Aconitum heterophyllum* Wall.

This tall herb, common in alpine or subalpine Himalayas possesses paired white tuberous roots, variable leaves (lower long-petioled, upper amplexicaul), racemes bearing blue flowers and follicles enclosing blackish obpyramidal seeds.

Alkaloid content of roots is upto 0.8%. The principal alkaloids are atisine (50%), heteratisine, heterophyllisine, heterophylline and heterophyllidine. Other alkaloids present are atidine, hetidine, benzyl heteratisine and hetisinone

Atisine exhibits lesser toxicity than aconitine. The whole aqueous extract showed marked hypotensive activity. Roots act as antifertility agents. Tubers are used against dyspepsia, acute inflammation, loss of memory, hysteria and piles. It is also an efficient febrifuge and a tonic to remove debility after malaria. The tubers are sometimes used as a vegetable.

4. *Aconitum chasmantham* Stapf.

Patisa, Indian Napellus

A plant very similar to *A. napellus* but differing in having deeply grooved and wrinkled tubers. The alkaloid content is about 3% and the major components are indaconitine, chasmaconitine, chasmanthinine, chasmanine and homochasmanine. The insecticidal uses of this plant are same as *A. ferox*.

5. *Adhatoda zeylanica* Medic. (Acanthaceae)

Syn. *A. vasica* Nees

Vasaka, Malabar nut

A tall perennial shrub with elliptic – lanceolate hairy leaves, large white flowers in large bracteolate spikes and clavate capsules containing globular seeds. Found naturally or cultivated throughout India.

The major alkaloids of the leaves are quinazoline compounds such as vasicine (peganine, upto 1.1%), vasicinone and vasicinol. More than half of vasicine occurs in the *l*-form and the rest in its racemic form. Vasicinone also occurs in *l*-form but racemises on standing. The other components of leaves are a steroid vasakin, betaine, β-sitosterol and an essential oil (containing limonene). Flowers also are found to contain vasicine, vasicinine, β-sitosterol, β-amyrin, essential oil, luteolin, quercetin and kaempferol. Roots yield vasicine, vasicinol and tritriacontane. Seeds yield upto 25% of a fatty oil rich in oleic and linoleic acids. This oil contains appreciable amounts of behenic, lignoceric and cerotic acids. β-Sitosterol also is obtained from seeds.

Leaves are antifeedant, insecticidal, repellant and effective against rice blast. The leaves possess weedicidal properties also.

6. *Amianthium muscaetoxicum* A. Gray (Melianthiaceae/Liliaceae)

Syn. *Zigadenus* sp.

Fly – poison

This is a toxic perennial herb with a basal bulb, radical long linear leaves, white flowers in scapaceous bracteate raceme and a spherical capsule containing ellipsoid black seeds. A native of USA.

The whole plant, especially bulbs, contains veratrum-like alkaloids, jervine and aminanthine. The bulbs, mixed with honey, are used as a fly poison.

Zigadenus densus Fernald (Crow poison), a related species containing veratrum – like alkaloids, is used similarly.

7. *Anabasis aphylla* Linn. (Chenopodiaceae)

Syn. *A. tartarica* Pallas

This is a sub-shrub with much branched woody grey-white stem, broadly triangular obscure leaves, small yellow/pink flowers in spikes and subglobose fleshy dark red fruits. A native of China and Russia.

The branches yield a piperidine alkaloid anabasine (an isomer of nicotine) and lupinane alkaloids like lupinine, aminolupinine, lupanyl acid, pachikarpine and N-nitrosoaffilinic acid.

Anabasine is a botanical pesticide like nicotine.

8. *Argemone mexicana* Linn. (Papaveraceae)

Swernakshiri/Bharband, Mexican poppy

This herbaceous annual bears sessile, pinnately lobed leaves with toothed margins. The beautiful yellow flowers occur singly on erect peduncles. Capsule is prickly, containing numerous seeds resembling mustard. A native of Mexico and S. America, *Argemone* occurs as a weed everywhere now.

The leaves and stem contain a number of alkaloids like berberine, protopine, retundine, munitagine, argemonine, nor-argemonine, codeine, morphine, toddaline, sanguinarine etc. Most of these alkaloids are present in the seeds also. Apart from alkaloids, seeds yield 15-30% fixed oil.

The seed extracts have been reported to show antinematodal activity against *Meloidogyne incognita* and insecticidal properties.

9. *Aristolochia bracteolata* Lam. (Aristolochiaceae)

Syn. *A. bracteata* Retz.

Kidamari, Dhumapatra, Bracteated Birthwort

This is a glabrous slender twining herb with striate stems, variable (reniform/broadly ovate) long-petioled leaves having a cordate base and crenulated margins, solitary flowers with sub-globose base, trumpet-shaped mouth and purple hairy lip, and oblong-ellipsoid 12-ribbed capsules containing deltoid compressed seeds which are black on one side and white on the other side. A native of Central Asia.

Roots yield an alkaloid, aristolochic acid. Seeds also contain aristolochic acid besides a fatty oil.

The whole plant is traditionally used to destroy insect pests. Aristolochic acid also is insecticidal and insect repellent. The plant is used for its purgative and anthelmintic properties.

10. *Bocconia cordata* Linn. (Papaveraceae)

Celandine Tree

A native of Tropical America, this plant is a annual/perennial herb with apetalous flowers and arillate seeds. The roots of this plant are found to contain isoquinoline alkaloids sanguinarine, chelerythrine and bacconine. These alkaloids are found to exert nematicidal activities against *Rhabditis* sp. and *Panagrolaimus* spp.

11. *Cassytha filiformis* Linn. (Lauraceae)

C. filiformis is a parasitic leafless thin twiner with yellowish green stem, small white flowers on lateral spikes and globose black drupes enclosed by fleshy perianth lobes. It parasitizes a number of plants and is seen throughout India.

The plant contains alkaloids cassyfiline, cassythine, cassythidine, nantenine, neolitsine, dicentrine, launobine, nuciferine etc.

The plant is said to possess piscicidal and insecticidal properties and yields a brown dye.

12. *Catharanthus roseus* G. Don (Apocynaceae)

Syn. *Vinca rosea* Linn.

Vinca pusilla Hook.f.

Lochnera rosea Reichb.

Baramasi, Vinca/Periwinkle

This plant is an erect much branched perennial with glossy oblong-elliptic leaves, fragrant white to pinkish purple flowers (blossoming in all seasons) in terminal or axillary cymose clusters and a pair of long slender follicles containing many oblong small seeds. A native of Malagasy, found wild or as cultivated throughout.

More than 110 indole alkaloids have been reported from the plant. But the more important are the 20 indole-indoline dimeric alkaloids such as vinblastine (vincaleukoblastine, VLB), vincristine (leurocristine, vincaleurocristine, VCR) are mainly present in aerial parts. Other dimeric alkaloids present are vincarodine, vincoline, leurocolombine, vinamidine, vincathicine, vincubine and isositsirikine. Roots are the main source of alkaloids where the alkaloids amounts to 0.13%. Ajmalicine (raubasine), serpentine and reserpine (the principal alkaloids of *Rauwolfia*) are the main alkaloids here wherein their concentration exceeds that of *Rauwolfia*. Other important alkaloids are vincolidine, lochrovicine, catharanthine, vindoline, leurosine, lochnerine, tetrahydroalstonine, vindoline and vindolinine. Iridoids such as loganin, deoxyloganin, dehydrologanin, sweroside, secologanoside and roseoside form another group of compounds present in *C. roseus*. Also present are oleanolic acid and α-amyrin acetate. Flowers contain the alkaloids, coronaridine 11-methoxy tabersonine, tetrahydroalstonine, ajmalicine, vindorosidine, vincristine etc. along with α-amyrin, cycloartenol, lupeol, campesterol, β-sitosterol, petunidin, malvidin, hirsutidin, kaempferol and quercetin. Leaves contain flavonols such as quercetin and kaempferol and phenolic acids, vanillic, syringic and *p*-coumaric acids. Leaves yield an oleoresin also containing lochnerol and lochnerallol, sterols and ursolic acid and small amount of a volatile oil consisting of citronellyl acetate, cadinene and 2-heptanol. Seeds yield 31% of a fatty oil and loganic acid.

The alkaloids showed sterilant activity on *Dysdercus cingulatus* Fabr., (the red cotton bug) and are considered useful for sterilising male population. Alcoholic extract of the shoot showed larvicidal activity against the mosquito *Aedes aegypti* Linn. at 1000 ppm after 7 days. Different bioactivities such as antifungal and molluscicidal have also been exhibited by the extracts of this plant.

13. *Cimicifuga foetida* Linn. (Ranunculaceae)

Bugbane

A robust tall perennial found in temperate Himalayas having long pinnately compound leaves (leaflets ovate, toothed, terminal leaflets 3-lobed) white/yellow flowers in racemes and 6-8-seeded follicles. The flowers and fruits possess a foetid smell.

The whole plant or roots are insect-repellent, insecticidal and used against bugs and fleas.

14. *Cocculus hirsutus* Diels. (Menispermaceae)

Syn. *C. villosus* DC.

Vasen-vel

A scandent, straggling densely villous herb with small alternate deltoid / hastate densely pubescent leaves, small greenish male flowers in axillary cymose panicles, small greenish female flowers in groups of 2-3 and a pea-sized keeled purplish black drupe containing horse-shoe shaped seeds. Common on the fences.

Roots and stems contain alkaloids coclaurine, magnoflorine, trilobine and cocculolidine.

Cocculolidine and related alkaloids are insecticidal in nature.

15. *Conium maculatum* Linn. (Apiaceae)

Poison Hemlock

Poison hemlock is a much branched herb with purplish red speckled stem, pinnately decompound leaves, small white flowers in large, many-rayed umbels and laterally compressed 5 ridged cremocarp devoid of oil tubes. Hemlock is native of Europe, occurring as a weed in America and some parts of Asia.

All parts contain liquid alkaloids, highest in fruit. The dry unripe fruits, which form the drug, contain 1-2.5% of the liquid alkaloid coniine, along with other alkaloids like N-methyl coniine, conhydrine, pseudoconhydrine and γ-coniceine. The alkaloid content decreases with the age of the fruit. Seeds contain coumarins such as bergaptene and xanthotoxin also

The unripe dry fruits - the conium of commerce - are used as antispasmodic, sedative, anodyne, in epilepsy, acute mania, and whooping cough and for treating obstruction of lymphatic system. In homeopathy it is used to prevent cataract.

16. *Consolida ambigua* Ball & Heywood (Ranunculaceae)

Syn. *Delphinium ambiguum* Linn.

D. pauciflorum D. Don

D. ajacis Linn.

Common Larkspur

A beautiful annual herb of 1m with much branched leaves (segments linear, mucronate), spurred blue/rose/white flowers born on raceme-like terminal cymes and beaked follicles containing black seeds. A native of Europe, now cultivated in gardens for the beautiful flowers.

The plant contains diterpenoid alkaloids delcosine and delsoline. Seeds yield an oil rich in oleic (60%), galoleic (18%) and linoleic (15%) acids.

Seeds are used as insecticides. The alkaloids are allelopathic to nearby plants and inhibit them.

17. *Coptis japonica* Makino (Ranunculaceae)

Berberine, palmatine and coptisine, the three isoquinoline alkaloids isolated from the roots of this Japanese plant are found to exhibit antifeeding activity against 4th – instar larvae of *Hyphantria cunea* and *Agelasia coerulea*. The total alkaloids show a more pronounced effect than individual alkaloids administered indicating a synergistic effect (Park *et al.*, 2000).

18. *Crotalaria retusa* Linn. (Fabaceae)

Dingala

C. retusa is a robust undershrub of 1m, with simple oblanceolate oblong silky pubescent (beneath) leaves (stipules subulate), numerous large and showy yellow (with a purple tinge) flowers borne in erect terminal racemes and linear oblong pods containing 15-20 seeds. Common in Central India.

Seeds contain 1% pyrrolizidine alkaloids such as monocrotaline, retusine and spectabiline.

The seeds and leaves are used as pesticides.

19. *Cuscuta reflexa* Roxb. (Convolvulaceae)

Amarvel

This is a slender leafless yellow twining parasite found on a number of host plants. The plant produces small white flowers in clusters of 2-4 and small globose circumscissile capsules containing 2-4 large black seeds. A common plant throughout India.

The plant contains scoparone, melanetin, quercetin, hyperoside and phenolic acids along with cuscutalin and cuscutin.

The aqueous extract of the plant exhibits nematicidal activity against second stage larvae of *Toxocara canis* and *Meloidogyne incognita*. Incorporation of chopped stem in soil reduced the population of root-knot nematodes in the soil and root-knot (gall) formation in tomato.

20. *Cynanchum arnottianum* Wt. (Asclepiadaceae)

This native of E. Asia is a perennial climber.

The whole plant, especially leaves, are insecticidal, useful to kill maggots in the wounds of animals.

21. *Datura stramonium* Linn. (Solanaceae)

Dhatura, Stramonium/ thorn apple

This Asian plant is an annual, 1-2 m in height. The stem shows dichasial branching with alternate slightly pinnatifid leaves. The large white flowers are solitary in the axils of leaves. The quadrilocular erect capsule is very prickly and contains numerous brown reniform seeds. Stramonium is often cultivated in U.S. and Europe, but occurs as a weed all over the world.

The plant is considered as one of the most poisonous ones. The drug stramonium constitutes the dried leaves and flowering tops. The total alkaloid content of the drug is between 0.25-0.5% of which the major alkaloids hyoscyamine and hyoscine occur in the ratio 2:1. But it is found that in young shoots hyoscine is in higher concentration. The seeds yield about 0.2% alkaloids and 15-30% fixed oil. Stramonium root contains ditigloyl esters of 3, 6-dihydroxy tropane and 3, 6, 7-trihydroxy-tropanes and some alkamines along with hyoscyamine & hyoscine. Leaves contain flavonoids such as chrysin, liquiritigenin, naringenin, kaempferol, quercetin and a withanolide- withastramonolide (Glotter, 1989; Lakshmi and Krishnamoorthy, 1991).

Aqueous extracts of seeds and leaves are found to kill root-knot nematodes like *Meloidogyne incognita* and *M. javanica*. Extracts of *Eclipta alba, Shorea robusta* and *Datura metel* are found to reduce the galls caused by *M. javanica,* in tomato roots (Goswami and Vijayalakshmi, 1986). It is also observed that the extracts of various species of *Amaranthus, Calotropis procera, Datura metel* and *Eclipta alba* caused an increase in overall performance of tomato infected with *Meloidogyne incognita.* Water and methanolic extracts are found to kill 2nd stage larvae of *Tylenchulus semipenetrans* and *Anguina tritici.*

22. *Delphinium brunonianum* Royle (Ranunculaceae)

Musk Larkspur

A musk-scented annual of 50cm with glandular pubescent stem, palmately 5-partite leaves (lobes cuneate-ovate, petioles very long), large pale blue spurred flowers in corymbs and viscidly pubescent follicles (5-6) containing many wrinkled seeds. Seen in Alpine Himalayas.

All the *Delphiniums* are found to contain benzyl isoquinoline alkaloids. The leaves and seeds of this plant and of *D. caeruleum* are irritants. Due to the poisonous nature they are used as insecticides

23. *Duboisia myoporoides* R. Br. & *D. leichhardtii* F. Muell.

Corkwood tree

D. myoporoides is a small tree with ovate-oblong leaves, white/ pale lilac flowers in pyramidal/corymbose panicles and globose berries. This and *D. leichhardtii,* both endemic to Australia, are some of the chief sources of atropine and hyoscine. Apart from these two compounds these plants possess other alkaloids like nor-hyoscyamine, tigloidine, valeroidine, poroidine, isoporoidine, valtropine, tigloyltropeine, lutropine and apohyoscine. Some other species of *Duboisia, i.e., D. hopwoodi* F. Muell. produce nicotine and nor-nicotine in the leaves and tropane alkaloids in the roots. The total alkaloid content and the proportion of individual alkaloids vary with chemotypes and ecological factors. The maximum content of alkaloids reported is 5.0%. All these alkaloids are insecticidal in nature.

24. *Equisetum arvense* Linn. (Equisetaceae)

Horsetail

A bushy rhizomatous perennial with hollow jointed grooved stems bearing up to 20 whorls of slender branches (terminating in long naked points), scale like deciduous leaves in whorls at the nodes and yellow fertile shoots terminating in blunt-tipped oblong cones. Cosmopolitan in distribution.

Contains silica (highest amount), alkaloids (nicotine, palustrine and palustrinine), flavonoids, saponins (equisetonin) tannins and sulphur.

This plant is a general purpose insecticide, due to the silica particles and nicotine, against plant insect pests and diseases.

25. *Euonymus verrucosides* Siebold (Celastraceae)

The root barks of *E. verrucosides, E. fortunei* and *E. phellomana* yielded three sesquiterpene pyridine alkaloids with a β-dihydroagarofuran sesquiterpene skeleton such as euoverrine, euophelline and euojaponine C. All these alkaloids are found insecticidal (Jinbo *et al.,* 2002).

26. *Gloriosa superba* Linn. (Liliaceae)

Bachnag, Gloriosa

G .superba is a climbing herb with a root stock of arched solid fleshy cylindric bifurcately branched tubers pointed at both ends, annual stems, arising from the angles of tubers, sessile ovate-lanceolate leaves having a tendril-like spiral tip, large axillary, solitary, showy flowers (petals linear-lanceolate with crispy wavy margins, color yellow passing through orange and scarlet to crimson) on a pedicel reflexed at the tip and a large linear-oblong coriaceous septicidal capsule containing many subglobose winged seeds. A native of tropical Africa, now common in India.

Roots contain up to 0.37% colchicine and seven colchicine derivatives, sitosterol and β-& γ-lumicolchicines. Seeds and flowers also contain colchicine and its derivatives. Flowers, in addition, contain luteolin, N-formyl-de-Ac-colchicine, β-& γ-lumicolchicines, 3-demethyl colchicine and 2-methyl colchicine.

Methanolic extract of the defatted seeds showed in vitro nematicidal activity against root knot nematode *Meloidogyne incognita* due to colchicines.

27. *Haplophyton cimicidum* A. DC. (Apocynaceae)

Arizona cockroach plant

A small shrub of 1m high with slender stems, small linear ovate pubescent leaves, bright yellow flowers and a pair of follicles containing black seeds. A native of Arizona.

The plant contains dimeric indole alkaloids such as haplophytine and cimicidophytine and aspidospermine alkaloids haplocine and haplocidine.

The latex of the plant mixed with molasses is used as a cockroach poison. As a lotion or infusion, latex is used to repel mosquitoes or fleas. The alkaloids are found to be the active insecticidal compounds.

28. *Lagarostrobos colensoi* Quinn.(Podocarpaceae);

Silver pine

This tree yields a group of homoerythrina alkaloids with dyshomoerythrine as the major component. These alkaloids are insecticidal (Bloor *et al.*, 1996).

29. *Lycopersicon esculentum* Mill. (Solanaceae)

Rakthamachi, Tomato

This is a pubescent spreading annual with pubescent unevenly pinnate curled leaves, yellow flowers in extra axillary cymes, large orange/red globose berries containing many flat kidney shaped seeds on a fleshy placentum. It is a native of Peru-Ecuador, now widely cultivated throughout the world.

The fruits contain protein (2%), carbohydrates (4.5%) minerals, pectin, citric, oxalic and malic acids, carotenoids and glucoalkaloids tomatine and solanine (in traces). The carotenoids present include lycopene, β-carotene, lutein and related compounds.

Leaves, branches and stems are insecticidal and repellant. Used for aphids, corn earworm, diamond black moth. Tomato is a carminative, blood purifier and general tonic.

30. ***Maytenus*** **spp. (Celastraceae)**

(Syn. *Gymnosporia* spp.)

This genus of laticiferous spiny shrubs / trees is found to contain sesquiterpene nicotinoyl alkaloids of reputed insecticidal activity. Putterine A & B and mayteine are some of the important alkaloids isolated from *Maytenus buchanani* and *M. putterlickoides* (Schaneberg *et al.*, 2001.

31. ***Mucuna pruriens*** **DC. (Fabaceae)**

Syn. *M. prurita* Hook.

Atmagupta/Kavach, Mucuna

A twinning annual with pubescent branches, pinnately trifoliate leaves with ovate leaflets and large purple flowers on drooping racemes. The fruit is an S-shaped legume densely clothed with brown, stinging hairs. It is a native of India and seen in many parts of S.E.Asia.

Seeds and roots contain a variety of alkaloids like physostigmine, mucunine, mucunadine and nicotine along with 3-4% of a red viscous oil, bufotenine and DOPA (Dihydroxy phenylalanine). The hairs on the fruit contain 5-hydroxy tryptamine (serotonin), which is said to be responsible for the irritation caused by them.

Bufotenine, DOPA, mucunine and serotonin are pesticidal in nature (Morris, 1999). Roots are used against cholera, as a diuretic and for dropsy. Seeds are powerful aphrodisiacs and are known to promote sexual vigour and semen. Seeds are also used in cholera, delirium, impotence, leucorrhoea, and spermatorrhoea and as a diuretic. It is also found to exert an antidepressant effect in cases of depressive neurosis.

32. ***Murraya paniculata*** **(L.) Jack (Rutaceae)**

All extracts, especially of petroleum ether of leaves, are found to cause mortality of *Callosobruchus maculatus* (Mollah and Islam, 2005).

33. ***Nicotiana tabaccum*** **Linn. (Solanaceae)**

Tobacco

Tobacco consists of the dried and cured leaves of an annual, indigenous to tropical America. It is a viscid pubescent herb with a woody base and large oblong lanceolate sessile leaves (lower leaves are decurrent and stem clasping), large rose-colored flowers borne on spreading panicles and ovoid capsules containing numerous seeds. Normally in the field the terminal bud is removed after the plant attains a particular height, so that the leaves become strong and healthy.

The leaves contain up to 16 pyridine alkaloids, nicotine being the principal one (1-10%). Other alkaloids are nornicotine, nicotelline, nicotyrine etc. Other compounds present are flavonoids like rutin, quercitrin, isoquercitrin, kaempferol, flavones, coumarins such as scopoletin, its glucoside scopolin, aesculetin, its glucoside cichoriin and simple phenols like chlorogenic acid, caffeic acid, quinic acid, phenol, guaiol, eugenol etc. Also present are organic acids (20%) mainly malic, citric and oxalic acids and smaller amounts of fumaric, malonic, acetic acids etc. The roots yield anabasine (an isomer of nicotine), anatabine and related compounds. The characteristic smell of tobacco is due to nicotianin (tobacco camphor) which is a mixture of terpenes.

Nicotine is best extracted by steam distillation. It is also extracted with water or hydrocarbon solvents. Free nicotine is very unstable, but its salts like sulphate, benzoate, oxalate, tannate or bentonite are more stable.

Nicotine is an acetylcholine mimic that binds to postsynaptic receptors. It is very toxic to humans also. It is used as a fumigant or a contact spray to destroy soft-bodied, sucking and rasping pests like aphids, thrips and mites

Nicotine is absent in seeds. Nicotine is generally manufactured from tobacco waste and is used for combating lice, gad flies and ticks. Nicotine is a contact insecticide but act principally as a fumigant and a stomach poison for vegetable and fruit pests which are soft-bodied and minute, such as aphids, white flies, leaf hoppers, thrips, red spiders, snails, slugs and cabbage butterfly larvae. Nicotine is used in the form of sprays (0.6-1.0%) and dusts (4%). Nicotine sulphate also is an effective insecticide.

Seeds are found to inhibit the nematode, *Meloidogyne incognita.*

34. ***Pachygone ovata*** **Miers (Menispermaceae)**

Syn. *Cissampelos ovata* Poir

A climbing dioecious shrub with ovate-oblong or sub-trapezoid leaves (petiole with a basal ring of hairs), male flowers minute in axils of subulate bracts in racemes exceeding the leaves, small male and female flowers and reniform drupes containing a horse-shoe shaped seed. A native of S.E. Asia.

This plant is said to have insecticidal properties.

35. ***Peganum harmala*** **Linn. (Zygophyllaceae)**

Harmal, Peganum

The seeds of *P. harmala,* a native of W. Asia and India, have been known for the psychotropic effect they produce. This plant is a dichotomously branched bush of 1 m in height with a dense foliage consisting of leaves having pinnately cut liner narrow acute spreading lobes, small solitary axillary white flowers and globose capsule enclosing numerous angular seeds.

The seeds contain β-carbolines like harmine (the main constituent), harmalol and harman.

The plant is used as an insecticide. It is also a hallucinogen where the active principle is harmine, which exhibits much more activity than mescaline. The seeds find use as a spice and a valuable aphrodisiac.

36. ***Piper nigrum*** **Linn. (Piperaceae)**

Kalimirich, Pepper

Indigenous to India and Indo-Malaya region but cultivated widely in the Eastern tropics, this perennial woody climber, the dried unripe fruits of which are known as black pepper, possesses simple, ovate, coriaceous and evergreen leaves. The small flowers are borne in catkins, and the fruits are small, red, one-seeded drupes, about 50 per catkin. Dried, full-grown but unripe fruits constitute the black pepper of commerce. White pepper is the ripened fruit from which the outer portion of the pericarp has been removed.

Chavicine, pipercide and piperolein A, (acid amides), free piperine ($C_{17}H_{19}O_3N$), piperettine ($C_{19}H_{21}O_3N$) and piperidine are the alkaloids contained in the fruits. The pungency of pepper is due chiefly to chavicine (less than 1%) and a lesser extent to piperine (present in amounts 10 times more than those of chavicine) and a resin. A volatile oil (1-2.5%) occurs in the fruit, which comprises of α-, and β-pinenes, α-phellandrene, citral, limonene, piperonal etc. The volatile oil is restricted largely to the pericarp and the pungency to the seed. A fixed oil (<1%) and starch (50%) are also present in the fruits.

The seeds along with groundnut oil have been reported to have action against tapeworm and other worms. Piperine acts as a synergist in some of the commercial insecticides. Pipercide and piperolein A are found insecticidal (Strunz and Finlay, 1994).

37. *Punica granatum* Linn. (Punicaceae)

Anar, Pomegranate

For a description of this plant please refer the section of Natural Dyes

The yield of volatile liquid alkaloids from the root and stem barks is about 0.5-1%.The principal alkaloid is pelletierine which is accompanied by isopelletierine, methyl isopelletierine and a solid alkaloid, pseudopelletierine. All these alkaloids occur in the form of tannates. The tannin content of the root bark amounts to 22%. Also present in bark are mannitol, steroids like friedlin, betulic acid and ursolic acid.

The bark is used as an insecticide in Japan.

38. *Ryania speciosa* Vahl (Flacourtiaceae)

This is a South American pubescent shrub with simple oblong acuminate leaves, white flowers and a globose capsule.

Stem and roots contain a pyrrole alkaloid-ryanodine, the major compound, and four minor ryanoids like 8, 9-didehydro ryanodine, 10-O-acetyl ryanodol, 3-O-benzl ryanodol and a deoxyryanoid. Other six compounds with insecticidal activity also are reported. But the major and insecticidally active compounds are ryanodine and 9, 11 - dehydroryanodine.

Ryanodine affects the Calcium 2+ ion release channel in muscle resulting in a sustained contraction of skeletal muscle and eventual paralysis. Ryanodine is mostly used for control of caterpillar pests of fruits and forage and persists for more time (3-5 days) in the field.

Ryania is used as a crude dust or as alkaloids extracted with water, acetone, ether or $CHCl_3$. It is a stomach poison used against codling moths, corn ear worm, corn borer and citrus thrips. But it is found toxic to man, fish and birds and very toxic to dogs.

39. *Sarcostemma acidum* Voigt. (Asclepiadaceae)

Somlatha

This is a leafless East Indian succulent wine, (leaves reduced to minute scales), flowers in umbellate cymes and thick follicles. It is also used as an insecticide.

40. *Schoenocaulon officinale* A. Gray (Liliaceae)

Cevadilla, Sabadilla, Veratrum

Seeds of this native of Peru contain up to 0.8% alkaloids like cevadine, veratrine (both most active), sabadinine, sabadiline and sabadine.

To get active extract, the seeds are aged, heated or alkali – treated to form or activate the insecticidal alkaloids. Alkaloids are extracted in kerosene, or the powdered seeds are used as such. The extracts are light sensitive and therefore stored in the dark.

Sabadilla action is similar to that of pyrethrum, by affecting the voltage dependent sodium channels of nerve axons. This nerve - poisoning results in instant death or several days of paralysis before death. Sabadilla or alkaloids are toxic to humans on ingestion. They are used for control of insects in crops, animals or even humans. They are particularly active against "true bugs" (Hemiptera). They are also highly toxic to honey bees.

41. *Solanum americanum* Mill.(Solanaceae)

Syn. *Solanum nigrum Linn.*

Kakamachi

This plant is described in Natural Dyes

The ethanolic leaf extract of this plant is found larvicidal to *Anopheles culicifacies, Culex quinquefasciatus, Aedes aegypti, Aedes caspius* and *Culex pipiens.*

42. *Stemona collinsae* Craib (Stemonaceae)

This plant contains alkaloids, isostenine, neotuberostemonine, bisdehydrone, 16,17-didehydro-16(*E*)-stemofoline and its isomer at C-4, 16,17-didehydro-4(*E*)-16(*E*)-stemofoline and stemofoline of which the latter three were found to show insecticidal, and antifeedant activities against the diamond black moth larvae (Jiwajinda *et al.*, 2001). This plant is a rich source of stilbenoids including 11 phenylbenzofurans and the stemofurans which are antifungal.

43. *Stemona kerrii* Craib.(Stemonaceae)

Stemona kerrii, S. curtisii; S. cochinchinensis and another unidentified species of *Stemona,* which are found in China, are found to contain pyrido- and pyrroloazepine alkaloids such as the pyrido [1,2-*a*] azepines stemokerrin, methoxystemokerrin-*N*-oxide, oxystemokerrin, oxystemokerrin-*N*-oxide, and pyridostemin, along with the pyrrolo[1,2-*a*]azepines dehydroprotostemonine, oxyprotostemonine, stemocochinin , protostemonine, stemofoline, 2'-hydroxystemofoline, and parvistemonine. The crude extracts were found to have strong insecticidal activity against *Spodoptera littoralis.* The activity was more in the first two species (Kaltenegger *et al.*, 2003).

44. *Stemona tuberosa* Lour. (Liliaceae)

Syn. *S. sessilifolia* Miq.

S. japonica Miq.

Roxburghia gloriosoides Roxb.

Wild asparagus

This is a herbaceous wine found in Central China and India. It bears tufts of yellow hollow cylindrical tubers opposite long-petiolate ovate-lanceolate leaves, axillary inflorescences, the peduncle borne on base of petiole, yellow green flowers and ovoid capsules containing oblong ribbed bearded seeds.

The tubers contain tropolone alkaloids (Stemona type, 14 in no.) such as stemonine, tuberostemonine, isotuberostemonine, stemonidine and sinostemonine, glycosides, proteins (9%), organic acids like malic, formic, citric and succinic acids and bibenzyls like stemonine and stemonal.

The aqueous and 70% alcoholic extract are found lethal to *Pediculus capitis, P. vestimenti, P. pubis* and lice ova. It is used for killing insects, worms, bedbugs and louses. Tubers (aqueous extract) are used successfully against caterpillars, crickets, flies, screw worms and weevil.

45. *Tinospora cordifolia* Miers (Menispermaceae)

Amruta/Guduchi, Tinospora

T. cordifolia is a woody twiner often producing filiform green aerial roots. Stem has a warty surface (due to lenticels) and leaves are thick, heart shaped and acuminate. Flowers are small unisexual borne on racemes in the axils of leaves. Fruit is a globose drupe. This plant is a native of India, common in tropical and subtropical parts.

Stem and aerial roots contain clerodane furanoditerpenes *viz.* columbin and tinosporaside; a lignan 3,4-*bis* (4-hydroxy-3-methoxy benzyl) tetrahydrofuran; alkaloids such as jatrorrhizine, palmatine, berberine and timbeterine; a sesquiterpene glycoside tinocordifolioside; phenylpropene disaccharides cordifolioside A & B; choline, tinosporic acid and tinosporon (Pachey and Schneidir, 1981; Wazir *et al.*, 1995; Maurya *et al.*, 1996, 1997).

Roots and stems of this vine also are used as insecticides.

46. *Tinospora rumphii* Merr.

T. rumphii contains bitter principles columbine and picroretine and alkaloids tinosporine, tinosporidine and berberine. These vines are used for making pesticides in Philippines. Roots and stems are used against diamond black moth, rice black bug, rice green leaf hopper, and rice stem borer.

47. *Veratrum viride* Ait. (Liliaceae)

American Hellebore

This is a perennial herb with an erect fleshy tunicated rhizome, dimorphic leaves, the lower being large and oval and upper being small and narrow, greenish yellow flowers borne in panicles and the capsular fruit containing numerous seeds. A native of America.

Rhizome contains two categories of alkaloids (1) jerveratrum alkaloids and (2) ceveratrum alkaloids. Jerveratrum alkaloids contain

1-3 oxygen atoms, occurring in plant both as free alkamines and as glycosides and are similar to saponins in possessing haemolytic properties. E.g. jervine, veratramine (free alkamines), pseudojervine (jervine and glucose) and veratrosine (veratramine+glucose). Ceveratrum alkaloids are highly hydroxylated containing 6-9 oxygen atoms and occur as free and esters with two or more acids e.g. germine and protoverine (free alkamines), germinitrine and protoveratrine A.

The rhizomes have been used for insecticidal purposes to a limited extent in America. They are used both in the form of sprays and in dusts which are lethal to the larvae of the houseflies breeding on manure. The insecticidal preparations are light sensitive, deteriorate rapidly on exposure of light and therefore recommended for the control of chewing insects on ripening fruits.

48. *Zanthoxylum hamiltonianum* Wall. (Rutaceae)

A large scandent shrub armed with short stout crooked prickles, pinnate leaves (petiole very prickly, leaflets 2-3 pairs, suddenly narrowed into a broad notched apex), small white flowers in axillary tomentose cymes and a fruit of 1-5 globose 1-seeded carpels. A native of S.E. Asia.

The plant yields a benzophenanthridine alkaloid, nitidine and 8-hydroxy flavonols and lignans.

The alkaloids and lignans are insecticidal in nature.

4.2.13 Alkamides

1. *Acmella oppositifolia* Lam. (Asteraceae)

Spot flower

A. oppositifolia is found to contain olefinic alkamides with affinin as the main lipidic component and *N*-2-methylbutyldeca-2*E*,6*Z*,8*E*-trienamide 2 and *N*-isobutyl-dodeca-2*E*,4*E*,8*Z*,10*E*-tetraenamide. These components can be found in roots and aerial parts also. All these compounds are insecticidal (Molinatorres *et al.*, 1996).

2. *Anacyclus pyrethrum* DC. (Asteraceae)

Syn. *A. officinarum* Hayne

Agragrahi, Spanish Pellitory

A procumbent perennial with numerous stems and pubescent branches, radical pinnatisect (lobes linear subulate) leaves and heads containing white ray flowers (purple beneath) and yellow disc florets. A native of North Africa, now introduced to Europe and Asia.

The roots contain various isobutylamides of which pellitorine and anacyclin are the principal ones, tyramine amides, polyacetylenes, sesamin and inulin.

Pellitorine is toxic (neuro-muscular toxicity) to molluscs, crustacea and fish. Anacyclin exhibits insecticidal activity against grain insect, *Tenebrio molitor*.

3. *Artemisia dracunculus* Linn. (Asteraceae)

Tarragon

The aerial parts of this plant contain alkamides, pellitorine, neopellitorine A and neopellitorine B, and one coumarin herniarine. All these compounds showed insecticidal activity against *Sitophilus oryzae* and *Rhyzopertha dominica* at 200 μg/ml concentrations (Saadali *et al.*, 2001).

4. *Heliopsis longipes* S. M. Blake (Asteraceae)

Scabrin

This species contains, in roots, olefinic alkamides with affinin 1 as the main lipidic component and *N*-2-methylbutyldeca-2*E*, 6Z,8E-trienamide 2, in lower quantities. All these compounds are insecticidal (Molinatorres *et al.*, 1996).

5. *Spilanthes acmella* Linn.(Asteraceae)

Akalkaaro

An annual erect hairy herb with opposite, toothed ovate pubescent leaves, homogamous long-peduncled solitary yellowish red conical heads in which the disc flowers are arranged in an elongated receptacle (pappus absent) and oblong much compressed achenes. The flower heads when chewed cause a hot burning taste and profuse salivation. A native of Tropical America, common as a weed in India.

Flower heads contain isobutylamides such as spilanthol (N-isobutyl deca-2,6,8-trienoamide), undeca-2*E*,7*Z*,9*E*-trienoic acid isobutylamide and undeca-2*E*-en-8,10- diynoic acid, fatty acid esters of α- & β-amyrins, β-sitosterol and sitosterol glucoside. The plant is employed as a fish-poison. Ether extract of the fresh flower tops of the plant is effective against anopheline mosquito larvae even in great dilutions with water. Spilanthol is a powerful insecticide. It is highly toxic to adult house-flies and anophiline and culicine larvae.

4.2.14. *Non-Protein amino acids*

When incorporated into artificial diets, powdered seeds of several Central American legume species containing albizzine; canavanine were found to be toxic or repellent to larvae of the southern armyworm (*Prodenia eridania*)(Rehr *et al.*, 1973).

1. *Calliandra* sp. (Mimosaceae)

Five non-protein amino acids such as *S*-(ß-carboxyethyl)-cysteine present in the young leaves and stems of *Calliandra* spp. were found insecticidal against the aphid, *Aphis fabae* (Simmonds *et al.*, 1988).

4.2.15. *Phenolics*

a. simple phenols

1. *Anacardium occidentale* Linn. (Anacardiaceae)

Kaju, Cashew nut

A. occidentale is a small spreading evergreen tree with obovate to ovate oblong rough glabrous leaves, small yellow flowers (with pink stripes) borne in long terminal panicles (flowers both male and bisexual) and a kidney shaped cashew apple. The plant is a native of America, now cultivated in Kerala, Goa etc.

Cashew nut is known for the kernels and the cashew apple, the former being a nutritious delicacy, and the latter edible and is a source of beverage. The shell of the nut yields up to 30%, a vesicant juice known as cashew nut shell liquid (CNSL) consisting of phenols such as anacardol (72%), cardol(18%), cardanol, and phenol. CNSL is an efficient mosquito larvicide. It enhances the antimosquito activity of kerosene or high speed diesel oil. A number of new pesticides are prepared from the phenolic components of CNSL.

2. *Aglaia basiphylla.* A. Gray and *A. gracilis* A. C. Smith (Meliaceae)

A. basiphylla and *A. gracilis*, two species native to Fiji were found toxic to *Spodoptera littoralis*. The insect toxicity of the former plant was caused by the well known benzofuran flavaglines rocaglamide, desmethylrocaglamide and aglafoline. In contrast, *A. gracilis* contained four related pyrimidinone derivatives in the root and stem bark, including two new congeners named marikarin and 3'-hydroxymarikarin as well as new putrescine bisamides, secoodorine and secopiriferine, a benzopyran flavagline, desacetyl aglain A, and a norsesquiterpene (Greger *et al* ., 2001).

3. *Aglaia duperreana* Pierre (Meliaceae)

Thirteen insecticidal cyclopentatetrahydrobenzofuran derivatives of the rocaglamide type were isolated from flowers of this plant, a native of Vietnam. All these compounds exhibited strong to moderate insecticidal activity against larvae of *Spodoptera littoralis.* The most active compounds were similar with regard to their bioactivity to azadirachtin (Chaidir *et al.*, 1999).

4. *Aglaia elliptica* Blume (Meliaceae)

Fruits of this plant contained seven insecticidal rocaglamide derivatives which exhibited strong insecticidal activity towards neonate larvae of *Spodoptera littoralis.* Didesmethylrocaglamide was the most active compound comparable to that of azadirachtin. (Nugroho *et al.*, 1997).

5. *Aglaia harmsiana* Perkins (Meliaceae)

The leaves of this tree also yield the seven insecticidal rocaglamide derivatives as that of *Aglaia elliptica* (Meliaceae) including didesmethylrocaglamide and are similar in properties (Nugroho *et al.*, 1997).

6. *Aglaia odorata* Lour. (Meliaceae)

The twigs and leaves of *Aglaia odorata* yielded eight insecticidal cyclopentatetrahydrobenzofuran rocaglamide derivatives, four cyclopentatetrahydrobenzopyran aglain derivatives, aminopyrrolidines odorine and odorinol, syringaresinol and flavonoid derivatives. The rocaglamide derivatives exhibited strong insecticidal activity towards neonate larvae of *Spodoptera littoralis.* The most active compounds were comparable to azadirachtin, which was used as a positive control (Nugroho *et al.*,1999).

7. *Aglaia spectabilis* Roxb. (Meliaceae)

Bark of *Aglaia spectabilis* in Vietnam yielded cyclopentatetrahdro-benzofurans of the rocaglamide type. All these compounds are found to be strongly insecticidal against larvae of *Spodoptera littoralis*. Methylrocaglate and C-3'hydroxy methyl rocaglate were the strongly active compounds, the activity being equal to that of Azadirachtin. (Schneider *et al.*, 2000).

8. *Capsicum annum* Linn. (Solanaceae)

Syn. *C. frutescens* C.B.Clarke,
C. purpureum Roxb.
C. minimum Roxb.

Lalmirich, Red Chilli

For a description of this plant please see section on Natural Dyes.

Chillies are found to contain an oleoresin (16%), ascorbic acid, protein (16% dry wt.), fat (6%, mainly in seeds), carbohydrates (32%), carotenoids, a volatile oil and vit. E. The oleoresin, yield varies from 15-18%, extracted by solvents, consists of capsainoids and some carotenoids. The most important component of capsainoids is a phenylamide – capsaicin- that is generally responsible for the pungency of chilly. Other capsainoids are dehydrocapsaicin, homocapsaicin II, bishomo-and trishomocapsaicin, homodihydrocapsaicin I and II, nordihydrocapsaicin, vanillyl amides of caprylic, nonylic and decylic acids and capsiamide. Capsaicin is maximum before the ripening stage of the fruit. Capsanthin and capsorubin are the carotenoids responsible for the red color, are absent in green chilies, which contain lutein, neoxanthin, violaxanthin and ß-carotene. Red pepper contains capsanthine-5, 6-epoxide, capsochrome and capsanthin 3,6-epoxide also along with ß-carotene, violaxanthin, cryptoxanthin and cryptocapsin. (Lutein is absent in red peppers.) Vitamin C is more in large chillies. The volatile oil, which varies from 0.1 to 2.6%, consists of 2-methoxy-3-isobutylpyrazine, *trans*-ß-ocimene, limonene, methyl salicylate, linalool etc. Fruit pericarp yielded glycosides of luteolin, quercetin, and apigenin and *p*-coumaric and ferulic acids (Materska *et al.*, 2003). Seeds yield a fatty oil rich in linoleic acid and many steroids such as 4α-methyl-5α-cholest-8(14)-en-3ß-ol, fucosterol, isofucosterol, citrostadienol and derivatives of lanosterol. Seeds contain free sugars and capsicoside.

The green fruits contain many polyphenols (which are found to be absent in mature ones). They are petunidin, luteolin and quercetin glycosides, esters of *p*-coumaric and quinic acids, acyclic diterpene glycosides such as capsiansides I to V and A-E. All parts except flowers and stem contain solanine and solanidine.

Chillies are used as stored grain protectants. The pungency of the oleoresin and capsaicin, due to which a burning sensation is produced, acts on the insects especially the soft-bodied ones. Chilli is a very popular spice and in small doses acts as a powerful stimulant and carminative. It is a well-known remedy in atomic and flatulent dyspepsia and dipsomania. A regular use is beneficial in varicose veins, anorexia and liver congestion.

9. *Curcuma longa* Linn. (Zingiberaceae)

Haridra/Haldi, Turmeric

The rhizome contains curcuminoids, volatile oil, sterols, sugars, starch and other polysaccharides. Curcuminoids are the principal coloring materials (upto 6%) of which curcumin amounts to 60%, with desmethoxy curcumin, *bis*desmethoxy curcumin and dihydrocurcumin forming the rest. Volatile oil contains high amounts of bisabolene derivatives along with borneol, camphene, linalool and α-phellandrene. The polysaccharides reported are ukonans A-D.

Turmeric has now been found to be a natural antibiotic, inhibiting the growth of bacteria, fungi and parasites. Both *C. longa* and *C. domestica* are repellants and insecticidal.

Aqueous extract is used for aphids, caterpillars, red spider, mites and powdery mildew. Turmeric rhizomes contain a nematicidal principle which is found effective against the root-knot nematode larvae. Turmeric exhibits insecticidal properties also and is found effective against mosquitoes.

10. *Curcuma raktakanda* Mangaly & Sabu (Zingiberaceae)

Tuber and leaves of this plant common in Kerala, are found larvicidal to *Culex quinquefasciatus, C. sitiens, Aedes aegypti* and *Anopheles stephensi* (Latha and Ammini, 2000).

11. Curcuma *zedoaria* Rose (Zingiberaceae)

Kachura, Zedoary

This is tuberous erect annual (of 0.5m) with large ovoid root stock, many sessile yellow tubers, large oblong leaves, yellow flowers overtopped by red bracts in spikes and ovoid trigonous spikes containing arillate seeds. A native of Eastern Himalayas.

Rhizomes are rich in starch (72%) and protein (15%), sesquiterpenoids like β-curcumene, *ar*-curcumene, curcumenone, curcumanolide A & B, furanogermenone, germacrone, zederone etc and also a volatile oil (1-1.5%) containing sesquiterpene alcohols, cineole and camphor.

β-Curcumene, *ar*-curcumene, germacrone, dehydrocurdione, furanogermenone and zederone are found to be highly toxic against *Spodoptera littoralis* (Pandji *et al.*, 1993).

12. *Eclipta alba* Hassk. (Asteraceae)

Bhringaraja/Bhangra

Please see the section on Natural Dyes for a full description and chemistry of this plant.

The extracts of roots, shoot and the whole plant showed nematicidal activity against nematode species like *Meloidogyne graminicola, M. incognita* and *Rolylenchulus* sp. due to the presence of thiophenes and thienyls. Extracts of *Eclipta alba, Shorea robusta and Datura metel* are found to reduce the galls caused by *M. javanica,* in tomato roots (Goswami and Vijayalakshmi, 1986). It is also observed that the extracts of various species of *Amaranthus, Calotropis procera, Datura metel* and *Eclipta alba* caused an increase in overall performance of tomato infected with *Meloidogyne incognita.*

E. alba is also hepatoprotective (antihepatotoxic) due to wedelolactone and demethylwedelolactone. It is also antimyotoxic and antihaemorrhagic in nature.

13. *Gaultheria procumbens* Linn. (Ericaceae)

Wintergreen

The plant is an evergreen shrub, with a creeping stem and erect leafy branches. The oval leaves are 3-6cm long, apiculate, glabrous and shining. These are the sources of wintergreen. The white flowers are solitary and the fruits, scarlet in colour. This is a native of North America.

The leaves contain gaultherin, a phenolic glycoside that, in the presence of water, is acted upon by an enzyme gaultherase to produce a volatile oil consisting of methyl salicylate (98%), and a sugar, primeverose. The other constituent of the oil is an ester of enanthic alcohol.

Oil of wintergreen (methyl salicylate) repels moths and beetles-used for ornamentals, both indoors and outdoors (may be toxic to humans in larger quantities). Wintergreen oil is used in medicine and flavoring. However, the chief source of the oil of wintergreen today, is the sweet birch- *Betula lenta* L. that contains gaultherin in the bark.

14. *Kaempferia rotunda* Linn.(Zingiberaceae)

For a description of this plant please see section on Natural Dyes.

Tubers containing crotepoxide and other cyclohexane diepoxides exhibits considerable insecticidal activity (Nugroho *et al.*, 1996).

15. *Melicope subunifoliolata* T. G. Hartley (Rutaceae)

This plant yields meliternatin (3, 5-dimethoxy-3',4',6,7-bismethylenedioxyflavone) and six other minor polyoxygenated flavones. The first compound is found to exhibit strong feeding deterrent activity against *Sitophilus zeamais* and very good larvicidal activity against *Aedes aegypti* (Ho *et al.*, 2003).

16. *Securidaca longepedunculata* Fers. (Polygalaceae)

Violet tree

This is a small tree of 6m with oblong leaves crowded at tips of branches, pink-purple flowers in clusters and a round winged nut. A native of tropical America, the roots and tree bark are used against coughs, rheumatism and constipation and as a contraceptive.

Roots contain a volatile oil rich in methyl salicylate. Roots and volatile oil exhibit repellant and toxic properties to storage pests like *Sitophilus zeamais, Rhizopertha dominica, Prostephanus truncates* and *Callosobruchus maculatus*. It is toxic to *Culex* larvae also.

17. *Zingiber cassumunar* Roxb. (Zingiberaceae)

The rhizomes contain two phenylbutanoids which are insecticidal (Nugroho *et al.*, 1996).

18. *Zingiber officinale* Rosc. (Zingiberaceae)

Adrak, Ginger

This erect perennial, a native of Southeastern Asia, has a much branched horizontal rhizome, oblong lanceolate leaves having a midrib and stem-clasping sheaths and yellow flowers having purple lips and greenish yellow bracts borne on a spike. Cultivated for the pungent rhizomes.

The rhizomes yield a pungent oleoresin containing phenolic derivatives such as gingerol, shagaol, zingerone etc. Also present in the rhizome are starch and a volatile oil containing camphene, β-phellandrene and sesquiterpenes like bisabolene, zingiberene and zingiberol.

Ginger powder and aqueous extract are used against aphids, thrips and plant hoppers, powdery mildew, root rot and fungal leaf blight. They are also found inhibiting the activity of the root-node nematode *Meloidogyne javanica.*

b. Lignans, Chromenes, Coumarins and Stilbenes

1. *Ageratum conyzoides* Linn. (Asteraceae)

Aerial parts of *Ageratum conyzoides* contained 11 chromenes, nine methoxyflavones (like 5,6,8,3',4',5'-hexamethoxyflavone and 8-hydroxy-5,6,7,3',4',5'-hexamethoxyflavone), sitosterol, (+)-sesamin and caryophyllene oxide (González *et al.,* 1991).Other components of this plant are alkaloids, flavonoids, benzofurans and terpenoids. The plant shows insecticidal properties (Okunade, 2002).

2. *Ageratum houstonianum* Garden(Asteraceae)

This plant contains chromenes precocene I and II, benzofurans, polysubstituted flavones and four pyrrolizidine alkaloids. The chromenes are found insecticidal to *Melanoplus sanguinipes* (Proksch *et al.,* 1987).

3. *Encelia farinosa* A. Gray (Asteraceae)

Brittle bush

This plant, the stem resin of which is used as an incense in Spain, yielded four analogous chromenes and benzofurans. The chromene encecalin has both antifeedant and insecticidal properties towards larvae of the noctuid pests *Peridroma saucia* and *Plusia gamma.* In addition to reducing larval growth and decreasing survivorship of neonate larvae fed artificial diets containing encecalin, this chromene was directly insecticidal to neonate larvae after 4 hr using a residue contact bioassay. Topical application of encecalin to third instar larvae of *Peridroma* led to a reduction in dietary utilization, indicative of a metabolic cost of handling the compound. The benzofurans euparin and methyleuparin, all common constituents of a number of *Encelia* spp., were only weakly active in all bioassays against the insects (Isman and Proksch, 1985).

4. *Eucalyptus ficifolia* F. Muell (Myrtaceae)

Red-flowering gum

This native plant of Australia is found to yield ficifolidione in the hexane extract of the aerial parts. Ficifolidione, is (1'*R*,2'*R*,4*R*)-4-isobutyl- 6,6,8,8-tetramethyl-2,3,4,8-tetrahydrochromene-5,7-dione-2-spiro-2'-6,6-dimethyl-bicyclo[3,1,1]heptane and is a moderately insecticidal compound (Khambay *et al.*, 2003).

5. *Kunzea ericoides* Burgan (Myrtaceae)

K. ericoides also is found to yield ficifolidione, in the hexane extract of the aerial parts which is a moderately insecticidal compound (Khambay *et al.*, 2003).

6. *Lyonia ovalifolia* Drude (Ericaceae)

Syn. *Pieris ovalifolia* D. Don

This is a deciduous pubescent shrub up to 2 m in height, with ovate lanceolate acuminate leaves having wavy margins, white drooping flowers in axillary racemes (lower leaves in axils of leaf like bracts) and globose capsules containing spindle shaped seeds. Native of China.

Lyonia-xyloside is a lignan glucoside obtained from this plant.

Leaves are insecticidal.

7. *Magnolia virginiana* Linn.(Magnoliaceae)

The leaves of *Magnolia virginiana* yielded neolignans, 4,4'-diallyl-2,3'-dihydroxybiphenyl ether, 3,5'-diallyl-2'-hydroxy-4-methoxy-biphenyl and 5,5'-diallyl-2,2'-dihydroxybiphenyl (magnolol). All these compounds and their methoxy analogues were very toxic to brine shrimp and mosquito larvae and showed strong anti-fungal and anti-bacterial activities (Nitao *et al.*, 1991).

8. *Mammea americana* Linn. (Clusiaceae)

Mammee apple

A tall tree indigenous to Surinam, with leathery oblong leaves crowded at the tip of branches, white fragrant flowers on older parts of branches and large (up to 3 kg) fruits containing edible orange pulp enclosing 4 brown ovoid seeds.

The fruit pulp and seeds contain isoprenoid coumarins such as mammein, mammeisin and related 4-alkyl coumarins. Also present are xanthones and phloroglucinol derivatives.

Seed powder or aqueous extract of seeds, fruit rind or gutta are found effective against aphids, cucumber beetle, cabbage worm, melon worm, mites and rice weevil.

9. ***Mammea longifolia*** **Planch.**

Nagkesar

This species yields a 4-propyl coumarin, surangin B-another insecticide.

10. ***Neorautanenia mitis*** **Verdc. (Fabaceae)**

Tubers of *Neorautanenia mitis* yielded the coumarin, pachyrrhizine, the isoflavonoids neotenone and neorautanone, and the pterocarpans, neoduline, nepseudin and 4-methoxyneoduline as the larvicidal and mosquitocidal constituents against the malaria- and filariasis-transmitting mosquitoes, *Anopheles gambiae* and *Culex quinquefaciatus*, respectively. These tubers are also used to kill bilharzias-carrying snails (Joseph *et al.*, 2004).

11. ***Yucca periculosa*** **Baker (Agavaceae)**

The bark of this plant yielded 4,4'-dihydroxstilbene, resveratrol and 3,3',5,5'-tetrahydroxy-4-methoxystilbene which exhibited growth regulatory activity against the Fall Army worm (*Spodoptera frugiperda*). 3,3',5,5'-Tetrahydroxy-4-methoxystilbene was the most active compound with significant effects at 3 μg/g in diets. It is inferred that these compounds could be involved in interference of sclerotization and moulting and appear to have selective effects on the pre-emergence metabolism of the insect. The results were fully comparable to known natural insect growth inhibitors such as gedunin and *Cedrela* extracts (Torres *et al.*, 2003).

c. Rotenoids and Flavonoids

Rotenoids are *cis*-fused tetrahydrochromeno [3, 4-b]- chromene nucleus. Rotenoids are known cancer chemopreventive compounds. Rotenone is a respiratory enzyme inhibitor. Rotenones are found in genera *Derris, Lonchocarpus, Tephrosia* and *Amorpha*. They were also known fish poisons. All are lipids soluble in hexane, $CHCl_3$ or petroleum ether. They may make up to 20% of dry weight of roots.

Extracted in acetone, ether or other organic solvents, rotenone disrupts energy metabolism in cell mitochondria either by inhibiting the electron transport systems or by uncoupling the transport system from ATP production. Rotenone is toxic to fish and is moderately toxic to mammals. Insects poisoned by rotenone cease feeding and lose locomotor function leading to paralysis and death within a few hours to a few days. Rotenone persists on the surface for 3-5 days. Rotenone is effective against a large variety of insects like aphids, saw flies, maggots, bagworms, codling moths, Japanese beetles, leaf hoppers, Mexican been beetles, cabbage worms, thrips, stinkbugs, flea beetles, vegetable weevils, mites, spiders and fish.

One of the oldest known biopesticides, rotenoids have been in use for centuries (at least since 1649) in South America to paralyze fish. But the use as crop insecticides was since 1848. Rotenoids are least toxic to man, but exert some toxicity to rats, dogs and pigs. They break down in soil and water within three days.

Rotenoids are reported from roots of *Clitoria fairchildiana* also.

1. *Buddleja lindleyana* Fortune (Buddlejaceae)

Butterfly bush

This is a perennial shrub of 1.5 to 2.5m in height, with ovate long leaves and deep violet flowers (hanging like grapes) in axillary racemes. A native of America.

The plant is found to contain diadzein, vanillic acid and glycosides of β-sitosterol, stigmasterol, spinasterol and betulinic acid.

Water extract is found useful against black bean aphids, caterpillars in cabbage and plant hoppers.

2. *Dalbergia paniculata* Roxb. (Fabaceae)

Rotenone is present in *Dalbergia paniculata* also. Therefore the pods are insecticidal and used as a fish poison.

3. *Dalbergia sissoo* Roxb.

A tall tree of 20 m, with pubescent young branches, alternate imparipinnate leaves (leaflets 3-5, sub orbicular, terminal leaflet largest) pale yellow sessile flowers in axillary panicles and strap-shaped legumes containing 1-4 reniform seeds. A native of India.

Leaves contain an isoflavone sissorin. Flowers yield tectorigenin. Stem bark contain dalbergenone, dalbergin and dalbergichromene. Heartwood also yields the same compounds along with biochanin A and stilbene.

Essential oil is reported larvicidal and repellent to *Culex quinquefasciatus* and *Anopheles stephensi.*

4. *Derris indica* Bennet (Fabaceae)

Syn. *Pongamia pinnata* Vent

P. glabra Pierre

Karanj, Pongam

Karanj is a tall tree with soft bark, alternate imparipinnate compound leaves (leaflets usually 5, elliptic, acuminate), pinkish white flowers in axillary racemes and a woody compressed indehiscent pod narrowed at the base and a short decurved mucro at apex enclosing a single reniform seed. A common tree in western India growing wild or cultivated.

Stem bark contains chromenoflavone, pongachromene, glabrin, glabra-I, glabra-II, karanjin, pogapin, kanugin and demethoxy kanugin. The wood also contains same compounds. Leaves yield a furanoflavone, 3'- OMe pongapin, karanjin, kanjone and other known flavonoids. Flowers contain another furanoflavone, pongaglabol, aurantiamide-OAc and ß-sitosterol. The fruits yield furanoflavonoids, pongapinnol A-D and a coumestan, pongacoumestan. The seeds yield a fixed oil rich in oleic (70%) and linoleic acids and flavonoids such as dimethyl chromenoflavanone, isolonchocarpin, demethoxy kanugin, pongol, a ß-diketone karanjin, pongapin, lanceolatin B etc. (Yadav *et al.*, 2004)

All parts of this plant exhibit insecticidal properties. Roots are piscicidal, insecticidal, antifeedant and repellant. Leaf extract is used for leaf eating caterpillars, aphid, armyworm, plant hoppers, citrus leaf miner, corn stem borer, green leaf hopper, potato tuber moth, rice caseworm and white fly.

5. *Derris elliptica* Benth. (Fabaceae)

A native of S. E. Asia, *D. elliptica* is a large handsome twiner with densely pubescent branches, long-petioled pinnate leaves having 9-13 subcoriaceous oblong-ovate leaflets (silky below), bright red flowers (petals silky outside) on lax racemes and 1-3 seeded thin flat pods winged on the upper suture only.

Roots contain rotenone, elliptone, deguelin, malacol, d-toxicarol, tephrosin, dehydroguelin and dehydrotoxicarol.

Root extracts are used against Adzuki bean beetle, army worms, American bollworm, fruit flies, Melon aphid and *Pyricularia oryzae*. A crude extract of root, leaves and branches was found to be lethal to mosquito larvae under laboratory conditions. No toxicity to human beings has been reported. The extracts also showed toxicity to brown plant hopper and root knot nematodes at a dose of 10% w/v.

6. *Derris scandens* Benth.

Five isoflavone glycosides, named derriscandenosides A-E (1-5), were isolated from the stems. Roots contain 3 aryl coumarins, osagin, scandenin and lonchocarpic acid, and isoflavones, warangalone, 8-γ-γ-dimethylallylwighteone, 3′-γ-γ-dimethylallylwighteone, and nallanin and chandalone. Warangalone, 8-γ-γ-dimethylallylwighteone, 3′-γ-γ-dimethlallylwighteone and nallanin are found selective and potent inhibitor of rat liver cyclic AMP-dependent protein kinase. Used as insecticide.

7. *D. philippensis* Lour.

Roots of this plant also contain the same components of *D. elliptica* and are used thus.

8. *Derris trifoliata* Lour.

Stem contained 6 α–12- hydroxy elliptone, deguelin and α-toxicarol.

9. *Enterolobium saman* Prain (Mimosaceae)

Syn. *Samanea saman* Jacq.

The aqueous extract exhibited inhibitory activity against *Drechslera oryzae*.

10. *Gliricidia sepium* Walp. (Fabaceae)

This native of Central America is a small tree with pinnate leaves (leaflets ovate) and reddish flowers at the end of branches without leaves.

Leaves contain tannins (up to 40%), isoflavonoids such as aformosin, formonetin, glyricidin A & B and medicarpin (a pterocarpan). Heartwood contains robinetin, sepiol, glyricidin, gliricidol and isomucronulatol.

The plant is insecticidal, repellant and rodenticidal and used against army worm, cabbage looper, ease worm, tobacco budworm, termites, aphids, etc. The leaf mulch effectively controls weeds and the leaf extract was found lethal to the nematode *Radopholus similis*.

11. *Haematoxylum campechianum* Linn. (Caesalpiniaceae)

Logwood

This is a small tree with a crooked trunk, spiny branches, pinnate leaves having 3-4 pairs of sessile ovate emarginate leaflets, small yellow fragrant flowers in spikate clusters from the leaf axils. A native of Central America.

Wood contains haematoxylin, haematoxylol A, tetra-D-methyl haematoxylol B, myricetin, quercetin, gallic acid etc.

The wood extract is used as an insecticide.

12. *Indigofera tinctoria* Linn.

The various plant parts were found to contain six rotenoids deguelin, dehydrodeguelin, rotenol, rotenone, tephrosin and sumatrol. The extracts were found effective against larvae of *Anopheles stephensi* (Kamal and Mangla, 1993).

13. Lamiaceae members

Five species of Lamiaceae showing insecticidal effects on *Acanthoscelides obtectus* (Say) (Coleoptera: Bruchidae) was analysed to determine if polyphenols were involved in the toxic effect. Chromatographic analyses indicated that the polyphenols were mainly cinnamic acids and flavones, the most abundant being rosmarinic acid, luteolin-7-glucoside and derivatives. A bioassay to assess the effect on the behaviour and survival of beetles was conducted with these main compounds compared to other polyphenolic acids and flavonoids (flavones, flavanones, flavonols and flavanols), aglycones and glycosides. The polyphenols tested were toxic to the beetles in different degrees. Caffeic and ferulic acids, vanillin and luteolin-7-glucoside induced a knockdown effect on the first day and gallic acid on the second day. Quercetin significantly decreased natural mobility from the first day, naringin, syringaldehyde, vanillic acid and gallic acid after the 4th day. On the 8th day all compounds caused significant mortality. Rosmarinic acid and luteolin-7-glucoside were the most active compounds. (Regnault-Roger *et al.*, 2004).

14. *Lonchocarpus* spp. (Fabaceae)

The roots of various species of *Lonchocarpus* such as *L. urucu* K. & S. (Brazil), *L. utilis* K. & S. (Peru) and *L. nicou* DC. (Guiana) and *L. fluvialis* F. & P. are the sources of rotenone. These plants are bush-like in the beginning, but later grow to vines and climb tall trees. The roots contain up to 12% rotenone occurring as solids.

Rotenone is unstable in light and heat and the toxicity is lost within 2-3 days in summer. But the plant powder is active for more time. Rotenones are stomach and contact poisons.

Stem bark of *Lonchocarpus fluvialis* F. & P. yields rotenoids sumatrol and villosinol, a pterocarpan and one chalcone.

15. *Milletia* spp. (Fabaceae)

A genus of trees or shrubs with imparipinnate leaves (leaflets 11-15, oblong/ovate, stipellate), flowers in axillary and terminal panicles and linear/oblong pods containing 1-5 seeds. A native of Tropics.

The root of *Milletia extensa* is reported to possess insect-repellent properties. It is used as fish poison and applied to sores of cattle to kill vermin. Extracts and aqueous suspension of finely ground seeds of *Milletia extensa* are effective when used in sprays against a variety of insects, like houseflies, bean aphids, pentatomids, leaf beetles and cabbageworms.

The rotenoids deguelin and tephrosin isolated from the seeds of *Milletia dura* were active against larvae (larvicidal) of *Aedes aegypti* but related milletone and milletosin were inactive (Yenesew *et al.*, 2003).

16. *Millettia racemosa* Benth. (Fabaceae)

The stems (without bark) of this tree contain isoflavans- 3*R* (+)-millinol, 3*R* (+)-millinol-B and 3*R* (+)-cyclomillinol. All these compounds showed promising insecticidal activity (Kumar *et al.*, 1989).

17. *Pachyrrhizus erosus* Urban (Fabaceae)

Yam bean

This is a vigorous spreading prostrate vine with an edible fleshy turnip like tap root, trifoliolate (leaflets broad) leaves, blue/purple flowers in axillary racemes and a flat legume containing many seeds. A native of S. America.

Seeds are found to contain five rotenones (dolineone, pachyrrhizone, 12-hdroxy rotenone, 12-hydroxy dolineone and 12-hydroxy pachyrrhizone), a furocoumarin, pachyrrhizin (furocoumarin with 3-aryl constituent) and erosnin (a furocoumestan).

Both pachyrrhizin and erosnin are insecticidal as also rotenones. Seeds also contain 2 amorphous saponins, achsaponin A and B. On hydrolysis these substances give, pachysapogenin B and A. Both saponins are poisonous to fishes.

The powdered seeds are useful as insecticide and fish poison in tropical countries. They act as contact and stomach-poisons and are reported to be highly effective against many insects such as silkworm larvae, aphids, cabbageworms, melon worms, beetles, leaf hoppers, cattle lice etc. A process has been patented which employs ground seeds in various formulations with pyrethrins, for use in insecticidal sprays and dusts. It is advantageous to use defatted seeds. The leaves, stem, roots and ripe pods also show insecticidal activity and leaves are toxic to mosquito larvae.

18. *Sophora griffithii* Ali (Fabaceae)

Powdered seed mixed with oil is applied to the hair to kill lice.

19. *Sophora mollis* Grah. (Fabaceae)

A silky hairy shrub having imparipinnate leaves (leaflets 21-25, emarginated, densely silky) appearing after flowering along with fruits, small yellow flowers in short crowded racemes and glabrous 4-winged pods containing 5-6 seeds. Native of Himalayas.

The plant is insecticidal.

20. *Sophora secundiflora* DC.

Seeds posses intoxicant and insecticidal properties and in large quantities they cause nausea, convulsions and finally death by asphyxia. Leaves also possess insecticidal properties.

21. *Tephrosia candida* DC.

White Tephrosia

The roots contain rotenone, tephrosin and three spirorotenoids such as tephrospirolactone and tephrospiroketone I & II. (amorphispiranone was the only spirorotenoid described from *Amorpha fruticosa* earlier) (Andrei *et al.*, 2002).

The plant serves as a cover crop, pesticide, piscicide, ascaricide and antifeedant.

22. ***Tephrosia major*** **Micheli**

The roots and aerial parts yield two rotenoids such as sumatrol and α-toxicarol, flavonoids like glabranin, tephrowastin, obovatin, tephrobbotin and quercetin, 2′,6′-dihydroxy-3′-prenyl-4′-methoxy-β-hydroxy chalcone, β-sitosterol, stigmasterol and lupeol.

23. ***Tephrosia purpurea*** **Pers. (Fabaceae)**

Sarpunka

A much branched herbaceous perennial with imparipinnate leaves having 11-12 oblanceolate mucronate leaflets (stipules linear subulate) clothed with fine appressed silky hairs beneath, purple flowers in leaf opposed lax racemes and linear (slightly curved) flat mucronate pods containing 5-6 flattened seeds. A common weed throughout.

The roots, stem and leaves contained tephrosin, deguelin and quercetin. Roots contain rotenone and isotephrosin.

The plant is antifeedant, pesticide, piscicide and ascaricide.

24. ***Tephrosia vogelli*** **Hook.f.**

Fish poison bean

This is a shrub of 3 m in height with woody stem & slender branches covered with hairs, imparipinnate leaves, white/purplish flowers and rectangled seeds. Common in Africa.

The plant is found to contain rotenoids deguelin, α-toxicarol and tephrosin. Tephrosin is maximum in flowers.

Tephrosin is insecticidal. Leaf extracts esp. of hexane extract was found to be highly effective against stored legume seed pests like *Acanthoscelides obtectus, Callosobruchus maculatus* and *C. chinensis* (Koona, 2005). In Zambia the most widely used pesticidal natural products were *Swartzia madagascariensis* Desv., *Tephrosia vogelli* Hook. f. and *Euphorbia tirucalli,* wood ash and cow dung.

d. Quinones

1. ***Aloe barbadensis*** **Mill. (Liliaceae)**

Syn. *A. vera* Linn

Kumari, Indian Aloe

Description is given in Natural Dyes.

The leaves contain mucilage, anthraquinones and chromones. Mucilage on hydrolysis yields glucose, galactose, mannose and galacturonic acid. The quinones present are barbaloin and isobarbaloin (C-glycosides) and their aglycone, aloe-emodin. Aloesone (7-hydroxy chromone) and its C-glycoside-aloesin are the chromones located. Also present are glycoproteins, alocutin A and B

Along with *Vitex,* aloe is used for armyworms, hairy leaf caterpillar, rice leaf roller, rice stem borer, semi-looper, and bacterial and fungal diseases. Aloe extract mixed with castor oil cake is used as a moth attractant. Aloe is said to be effective against white ants and soil nematodes and therefore is used as an intercrop.

2. *Arnebia nobilis* DC. (Boraginaceae)

Ratanjoth

A native of Afghanistan, this plant is a hairy herb with woody deep-red roots, lanceolate leaves covered by white trichomes, pink/ white flowers in terminal spikes and tuberculate nutlets.

The roots and root stocks are purple brown (up to 10cm long and 6cm in dia.) covered with several layers of thin scaly bark of same color. The color is due to naphthoquinones (seven in No.) such as arnebin-1 (alkannin-ß, ß-di methyl acrylate), arnebin-2 (ß, ß-dimethyl acrylic ester of hydroxy alkannin), arnebin-3 (alkannin acetate), arnebin-4 (arkannin), arnebin-5, arnebin-6 and arnebin-7. Also present are waxes and sitosterol.

It is used as a bioactive protectant to stored grains and a dye. Arnebins are anticancer in nature against walker carcino-sarcoma-256 in rats and each of the arnebins inhibited tumour cells in vitro. These naphthoquinones are found antimicrobial against gram positive bacteria, and *Candida albicans,* arnebin-2 being the most potent. Ratanjot is used as a medicine for enlarging liver and against worms.

3. *Calceolaria andina* Linn. (Scrophulariaceae)

The leaves and stem of this plant yielded two naphthoquinones, 2-(1,1-dimethylprop-2-enyl)-3-hydroxy 1,4-naphthoquinone and 2-acetoxy-3-(1,1-dimethyl-prop-2-enyl),1,4-naphthoquinone which are found to be active against sucking pests of cereals, cotton, vegetables and ornamentals and also against fruit destroying mites. These compounds are found to be weak against beneficial sp. and show very low toxicity to mammals (Khambay *et al.,* 1999).

4. *Cassia alata* Linn. (Caesalpiniaceae)

Dadmurdan, Candle bush

C. alata is a large handsome shrub reaching a height of 2-3 m, with thick branches, long paripinnate leaves having persistent stipules, 8-12 pairs of oblong leaflets, bright yellow flowers in dense paniculate racemes and long winged pods containing 50 or more triangular flattened seeds. A native of Africa and Asia.

Roots contain anthraquinones and steroids. Leaves yield cassiaxanthone, aloe-emodin, chrysophanol, isochrysophanol, rhein, physcion, kaempferol and β-sitosterol.

The leaves are insecticidal and piscicidal. In Africa the plant is used as an ant-repellent.

5. *Cassia occidentalis* Linn. (Caesalpiniaceae)

Kasamarda, Coffee senna

The description and chemistry is given elsewhere (See Natural Dyes).

All parts of this plant have insecticidal properties.

6. *Cassia siamea* Linn.

Kasid, Siamese Senna

A medium sized evergreen tree with paripinnate leaves of 6-14 pairs of elliptic oblong leaflets, bright yellow flowers in axillary / terminal panicles and slightly curved long fattened pods containing many brown seeds. A native of S.E. Asia.

The heartwood is found to contain cassiamine, chrysophanol and stilbene. Leaves yield anthraquinones like sennosides, cassiamine, physcion, chrysophanol and tannins. Bark also contains tannins & anthraquinones of wood.

C. siamea is found to provide disease resistance to sandal wood growing on them against spike disease. It is also used as a nurse crop to *Sweitenia mahogani* to reduce or control the attack of borers. It is also found to reduce incidence of *Lantana.* Mulching of leaves prevents termites in the soil.

7. *Cassia sophera* Linn. (Caesalpiniaceae)

Janglitakla

C. sophera is a perennial shrub of 3m with paripinnate leaves (leaflets 6-10 pairs, lanceolate), yellow flowers in axillary few-flowered corymbose racemes and slightly recurved pods containing 30-40 brown flattened seeds. A native of Africa, occurring as a weed in India.

Flowers are found to contain anthraquinones and flavonol glycosides. Leaves also are found to have flavonols and sennosides. Root bark also is rich in anthraquinones.

Leaves of this plant was found to be very effective against storage pests like *Callosobruchus maculates* and *Rhyzopertha dominica,* and they reduced the F1 emergence of *Sitophilus zeamais* (Belmain *et al.,* 2001).

8. *Chamaecrista nigricans* Greene (Caesalpiniaceae)

Syn. *Cassia nigricans* Vahl

This is an unbranched annual herb of 1m with pinnate leaves (leaflets 7-15 pairs) and cream yellow flowers. This plant also is attributed with insecticidal properties.

9. *Embelia ribes* Burm.f (Myrsinaceae)

Vidanga

The description, chemistry and uses as dye are given in Natural Dyes.

Embelia acts as a protectant of wheat against stored product insect pests *Sitophilus oryzae, Rhizopertha dominica* and *Ephestia cantella.*

10. *Juglans regia* Linn. (Juglandaceae)

Akhrot, Walnut

The description, chemistry and uses as dye are given in Natural Dyes.

Juglone is an antifeedant to bark beetle *Scolylus multistriatus.* It is also allelopathic to nearby plants causing wilting and inhibition of seedling growth. Toxicity is more in poorly strained soils. Juglone is toxic to earthworms also.

11. *Lawsonia inermis* Linn. (Lythraceae)

The description and chemistryof this plant and uses as dye are given in the previous section on Natural Dyes.

The aqueous leaf extract of this plant is found to cause significant mortality of the 2nd stage larvae of *Tylenchulus semipenetrans* and *Aguina tritici.*

12. *Polygonum hydropiper* Linn (Polygonaceae)

Syn. *P. flaccidum* Meissn

Pani maricha, water pepper

Please see the section on Natural Dyes for a description of this plant.

The plant is rich in flavonoids such as quercetin, quercitrin, rutin, hyperoside, rhamnetin, isorhamnetin, etc. and tannins (15-36%) consisting of catechol, phloroglucinol, gallic acid and phlobaphenes. Also present are caffeic, chlorogenic and protocatechuic acids and ascorbic acid.

Seedlings contain a deep red pigment idacein. Roots contain tannins and anthraquinones.

The dried powder of herb is spread on clothes to guard against moths. It is also used as a fish poison. The greenish mucilaginous juice kills mosquito larvae.

13. *Roldana barba-johannis* H. Rob. & Brettell (Asteraceae)

The methanol extract from the aerial parts of *Roldana barba-johannis* yielded tocotrienols and hydroquinones such as sargachromenol, sargahydroquinoic acid, and sargaquinoic acid. These natural products and their corresponding acetylated and methylated derivatives showed insecticidal and insect growth regulatory activities against larvae and adults of the Fall Armyworm (*Spodoptera frugiperda*) which is an important insect pest of corn. Sargachromenol and its acetylated derivative; sargahydroquinoic acid and its acetylated derivative; and a mixture of sargachromenol, sargahydroquinoic acid, and sargaquinoic acid (6:3:1) and the acetylated form of this mixture all these compounds and mixtures had significant inhibitory effects between 5.0 and 20.0 ppm and diets. These compounds were fully comparable in activity to those known natural insect growth inhibitors such as gedunin and methanol extracts of *Cedrela salvadorensis* and *Yucca periculosa* (Cespedes *et al.*, 2004).

e. Tannins

1. *Acacia nilotica* Delile (Mimosaceae)

Syn. *Acacia arabica* Willd.

Indian gum arabic tree, Babul

This is a small spiny tree with bipinnate leaves (stipules long spiny, leaflets 10-25pairs, small, linear-oblong), small yellow flowers in globose heads in fascicles of 2-6, and narrow moniliform fruits containing 8-12 seeds.

Pods contain gallic acid, ellagic acid and condensed tannins. Aqueous extracts of fruits of this plant was found to be larvicidal to *Culex* mosquito.

2. *Butea monsperma* Taub. (Fabaceae)

Syn. *B. frondosa* Koenig.

Palas, Bengal kino tree

For the description and chemistry, please refer Natural Dyes

A decoction of flowers is sprayed in fields to control white ants. Maggots are killed when the seed powder is sprinkled over them. Seeds are effective on other insects also. Palasonin is an anthelmintic against *Ascaris*.

4.2.16. Lectins

1. *Canavalia ensiformis* DC. (Fabaceae)

Jack Bean

This is a stout perennial twiner with trifoliate leaves having triangular stipules, ovate apiculate, membraneous leaflets, white large flowers in curved axillary racemes and long 3-keeled pods containing 8-20 reddish brown seeds. Cultivated for edible pods.

Seeds contain toxic proteins like concanavaline A & B., canavaline, and canatoxin alongwtih betonicine and trigonelline. Seedlings are found to contain canaline, dimethyl homopterocarpin and 3-isoxazolidine

The proteins are haemagglutinins and therefore the seeds are used as a pesticide, bactericide and fungicide.

2. *Canna edulis* Ker-Gawl (Cannaceae)

Queensland Arrow root

This is a robust erect handsome perennial with much branched yellow/rose colored rhizomes, reddish brown pseudostems, long-petioled bronze-colored ovate leaves, red flowers in pairs in bracteate loose inflorescences, and obovoid capsules containing hard rounded seeds. A native of Peru, cultivated for starch yielding rhizomes.

Rhizome contains up to 24% starch.

Burning of the plant produces an insecticidal smoke (Rogers, 1984).

3. *Canna indica* Linn

A perennial with tuberous root stock with 1-1½m high stem, large oblong petiolate leaves, scarlet/yellow flowers and subglobose (obscurely 3-lobed) fruits containing 3 black spherical (pea-sized) shining seeds. Common garden plant.

The compounds reported from this plant are campesterol, ß-sitosterol, stigmasterol and ß-lectin.

Aqueous and ether extracts of the whole plant, methanol extracts of root, and leaves were found to have molluscidal activity. This may prove useful in control of *Schistosomiasis* by controlling the molluscs.

4. *Ricinus communis* Linn. (Euphorbiaceae)

Erandah, Castor

Castor plant is a tall soft hollow-stemmed glabrous laticiferous shrub with long-petioled peltate palmately 7-9 lobed (lobes serrate) leaves, monoecious inflorescences containing male apetalous flowers with branched stamens at the top and female flowers (also apetalous) towards the base. Fruit is a 3-valved spinous schizocarp (regma) containing 3 mottled and carunculate seeds. A native of tropical Africa, now extensively cultivated for oil.

Castor seed contains oil (45%) and protein (16%), enzymes like lipase, alkaloid ricinine and allergens. Ricin, a toxalbumin, is the principal haemagglutinin. Ricinine ($C_8H_8O_2N_2$, 1-methyl-3-cyano-4-methoxy-2-pyridone), a water soluble alkaloid, is present mostly in the seed coat. The oil contains mainly ricinoleic acid (up to 90%). Leaves contain ricinine, N-methyl ricinine and glycosides of Kaempferol and quercetin

The powdered leaves are used for repelling aphids, mosquitoes, white flies and rust mites- due to ricin or alkaloid ricinine. Roots and leaves are considered highly effective in rheumatism. Seed oil is purgative and used for fever, inflammation of scortum and colic.

4.2.17. Proteins/Enzymes

1. *Archidendron ellipticum* I.C. Nielsen (Mimosaceae)

From the seeds of this native tree of Great Nicobar Island, India, a serine proteinase inhibitor, AeTI, is isolated. This protein is found to be a potent insect antifeedant (Bhattacharya *et al.,* 2006).

2. *Phaseolus vulgaris* Linn.(Fabaceae)

Garden bean

An α-amylase isoinhibitor, α-AI-1 from this plant is found to inhibit digestive α-amylases from insects belonging to the orders Coleoptera, Hymenoptera and Diptera and of an annelid worm. The ingested inhibitors are found to suppress the development of insect larvae (Kluh *et al.,* 2005).

4.2.18. Other principles

1. *Abuta grandifolia* Sandwith. (Menispermaceae)

Syn. *Cissampelos pareira* Linn.

The dichloromethane extracts of this plant demonstrated high larvicidal activity against *A. aegypti,* with an LC_{50}=2.6 μg/ml (LC_{100}= 8.1 μg/ml), indicating an activity 2-fold higher than β-asarone, a natural botanical insecticide used as a positive control (LC_{100}=16 ìg/ ml) (Ciccia *et al.,* 2000).

2. *Acalypha indica* Linn. (Euphorbiaceae)

This plant is used as an insecticide.

3. *Andrachne cordifolia* Muell. Arg. (Euphorbiaceae)

Syn. *Leptopus cordifolius* Decne.

This is a small herb having ovate lanceolate leaves and unisexual flowers. Found in Central and Western Himalayas. Leaves contain hydrocyanic acid. Aqueous extracts of branches and leaves are insecticidal.

Aqueous extracts of branches and leaves are insecticidal.

4. *Arachis hypogea* Linn. (Fabaceae)

Groundnut, Pea nut

The seed kernel and oil cake containing di-*n*-butyl succinate, is found to be effective against nematodes such as *Meloidogyne incognita, M. Javanica, M. exigua, Heliocotylenchus erythrine, Hoplolaimus indicus,*

Aphelanchus avenae, Tylenchorhynchus brassicae and *Ditylenchus cypei.* The oil cake is amended into the soil to control various nematodes.

5. *Arisaema speciosum* Mart (Araceae)

Samp-Ki-Khumb, Cobra lily

This is a tall paradioecious tuberous perennial, with obliquely elongate creeping tubers, solitary trifoliolate leaves (leaflets elliptic-ovate with red margins), dark purple spathes which are longitudinally white-striped, pink spadices and fusiform-ovoid fruits. Found wild or cultivated in Himalayas at higher attitudes.

Tubers are rich in carbohydrates (76%), but contain irritating needle-like crystals. They are anthelmintic and insecticidal.

6. *Arisaema tortuosum* Schott.

Syn. *A. helleborifolium* Schott.

Cobra Flower

A monoecious tall perennial with depressed globose tubers, a tall pseudostem, 2-3 leaves having 5-17 variable leaflets (rhombic-ovate or lanceolate) green/purple spathes, bisexual spadices and fruits containing 3-5 ovoid seeds. Common in subtropical/temperate Himalayas.

The tubers contain n-alkanes, n-alkanols, stigmasterol, sitosterol, campesterol, cholesterol, choline chloride and stachydrine hydrochloride. The seeds contain lauric, myristic, palmitic, stearic, arachidic, eicosanoic and behenic acids in different proportions. They are credited with anti-fertility, insecticidal and insect-repellent property and are used to kill worms which infest cattle during the rains.

The tubers and seeds are used as insecticidal and insect-repellents. They are also useful for cold and cough.

7. *Aristolochia argentina* Gris. (Aristolochiaceae)

CH_2Cl_2 and MeOH extracts of this plant showed a significant activity against *Sitophilus oryzae* (Broussalis *et. al.,* 1999).

8. *Azolla pinnata* R. Br.

Syn. *Salvinia imbricata* Roxb.

Mosquito fern

Azolla is green/reddish small free-floating annual/perennial common in water bodies throughout India. The plant possesses a branched rhizome, fascicled roots, and oblong/deltoid fronds having sessile bilobed leaves. Sporocarp is on the underside of the plant between roots and is covered by imbricate scales. *Azolla* contains *Anabaena azollae*, a blue-green alga living as an endophytic symbiont in the inter-cellular spaces of basal leaves.

Azolla is known as mosquito fern because these plants, when profusely growing will cover the water surface completely to make an impenetrable layer so that mosquitoes are unable to lay eggs on the water bodies. *Azolla* is also found to be molluscicidal and nematicidal.

The ethanolic extract of *A. pinnata* was found to have marked nematicidal activity against blood fluke *Schistosoma mansoni miracidia* and *cercariae* as well as high molluscicidal activity against the snail, *Biomphalaria alexandrina* (Zidan *et al.*, 1998) Fresh extract of this plant is found to reduce egg hatching of *Meloidogyne javanica* and *M. incognita* (Thaker *et al.*,1988).

9. *Bambusa arundinacea* Roxb.

Syn. *B. bambos* Druce.

Vansa, Thorny Bamboo

B. arundinacea is the most common spiny bamboo with short stout knotty rhizomes, dense green (purplish when young), tall (24-30m) hollow culms, linear lanceolate large leaves, large yellow flowers produced once in 30-45 years (plant flowers only once in lifetime) and oblong grooved grains. Common near rivers and most deciduous forests in almost all parts of India.

The fresh young shoots are found to be lethal to mosquito larvae and thus are insecticidal in properties.

10. *Carica papaya* Linn. (Caricaceae)

Papeetha, Papaya

A quick growing tall soft-stemmed laticiferous tree having a hollow trunk bearing leaf scars, a crown of large long-petioled deeply palmatifid leaves and unisexual dioecious flowers. The male flowers are small yellowish-white, borne in many-flowered densely pubescent cymes. Female flowers are large, solitary or in few-flowered racemes.

Fruit is a large berry mostly obovate with a hollow center and enclosing many black reticulate seeds enveloped by a transparent aril. Papaya is a native of southern Mexico, now cultivated everywhere.

The ripe fruits are a good source of pectin, carotenoids like β-carotene, cryptoflavin and violaxanthin. Other carotenoids present are phytoene, γ-carotene, chrysanthemaxanthin and neoxanthin. Vitamin C, sugars, citric, malic and butanoic acids, volatile oil consisting of methyl butanoate, linalool, benzyl isothiocyanate and benzyl glucosinolate and alkaloid like carpaine are the other constituents. Raw papaya is a good source of pectin and proteolytic enzymes papain and chymopapain. The latex, generally collected from fruits, is the main commercial source of papain. Besides papain and chymopapain A, B & C, the latex contains sulfhydryl compounds, glutamine cyclotransferase, peptidase and lysozyme. Seeds yield protein (8%), fatty oil (9.5%, rich in oleic acid), phospholipids, carpaine, and glucosinolates like caricin, benzylglucosinolate, and glucotropaeolin and benzyl isothiocyanate. Roots contain carposide and myrosin and the leaves yield carpaine, pseudocarpaine, dehydrocarpezine, choline, vitamin C and vitamin E.

Leaves, seeds and unripe fruits are insecticidal. The leaf extract is used for leafy caterpillars, coffee rust, leaf rust, mosaic virus, powdery mildew, grass looper, white grub, flower thrips and fruit flies.

All parts of the plant are of medicinal value. Unripe fruit is laxative and diuretic. Ripe fruit is stomachic, galactogogue, diuretic and is effective in dysentery, chronic diarrhoea, bleeding piles, dyspepsia and as an emmanagogue. Seeds are emmanagogue, carminative and vermifuge. Carpaine shows antitumour and antitubercular activity.

11. *Carthamus tinctorius* Linn. (Asteraceae)

Kusumbha, Safflower

This is a much branched slender annual growing up to 50 cm to 1m in height with lanceolate spinulose-serrate leaves, orange red flower heads having spiny involucrate bracts, and obovoid 4-angled achenes. A native of Europe now cultivated throughout.

Polyacetylenes of safflower (flowers), 3-*cis,* ll-*trans*-trideca-1,3,11-triene-5,7,2-triyene and 3-*trans,* 11-trans-trideca-1,3,11-triene-5,7,2-triyene are found to have significant nematicidal activity against *Aphelenchoides besseyi*.

12. *Chenopodium multifidum* Linn. (Chenopodiaceae)

CH_2Cl_2 and MeOH extracts of this plant showed a significant activity against *Sitophilus oryzae* (Broussalis *et. al.*, 1999).

13. *Cochlospermum religiosum* Alston (Cochlospermaceae)

Syn. *C. gossypium* DC.

Girisalmalika, Cotton tree

A soft-stemmed deciduous tree with deeply furrowed bark containing gum, palmately 5-lobed tomentose leaves, golden yellow flowers in terminal panicles (appearing after leaf fall) and large pear-shaped 3-5 valved capsules containing numerous seeds covered by wooly hairs. A native of India.

The leaves are found to contain flavonoids, phytosterols, saponins and tannins.

This plant also is referred as insecticidal. The gum-kadira gum is used as an anodyne and sedative.

14. *Combretum pilosum* Roxb. (Combretaceae)

This is a large scandent hairy shrub found in the North-Eastern region of India, having large oblong-lanceolate leaves, flowers in densely crowded silky-villose spikes and elliptic fruits.

The leaves of *Combretum pilosum* and *C. wallichi* DC. are considered pest-repellant.

15. *Echinops echinatus* Roxb. (Asteraceae)

Utkatar, Echinops

A thistle-like much-branched spiny bush with white cottony pubescent widely spreading branches, sessile pinnatifid pubescent leaves (lobes of leaves triangular and spinescent (spines often 3 cm. long.), white disc florets in globose spiny heads (involucral bracts modified to spines) and obconical densely villous achenes. A native of S. Europe, Asia, and N. Africa.

The powdered roots are used to destroy maggots in the wounds of cattle. It is also applied to hair, alongwith *Acacia* gum, to destroy lice.

16. ***Flaveria bidentis*** **(L.) O.K. (Asteraceae)**

CH_2Cl_2 and MeOH extracts of this plant showed a significant activity against *Sitophilus oryzae* (Broussalis *et. al.*, 1999).

17. ***Gaillardia pulchella*** **var.** ***picta*** **Gray.(Asteraceae)**

Fire wheel, Blanket flower

This small herbaceous annual bears oblong-obovate (sometimes pinnatifid) leaves and large heterogamous purple (rays with yellow or white tips) heads. A native of America, often cultivated for the beautiful flowers.

The extracts were found to cause high mortality rate of the second stage juveniles of *Meloidogyne* spp.(Ali *et al.*, 1985).

18. ***Glycine max*** **Merril (Fabaceae)**

Syn. *G. soja* Maxim,

Soja hispida Moench.

Bhat, soybean

Soybean, a very important food and forage crop of Asia, is an erect annual having trifoliolate leaves with large ovate- lanceolate hairy leaflets, blue-violet mostly inconspicuous flowers in short axillary racemes and broad pods containing 2-5 compressed, spheroidal/ellipsoidal yellow/brown/white seeds. A native of China, now widely cultivated.

Seeds contain carbohydrates (35%), protein (45-50%), oil (20%), saponins, sterols, tocopherols, phospholipids, lectins and isoflavonoids. Oil consists of linoleic (50%), oleic (35%), linolenic (6%), palmitic (6%) and stearic (4%) acids. Six kinds of group A saponins Aa, Ab, Ac, Ad, Ae and Af giving soyasapogenol A as aglycone and 5 group B saponins, Ba, Bb, Bc (all three giving soyasapogenol B as aglycone), Bd and Be (soyasapogenol E) are reported from seeds. The steroids present are stigmasterol and β-sitosterol. Several isoflavone glycosides genistein, diadzin (glycitein 7-O-β-D-glucoside), glycitin, 6′′-O-acetyl genistein, 6′′-O-acetyl diadzin and their aglycones, and 6-O-malonylated derivatives of diadzin, glycitin and genistein also are found to be present. Genistein, diadzin and their derivatives form 0.25% of the defatted meal.

Soyabean oil kills mites, beetles and other insect pests and useful for food and feed crops and indoor and out door plants.

Soybean oil is known to decrease thrombosis and found useful in treating cholethiasis. Isoflavones are estrogenic and therefore are useful in treating breast cancer.

19. *Guizotia abyssinica* Cass. (Asteraceae)

Seed extract is found useful in controlling species of *Meloidogyne.*

20. *Heliotropium bacciferum* Forssk. (Boraginaceae)

The leaves are found to cause a reduction in oviposition of *Callosobruchus maculatus.*

21. *Hiptage benghalensis* Kurz. (Malpighiaceae)

Syn. *H. madablota* Gaertn.

This is a large woody pubescent climber with coriaceous elliptic oblong silky pubescent leaves; fragrant white flowers in tomentose racemes and 3- winged samaras (central wing longer) containing solitary globose seed. A native of tropical Asia.

Root bark contains mangiferin.

The leaf juice possesses insecticidal properties.

22. *Kalanchoe integra* Kuntze (Crassulaceae)

Syn. *K. spathulata* DC.

A stout succulent perennial herb of 1m with fleshy spathulate oblong crenate leaves, yellow flowers in dense many flowered cymose panicles and glabrous follicles containing numerous oblong-ellipsoid longitudinally ribbed seeds.

Leaves and flowers contain flavonols and sterols.

The plant possesses insecticidal properties.

23. *Kigelia africana* Benth. (Bignoniaceae)

The stem bark of *Kigelia africana* was found larvicidal to *Culex* mosquito (Taura *et al.*, 2004).

24. *Kunzia ambigua* Thom. and *K. baxterii* A. rich. (Myrtaceae)

A mixture of tetramethylcyclohexenedione isomers, all 4-[1-(5, 7-dihydroxy-6-methyl-4-oxo-2-phenyl-chroman-8-yl)-3-methylbutyl]-5-

hydroxy-2,2,6,6-tetramethyl cyclohex-4-en-1,3-diones, are isolated from the aerial partds of these trees. All of them are found to be insecticidal (Khambay *et al.*, 2002).

25. *Lagenandra toxicaria* Dalz. (Araceae)

A marshy herb with a thick creeping rhizome, long petioled large leaves (lamina elliptic-oblong, midrib stout and margins undulate) and spathe tubular at base and limb long-tailed. Male inflorescence is cylindrical with crowded yellow stamens. Female inflorescence has a spiral row of many ovaries crowded in a globose head with peltate stigmas. Fruit is a globose syncarp containing many slender furrowed seeds. Common in Western Peninsula and Ceylon.

This plant also is said to exhibit insecticidal properties.

26. *Parthenocissus quinquefolia* Planch. (Vitaceae)

Virginia creeper

This is a perennial vine, native to USA, with alternate palmate pentafoliolate leaves (leaflets ovate, serrate) small green flowers borne in clusters and blue black berries. A native of North America.

An enzyme with chitinase activity is found to be present in this plant. β-Amyryl hexadecanoate, a thrombin inhibitor also is present here.

27. *Plumeria rubra* Linn. (Apocynaceae)

Frangipani

This native of Mexico is a soft-stemmed laticiferous tree with obovate leaves having an acute base and acuminate apex and large red/white fragrant flowers borne in apical cymes.

The bark contains iridoids such as plumericin 1, isoplumericin 2 and plumieride, acetophenone etc.

The extracts of the leaves and bark of this plant were found to cause 100% mortality to *Meloidogyne javanica* within 12 hours (Ahmed *et al.*, 1990). Plumericin and isoplumericin are found molluscidal and antibacterial. Plumericin and plumieride are algicidal too.

28. *Pterospermum acerifolium* Willd. (Sterculiaceae)

Hathipaila

The description of this plant is given in section on Natural Dyes.

The flowers are sharply bitter and acrid in taste and are used as insect repellants and disinfectants.

29. *Rhazya stricta* Decne (Apocynaceae)

The leaves are found to be cause a reduction in oviposition of *Callosobruchus maculatus*.

30. *Rudbeckia hirta* Linn.(Asteraceae)

Gloriosa daisy

This plant is found to contain polyacetylene derivatives (PADs) such as 1-tridecene-3,5,7,9,11-pentayne (highly toxic), a thiarubrine, 3-(1-propynyl)-6-(3,5-hexadien-1-ynyl)-1,2-dithiacyclohexa-3, 5-diene, and a thiophene, 2-(1-propynyl)-5-(3,5-hexadien-1-ynyl)-thiophene. The thiarubrine was more toxic against mosquito larvae and also to larvae of a herbivorous insect, *Manduca sexta* (Guillet *et al.*, 1997).

31. *Synedrella nodiflorum* Gaertn (Asteraceae)

Node weed

An erect herb clothed by appressed hairs with opposite narrowly elliptic serrate leaves having the primary vein ending in an apical point, heterogamous yellow heads in the axils of leaves (pappus, a row of bristles) and black achenes. A native of Pacific Islands. This plant also is used as an insecticide.

32. *Tournefortia hirsutissima* Linn. (Boraginaceae)

Chiggery grapes

This is a high climbing vine (stems up to 10 m in length) with stinging hairs, rough lanceolate leaves, white flowers in scorpioid cymes and round white drupes. A native of Central and South America. It is said to be insecticidal.

33. *Viguiera annua* (M. E. Jones) Blake (Asteraceae)

The leaves of this herb contain two major polyacetylenes, *trans*-1, 3,5,11-tridecatetraen-7,9-diyne and 1,*cis*-3,*trans*-5,*trans*-11-tridecatetraen-7,9-diyne, possessing light-activated toxicity (phototoxicity) to many insects (Guillet *et al.*, 1997).

4.2.19. Preparation of insecticidal extracts

A number of methods of preparing extracts are available with the rural folk of India. The extracts are prepared either in water or in kerosene. In both cases the extracts are diluted in water before spraying. Addition of soap to this solution facilitates the oil-solubles to form emulsions with water. Even water soluble compounds also will be distributed in soap solution. Soap gives another advantage in that the sprayed solution will stick to the body of the insect as well as to the leaf/plant surface and will not drip down. Since soap itself possesses insecticidal properties (described later in this section) it will give added potency to the plant extracts. Almost all the data are from the book "Botanical Pesticides in Agriculture" (Prakash and Rao, 1997).

1. *Aloe* + *Vitex* extract

Aloe pulp (prepared by removing outer skin of leaves and then grinding) 2 litres + 5 kg *Vitex* leaves boiled in 10 l water → mixture made to 50 l. 50-60 ml soap added. Mixture sprayed in the morning or late in afternoon.

2. *Andrographis*

2 kg Fresh plant + 10 g of chilli powder - ground with 250 ml water - diluted with 10 l water - filter through cheese cloth. Sprayed in the early morning or late afternoon.

3. *Ocimum*

50g Fresh leaves ground, soaked in 3 l water overnight, 8-12 ml soap added - sprayed in the morning.

4. *Buddleja lindleyana*

4 kg Leaves ground; some water added and the extract squeezed out, 10 times water + 10-15 ml soap added.

5. Chilli

500g Chilli powder boiled in 3 l water, + 30 ml soap.

Chilli + neem leaves

20 Hot pepper + 2 kg of neem leaves - ground well and 1 l water added, kept overnight, extract strained/squeezed. 20 l water + 2 tbsp. powdered soap added.

Chilli + neem seeds

12 Pieces of pepper + 200 gm seeds of neem - ground. Soaked in 4 l water, strained.

Chilli + *Annona* + neem

a) 25 g Chilli powder soaked in 100 ml water
b) 100 g *Annona* leaves crushed, soaked in 500 ml water
c) 50 g Crushed neem fruits - soaked in 200 ml water

All the three extracts mixed and 5 l water + 20 ml soap powder added.

6. Coriander

200g Seeds crushed, boiled with 1 l water for 10 min + 2 l water added, sprayed in the morning.

7. *Annona*

a) 500 g Leaves boiled in 2 litres of water for 30 min, strained and filtrate diluted with 15 l water.

b) 500g Seed powder, kept in water for 2 days, strained, diluted with 15 l water.

c) Seeds ground and oil crushed out, 1 part of oil mixed with 20 parts of water.

8. *Euphorbia tirucalli*

Young branches pound to pulp, water added, strained.

OR

10 drops of latex + 1litre water.

9. Garlic

a) 85g Chopped garlic + 50 ml kerosene/vegetable oil - allow to stand for a day, soap added (10ml) and 100 ml of this emulsion mixed with 1900 ml water.

b) Ground garlic + soap + water - strained, 100 ml of emulsion + 900 ml water.

10. Ginger

50g Ginger ground to paste, water (3 l) added with 12 ml soap. Also used with garlic + chilli in equal proportions.

11. *Gliricidia*

1kg Fresh leaves, ground, soaked in 3 litres water, strained, 20 litres water added; used with neem leaves, chilli or onion.

12. *Petiveria*

10 kg Roots + 3 spoon salt + 30 gallons of water: kept for 8 days, strained. 1 litre filtrate is mixed with 20 l water for spraying.

13. *Equisetum*

50g Fresh chopped stems + 4 gallons of water – boiled for 30 min, strained.

14. Lemongrass

50g Ground leaves + 2 l water – soaked for 3-4 h – strained. Also used with chilli, bitter wood, tobacco, *Gliricidia* and *Tinospora*.

15. *Lansium*

500g Seeds ground, soaked in 20 l water, strained.

16. *Adhatoda*

1kg Leaves + 10 l water, boiled for 30 min., strained, diluted to 50 l with water.

17. *Mammea*

Ground 1 kg seeds + 100 l water + 100 ml soap for 2 h., strained. Also used as seed powder mixed with saw dust or ground dry leaves.

18. Marigold

a) Whole plant, 5-10kg, kept in a drum filled with water for 5-10 days, strained and 1 ml soap added to 1 l extract.

b) Pounded leaves soaked in hot water for 24h., strained.

Also used with tomato leaves (equal proportions) or with chilli, garlic and onion.

19. Mint

250g Ground fresh leaves + 2 l water, stirred for 2 h., strained.

20. Neem

a) *Leaf.* 2 kg Fresh pound leaves + 4 l water, kept for 3 days. Strained, diluted 1 l extract with 9 l water + 100 ml soap

b) *Seed.* Seeds pound, mixed with water, stirred, kept for 6-10 h. add soap, strain.

c) *Seed oil.* Seed powder, mixed with water slowly to knead the mixture to get oil. The oil is mixed with soap + water.

21. Onion

Onion extract is prepared water based or oil based. For preparing oil based extract 85gm chopped onion is mixed with 50ml kerosene for 24 h. To this 450 ml water and 10 ml soap is added and mixed. Water extract is prepared by pouring boiling water (1 l) to a container having 1 kg chopped onions. It is allowed to stand for 24 h., strained and diluted 10 times.

22. Papaya

Leaf extract is prepared by soaking 50 g shredded leaves in 100 ml water, stirring well and keeping overnight. The extract is squeezed through muslin cloth. The extract diluted with 3 l water and 10 ml soap added.

23. *Schinus molle*

3 kg Shredded leaves soaked in 15 l water for 3 days, strained, 15 ml soap added.

24. Pyrethrum

1. 100 g Fresh flowers soaked in 30 ml of alcohol overnight, strained, diluted with 3 l of water.
2. 1 Kg of dry pyrethrum flowers added to a drum containing 100 l water, stirred vigorously, strained and 3 kg liquid soap is added. Spraying is done preferably in the evening.
3. Pyrethrum flower powder (3g) + 1 tsp. Soap added to 1 l water, stirred well and applied immediately.

25. Pongam

a) *Leaf extract* is prepared by keeping 1 kg shedded leaves in water (to cover the leaves) overnight, grind the leaves, strain and dilute the extract with 5 l water, add soap (5 ml) and apply.

b) *Seed extract.* 50g seeds ground and the powder is tied in a muslin cloth and soaked in 1 l water overnight. The pouch is squeezed to get the extract. 1 ml soap is added and sprayed.

c) *Pongam cake extract:* 100g seed cake is wrapped in a muslin cloth and soaked in 1 l water overnight, extract squeezed out, soap added. Powdered cake can be used as soil manure also.

d) *Seed oil:* 1g seed oil is mixed with 1 ml soap, dissolved in 1 l water.

e) *Pongam oil + neem seed oil:* 10 ml pongam oil + 50 ml neem oil + 1 ml soap + 1l of water.

f) *Pongam + aloe + neem:* 1 kg pongam seed cake + 1 kg neem seed cake, soaked in a muslin pouch in water overnight. The extract is squeezed and 250 ml *Aloe* juice added. The whole mixture is diluted with 15 l water.

26. *Quassia*

50 g Bark chips, ground and added to 2 l water, boiled, strained and sprayed.

27. *Toona*

1 kg Fresh leaves pounded, soaked in 2 l water, extract squeezed out and sprayed.

28. *Bidens*

a) 100g Seeds + 1 litre water – boiled for 5 min., filtrate strained, 1 ml soap added.

b) Whole plant (powdered) soaked in water overnight, strained, soap added.

29. *Acorus*

Pound rhizome (20gm) + 1 litre water, allowed to stand for a day, strained, soap added.

30. *Stemona*

Dried root (200g) soaked in water overnight, strained and used as a spray.

31. *Tripterygium*

Powdered roots (1kg) stirred with 200 l water for 2h., strained and sprayed.

32. *Tinospora*

Chopped vine (200g) + water (1 l), stirred for an hour, rice seedlings are soaked overnight or sprayed on seedling.

33. Tomato

Tomato leaves (1 kg) pound, mixed with 17 l water for 2 h., filtered, soap added.

34. Turmeric

20g Shredded rhizome in 200 ml water, kept overnight, filtered, 10 ml soap added.

35. *Vitex*

Vitex leaves (2kg) soaked overnight in 5 l water. Mixture boiled for 30 min. in the morning, strained, diluted with 10 l water, soap (10 ml) added.

36. *Chenopodium ambrosioides*

Leaves (1-6 kg) ground, soaked in (5-30 l) water, strained, filtrate diluted with 5 times water.

37. *Pachyrrhizus*

Seeds (500g) ground, soaked in water for 1-2 days, strained.

4.2.20. Other plant derived substances

1. Diatomaceous Earth

Diatomaceous earth (a geological deposit consisting of the fossilized skeletons of numerous species of siliceous marine and fresh water unicellular organisms, particularly diatoms and other algae) also is used as an insecticide because of its abrasiveness to the insect cuticle and /or their absorption of cuticular fats. After quarrying, crushing and milling, a fine light dust is obtained, containing porous particles with certain abrasive properties and the ability to absorb lipids to about three or more times the particle mass. Any diatomaceous earth with high oil absorbing capacity is a potential insecticide. Beyond the absorbing capacity, the size of particles, uniformity and shape of the particles, pH, and the purity of formulation affect the compound's insecticidal efficacy. Insecticidal diatomaceous earth should be a highly pure amorphous silica, having particles of equal diameter (< 10 μm), pH < 8.5, containing the least possible number of clay particles and less than 1% crystalline silica. The particles of diatomaceous earth are easily picked up by rough bodied insects. These particles damage the cuticle through hydrocarbon absorption and abrasion making the cuticle permeable to water which rapidly leaves the insect's body causing death from desiccation. This dehydration causes the death of insects. It is used as a dust on livestock animals or added to stored grains or seeds to reduce losses due to stored product insects (Korunic, 1998).

2. Soaps

These are not grouped as botanical pesticides, but included here because oils which produce soaps are of plant origin. Fatty acids and their salts (soaps) have been known for centuries as insecticidal and used widely for insect control in US and other developed countries since early 19th century. The most toxic insecticidal soap is potassium oleate and the activity is attributed to the caustic action due to the alkali released by hydrolysis, entry into tracheae and subsequent lysis of hemocytes and body wall and disruption of the nervous system. The fatty acids disrupt cell membranes and uncouple oxidative phosphorylation. Even the cuticle of insects also is disrupted.

Soaps are used to kill soft bodied pests such as aphids, thrips, scale crawlers, whitefly and leaf hopper nymphs and mites. They are less effective against the adults of hard-bodied insects. They are preferred on ornamental landscape plants and house plants. The recommended concentration is 1-2% solution in water. Laundry detergents are not used because they dissolve the cuticle of plant and thus cause injury.

3. Vegetable oils

Cottonseed oil, and oils of corn, peanut, safflower and soybean are found to be toxic to several soft-bodied insect pests. These oils block the spiracles of the insects causing suffocation and killing. Since oils prevent gas exchange through the egg membrane, the insect eggs also get destroyed. The fatty acids of oil disrupt cell membranes and uncouple oxidative phosphorylation. These oils kill only the insects on which they are applied (no residual effect) They affect relatively immobile soft bodied pests such as aphids , scales, leafhopper nymphs, caterpillars and mites. Jojoba oil is used for all crops, this works as a physical barrier and kills/repel, whiteflies and useful for powdery mildew on grapes and ornamentals.

4. Acetone

Acetone vapours are found to achieve 100% mortality with a dose of 123 μl acetone/l air in 1–3 days for *Tribolium confusum* du Val adults and eggs, and *Ephestia kuehniella* Zeller larvae and eggs. However a lower dose, 61.5 μl acetone/l air achieved 99% mortality in 3–6 days, in these life stages except for the larvae of *E. kuehniella* (Tunç, *et al.*, 1997).

5. Ashes

Grains treated with ashes are foúnd to get protected from insect pests. The ashes of cattle dung, goat droppings, *Cochlospermum mopane* or *Combretum imberbe* are found particularly useful. The ashes are either mixed thoroughly with the grain or added to the stored product in various layers. The insects are found suffocated or desiccated by ashes. Ash is found effective against bruchids in cowpea, stem borers in maize, aphids, banana weevils and blight in tomatoes. However in the field, ash is of limited value because though it reduced the infestations, it is not totally protective to the crops.

4.3. Parasitoids and Predators

These are the natural enemies of insects which are used to control insect populations. Biological control deals with the use of parasitoid, predator, antagonist or competitor populations to suppress a pest population, making it less abundant and thus less damaging than it would other wise be (Van Driesche and Bellows,1996). There are three aspects of biological control (a) importation, (b) augmentation and (c) conservation of natural enemies. Importation and establishment of non-native natural enemy populations is the first step followed by augmentation by increasing the number (by mass culture, periodic release and colonization). Conservation is the study of modification of human influences that allow natural enemies to realize their potential.

Some of the natural enemies which were used in the past are the following.

1. *Oecophylla smaragdina* F., an ant which Chinese growers used to protect *Citrus* from insets.
2. *Cotesia glomerata* L., a parasitic wasp used to protect cabbage
3. *Gracula religiosa* L. a mynah bird to control red locust, *Nomadacris septemfasciata* in Mauritius.
4. *Picromerus bidens* (L), stinkbug to control bedbugs
5. *Rodolia cardinalis* Mulsant, an insect against cottony cushion scale *Icerya purchasi.*
6. *Cryptochetum iceryae*, a parasitic fly etc.

Since this involves importing a living population to another country or continent, a lot of legislative procedures are needed for the same .

A detailed account of Parasitoids and Predators is given by Orr. and Suh (1998).

4.4. Pheromones and other Semiochemicals

Pheromones are the volatile compounds which are the agents released by an organism into the environment so as to convey a signal to another organism. They are used to attract mates, to mark trails and to promote social cohesion and co-ordination in colonies. The odors that trigger specific behavioral responses in an organism together are called semiochemicals and pheromones are one class of them. A review on the role of pheromones in pest management is given by Shani (1993) in which it was reported that over 1.3 million ha of crops (1% of cultivated area) were being treated by such chemicals in 1990.

More than 1000 sex pheromones are reported. They are volatile and specific to one species or a small number of related species and are very potent over considerable distances. Using pheromones (mostly synthesized) the particular insect population can be attracted and trapped. Some of the pheromones used thus are *p*- anisaldehyde (used to attract *Bactrocera dorsalis*), myrcene (to attract *Dentroctonus ponderosae*), isothiocyanates (for brassica pod midge *Dasyneura brassicae*) etc. Pheromones are also used for "mating disruption" of insects. By releasing a large amount of sex pheromones in an area, the behaviour of male insects are disturbed.

Floral fragrances are either attractant or repellant to insect. For e.g. pine oil has been found to have repellant or ovipositor deterrent properties against some insects (Ntiamoah, 1996). Linalool is found to reduce feeding and oviposition of houseflies. α-Pinene is found as an oviposition repellent for cherry bark tortrix *(Enarmonia formosana)*. Even palmitic and oleic acids are seen to reduce oviposition of European grape berry moth (*Lobesia botrana*).

4.5. Ecdysteroid Agonists

A number of ecdysteroid agonists have been used to control insect pests. A good number of non-steroidal bisacylhydrazine ecdysteroid agonists, which are more potent than the native hormone, 20-hydroxy ecdysone in inducing molting, are found to cause reduced feeding and weight gain followed by a lethal molt causing death of the insect trapped within the exuvial cuticle. The larvae of Lepidoptera, Diptera and Coleoptera are found affected by ecdysone agonists. A number of plants like *Pteridium* and *Podocarpus* contain ecdysones in high concentrations. They are also reported from *Sesuvium, Achyranthes, Gomphrena* etc and the role of these plants as insecticides are to be explored.

4.6 Brassinosteroids

These compounds as an aerial spray are found to increase the resistance of plants against pathogens (like *Phytophthora* on potato), and viral infections. They also do not initiate co-evolution of pests (Khripach *et al.,* 2000).

4.7 Nematodes

(1) Heterorhabditis bacteriophora: This entomopathogenic nematode is characterised by its symbiotic association with bacteria, *Xenorhabdus* and *Photorhabdus*. The juveniles of this nematode which are free-living in the soil enter the host insect through mouth, anus or spiracula. Once inside the hemocoel, the nematode releases the symbiotic bacteria which secrete toxins that kill the insect within one or two days. This is effective against Black vine weevil, turf grubs and other soil insect pests as well as wood borers, scarabs, billbugs etc. It is also found effective against many types of lepidopteran larvae in USA.

(2) Steinernema carpocapsae: The mode of action of this nematode is similar to that of *Heterorhabditis.* This is used against Black vine weevil, strawberry root weevil, cranberry girdler and many other larval insects.

4.8 Other Substances

A number of other substances like alcohol, ammonia, baking soda, bleach, boric acid, copper, flour spray, horticultural oil and sulphur also are used as pesticides. Even bug juice also repels insects.

Bug juice spray

It is found that pests of the same species will not feed on the plants if members of their colony are found dead on those plants. The smell serves as a repellant and as a warning to other incoming pests. To prepare this the pests are pound or ground, extracted in water and sprayed on plants.

□□□

5

Biofertilizers

They are biologically active organisms/products including bacteria, algae or fungi, with the ability to provide nutrients to plants. They form an important alternative source of plant nutrition. They belong to two types; (1) nitrogen fixing and (2) phosphate solubilising. Nitrogen fixing biofertilizers fix atmospheric nitrogen into forms which are readily usable by plants. These include *Rhizobium, Azotobacter, Azospirillum,* blue green algae and *Azolla.* While *Rhizobium* requires symbiotic association with the root nodules of legumes to fix nitrogen, others can fix nitrogen independently. Phosphate solubilising micro-organisms (PSMs) secrete organic acids which dissolve rock phosphate and tricalcium phosphates which are readily absorbed by plants. PSMs are particularly valuable as they are not crop specific and can benefit all the crops.

Fungal biofertilizers are not well suited to European agriculture where large-scale automated planting methods are not amenable to inoculating soil around plant roots. By contrast, in developing countries the crops are usually planted by hand and hand-planting makes it easy to introduce small amounts of inoculum into soil.

The potential demand for biofertilizers in India is of the order 3.4 lakh tonnes and the country's annual production has reached only 13,000 tonnes (Hindu, Feb 12, 2003). Though the awareness among Indian farmers is very high, many constraints such as quality of the product, poor storage facility and inconsistent field performances of inoculates are impeding their production.

The sharp increase in the use of chemical fertilizers in last three decades caused grave concern on the quality of soil and ground water. The large scale production of biofertilizers was initiated only in 1990 mainly due to the initiative of National Biofertiliser Development Centre at Ghaziabad. India's total production was reported to be 16,000 tonnes in 1998-99. *Rhizobium* accounts for 40% followed by *Azotobacter*. *Azotobacter* is found to increase the yield of rice (5%), wheat (8-10%), maize (15-20%), Sorghum (15-20%), Potato (13%), Carrot (16%), Cauliflower (40%), Tomato (24%), Cotton (up to 27%) and sugarcane (9-24%) (Das, 1991).

Biofertilisers are defined as a living fertilizer compound of microbial inoculants or groups of microorganisms which are able to fix atmospheric nitrogen or solubilise phosphorus, decompose organic material or oxidize sulfur in the soil. On application it enhances the growth of plants, increases the yield and also improves soil fertility and reduces pollution. Chemical fertilizers supply limited number (NPK) of minerals, but biofertilisers provides certain growth promoting substance like hormones, vitamins, amino acids also. The nitrogen supplied by chemical fertilizers depletes and therefore are to be supplied repeatedly. In contrast the biofertilizers supply the nitrogen continuously throughout the entire period of crop growth in course of time and there is no toxicity at higher doses (in contrast to chemical fertilizers).

Transgenosis

By this process transgenic biofertilisers are produced by increasing the number of *nif* genes, incorporation of *hup* genes wherever hydrogenase enzyme is not produced and knocking of negative gene from the regulatory operon.

Azospirillum was incorporated with phosphoenol pyruvate carboxylase (pqq.) synthase gene. This transcojugant was found to possess both mineral solubilising (MPS) and nitrogen fixing properties. The pqq synthase gene was first isolated from *Buckholderia cepacia* via PCR amplification of primers designed using the consumed sequence across the gene reported in different species and cloned to *E. coli*. The *E. coli* containing pqq synthase gene now showed MPS activity.

5.1 Nitrogen-fixing Bacteria

a. *Rhizobium*

Rhizobium is a soil bacterium producing root nodules in leguminous plants. They live symbiotically inside the nodules and provide up to 90% nitrogen requirements of legume crops. The increase in crop yield is 15-30%. *Bradyrhizobium* is a small-sized bacterium found in the root nodules of soybean, chickpea and some other legumes. Both these bacteria lose their outer membranes on entry into the root cell, and become irregular in shape - "bacteroids". The fixation of nitrogen takes place in a strictly controlled microaerophilic environment. The oxygen level in root nodule is regulated by a special hemoglobin - leghemoglobin which is produced only when the host plant is infected with bacteria. The globin protein is encoded by plant genes but the heme cofactor is supplied by the bacteria. The plant cells convert sugar to organic acids and supply to bacteroids. In exchange, the plant receives amino acids rather than free ammonia. Nodules on some other leguminous plants are formed by yet other genera. In all these cases the bacteria "invade" the plant and cause the formation of a nodule by inducing localized proliferation of the plant host cells. Yet the bacteria always remain separated from the host cytoplasm by being enclosed in a membrane, a necessary feature in symbiosis. *Mesorhizobium* and *Sinorhizobium* are other two strains of *Rhizobium* isolated from different plants.

Specific strains of bacteria are found inhabiting specific plant species. *Rhizobium trifoli* is such an example found in clover (*Trifolium*). The lectins on the surface of root cells of clover, binds specifically to the lipopolysaccharide of *Rhizobium* which contains 2-deoxy glucose. The bacteria then enter the root cells and produce cytokinins which promote rapid cell division resulting in the formation of nodules. *Rhizobium* is recommended for pea, green gram, black gram, peanut and cow pea.

b. Azotobacter

One of the non-symbiotic nitrogen fixing bacterium, this is a gram negative, oval, flagellate pleomorphic organism. It occurs free in soil and water. It does not form root nodules or associate with leguminous crops. The increase in crop yield is 5-15%. *Azotobacter* is found to secrete thiamine, riboflavin, indole-3-acetic acid and gibberallic acid to environment which benefits the plants around. The incidence of diseases in *Azotobacter* containing fields are said to be very low probably due to the secretion of antibiotics by this fungus. This is recommended for upland crops (cannot survive wetland contains). In soils of poor fertility and organic matter, *Azotobacter* is to be regularly applied. It is seen that addition of 20 kg biofertiliser containing *Azotobacter* provides 50% of nitrogen needed to the crop.

c. *Beijerinkia*

This is a gram-negative free living aerobic bacterium found in sugarcane soil. It is a nitrogen fixing bacterium generally occurring alongwtih mycorrhiza of dipterocarps. The genus was earlier called *Azotobacter indicum*. This genus is characterized by cohesive colonies, the dumbbell shaped cells with fat inclusions, reddish pigmentation, slow growth of isolates and the ability to fix nitrogen at a pH well below 6.

d. *Azospirillum*

This bacterium thrives on the root zones of rice and fixes nitrogen which is readily absorbed by plants. They do not produce root nodules or associate with leguminous crops. They are not free-living and live inside plant roots where they fix nitrogen. The leachates or exudates of the roots provide nourishment to bacteria which flourish. This organism can solubilise phosphorus and silicon also. *Azospirillum* can be used in wetland conditions and thrives well in soils with a good amount of organic matter. It requires a pH level above 6.0. It produces other beneficial growth substances also.

e. *Bacillus cereas*

BP 01 strain of this bacterium is a microbial plant regulator that is used on cotton to produce higher quality cotton at lower cost. This is applied by aerial or ground spraying in USA.

Clostridium, Desulphovibrio, purple sulphur bacteria, purple non-sulphur bacteria and green sulphur bacteria are the anaerobic nitrogen fixing bacteria.

The **nitrogenase enzyme complex** is highly sensitive to oxygen. It is inactivated when exposed to oxygen because this reacts with the iron component of proteins. Although this is not a problem with anaerobic bacteria, it could be a major problem for aerobic bacteria such as Cyanobacteria (which generate oxygen during photosynthesis) and the free living aerobic bacteria of the soils, *Azotobacter* and *Beijerinkia*. These organisms have various methods to overcome this problem. For e.g. *Azotobacter* has the highest known rate of respiratory metabolism of any organism, that they might protect the enzyme by maintaining very low levels of oxygen in their cells. *Azotobacter* and *Rhizobium* also produce copious amounts of extracellular polysaccharides and by maintaining water within the polysaccharide slime layer these bacteria can limit the diffusion of oxygen in the cells.

In symbiotic nitrogen fixing organisms like *Rhizobium,* the root nodules contain oxygen scavenging molecules such as leghemoglobin (which shows a pink color when the active nitrogen fixing nodules are cut open). Leghaemoglobin may regulate the supply of oxygen into the nodule tissues in the same way haemoglobin regulates the supply of oxygen to the mammalian tissues. Some of the cyanobacteria have another mechanism to protect nitrogenase in that nitrogen fixation occurs in special cells, the heterocysts (which possess only PSI used to generate ATP by light mediated reactions and no photolysis of H_2O) while other cells have both PSI and PSII (which generate $O_{2)}$. Compounds like luteolin, genistein, stachydrine, trigonelline and erythronic acid are signals used by *Rhizobium. Rhizobium* produces nod-factors, which are lectins containing oligosaccharides, which go and interact with roots. It is seen that roots of legumes contain many genes which favour rhizobia.

Symbiosomes are organelle-like structures of plant plasma membrane vescicles containing bacterioids fixing N_2. The membranes contain plant proteins and bacterioid membrane proteins. Bacterioids transfer ammonia to plants and takes up amino acids from plants.

f. *Frankia*

Frankia belongs to the Actinomycetes, a group of filamentous bacteria that are characterized by the production of airborne spores. *Frankia* spp. are slow-growing in culture and require specialized media. This genus (sometimes called *Actinorhizae*) is specific to trees and found in symbiotic association with 200 dicots of 25 genera belonging to 8 families such as Betulaceae, Casurinaceae, Coriariaceae, Datiscaceae, Eleagaenaceae and Myricaceae etc. They form nitrogen fixing root nodules (multilobed) with several woody plants such as alder (*Alnus* sp.), sea buckthorn (*Hippophae rhamnoides*) and *Casurina.* These nodules contain vescicles with very thick walls which are similar to heterocysts to reduce the entry of oxygen and cause finger-shaped extensions. The type of vescicles varies from plant to plant. They may be club-shaped, spherical or filamentous. Alder and other woody hosts of *Frankia* are typical pioneer species that invade nutrient poor soils. *Frankia* is said to be responsible for 15% of the biologically fixed nitrogen in the world.

Frankia deforms the root hairs of host plants by going into the cortical cells and causing the formation of nodules. Vegetative hyphae colonize the nodules and differentiate into diazovesicles which are thick-walled spherical cells in which reductive nitrogen fixation occurs. The membranes of *Frankia* contain lipid components called hopanoids (amphiphilic pentacyclic triterpenoid lipids). It is these hopanoids which thicken and stabilize the walls which help to keep oxygen away from nitrogenase.

5.2 Cyanobacteria

Cyanobacteria most commonly used as a biofertiliser are *Anabaena, Nostoc* and *Tolypothrix*. Other nitrogen fixing cyanobacteria, used thus are (1) unicellular forms like *Gloeocapsa* and *Aphanotheca,* (2) filamentous heterocystous forms such as *Rivularia* and (3) filamentous non-heterocystous forms like *Oscillatoria, Lyngbya* and *Spirulina.* It is estimated that these algae can provide up to 20-30 kg N/ha per season. They are found to elaborate vit. B and other growth factors which are then available to the plant. The oxygenation of water by these microbes is another advantageous factor. The algal colonies (normally in the form of a mat) prevent loss of moisture from the soil also. Thus the plants can tolerate draught when irrigation or rainfall is delayed.

Tolypothrix is found to increase paddy yield by 15-25% in Japan. A mixture of *Anabaena, Aulosira, Tolypothrix, Nostoc* and *Plectonema* used in paddy fields of Tamil Nadu, resulted in substantial increase in yields. In places where other inorganic fertilizers are used these microbes are found to augment the crop yield.

The application of the blue-green algae was found to benefit vegetables like lettuce, chillies and tomato as well as other crops like wheat. *Nostoc, Scytonema, Camptylonema, Aulosira* and *Cylindrospermum* were used to reclaim alkaline soils and it is found that within a year the pH of the soil has been reduced from 9.5 to 7.26.

5.3 Phosphate Solubilising Microorganisms (PSMs)

Phosphorus is one of the thirteen essential elements required by plants and is responsible for disease resistance and higher yield. PSMs include bacteria such as *Bacillus megatherium* var. *phosphaticum* and *Pseudomonas fluorescens,* fungi like *Penicillium digitatum* and *Aspergillus niger* and mycorrhiza. Bacteria work well in neutral or alkaline soil conditions. Fungi are better suited to acidic soils.

5.4. Vascular Arbuscular Mycorrhiza (VAM)

Certain fungi belonging to Glomalean group remain in association with roots and form vescicles and arbuscules within the root. They are known as VAM, AM (arbuscular mycorrhizae), endomycorrhiza or endotrophic mycorrhizae. More than 80% of plant families form relationships with VAM. They take sugars from the plant and the plant gains nutrients, usually phosphate from the fungus. The mycelia are seen within and around the roots. The hyphae are non-septate when young and ramify within the cortex and even in pith region. Arbuscules are intricately branched haustoria in cortical cells. Vesicles are globular storage structures found in cortex. A network of hyphae is formed in the soil with thicker hyphae function as conduits and thin branched hyphae are absorptive in function. Spores are asexual spherical structures (20-1000 μm in diameter) that form on hyphae in soil or roots.

Based on the thickness of hyphae and nature of spores, various genera such as *Glomus, Scutellospora, Acaulospora* and *Gigaspora* are recognized. The wall layers are one of the important characters to identify the species.

VAM fungi are symbiotic in nature and colonise the roots of approximately 99% of plant families. The host plants are benefited by increased uptake of phosphorus and other immobile nutrients from the soil, enhanced nutrient acquisition, water relations, pH tolerance and disease and pest resistance, resulting in increased growth and yield. The extra radical mycelium of the fungus acts in effect as an extension of root system which extends up to 8 cm or more beyond the roots. A number of other microbes also are found to colonise the spaces among the mycelia of AM or the spores. The spores of AM fungi are known to harbour a large population of bacteria like organisms (BLO), characterized as *Burkholderia capacia.*

AM is found to interact with other microbes in the soil such as nitrogen fixing bacteria and saprophytic and free-living microbes. When it is in association with nitrogen fixing bacteria, it helps in better nodulation and nitrogen fixation through improved phosphorus uptake. With *Rhizobium,* AM shares the signal molecules such as flavonoids and the bacterium is found to stimulate AM colonization by *nod* factors. Dual inoculation of VAM with *Rhizobium* or *Azotobacter* or *Azospirillum* is found to provide the plants with a positive growth response and nutrient status (Sharma & Joshi, 2002).

Another positive factor of AM is that it checks the growth of soil and root borne pathogenic fungi. AM is a good control agent for root knot nematodes belonging particularly to *Meloidogyne*. The improved resistance of plants to plant pathogenic fungi and nematodes in presence of AM fungi is due to improved production and secretion of factors such as phytoalexins, lytic and other enzymes, pathogenesis-related (PR) proteins, callose formation, hydroxyproline-rich glycoproteins (HRGP) and phenolics. AM is found to increase the actinomycetes, bacteria and saprophytes (all beneficial) and decrease pathogenic fungi in the vicinity of roots.

The life history of AM fungi includes a brief free-living phase prior to root colonization and an extended biotrophic phase. AM is best useful in conditions of both abiotic and biotic stress. They thrive best in acidic soils, but can thrive successfully (which benefits the host plants) in arid zones, salt affected soils, sand dunes etc. The capacity of mycobiont (fungi), to bind soil particles to form water stable macroaggregates, is found useful in raising AM colonized grasses and other plants in stabilization of sand dunes of coastal areas. The extracellular hyphae by their polysaccharides on the hyphal surface help in binding the soil particles and stabilize them further by the formation of larger lumps (macroaggregates). These lumps are bound further by the spread of hyphae (up to 5 cm) and humus in the soil. The soil aggregation is found to be up to four fold times.

Mychintec [Mycorrhiza technology for staple food production in small scale sustainable agriculture in China.] is a three year project started in 2000 in China to use soil enhancing abilities of Arbuscular Mycorrhizal Fungi (AMF). In China the crops studied are sweet potato, maize and cassava. Thirty one fungal isolates from 3 different Chinese regions were collected and scientists at INRA (Institute National de la Recherché Agronomique) in France analyzed their genetic sequences and devised fungus specific DNA probes that could track the fate of individual fungal types in soil and plant roots. On testing these fungi for their ability to provide extra phosphate to the plants at least 15 isolates were found to have really interesting plant fertilizer abilities. Then these fungi were cultured into a soil based inoculum usable by farmers. The trials in 2002 and 2003 proved that crop yields of sweet potato and maize increased by 11% and there was a quality increase up to 26%.

5.5 *Azolla pinnata* R. Br.

Syn. *Salvinia imbricata* Roxb.

Description of this plant is given in Biopesticides

Azolla multiplies rapidly and is used as green compost also. It has been found to increase the yields of paddy by 300%, when incorporated in the soils of transplanted paddy. The number of fillers and panicles and the weight of panicle also were found to be increased substantially. The presence of *Azolla* was also found to be reducing the angiosperm weeds in the filed. *Azolla* is found to be rich in proteins and minerals and therefore is used as a cattle and poultry feed also.

List of commonly produced biofertilizers in India

Name	Crops	Benefits	Remarks
Rhizobium strains	Legumes like pulses groundnut, soybean	10-35% increase 50-200 kg N/ha	Fodders give better results leaves residual N in soil.
Azotobacter	Non-legume crops including dry land crops	10-15% incr. Adds 20-25 kg N/ha	Also control certain diseases.
Azospirillum	Maize, barley, oats, sorghum, millet, sugarcane, rice	10-20% incr.	Fodders give higher/ enriches fodder response, produces growth-promoting substances. It can be applied to legumes as co-inoculant.
Phosphate solubilisers include 2 bacterial and 2 fungal species	All crops	5-30% incr.	Can be mixed with rock phosphate
Blue-Green algae and *Azolla*	Rice/wet lands	20-30kg N/ha, *Azolla* can give biomass up to 40-50 tonnes and fix 30-100 kg N/ha	Reduces soil alkalinity can be used as a feed for fishes, show growth promoting hormonal effects TNAH has developed high yielding *Azolla* hybrids.
Mycorrhizae (VAM)	Many trees, some crops and some ornamental plants	30-50% incr. Enhances uptake of P, Zn, S. and water	Usually inoculated to seedlings.

5.6 Mixed Cropping

Mixing fallows of fast growing, nitrogen fixing trees such as *Sesbania sesban, Tephrosia vogelli, Gliricidia* etc. the soils will become a physically changed medium with reduced crusting, better water infilteration and holding capacity. The defoliating beetle, which a serious threat to *Sesbania,* can be destroyed by planting *Gliricidia sepium* in adjacent plots. The adult beetles died after feeding on *Gliricidia* and therefore, mixed planting is found better than single tree fallows.

Other trees used similarly are *Faidherbia albida, Acacia polycantha* and *A. galpinii. Tephrosia* or *Cajanus cajan* may be under-sown to increase soil fertility. The leaves of *Cassia alata* are used as a green manure in Tripura where acidic nature of the soil is not suitable for chemical fertilizers. The red glands on the petioles of leaves of *Cassia occidentalis* are found to contain *Bacillus megaterium* (a phosphate dissolving bacterium) and using the leaves as green manure increased the grain and straw of wheat. The extracts of *Jatropha curcas, Azadirachta indica* and *Derris elliptica* have been found to enhance cotton yield.

□□□

6

Biofuel

6.1 Petroleum

Petroleum is a heterogeneous assemblage of organic compounds predominantly containing hydrocarbons. The hydrocarbon fraction consists of straight chain alkanes of 1-30 carbons (alkenes are almost absent) and cyclic compounds- naphthenes-like cyclohexanes and cyclopentanes and aromatics such as benzene, toluene, xylene and naphthalene; the relative proportion of which varies with the region of origin. For example, oil from Pennsylvania is rich in aliphatic hydrocarbons; that of Borneo in aromatics and of Caucasus in naphthenes. Heavier asphaltic fractions contain polycyclic aromatics and cycloalkanes of high molecular weight. Besides hydrocarbons, there are many complex components containing oxygen, nitrogen and sulphur. Oxygen is present as carboxylic acid derivatives of alicyclic hydrocarbons and fatty acids, while nitrogen occurs in the form of heterocyclic compounds such as pyridines, quinolines, isoquinolines and their alkyl derivatives. Thiols, sulphides and thiophenes form the different sulphur compounds in petroleum.

There are different theories about the formation of petroleum. It might have formed in the primaeval earth, when there was a great abundance of hydrogen in the atmosphere by (1) reactions like carbonium ion mechanisms catalysed by minerals in oil bearing rocks or (2) by thermal decomposition supported by natural radioactivity or (3) like coal, by the accumulated decay brought about by the enzymatically catalysed reactions of anaerobic microorganisms. Though the first two theories cannot be ruled out, the overwhelming evidences suggest a biological origin of petroleum.

The debris of the marine and land life, collected at the depths of primordial oceans, is believed to be the source of petroleum. The various chemical compounds like carbohydrates, proteins and lipids might have been reduced to a hydrocarbon conglomerate by the enzymatic reactions of microorganisms and by the heat and pressure of rock formation over periods of millions of years resulting in a syrupy liquid - the petroleum.

The optical activity exhibited by the petroleum, occurrence or porphyrin skeletons, aromatics and the predominance of alkanes with an odd number of carbon atoms are the evidences cited in support of its biological origin. Optical activity is associated with almost all the components of living organisms. A complex skeleton like porphyrin is easily derived from the chlorophyll molecules.

Pristane (2, 6, 10, 16-tetramethyl pentadecane), which amounts to 0.5% (of East Texas) or 0.2% of (Mid-Continent) oils, is believed to be a derivative of phytol, a terpenoid component of chlorophyll molecule. It is assumed that phytol (C_{20}) on decarboxylation gives rise to pristane (C_{19}). It is suggested that isoprenoids like carotenes undergo low temperature reactions to produce napthenic and aromatic hydrocarbons similar to those of petroleum (Erdman, 1961; McMarry and Thornton, 1952). The abundance of carotenoids in many of the marine algae like brown algae which constitute the major share of ocean flora correlates well with this idea of carotenes giving rise to naphthenes and some aromatics.

The aromatics might have also been derived from the phenols and simple polyphenols abundant in the early vascular plants like Pteridosperms. Lignin present in them, in whatever primitive form, also would have contributed to the aromatic fraction of the petroleum.

The principal components of many plant waxes are odd chain alkanes presumably formed by the decarboxylation of fatty acids. The presence of a heavy waxy cuticle in the earlier giant vascular plants-as also in the ferns of today-leads us to trace the origins of alkanes to these plant components. In addition, the fatty acids-abundant in plants in the combined form as triglycerides, waxes, cutins, suberins, phospholipids and lipoproteins, may on decarboxylation give rise to alkanes or in unaltered form remain in petroleum to form the fatty acid fraction. Cyclohexane carboxylic acids like shikimic and quinic acids - important intermediates in carbohydrate-phenol interconversion-are present in almost all tissues. The amino acids like cysteine, cystine and methionine and thiols, sulphides etc. might have contributed to the sulphur containing compounds in petroleum.

Petroleum varies in appearance and in composition depending upon the places of origin. These variations might have been due to differing flora and fauna of every region. It is therefore logical to assume that the crude oil, obtained from a place where the source materials are marine algae containing abundant carotenoids, consists of naphthenes; while those derived from a vascular flora contain more of aromatics and alkanes.

Petroleum derived from rocks of very great age (ca 1,000 million years old or older) does not contain any plant derived compounds.

Natural gas is found alone or in conjugation with petroleum deposits. Though the principal component is methane, it also contains progressively lesser amounts of other alkanes of increasing chain lengths (e.g., C_1=80%, C_2=10%, C_3=5%,C_4-C_7=5%).

The fractional distillation of petroleum to separate the different ingredients is known as refining. The various fractions collected are: 1. Light petroleum or ligroin (b.p. 35-100°C), 2. Gasoline or petrol (Bp 50-220°C), 3. Paraffin oil or kerosene (b.p.180-325°C), 4. Heavier oil fractions (all boiling above 300°C) like diesel, furnace oil, lubrication oils, greases and waxes and (5) a black residue, bitumen or asphalt. The low boiling fractions are used as fuels and the high boiling fractions, as lubricants, greases and waxes. The gases accompanying the petroleum are liquified under pressure and used as heating and cooking gas. The petroleum ether and ligroin are used as solvents. Gasoline is the motor fuel; kerosene (C_{22}-C_{15}) is the fuel for the jet engines and furnaces. The lubricating oil contains different fractions and after removal of these fractions yield paraffin wax, mainly used for candles.

The amount of petrol in crude oil being insufficient to meet the growing demands, the higher boiling fractions (400°-600°C) are heated in presence of catalysts like aluminum oxide or silica. The free radicals that are formed decompose or recombine in many ways to produce low boiling petrol. This process is known as *cracking* or *pyrolysis*. The yield of petrol is about 60%. The branched chain alkanes burn smoothly than unbranched and are preferred. The linear alkanes when heated above 350°C in presence of borontrifluoride or aluminium chloride are converted to branched chains by *isomerisation*. Also by *alkalating* or *polymerizing* the unbranched alkanes, the branched ones are obtained. In *reforming* process, the aromatic hydrocarbons are produced from alicyclic compounds by catalytic dehydrogenation over platinum.

Synthetic Petroleum

Synthetic petroleum is an oil resembling petroleum produced by the hydrogenation of coal at 400-500°C and 200-700 atm pressure. Coke also yields a similar product by a process known as Fischer Tropsch process in which the carbon monoxide and hydrogen produced by treating coke and steam at high temperatures are enriched with more hydrogen and passed over a cobalt or iron catalyst at 200°C and 5-35 atm. The oil thus obtained may be fractioned like petroleum.

Petroleum substitutes

Petroleum is non-renewable source of energy, but man is totally dependent of this energy today. The sources of petroleum are dwindling very fast and it is seen that oil is not available at all times due to politico-economic considerations. Though coal and natural gas yield many of the compounds present in petroleum, they cannot fully substitute petroleum. Man has already started looking for alternative sources of energy which can be periodically renewed and choice obviously falls on living plants, the fossilized ancestors of which procured us this valuable fuel. This does not undermine the application of nuclear energy and solar energy anyway.

But there exists a number of difficulties in the production and application of these forms of energy within the reach of common man. The various ways by which the plants can be tapped of their energy are: (1) petroleum from plants (petro-crops), (2) vegetable oils, (3) biogas, (4) alcohols, (5) biological hydrogen, (6) wood and (7) coal.

6.2 Petro-crops

As described earlier, the plants manufacture a variety of hydrocarbons, the type and relative amounts varying with the groups of plants. Thus some plants are rich in waxes while some others contain oils stored in the plant parts and still others possess terpenes. The xerophytes and advanced pteridophytes are covered by a very heavy waxy coating. Pines and resinous trees manufacture oleoresins which are rich sources of mono-, di- and triterpenes. The plant families like Rutaceae, Lamiaceae, Myrtaceae and Asteraceae contain volatile oils in their leaves. The laticifers belonging to Euphorbiaceae, Sapotaceae,

Apocynaceae, Moraceae and Asteraceae store large polymeric hydrocarbons like rubber and gutta in the latex. The cutins of leaves, suberins of bark and the seed oils are other sources of hydrocarbons in plants.

Though the idea of obtaining petroleum like fuel directly from living plants apparently is simple, a lot more is to be desired in the technology to do it. It is to be borne in mind that the petroleum stored within the deeper strata of earth is formed from countless trillion tons of plant materials (both terrestrial and marine) which existed for many millions of years and were destroyed by natural calamities occurred one after another and lied buried under the earth crest for a markedly long period of time. This epic cannot be repeated now in laboratories and industries. The petroleum sources formed from such vast amounts of plant material during equally long period of time will be consumed by man within a period of one century or less. If a supplementary source is procured in time, we may prolong the availability of petroleum for a few decades more by reducing the stress on the same.

The plants manufacturing polyterpenes like rubber and gutta are looked upon as the sources of hydrocarbons today. Though all the related (taxonomically) plants of known rubber and gutta plants do not produce such macromolecules, they are bound to possess similar biosynthetic mechanisms and may be producing low molecular weight hydrocarbons. The most promising option for producing liquid fuels from biomass seems to be the direct extraction of low molecular weight non-polar constituents from plants. This material, termed *biocrude* (McLaughlin and Hoffmann, 1982) is a complex mixture of lipids such as triglycerides, waxes, terpenoids, phytosterols and other modified isoprenoids which can be 'cracked' for use as liquid fuels and chemical feed stocks. It is demonstrated (Weisz *et al.* 1979, Haag *et al.* 1980) that plant hydrocarbons can be converted by a zeolite type catalyst (ZSM-5 type) directly to gasoline and closely related materials. Johnson and Hinmann (1980) concluded that the most successful energy crops are available in the arid lands. The xerophytic environment induces the higher production of waxes as an adaptation to preserve the precious water there. A laticiferous plant growing in the same area is having an added advantage of having waxes spread over the surface and polyterpenes distributed within. As a number of *Euphorbias* are found in arid or semi-arid conditions they are looked upon as the most potential sources of hydrocarbons. Some of the petrocrops are the following.

1) *Chrysothamnus nauseosus* Britt. (Asteraceae)

Rubber rabbit brush

This is a perennial shrub of 2 m. with several fibrous barked main stems from the base, densely leafy ill-scented twigs, linear lanceolate leaves and yellow flowers in terminal heads. A native of America.

The plant yields 15% (cyclohexane) or 20% (ethanol) extract. The yield of plant is 4.5 mt/ha. The waxy leaves contain 12.8% protein, 3.2% fat and 75% carbohydrates also.

2) *Cryptostegia grandiflora* R. Br. (Asclepiadaceae)

The biocrude obtained from the plant on hydrocracking yield gases, naphtha, kerosene and gas oil along with coke and residue rich in hydrocarbons (65%). Some of the main constituents of gases are methane, n-heptane, ethane and n-octane. The naphtha contains alkanes and naphthenes and can be used for production of benzene, toluene, xylene or high octane gasoline pool. (Anon, 2001)

Hexane is the preferred solvent for biocrude extraction. The biocrude can be used as a chemical feed stock.

3) *Euphorbia caducifolia* Haines. (Euphorbiaceae)

This is a pale green dense fleshy dendroid shrub reaching a height of 1m with numerous branches arising from the base. The leaves appear only in rainy season which fall off later. Flowers and fruits appear long after the leaves fallen. Seen in the dry regions of Gujarat, Rajasthan, Orissa etc.

The latex contains triterpenoids like 3-epicyclolaudenol, cyclolaudenol, cyclocaducinol, cycloartenol, euphol, tirucallol, cyclotrichosanthol and cycloeucalenol and hydrocarbons such as nonacosane and dotriacontane. The thick waxy covering on the stem yields fatty acids, hydrocarbons and free alcohols.

The ethyl acetate extractable material is about 10% which can be used as biocrude.

4) *Euphorbia lathyris* Linn.

Gopherweed

This is a glabrous perennial (annual in cultivation) 1 m tall with opposite/whorled lanceolate leaves, greenish yellow cyathia (flowers) and 3-lobed subglobose fruits containing 3 flattened ovoid seeds. A native of Mediterranean.

The latex and the whole plant contain rubber, L-DOPA, flavonols like quercetin, kaempferol, their glycosides, ß-sitosterol, ingenol hexadecanoate, *p*-coumaric and ferulic acids. Total hydrocarbons amount to 6.2%. The plant contains 14% glycerides and 8.3% terpenoids. Seeds yield 40-47% oil. These plants yield up to 20% sugars also, which are fermentable to alcohol

There are varying reports on the biocrude from this plant ranging from 10.3 to 22 mt/ha/yr. Seed yield is 3 mt/ha. The oven dried and powdered plant material contained about 8% terpenes which are converted to "high grade transportation fuel" with the help of zeolite catalysts (Nemethy *et al.*, 1980).

5) *E. neriifolia* Linn.

The latex containing triterpenoids is considered to be a good source of biocrude.

6) *E. tirucalli* Linn.

E. tirucalli is a perennial growing very fast and can be easily cultivated. The yield and properties of the oil are similar to that of *E. lathyris*.

The latex and stem bark contains triterpenoids such as ingol, acetyl and isovaleryl ester, ephorbinol, cycloeuphorbinol, euphorcinol, euphorginol, glutenol, cycloart-23-ene-3ß,25-diol, α-taraxerol, cycloartenol, α-amyrin, isoeuphorbol, euphorbol, euphol, its acetate, taraxerol and taraxerone. Also present are diterpenoids like ingenol and phorbol esters and ellagic acid. Rubber and n-hentriacontane are the other components of stem.

Hexane is the preferred solvent for biocrude extraction. The biocrude can be used as a chemical feed stock.

The only practical difficulty with *Euphorbias* is them containing many carcinogenic and skin irritant materials which on contact with the skin causes great harm.

It is to be noted that as far back in 1935, *Euphorbia abysinica* Gmel. was adjudged as the source of "vegetable gasoline" by Italians who even planned to set up a "Euphorbia gasoline refinery", when they met with a fuel crisis. *E. resinifera* Berg. was chosen as the petroleum source by French in 1940 which yielded upto 30 metric tons of oil per hectare. Both these projects were not pursued further because of the higher cost involved and the availability of oil, once the world war was ended (Khoshoo, 1982).

7) *Pedilanthus tithymaloides* Poit.(Euphorbiaceae)

The hydrocarbons extracted from this plant are comparable to gasoline (De *et al.*, 1997).

A systematic programme for the detection of petroleum plants - petrocrops - is initiated in all the developed and developing countries. According to Bucchanan *et al.*, (1978), a petro-crop must be (1) fast growing, (2) highly adaptable to any environment, (3) easily cultivated (4) have low harvesting cost and (5) high oil and hydrocarbon fractions. The presence of high quantities of fibre, protein or carbohydrates would increase their utility. Such multiuse plants are always preferred in an agricultural endeavour.

6.3 Biocrude

Biocrude is the hydrocarbon and hydrocarbon–like chemical fraction of plants that may be extracted by organic solvents and upgraded to liquid fuel and chemical feed stocks.

Biocrude produced from the materials having some wood component is undesirable as a fuel because of the presence of some thermally unstable components which can lead to gum formation, low energy density because of dissolved water and highly oxygenated compounds, a corrosive organic acid component and phase instability with a tendency towards phase separation. These problems, to some extent, are solved by high severity hydrotreating involving complete hydrodeoxygenation, some hydrocracking and minimal hydrogenation to produce an aromatic hydrocarbon fuel. But these processes are highly expensive and therefore low severity hydrotreating is resorted to, wherein partial hydrodeoxygenation, minimal hydrocracking and effective hydrogenation is done to stablise the biocrude and improving its energy density and handling properties.

Hydrotreating of the biocrude is often hampered by the instability of biocrude. By hydrodeoxygenation by catalytic processing involving low temperature hydroprocessing, hydrocarbon products with gasoline and light fuel oil distillation ranges are obtained. The hydrogen consumption for this processing is high and only 30% of the biocrude is separated by this process. The high costs involved in this process caused the emphasis to be shifted to lower cost processing to produce heavier products such as turbine fuel for electricity generation or as a first step toward cofeeding to a conventional petroleum refinery. Low-severity hydrotreating is then practised to upgrade biocrude to a useful fuel oil product with a minimum of cost through catalytic hydroprocessing. The catalysts used are NiMo or CoMo (Elliot and Neuenschwander, 1996).

Biofuel is generated from sewage sledge also. Sewage sledge is considered a waste product, does not need milling and is more or less homogenous. This is fed into a reactor; a pressure of 20-180 bars is applied and heated up to 350°C. 25% Of the biomass is converted to CO_2 and is lost, half of the biomass is converted to biocrude and 25% remains in the water fraction. After the conversion and removal of CO_2, the water is separated from the biocrude. The biocrude is then separated into a light fraction and a heavy fraction. The light fraction goes to the hydrodeoxygenation process and the heavy fraction of the biocrude is used to generate energy. The biogas produced at the time of initial treatment of waste water (anaerobic digestion followed by aerobic digestion) is used as a fuel in the heater.

6.3.1. Biomass conversion

There are two main ways to convert biomass to biofuel. They are (a) thermal or thermo-chemical process and (b) biological conversion process.

a. Thermal and thermochemical processes

The thermal conversions are in two forms, gasification and pyrrolysis or carbonization.

Gasification is a process in which the soil fuel is converted to a gaseous one through reaction with hot steam and oxygen. The process of heating biomass is carried out in presence of some air or oxygen which is insufficient to fully combust biomass. The fuel gas developed through this process is a mixture of methane, CO_2, N_2, CO and H_2, the rate of which is determined by the technology. The process begins by feeding the feedstock to the gasifier in dry or slurred form. It reacts in the gasifier with steam and oxygen at high temperature and pressure in an oxygen starved atmosphere. This produces synthetic gas (*syngas*) made up primarily of CO and H_2 and smaller quantities of CO_2 and CH_4. This gas is cleaner than biomass because the chemical pollutants are removed during the process. The gas thus produced can be used for direct combustion in, internal combustion engines, gas turbines or for producing hydrocarbons like methane or methanol.

Pyrrolysis or carbonization is a form of incineration, wherein the biomass is made to decompose producing a mixture of solids (char), liquids (oxygenated oils) and gases (methane, CO and CO_2). The exact ratio of components varies with the reaction conditions and the biomass feedstock employed. Normal pyrrolysis is done by heating at a temperature between 300 - 350°C while fast pyrrolysis involves high speed heating (at a temp. of 800 - 900°C) resulting in less char, more gas and oil with a higher CO content.

With fast pyrrolysis techniques (flash pyrrolysis techniques), the liquid fraction can be up to 70% of the thermal biomass input. Bio-oil contains 40% oxygen and is corrosive and acidic. This oil can be upgraded to reduce oxygen content, but this process is economically not viable.

Hypothermal upgrading (HTU), originally developed by Shell, converts biomass at high pressure and at moderate temperatures in water to biocrude. Biocrude contains far less oxygen then bio-oil by pyrrolysis.

The combustible gases are treated in a secondary combustion chamber, burnt and partially condensed. Several types of pyrrolysis units are available including rotary kiln, rotary hearth furnace or fluidised bed furnace.

b. Biological conversion processes

There are two types of these processes. Anaerobic digestion produces methane gas (which can be used as natural gas), while fermentation produces ethanol.

Anaerobic digestion is a process by which the plant material is digested by bacteria in the absence of oxygen. Hydrolytic microorganisms (including the food - spoiling bacteria) break down complex organic waste to smaller units which are then fermented to short chain fatty acids, CO_2 and H_2. Syntrophic microorganisms then convert the fatty acids to acetic acid (with release of CO_2 and H_2). Methanogenesis is the final step in which biogas (consisting of methane - 50 - 60% and CO_2) is produced.

Similar anaerobic digestion occurs naturally when wet organic matter accumulate (in the absence of oxygen) as in the case of sediment of ponds & lakes, swamps, peat bags etc. The decay of biomass in landfill also produces methane which can be collected and used profitably.

Fermentation is the process in which microorganisms break down the sugars or starches to produce ethyl or methyl alcohol. When starch is the raw material, it is first broken down to simple sugars before fermentation. Yeast is the microbe used here.

Microbial Fuel Cell (MFC) is a device to convert the chemical changes taking place in sewage into electricity. Designed by Bruce Logan of Pennsylvania State University, this cell collects electrons released by the oxidation of organic matter by sewage bacteria. Normally these electrons enter into the respiratory reaction of the bacterial cells and are eventually combined with oxygen to form water. By depriving the bacteria of oxygen on one side of the MFC, the electrons can be wrested from them and used to power a circuit.

The MFC comprises of a sealed 15 cm long can with a central cathode rod surrounded by a proton exchange membrane (PEM), which is permeable to protons. Eight anodes are arranged around the cathode. The long slender graphite anodes increase surface area in contact with organic waste. Bacterial enzymes break down the food liberating protons and electrons. Electrons pass into the anodes, while the protons migrate to the cathode. At the central cathode an outer membrane allows only proton to pass through. There they combine with electrons from the cathode and oxygen from the air to form water (Biever, 2005).

c. Biocrude projects

Here a few Biocrude projects are cited; all managed by US. Department of Energy (DOE), Washington.

1) Lignin to reformulated gasoline

Because lignite can be easily converted to liquid fuels via a two-step process in which the lignite is first subjected to a mild hydrotreatment, followed by a base-catalysed hydrolysis at high temperature, a similar process on lignin is attempted. In this process lignin is to be depolymerised to low molecular weight aromatics and to catalytically upgrade these cyclic compounds via hydrogenation, mild hydrocracking and alkylation. A mixture of alkylated cyclohexanes has been obtained in good result (Diebold *et al.* 1995).

2) Hemicelluloses to tetrahydrofural ethyls

Acids convert pentoses to furfural and hexoses to hydroxy methyl furfural. Hemicelluloses consisting of xylans, arabans, mannans etc are first converted to furfural and 5-hydroxy methyl furfural using acid catalysis. Furfurals are further hydrogenated to tetrahydrofurfuryl alcohols (THFAS). The latter can, then be etherified to tetrahydrofuryl ethers (THFES) having high octane numbers. National Renewable Energy Laboratory (NREL), USA has identified a new catalyst which generates furfural very efficiently from xylose at high temperature under dry conditions.

2) Biomass pyrrolysis

This involves pyrrolysing biomass to oxygenated organic vapours which can be condensed to form a high yield of an acidic fuel oil having very low sulphur and ash. The oxygenated vapours are then being upgraded by removing oxygen to produce hydrocarbon-rich oil. This can be used directly in diesel engines or integrated to a petroleum refinery.

3) Biocrude derived diesel

Straight biomass pyrolysis oils are unstable, acidic, corrosive, and viscous and have high water and ash contents. These oils do not lend themselves readily to upgrading in conventional petrochemical processes. These oils are modified to improve their properties for co-feeding with heavy and light cycle oils in a Fluid Catalytic Cracking (FCC) unit. The upgrading of oil includes partial hydrodeoxygenation, catalytic derivatisation and minimal deoxygenation. Using NREL'S bench scale fluid bed reactor, which contains HZSM – 5 and modified zeolite catalyst, biomass feed stocks are catalytically pyrolysed to produce partially deoxygenated biocrude oils. In this approach, the pyrolysis and catalysis are carried out simultaneously, reducing coke yield and partially deoxygenating the pyrolysis products. The partially deoxygenated products can either be catalytically derivatised to methyl aryl ethers or co-fed with heavy cycle oils to FCC units to produce oxygenated transportation products.

6.4 Liquid resins

The trees containing large amounts of oleoresins/hard resins are also promising sources of liquid fuel. The volatile oils and accompanying resins are pure lipids containing hydrocarbons or their oxygenated derivatives and have high fuel efficiency. The spp. of *Copaifera, i.e., C. longdorfii* and *C. multijuga* produce appreciable quantities of oleoresins in the stems. These trees are tapped by drilling a hole into the wood (the wood is having a system of canals) and a tree of 90-120 cm diameter yields about 20 litres of resin within 2-3 hours. Once the oil extraction is over, the hole is closed by a wooden plug and clay, which may be reopened for the next tapping after about 6 months. The oleoresin is mostly of sesquiterpenes and can be used as fuel in diesel engine directly. The cracking pattern of this oil is similar to that of *E. lathyris* (Calvin, 1980). A number of related genera and spp. are under investigation.

1) *Copaifera spp.* (Caesalpiniaceae)

This genus, a native of S.America, consists of large trees producing an oleoresin, Copaiba, from the wood. These trees contain large resin ducts in the heartwood and the resin flows out due to the spontaneous bursting caused by the pressure developed by the secreted resins in these ducts. The resin is also obtained by boring holes with the heartwood. The clear colorless liquid resin possessing a specific aroma and a bitter taste turns viscous yellow on storing.

The oleoresin of *C. langsdorfii* contains sesquiterpenes like caryophyllene, copaene, ß-bisabolene, diterpene acids like polyalthic acid, (-)-kaur-16-en-19-oic acid, (-) 16 ß-kaurene-19-oic acid, and eperu-8 (20)-ene-15,18-dioic acid, copalic acid, enantioagathic acid and enantio-3-hydroxy, labda-8(20), 13-dien-15-oic acid

Copaifera multijuga yields up to 25 litres of resin from the wood in a period of six months. This resin is directly used in diesel engines. *C. officinalis* is another source of liquid resin.

2) *Dipterocarpus turbinatus* Gaertn.f. (Dipterocarpaceae)

Gurjan balsam

This is a tall resinous tree with a clear bole, deeply cracked bark, simple ovate acute, glabrous shining leaves, large white flowers in axillary 3 – 8 flowered racemes (stamens numerous, terminated by a long slender bristle) and a woody nut like (rarely 2) fruit, enclosed in an enlarged calyx in which two of the calyx lobes are enlarged to long wings. A native of tropical Asia.

The oleo-resin consists of 18% sesquiterpenes such as caryophyllene, humulene, humulene epoxide and caryophyllenol I and

II, and triterpenes like damamar-24-ene-3, 20 –diol, dammar–20, 24-dien–3-one, dipterocarpol, dipterocarpolic acid, asiatic acid, acetyl asiatic acid and 2- α- hydroxy ursolic acid.

3) *Grindelia camphorum* Green (Asteraceae)

Gumweed

This species produces large amount (10%) of extractable diterpene resin acids such as grindelic acid and several of its derivatives.

4) *Kingiodendron pinnatum* Harms.

Syn. *Hardwickia pinnata* Roxb.

The tree yields, on tapping, a dark reddish brown oleoresin resembling copaiba balsam in odor and taste. The tapping is done by boring a hole of 2cm diam-in trees of 2 m or more in girth. The hole which reaches the pith is placed 1m above the ground and slopes downwards from the pith to the bark. The exudate is collected and the hole is plugged with a piece of wood. The tree is rested for 10 years before tapping again. A healthy tree of 3 m girth yields about 12 gallons of oleoresin. The maximum yield reported is 40 gallon. The main constituent of oleoresin is β-caryophyllene (84%) and cadinene. The resin left after the volatile oil distilled is a greenish yellow hard brittle mass.

The resin is highly inflammable and causes wild fire when the drilled hole is left uncovered after extraction. (Illegal extraction is rampant in forests containing these trees).

5) *Knema angustifolia* Warb. (Myristicaceae)

Syn. *Myristica longifolia* Hook .f.

An evergreen tree of 25m, found in Eastern Himalayas with lanceolate coriaceous leaves, unisexual flowers and oval fruits borne on axillary woody tubercles.

The resin, a red fluid containing tannin (33%), is used as a varnish.

Algae as sources of hydrocarbons

In addition to the higher plants mentioned above a few algae also are known sources of hydrocarbons.

1) *Botryococcus braunii*

This fresh water green colonial microalga belonging to Chlorococcales, inhabiting freshwater ponds and lakes, has the peculiar ability to synthesize hydrocarbons in large quantities. It is known to be a rich source of lipids including fatty acids, epoxides, alkylphenols, ether lipids and also hydrocarbons sometimes in very high amount (Metzger *et al.* 1991). The algal scum occurring near southern Australia and Soviet Kazakhstan contains upto 70% hydrocarbons very similar to crude petroleum. In the intact cell, the hydrocarbons occur mostly (upto 95%) in the cell walls and have carbon number C_{25}-C_{31}. The location of hydrocarbons in the outer cell makes the recovery of hydrocarbons easy. According to the type of hydrocarbons produced 3 chemotypes are recognized (1) A-race producing odd carbon numbered n–alkadienes and trienes (2) B-race, producing C_nH_{2n-10} triterpenes (n=30-37) called botyrococcenes and (3) L-race yielding an acyclic tretraterpene, *trans, trans,* lycopadiene.

In addition to botyrococcenes, the B-race produces small amount of squalenes and methylated squalenes. Unusual carotenoids with a tetramethyl squalene skeleton and C_{31} - C_{34} methylated squalenes also are reported from B-race (Achitouv *et al.,* 2004)

This alga is very slow growing and measures are in progress to increase its rate of growth (Largeau *et. al.,* 1980).

2) *Chlorella pyrenoidosa*

This is another fresh water alga of great promise. This algal biomass can be catalytically hydrogenated at high temperature and pressure (about 400°C and 1200 psig pressure) in presence of catalysts like cobaltomolybdate. About 50% of the algal mass is converted to clear golden oil which can be processed further to get different fractions. Ammonium carbonate is the byproduct (Chin and Engel, 1981).

6.5 Vegetable oils, Waxes and Plant oil Methylester (PME)

The vegetable oils, consisting almost entirely of triglycerides and some free fatty acids, are easily ignited and act as an energy source of high fuel efficiency. The high viscosity and molecular weight are the two impediments to overcome before their use as fuel in engines. These oils are about 11-20 times viscous than diesel. The viscosity can be reduced to the level of diesel by methylating or ethylating these oils, which make them more volatile and thereby increase the thermal efficiency. A further step of cracking would produce gasoline.

The edible oils, which are successfully experimented, are from sunflower, safflower and palms. In Brazil, a mixture of 80% diesel & 20% sunflower oil is used in tractors and another mixture of 73% diesel, 20% palm oil and 7% ethyl alcohol is used in buses. In Australia, various oils in combination with diesel or alone were used in different engines. It is found that no modification was necessary in the engines and their output was near the same. But vegetable oils being heavier, generated less energy per unit weight than diesel (Peterson *et al.* 1980). Other edible oils used similarly are those of maize, soybean, olive, castor, rapeseed, eucalyptus etc. Though the cost factors weigh heavily against their use as fuels, they can completely replace diesel in times of emergency.

In contrast to edible oils where the cost is high, non-edible oils like those from neem *(Azadirachta indica* A. Juss.*)*, Karanj *(Pongamia pinnata* Pierre*)*, Kusum *(Schleichera oleosa* Oken*)*, Khakan *(Salvadora oleoides* Decne.*)*, Undi *(Calophyllum inophyllum* Linn.*)* etc. form alternative propositions. All these plants are tall trees and the harvesting is often difficult. But an organized effort would make these oils easily available.

The liquid waxes like Jojoba *(Simmondsia chinensis* Schneid.*)* and Carnauba (*Copernicia cerifera* Mart.) also are considered as alternative fuels. These waxes can replace petroleum products like lubricating oils in cosmetics and pharmaceutics.

Chemical conversion of oils to PME (biodiesel)

Biodiesel is the plant oil methyl ester (PME) which is produced by transesterification of vegetable oils or animal fats. By the addition of methanol and a catalyst to oil, the fatty acid methyl esters -the biodiesel- and glycerin are produced. The glycerin is removed and processed to methanol. The waste cooking oil also can be profitably used for biodiesel.

The main raw materials used in beginning was rape seed or canola containing high content of oleic acid (60%) and low levels of saturated fatty acids (palmitic and stearic acid < 6%) and iodine value of < 115. Other raw materials used were palm oil (>40% palmitic acid) in Malaysia (no problems due to high temp. there and therefore operable in winter also), sunflower oil in Italy, Australia & France and soybean oil in USA. Both these have linoleic acid content of > 50% and iodine value of > 115%. Frying oil left after using in restaurants and household are found to reduce costs.

The advantage of Biodiesel is that it is sulphur- free, has higher flash point (110°C) and the same Cetane number (>49) as diesel. The oxygen content of biofuel is about 10% in the molecule which varies by fatty acid chain length. The fuel properties of biodiesel, compared with those of diesel, are given in Table I.

Table I. Fuel properties of diesel and biodiesel (Walter, 1992)

Physical- chemical properties	Unit	Diesel	Biodiesel
Density at 15°	Kg/m3	820-860	875-900
Viscosity at 40°	mm^2/s	2.00 - 4.50	35 - 50
Flash point	°C	>55	>110
Sulphur	%w	<0.20	<0.01
Cetane number		> 49	>49
Oxygen	%w	0.0	10.9
Caloric values	MJ/dm^3	35.6	32.9
Efficiency degree	%	38.2	40.7

In addition, biodiesel has excellent lubricity properties and is considered in France as a lubricity additive for low sulphur diesel fuel. Biodiesel cause a reduction in green house gases also and has very low toxicity and is easily biodegradable (Biodegradability of more than 90% within 3 weeks).

Now biodiesel is used mixed with fossil diesel (5% in France, 20% in USA). Germany and Austria use 100% biodiesel.

The plants which can yield fatty oils (non-edible) in good amounts are as follows:

1. *Azadirachta indica* A. Juss. (Meliaceae)

Nimba, Neem

Seeds yield about 25% oil and a number of bitter principles both in oil and outside. The oil consists of oleic (56%), palmitic (16%), stearic (14%) and linoleic (9%) acids and 0.8% unsaponifiable matter.

2. *Brassica nigra* Koch. & *B. alba* Hook.f. (Brassicaceae)

Mustard

The oil obtained from the seeds of various spp. of *Brassica* is the Rape or colza oil. The yield of fixed oil is 30-45%. It contains glycerides of erucic (50-55%), oleic (20-25%) and linoleic (15%) acids.

The oil is an illuminant and the choice for PME in many countries.

3. *Copernicia cerifera* Mart. (Arecaceae)

Carnauba wax

This unbranched palm, a native of Brazil, yields a wax from the leaves. The leaves are cut before they are fully open, dried in the sun and the powdered wax is thrashed/melted out.

The wax is greenish yellow in color, tasteless and colourless, consists of myricyl cerotate (80 %) and 10% of free alcohols having a chain length of C_{28} - C_{36} (the higher alcohols predominate). Some resinous material, a lactone and aromatic acids like *p*– coumaric and chlorogenic acids also are present.

4. *Calathea lutea* Mey. (Marantaceae)

Cauassu wax

This rhizomatous perennial, a native of Amazon, produces a wax on the underside of the leaves. This wax is extracted by scrapping off from the dried leaves. It is very similar to carnauba wax and used similarly.

5. *Calophyllum inophyllum* Linn. (Clusiaceae)

Undi, Alexandrian Laurel

This is a handsome medium-sized glabrous tree with a crooked trunk, leaves having broadly elliptic striated blade, white fragrant large (3 cm diam.) tetramerous flowers and globose drupes containing a single ovoid seed. A native of Africa and Asia.

Seeds yield 55% of oil containing oleic (60%), palmitic (20%) and stearic (13%) acids.

6. *Derris indica* Bennet (Fabaceae).

Karanj

The seeds yield a fixed oil rich in oleic (70%) and linoleic acids.

7. *Euphorbia antisyphilitica* Zucc. (Euphorbiaceae)

Candelilla wax

This xerophytic shrub, a native of N. America, contains wax on the stems. This wax is extracted by using solvents or by boiling with water when the wax melts and floats on the surface of water.

The wax is rich in alkanes (more than 50%). Esters containing alcohols and acids form the rest.

8. *Glycine soja* S. & Z.

Soybean

Soybean seeds contain about 20% fixed oil consisting of glycerides of linoleic (50%), oleic (30-35%), linolenic (6%), stearic (4%), palmitic (6%) and traces of arachidic acid and stigmasterol.

9. *Jatropha curcas* Linn. (Euphorbiaceae)

Seeds contain 40-50% oil consisting of 80% unsaturated acids (oleic, 47% and linoleic, 31%) and 20% saturated acids(palmitic, 14%, stearic, 6%). The seeds yield up to 4% hydrocarbons also.

This oil is methylated and used as the biodiesel.

10. *Madhuca indica Gmel. (Sapotaceae)*

The kernels yield 55-65% of a whitish yellow oil consisting of palmitic (23%), stearic (20%), oleic (45%)and linoleic (10%) acids. In some samples myristic acid replaces stearic acid.

11. *Salvadora oleoides Decne. (Salvadoraceae)*

S. oleoides is a small tree with a twisted or bent bark, white branches, whitish-green linear-lanceolate leaves, greenish small flowers in axillary spikes and a globose yellow berry. A native of Asia.

Seeds contain 40-50% of a greenish yellow fat consisting of lauric (35%), myristic (55%) and traces of other acids.

12. *Schleichera oleosa* Oken.

The kernels contain 60-70% oil consisting of oleic (52%), arachidic (20%), stearic (10%) and gadolic acids.

13. *Shorea robusta* Gaertn. (Dipterocarpaceae)

It is a large resinous tree with pubescent petioles and inflorescence, large ovate oblong leaves, cream colored pubescent flowers in axillary panicles and a one-seeded indehiscent capsule covered by the bases of accrescent sepals of which three are longer acting as wings. A native of Asia.

Seeds yield 20% oil composed of oleic (42%), stearic (35%), arachidic (12%) and palmitic (8%) acids.

14. *Simmondsia chinensis* Schneid. (Buxaceae)

Jojoba oil

This liquid wax is obtained from seeds of an evergreen shrub indigenous to S.W.USA and Mexico. The wax content of the seed is more than 50% and contains esters of higher aliphatic alcohols and acids. The acids present are n - Δ^{11}–eicosenoic acid (64%) and erucic acid (30%) and the alcohols are n-Δ^{13}-docosenol (70%) and n-Δ^{11} - eicosenol (30%).

Waxes are also extracted from sugarcane stem (containing n-octacosanol as the main component), saffea palms, licuri (*Syagrus*) leaves and esparto grass (*Stipa*) leaves.

6.6 Biogas

The term biogas includes all the gaseous products obtained by a partial microbial conversion of biomass in an anaerobic environment. All the agricultural wastes like weeds, straws, parts of stems, leaves, animal dung and even sewage are used for generating the biogas. Biogas consists of 60-70% methane and the rest oxides of carbon like CO_2 and CO. In the gasification process, the complex compounds like carbohydrates, proteins and most of the lipids are easily hydrolyzed to their monomers by the accompanying enzymes, acids or alkalies. Even the major plant constituent cellulose is not spared here which is hydrolyzed to simple sugars. These monomers are acted upon by acid-forming, facultative anaerobic bacteria, to produce simple organic acids like acetic acid. The next step, the formation of methane, is a strictly anaerobic process needing an oxygen free atmosphere. A group of anaerobic bacteria- methane producers-convert the organic acids to methane and carbon dioxide. The gas is piped out and used as a fuel. A mixture of 35% biogas and 15% diesel is successfully used in running tractors and other agricultural implements. The sludge remaining is a very precious fertilizer and an excellent soil conditioner. Animal dung being rich in methane-forming bacteria, mixing them with any other biomass gives higher yields of methane.

Methane is also produced by hydrogasification method, in which the farm wastes are burned with hydrogen at temperature of 500°C-600°C. A mixture of methane and ethane is produced.

6.7 Alcohols

Short chain alcohols like methanol and ethanol form another alternative for fossil fuel. Both these alcohols are blended with petrol in varying proportions and used directly in automobiles.

6.7.1. Ethanol

Ethanol is the most widely used fuel today. In the United States each year more than 1.5 billion gallons are added to gasoline as an oxygenate, to improve vehicle performance and reduce air pollution. Ethanol is produced by fermenting sugar or starch. Ethanol produced from cellulosic biomass materials instead of traditional feed stocks (starch crops) is called bioethanol. In the coming years it is believed that the cellulosic biomass will be the largest source of bioethanol. The agricultural residues, woods, weeds and almost all the plants form the raw material for production of ethanol. The amount of ethanol expected from cellulosic matter is 112 gallons from 1 tonne of bagasse, 113 gallons from 1 tonne of corn straw, 110 gallons from rice straw, 82 gallons from forest thinning, 101 gallons from saw dust and 116 gallons from waste paper. From the pulped material cinnamoyl esterases are used to remove phenolics especially lignin. Hydrolysis involves both of cellulose and hemicelluloses and a number of genetically modified strains of *E. Coli, Zymomonas, Saccharomyces* etc. have been produced that exhibit varying degrees of ability to ferment the full spectrum of available sugars to ethanol. Enzyme mixtures (mainly cellulases) are obtained from *Trichoderma reesei* and related microbes. Ethanol is used in the pure form (neat), as a blend with gasoline or as a fuel for fuel cells.

Brazil with its abundant sugarcane crop is the first to come up with the idea of using alcohol as a fuel. Petrol containing upto 20% anhydrous ethanol (gasohol) is used in all the vehicles without any change in the machinery. New engines are being designed which run on alcohol alone. The normal alcohol (rectified spirit-95%) contains 5% water in addition to the water it absorbs from the atmosphere causing corrosion in the metal containers. So alcohol is to be made anhydrous before mixing with petrol.

The cheapest way of manufacturing alcohol is from molasses in which the free sugars are abundant. It is also produced from other sugar sources such as sugar beet, sweet sorghum, mahua flowers. Commercial starches also are fermented to alcohol after being hydrolysed to sugars. The common sources of starch commercially utilized for this purpose are tapioca, sweet potato, maize, barley and other coarse grains. Because this is a two-stage process involving, 1) hydrolysis of polysaccharides to monomers and 2) fermentation of sugars, the cost of production of alcohol from starches is relatively high.

6.7.2. Methanol

This also is used as a transportation fuel. U.S. Industry produces approximately 1.2 billion gallons of methanol annually, 38% of which is used in the transportation sector. Currently produced from methane (natural gas), methanol can be produced from biomass through a two step thermochemical process. First the biomass is gasified to produce hydrogen and carbon monoxide. These gases are then reacted to produce methanol. Methanol can be used in its pure form (neat), as a feed stock for the gasoline additive methyl tertiary butyl ether (MTBE) or as a fuel for fuel cells.

Methanol is synthesized from natural gas or produced by the destructive distillation (pyrolysis) of wood and other agricultural products. In West Germany, 15% methanol-petrol blend is found quite satisfactory in its performance to pure petrol. Methanol, in presence of zeolite catalysts, can be converted to gasoline (85%). Looking to the net productivity, potential of resources, economics and pressure on agricultural land, methanol is preferred as the motor fuel while ethanol is the raw material of choice for chemical industry.

All these methods create a large number of jobs. Biofuels reduce CO_2 emission when they burn and the plants producing them use CO_2 from the atmosphere. Ethanol is used successfully on an oxygenate additive in gasoline to reduce emissions from the older, high – polluting vehicles. Biofuels degrade quickly in water as against the hazard of marine oil spills, ground water contamination from underground gasoline storage tanks and run off resulting from fuel spills. Biofuel does not emit most toxic parts of gasoline.

Waste disposal from agricultural and forest products industries produce huge amount of waste each year in advanced countries. Much of waste is of lignocellulosic material that can be converted to biofuels. This form of recycling relieves the growing waste problem also.

6.8 Biological Hydrogen

Hydrogen, an excellent fuel, can substitute petroleum very efficiently. The best source of hydrogen is water molecule from which it is obtained 1) electrolytically, 2) thermochemically, 3) radiolytically or 4) photolytically. The photolytic production of hydrogen occurs in all green plants during the process of photosynthesis. In normal photosynthetic pathway, the electrons, derived from the chlorophyll molecule or water, are accepted by an electron acceptor like NADP.

This reduced NADP accepts the hydrogen ion (H^+), which is formed by the photolysis of water, to form $NADPH_2$. The enzyme hydrogenase present in a variety of bacteria has the power to procure the high energy electrons to the H^+ ions (instead of NADP), producing gaseous hydrogen which is released. So a system of chloroplasts and hydrogenase in water, when illuminated in the absence of carbon dioxide, liberates hydrogen gas. Hydrogenase is present in algae like *Scenedesmus, Chlorella* and *Anabaena* that may be utilized for a similar process. As oxygen also evolves during photosynthesis, the two gases H_2 and 0_2 get mixed up. Techniques to separate these two gases or inhibiting oxygen production and also to stabilize the hydrogenase enzyme systems are under careful scrutiny. Because the yield of H_2 is very low, trials to enhance the yield also are under study. Some other microbes which can be involved in hydrogen production are *E. coli, Citrobactor intermedices, Closterium butyricum, Sirulina maxima, Bacillus macerens* and some holobacteria.

Hydrogen producing bacteria are everywhere. They can be segregated from hydrogen consuming bacteria by just boiling the soil at a temperature just above water's boiling point. The hydrogen consuming bacteria dies off leaving the heat resistant spores of hydrogen generating bacteria. If this soil solution is mixed with a solution of sugar in a reactor, hydrogen (up to 60%) is regenerated. (CO_2 from the remaining gas can be removed by bubbling through NaOH solution). But this hydrogen is to be removed continuously because the hydrogen accumulated suppresses hydrogen production.

A species of *Anabaena, A. azolla,* growing endophytically in a water fern *Azolla pinnata,* is found to release hydrogen from water. But this hydrogen combines with the nitrogen fixed by the same plant producing ammonia. Once somehow the N_2 fixation is suppressed, gaseous hydrogen can be liberated. Some of the leguminous plants containing nitrogen fixing bacteria liberate some hydrogen to atmosphere and are considered a potential source of this gas.

Chlamydomonas reinhardti, a unicellular biflagellated green alga, also is found to release H_2.

Biological hydrogen production, using hydrogenases from Cyanobacteria (blue - green algae), is undertaken by many R & D laboratories. The Hawaii Natural Energy Institute (HNEI) at the University of Hawaii is developing a microalgal indirect biophotolysis process in which water is converted into O_2 and H_2 in separate stages.

In this conceptual process, an initial photosynthetic stage, an open pond micro algae cultivation system, fixes CO_2 into starch (green algae) or glycogen (Cyanobacteria). This is followed by a second stage where the algae are concentrated as needed, become anaerobic (through respiration), include the hydrogenase enzyme and start evolving H_2 from their stored carbohydrates in a dark fermentation. Finally a light-driven H_2 production stage would complete the conversion of the biomass to H_2, using only part of the photosynthetic apparatus of algae. A single organism, a green alga or a cyanobacterium would carry out all these reactions and be reused through several such cycles of CO_2 fixation and H_2 evolution (with CO_2 recycled between stages). The cyanobacterium used was *Spirulina* (now called *Arthrospora*) because it does not fix nitrogen (and thus hydrogen evolution would not be dominated by this enzyme), but produce hydrogen under anaerobic conditions in the dark. The experiment was done in Photobioreactors (Rocheleau, 2000). *Nostoc, Plectonema* and *Anabaena* are the other cyanobacteria used in similar programmes.

Hydrogen is seen as the future energy source because it is renewable, does not evolve the "green house gas" CO_2 in combustion, liberates large amounts of energy per unit weight in combustion and is easily converted to electricity by fuel cells. The heating value of hydrogen is 61,110 Btu/lbs, much higher than that of methane (23879 Btu/lb). Since electrochemical hydrogen requires high energy solar batteries, the biological hydrogen production by photosynthetic micro organisms is preferred. There are two types of Biophotolysis of water.

(1) Hydrogenase-dependent hydrogen production and

(2) Nitrogenase-dependent hydrogen production.

Hydrogenase dependent reactions are observed in the case of a green alga *Scenedesmus* which produced molecular hydrogen under light conditions after being kept under anaerobic and dark conditions. Hydrogenase, the enzyme responsible for this hydrogen production, catalyse the following reaction:

$$2\ H^+ + 2\ X_{reduced} \rightarrow 6H_2 + 2X_{oxidized}$$

The electron carrier X is considered to be ferredoxin. Since ferredoxin is reduced with water as an electron donor by the photochemical reaction, green-alga are theoretically, water-splitting microorganisms. Further studies concluded that the reducing power (electron donation) of hydrogenase, does not always comes from water, but may sometimes originate intracellularly from organic compounds like starch. In anaerobic conditions, hydrogenase is thought to become active in order to excrete excess reducing power. It is also found that *Scenedesmus* produced hydrogen fermentatively under dark anaerobic conditions with intracellular starch as the reducing source. Although the rate of fermentative hydrogen production per unit was less than that obtained through light-dependant hydrogen production, the hydrogen production was sustainable due to absence of oxygen.

Cyanobacteria also are found to produce hydrogen gas auto-fermentatively under dark and anaerobic conditions. *Spirulina* species are found to have the highest activity in this group.

Nitrogenase dependant hydrogen production was discovered by Benemann and Weare (1974) when they observed that *Anabaena cylindrica,* a nitrogen fixing organism, produced hydrogen and oxygen simultaneously in an argon atmosphere for several hours. Nitrogenase, the enzyme responsible of nitrogen fixation, is common among prokaryotes including cyanobacteria, but not in Eukaryotes (Algae). Molecular nitrogen is reduced to ammonia with consumption of reducing power (e^- mediated by ferredoxin and ATP). However nitrogenase catalyses proton reduction in the absence of nitrogen gas (i.e. in an argon atmosphere). Hydrogen production catalysed by nitrogenase occurs as a side reaction at a rate of $1/3^{rd}$ to $1/4^{th}$ of that of nitrogen fixation, even in a 100% N_2 gas atmosphere.

Nitrogenase itself is oxygen labile and therefore is localized in heterocysts in filamentous cyanobacteria away from the photosynthetic vegetative (normal) cells. Hydrogen producing activity of cyanobacteria is stimulated by nitrogen starvation.

Synechococcus is another unicellular aerobic nitrogen fixer. This cyanobacterium, which has a mechanism to protect nitrogenase in non-heterocystous and oxygen evolving cells, has a much higher (3.5%) conversion efficiency.

Photosynthetic bacteria like *Rhodobactor* is another bacterium successfully used in converting waste from fruits and vegetables and sewage.

Photovoltaic cells are combined with a catalyser which acts as an electrolyser and splits hydrogen and oxygen directly from the surface of the cell. The American National Renewable Energy laboratory has attained an efficiency of over 12% with a cell consisting of a layer of gallium indium phosphide ($GaInP_2$) over a layer of gallium arsenide connected to a platinum electrode in water. Theoretically it is possible to obtain an efficiency of 24% and researches are now being conducted to replace the high priced materials with less expensive ones and to increase efficiency. This solar cell can reduce the cost of photolytic production of hydrogen to one fourth of what it costs for photovoltaic panels and electrolysis.

Instead of microorganisms, oxygenated catalysts such as a platinum catalyst on an aluminum oxide base can be used to make H_2 from glucose. The reaction also produces CO_2 and methane and the later might be burned to liberate more energy.

6.9 Wood

The fuel value of wood cannot be underestimated any time. Firewood is the chief cooking fuel in rural areas while fuelwood cater to the needs of many industries. Fast and continuous depletion of forests for timber as well as for fuel is going on for quite sometime, resulting in near extinction of some of the well-known forest ecosystems. To rectify this loss, the idea of energy plantations is mooted in which fast growing trees / shrubs, which give high / yield and incur minimum cost, are selected and harvested periodically. Two different methods, single stem trees and coppicing are adapted for growing trees for energy purposes. Single stem trees are usually produced for timber purposes. In coppicing the trees are encouraged to grow with several stems. Coppiced trees grow much faster than single stem trees and periodical harvesting can be done from the same plant. In India, deciduous trees with high coppicing ability and are easily multiplied, are selected and grown in marginally useful land. Those trees, which are of multiple uses like fodder, fibre etc., are always preferred. The plants utilized for energy plantations in India are *Populus* (Poplar bark contains medicinal phenols and tannins), *Nothofagus* (Southern beech), *Alnus* (alder, tannins in bark), *Salix* (Willow-tannins), *Eucalyptus* (essential oils from the leaves) and *Casurina* (bark, a tannin source). Leguminous trees have an added advantage of enriching the soil by the production of root nodules. *Leucaena leucocephala* (Lamk.) de Wit., a native tree of tropical America belonging to Mimosaceae, is a multipurpose plant introduced to this country and is getting naturalized in many places. In addition to the wood it supplies, the leaves can be used as a fodder and poultry rations because of their higher (6%) protein content. The seed, yielding about 9% of an oil and mucilage, are eaten after parching or used as a cattle feed. (When taken in excessive quantities, all parts of this plant are toxic to animals due to the non-protein amino acid, mimosine, contained in them and cause great loss of hair and fertility). Various spp. of *Lespedeza* and *Calliandra* are also introduced which are believed to increase the wood and fodder production in this country.

References

Abeysekara, *I. et al.*, (1990) *Fitoterapia*, **61:** 473

Achitouv, E. *et al.*, (2004) *Phytochemistry*, **65(23):** 3159

Ahmed M. U. *et al.*, (1990) *Intel. Nematol Network News*, **7:** 5

Akihisa, T. *et al.*, (1990) *Phytochemistry*, **29(5):** 1639

Akihisa, T. *et al.*, (1999) *J. Nat. Prod.* **62:** 265

Ali. S., Junaid, M and M. Mashkoor (1985). *Intel. Nematol Network Newsl*, **3:** 19.

Andrei, C. C. *et al.*, (2002) *Z. Naturforsch.*, **57c**: 418

Anonymous. (1958), *Agricultural Handbook*, **45,** USA Washington

Anonymous. (1975), *Agricultural Handbook*, **154,** USA Washington

Anonymous. (1991a) *Traditional veterinary Medicine in Srilanka*, APHCA Publn. Thailand.

Anonymous. (1991b) *Traditional veterinary Medicine in Philippines*, APHCA Publn. Thailand.

Araya, H (2004) *Bull. Natl. Inst. Agro. Environ. Sci.*, **23** : 77

Aruna K. and V.M. Sivaramakrishnan (1996) *Phytotherapy Research.*, **10:** 577

Aslan, I. *et al* (2006) *Industrial Crops and Products*, **23(2):** 171

Asseleiah, L.M., Hernandez, O and J.R. Sanchez (1990) *Phytochemistry*, **29:** 3095

Badar A. *et al.*, (2003) *Phytochemistry*, **62(8):** 1271

Bagchi, A. *et al.*, (1991) *Planta med.*, **57:** 96.

Belmain S.R. *et al.*, (2001) *Food Chem. Tox.*, **39:** 287

Benemann, J.R. and N.M. Weare (1974) *Science*, **184:** 174

Bhattacharya, A. *et al.*, (2006) *Phytochemistry,* **67(3):** 232

Bhutani, K.K. *et al.*, (1990) *Phytochemistry,* **29:** 969.

Biever, C. (2005) *New Scientist,* 10th March,2005

Blankenship, J. D. *et al.*, (2001) *Phytochemistry,* **58(3):** 395

Boar, R.B. *et al.*, (1984). *J. Am. Chem. Soc.,* **106**: 2476

Bohnenstengel, F.I. *et al.*, (1999) *Phytochemistry,* **50(6):** 977

Bringmann, K. *et al.*, (2003) *J. Nat. Prod.,* **66(1):** 80

Broussalis, A. M. *et al.* (1999) *J. Ethnopharmacol.* **67(2):** 219

Buchanan, R.S. *et al.*, (1978) *Econ. Bot.,* **32:** 131

Calderone, N.W. and M. Spivak (1995) *Jour. Econ. Entomol.,* **88:** 1211

Calvin, M. *et al.*, (1981) *LBL, Univ. of California Preprint,* 12413.

Cantin, A. *et al.*, (2001) *Tetrahedron: Asymmetry,* **12(4):** 677

Carpenter, I., Lockesley, H.D. and F. Scheimann (1969) *Phytochemistry,* **8:**2013

Carpita, N. and M. McCann (2000) In *Biochemistry and Molecular Biology of Plants* (Buchanan B.B. *et al* Eds) American Society of Plant Physiologists, USA. p.52

Céspedes, C. L. *et al.*, (2004) *Phytochemistry,* **65 (13):** 1963

Céspedes, C. L. *et al.*, (2005) *Phytochemistry,* **66(20):** 2481

Chadwick, L *et al.*, (2004) *J. Nat. Prod.* **67(12):** 2024

Chaidir, K. *et al.*, (1999) *Phytochemistry,* **52(5):** 837

Chiang, Y *et al.*,(2005) *Phytochemistry,* **66(4):** 495

Choi, W. *et al.*, 2003). *Jour. Eco. Entom.,* **34(8):** 231

Ciccia, G., Coussio, J. and E. Mongelli (2000) *J. Ethnopharmacol.* **72(1-2):** 185

Coats, J.R., Karr L.L. and C. D. Drew (1991) In *Naturally Occurring Pest Bioregulators* (Hedin P A Ed) ACS, Washington DC. p. 456

Collins, D. O. *et al.*, (2001) *Phytochemistry,* **57(3):** 377

Cortes, D., Figadere, B and A. Cave (1993) *Phytochemistry,* **32(6):** 1467

Crowden, R.K. and S.J. Jarman, (1974) *Phytochemistry,* **13:** 1947

Dahanukar, S. A., Kulkarni, R.A. and N.N. Rege (2000) *Ind.J.Pharm.,* **32:** S81

Da Silva, B.P. and J.P. Parente (2002) *Phytotherapy Research,* **16:** 87

Daniel, M., (1989) *Basic Biophysics for the Biologists,* Agrobotanical publishers, Bikaner.

Daniel, M.,. (1989a) *Curr. Sci.,* **46(14):** 472

Daniel, M.,. (1991) *Methods in Plant Chemistry and Economic Botany,* Kalyani Publishers, New Delhi.

Daniel, M. (2004) *Curr. Sci.,* **87(3):** 275

Daniel, M, (2006) *Natural Dyes, Scope and Challenges* (Ed.). Scientific publishers, Jodhpur.

Daniel, M. (2006) *Medicinal Plants: Chemistry and Properties,* Science Publishers, USA and Oxford & IBH, India.

Darwish, F. M.M. and M.G. Reinecke (2003) *Phytochemistry,* **62(8):** 1179

Das, B. and K. Chakravarthy (1991) *Ind. J. Chem.* **30B:** 1052

Dayal, R. (1988) *J. Sci. Ind. Res.,* **47:** 215

De, S., Bag, A and S. Mukherji (1997) *Bot.Bull Acad Sin.* **38:** 105

De Luca, V. and B. St. Pierre (2000). *Trends Plant Sci.,* **5:** 168

Deshmukh, B. *et al.,* (1982) *Pesticides,* **16(12):** 7

De Vries, J.A. *et al.* (1982). *Carbohydrate Polym.* **2:** 25

Diaz. F. *et al.,* (2003) *J Nat Prod,* **66(6) :** 865

Diebold, J.P. *et al.,* (1993). In *Advances in Thermochemical Biomass Conversion* (Bridgewater, A.V. and D.G.B. Boocock, Eds). Blackie, London. p. 1315

Duke J. A. (1997). *The Green Pharmacy* (Indian Edn.) Scientific Publishers, Jodhpur. p. 85.

Duke, J. A. and S. Foster (1989) *Herbalgram,* **20:** 20

Dunkel, F. V. and L. J. Sears (1998) *J. Stor. Prod. Res.* **34(4):** 307

Dutt, C.B.S., Daniel, M and S.D. Sabnis (1987) *Jour. M.S.U.* **31:** 27

Elliot, D.C. and G.G. Neuenschwander (1996) In *Developments in Thermochemical Biomass Conversion* (Bridgewater, A. V. and D.G.B. Boocock, Eds) Blackie, London. **I :** 611

Ellis, B. and F. Bradley (1996). *The Organic Gardener's Handbook of Natural Insect and Disease Control,* Rodale Press, Pennsylvania.

Fairburn, J. W. (1964) *Lloydia,* **17:** 79

Fang, Y.Z., Yang, S. and G. Wu (2002) *Nutrition,* 872

Farooqui, A.A. and B. S. Sreeramu, (2001) *Cultivation of Medicinal and Aromatic Crops,* Universities Press, Hyderabad

Fatope M.O. *et al.,* (2003) *Phytochemistry,* **62(8):** 1251

Fiehn, D (2002) *Plant Mol. Biol.*, **48:** 155

Fossen, T. and O. M. Anderson (2003) *Phytochemistry,* **62(8):** 1217

Fozdar, I. *et al.*, (1989) *Phytochemistry,* **28:** 2489

Frankel, E. N. and A. S. Meyer (2000) *J. Sci. Food Agric.*, **80:** 1925

Fujimatu, E., Ischikawa, T. and K. Kitajima (2003) *Phytochemistry,* **63(5):** 609

Fukuzawa, A. and T. Masamune (1981) *Tetrahedron Lett.*, **22(41);** 4081

Ghayur Alam (2000) *A study of Biopesticides and Biofertilisers in Haryana, India* Gatekeeper series No 93.

Gibbs, R. D. (1974) *Chemotaxonomy of Flowering Plants,* McGill Queens, Montreal

Gijibels, M. J. M., Scheffer, J.J.C. and S.A. Baerhiem (1982) *Fitoterapia,* **53 :** 17

Glotter J., (1989) *Nat. Prod. Rep.*, 415

Gonazalez-Coloma, *et al.*, (1993). *Phytochemistry,* **34(2):** 397

Gonzalez, A.G. (1991) *Phytochemistry,* **30(4):** 1137

Gonzalez, A G *et al.*, (1992). *Phytochemistry,* **31(5):**1549

Gonzalez A. G. et al., (1997) *Biochem. Syst. Ecol.* ,**25(6):** 513

Goswami B.K. and K. Vijayalakshmi, (1986) *J. Agric.Res.*, **7:** 363

Grainage, M and S. Ahmed (1988) *Handbook of Plants with Pest Control Properties,* John Wiley & sons, NY

Green, M.M. *et al.*, (1991) *Jour American Mosquito control Ass.*, **7(2):** 282

Greger, H et al., (2001) *Phytochemistry,* **57(1):** 57

Grundy, D. L. and C. C. Still (1985) *Pesticide. Biochem. Physio.* **23(3):** 378

Guillet, G *et al.*, (1997a) *Phytochemistry,* **45(4):** 695

Guillet, G *et al.*, (1997b) *Phytochemistry,* **46(3):** 495

Guillet, G. et al.,(1998) *Phytochemistry,* **49(2):** 423

Gupta, V. *et al.*, (1989) *Indian J. Chem.* **28B:** 282

Hagnauer, R. (1966) In *Comparative Phytochemistry,* (T. Swain ed.) Academic Press. London, p.211

Hameed, S.F. (1968) *Madras Agri. Jour.* **55:** 168

Hamerski, L *et al.*, (2003) *Phytochemistry,* **63(4):** 397

Handa, S.S. *et al.*, (1999) *Fitoterapia,* **63:** 3

Handa S.S. *et al.*, (2004) *Phytochemistry,* **65(7) :** 915

Harborne, J.B. (1964) *Biochemistry of Phenolic Compounds*, Academic Press. London,

Harborne, J.B. (1973). *Phytochemical Methods*, Chapman & Hall, London

Herzberg, S. and Liaaen-Jensen, (1971) *Phytochemistry*, **10:** 3121

Ho, S.H. *et al.*, (2003) *Phytochemistry*, **59(6):** 579

Hopkinson, S. M. (1989) *Quart. Rev.*, **23:** 98

Hopp, D.C. *et al.*, (1996) *Jour. Nat. Prod.* **59(2)**:97

http//oisat. org/control (Preparation of extracts)

Huang, Y. and S. H. Ho (1998) *J. Stor. Prod. Res.*, **34(1):** 11

Huang, Y. et al., (2002) *J. Stor. Prod. Res.*, **38(5):** *403*

Imanishi K. and T. Ishiguro (1981) *Experienta*, **37:** 1186

Iosi, F., Santini M.T. and K. Malorni (1993) *Anticanc. Res.*, **13:** 545

Isman, M. B. and P. Proksch (1985) *Phytochemistry*, **24(9):** 1949

Ito, C. *et al.*, (2003) *Phytochemistry*, **66(5):** 567

Jacobson, M and D.G. Crosby (1971), *Naturally occurring insecticides*, Marcel Dekker Inc. New York

Jain N. *et al.*, (1990) *Phytochemistry*, **29**: 3988

Jakupovic, J *et al.*, (1999) *Phytochemistry*, **26:** 1536

Janssen, O., Scheffler, A and D. Kabelitz (1993) *Arzneim Forsch / Drug Res.* **43:** 1221

Jinbo, Z. *et al.*, (2002) *Phytochemistry*, **61(6):** 699

Jimenez, A *et al.*, (1998) *Phytochemistry*, **49(7):** 1981

Jiwajinda, S. *et al.*,(2001) *Phytochemistry*, **56(7):** 693

Joseph, C. C. *et al.*,(2004) *Trans. Roy. Soc. Trop. Med. Hyg.*, **98(8):** 451

Joshi K and P Dev., (1988) *Indian Jour. Chem.* **27B:** 12

Joshi, M.C. (2006) *Handbook of Indian Medicinal Plants*. Scientific Publishers, Jodhpur.

Kahlos, K. *et al.*, (1994) *Phytochemistry*, **36(4):** 917

Kaltenegger, E. *et al.*, (2003) *Phytochemistry*, **63(7):** 803

Kamal, R. and M. Mangla (1993) *Jour. BioSci.*, **18(1):** 93.

Karazu, C. *et al.*, (1997) *Metabolism*, **46:** 872

Kawaguchi, Y *et al.*, (2004) *J. Nat. Prod.* **67(11):** 1893

Keinan, E. *et al.*, (1993) *Pure & Appl. Chem.*, **69(3):** 423

Kesari, A.K. *et al.*, (2004) *Phytochemistry*, **65(23):** 3125

Keville, K. and M. Green (1995) *Aromatherapy: A complete Guide to the Healing Art*, Crossing Press, USA

Khambay B.P.S. *et al.*, (1999). *J. Agric Food Chem.* **47**: 770)

Khambay, B.P.S. *et al.*, (2002) *Phytochemistry*, **59(1):** 69

Kambay, B. P. S. *et al.*, (2003) *Tetrahedron*, **59(36):** 7131

Khan A and A Malik, (1989) *Phytochemistry*, **28:** 2859

Khan, T. *et al.*, (2002), *Phytomedicine*, **9(8):** 749

Khanbay, B.B., Beddie, D.G. and M.S.J. Simmonds (1999) *J. Nat Prod.*, **62 :** 1423.

Khripach, V., Zhabinskii, V. and A.D. Groot (2000). *Annals of Bot.*, **86: 441**.

Kim, S. *et al.*, (2003) *J. Stor. Prod. Res.*, **39(3):** 293.

Kitajima, J. *et al.*, (2003) *Phytochemistry*, **64(5):** 1003

Klingauf, F. *et al.*, (1982) *Mitt. Deut. Ges. Allg. Ang. Ent.*, **4:** 123

Klocke, J. A. *et al.*, (1991) *Phytochemistry*, **30(7):** 2165

Kluh, I. *et al.*, (2005) *Phytochemistry*, **66(1):** 31

Koona, D. (2005). *Ann. Appl. Biol.*, **147 (1):** 43.

Korunic, Z. (1998) *J. Stor. Prod. Res.*, **34(2-3):** 234

Koyama, J. *et al.*, (1992) *Phytochemistry*, **31(8) :** 2907

Krasnoff, S. B, (1991) *J. Invert. Path.* **58(2):** 180

Krishnaveni, K.S. and J.V.S. Rao (2000) *Phytochemistry* ,**57(5):** 605

Kubo, I. et al., (1986)*Tetrahedron Letters*, **27 (36):** 4277

Kumar, R. J., Krupadanam, G, L. D and G. Srimannarayana,(1989) *Phytochemistry*, **28(3):** 913

Kwiligwa, E *et al.*, (1999) In *Meeting the Challenges of Animal Traction* (Starkey, P and P. Kaumbutho, Eds) Intermediate technology publishers, London.

Lakshmi K and V Krishnamoorthy, (1991) *Ind. J. Phar. Sci.*, **53:** 94

Lamparsky D and I. Linies (1988) *Perfumer and Flavorist*, **13:** 17

Latha, C. and J. Ammini (2000) *Pharma. Biol.*, **38(3):** 167).

Lee, S.M. *et al.*, (1991) In *Naturally Occurring Pest. Bioregulators* (Hedin P.A. Ed) ACS, Washington DC, p. 339

Lee, H. Park, C. and Y. Ahn, (2002) *Appl. Entom. Zool.*, **37(3)** : 459.

Li, X.H. *et al.*, (1990) *Jour. Nat. Prod.*, **53(1):** 81

Ling, B. *et al.*, (2003) *Ying Yong Sheng Tai Xue Bao,* **14(5):** 744

Liu, J. *et al.*, (1996). *Phytochemistry*, **40(3):** 841

Logan, B.E. *et al.*, (2002). *Environ Sci. Technol.*, **36:** 2530

Loo, G.(2003) *J. Nutr. Biochem.*, **14:** 64

Mansour, F., Ravid, U and E. Putievsky (1986) *Phytoparasitica,* **14:** 137

Marquina, S. *et al.*, (2001) *Phytochemistry*, **56(1):** 93

Marquez, B *et al.*, (2005) *Phytochemistry*, **66(15):** 1804

Marr K. L. and C. S. Tang (1992) *Biochem. Syst. Eco.*,**20(3):** 209

Materska M. *et al.*, (2003) *Phytochemistry*, **63(8):** 893

Maurya R. *et al.*, (1996). *Nat. Prod. Lett.*, **8:** 7

Maurya, R *et al.*, (2004) *Phytochemistry*, **65(7):** 915

Maurya, R. Dhar K.L. and S.S. Handa (1997) *Phytochemistry*, **44:** 749

Meltinger, C G *et al.*, (2004) *J. Nat. Prod.*, **67(12):** 2024

Metzger, P., Largeau, C and E. Casadevall (1991) In *Progress in the chemistry of Organic Natural Products* (Herz, W *et al.*, Eds) Springer, Vienna **57:** 1

Mier, N. *et al.*, (1996) *Phytochemistry*, **41(3):** 801

Miller, I.M. *et al.*, (1991) *Pharmazie*, **46:** 657

Mitsui, K *et al.*, (2004) *Phytochemistry*, **65(23):** 307

Miyake, J. (1984). *J. Ferment Technol.*, **62:** 531

Molinatorres, J. *et al.*, (1996) *Biochem. Syst. Eco.*, **24(1):** 43

Mollah, J.U. and W. Islam (2005) *Frontiers Nat. Prod. Chem.*, **1(1):** 209.

Morris J.B. (1999) *Perspectives on New Crops and New Uses* ASHS Press, Alexandria p. 196.

Muhammad I. *et al.*, (2003) *J. Nat. Prod.*, **66(6):** 804

Musago, L and G. Rodghiero (1972) In *Photophysiology* (Geisse, A.C. ed.) Academic Press, New York, **7:** 115

Niki, E. et al., (1995) *Am. J. Clin. Nutr.*, **62:**1322

Nitao, J. K. *et al.*, (1991) *Phytochemistry*, **30(7):** 2193

Ntiamoah, Y.A., Borden, J.H. and H.D. Pierce, Jr. (1996) *Entomol Exp. Appl.*, **79:** 219

Nugroho, B.W. *et al.*, (1996) *Phytochemistry,* **41(1):** 129

Nugroho, B.W. *et al.*, (1997) *Phytochemistry,* **45(8):** 1579

Nugroho, B.W. *et al.*, (1999) *Phytochemistry,* **51(3):** 367

Okunade, A. L.(2002) *Fitoterapia,* **73(1):** 1

Omolo, M. O. *et al.*, (2003) *Phytomedicin,* **12(3):** 241

Ono, M. *et al.*, (2004). *J. Nat. Prod.*, **67(12):** 2073

Orr, D.B. and C.P. Suh, (1998) In *Biological and Biotechnological control of insect pests* (Rechcigl, J.E. and N.A. Rechcigl, Eds) CRC Press, Florida, p.3.

Pachey, P and C. Schneidir, (1981) *Arch. Pharm.*, **314:** 251

Pandey R *et al.*, (2000) *Phytopathol Zeitschrift,* **148:** 501

Pandji, C *et al.*, (1993) *Phytochemistry,* **34(2):** 415

Park, B. *et al.*, (2002) *Crop Protection,* **21(3):** 249

Park, C., Kim S.I. and Y. Ahn. (2003) *J. Stor. Prod. Res.*, **39(3):** 333.

Park, I. *et al.*, (2000) *Jour. Eco Entomol.* **12(1):** 121

Park , I. *et al.*, (2003) *J. Stor. Prod. Res.*, **39(4):** 375

Pathak, K.N. Nath, R.P. and A.K. Jha (1989). *Nematol. Abs.*, **59:** 92.

Pemonge, J. *et al.*, (1997) *J. Stor. Prod. Res.*, **33(3):** 209

Peng. S. Y. and J. Normal (1991) *Mol. Biotherm.*, **3:** 79

Perry, N.B. *et al.*, (1996). *Phytochemistry,* **40(3):** 841

Phillip. O. and Y. Nkunika (2002) *Insect Science and its Applications,* **22(3):** 235

Pietta, P. G. (2000) *J. Nat. Prod.*, **63:** 1035

Prakash, A. and J. Rao (1997) *Botanical pesticides in Agriculture,* CRC press, NY

Prates, H. T. *et al.*, (1998) *J. Stor. Prod. Res.*, **34(4):** 243

Proksch, P *et al.*, (1987) *Phytochemistry,* **26(8):** 2227

Prot. J.C. and J.M. Kornprobst, (1983) *Rev. de Nematol.*, **6(2):** 330

Pungitore, C. R., *et al.*,(2005) *Jour. Stor. Prod. Res.*, **41 (4):** 433

Quesada-Moraga, E. *et al.*, (2006) *Biological Control,* **36(3):** 274

Raja, N. *et al.*, (2005), *Curr. Sci.*, **88(2):** 220

Ramirez, I. *et al.*, (2003) *Phytochemistry* ,**64(8):** 142

Regnault-Roger, C. *et al.*, (2004) *J. Stor. Prod. Res.*, **40(4):** 395

Rehr, S. S. *et al.*, (1973) *Biochem. Syst. Ecol.*, **1(1):** 63

Ren Y.J. *et al.*, (1995) *Med. Arom. Pl. Abstr.* 1166

Rice, P.J. and J. R. Coats (1994) *Jour Econ. Entom.*, **87:** 1172

Robinson, T. (1975) *The Organic Constituents of Higher Plants*, Cordus Press, Mass.

Rocheleau, R. (2000). *Biohydrogen Production, Final Summary Report.* U.S. Dept. of Energy Hydrogen Programme.

Rogers, A. (1984) *J. Arnold Arbor.*, **65:** 35

Rucker G. *et al.*, (1991) *Phytochemistry*, **31(1)** : 340

Rucker G. *et al.*, (1993) *Phytochemistry*, **33** : 141

Saadali, B. *et al.*, (2001) *Phytochemistry*, **58(7):** 1083

Saleh, M.A. (1984) *Phytochemistry*, **23(11):** 2497

Sarac, A and I Tunc (1995) *Zeitschrift fur Pflanzenkrankheiten und Pflanzenschutz*, **102:** 429

Saxena, D.B., Goswami B.K. and S.S. Tomar (1987) *Indian Perfumer*, **31(2):** 150.

Scartezzini P. and E. Speroni (2000) *J. Ethnopharmacol.*, **71** : 23

Schaneberg, B. T., Green, D. K. and A. T. Sneden (2001) *J. Nat. Prod.*, **64:** 624

Schmutterer, H. (1990) *Ann. Rev. Entomol*, **35:** 271

Schneider, C. *et al.*, (2000) *Phytochemistry*, **54:** 731

Seri (2002) *Colours From Nature. Silk Dyeing Using Natural Dyes.* Oxford & IBH, New Delhi.

Shani, A. (1993) *10BC WPRS Bulletin* ,**16:** 359

Sharma, A.K. and B.N. Johri (2002) *Arbuscular mycorrhizae*, Oxford & 1BH New Delhi

Siddiqui, B.S. *et al.*, (1994). *J. Nat. Prod.*, **57:** 27

Sikroia B.C., Srivasthava, S.J. and G.S. Niranjan, (1982) *J Indian Chem. Soc.*, **59:** 905

Simmonds, M. S. J., *et al.*, (1988) *Biochem. Syst. Eco.* **16(7-8):** 623

Singh P and S. Bhargava (1992) *Phytochemistry*, **31:** 2883

Smith, J. A. *et al.*, (2005) *Cancer Res.*, **65:** 1027

Smith T. J. *et al.*, (1995) *Ann. N. Y. Acad. Sci.*, **86(8)** : 1092

Soledade, M. *et al.*, (2003) *Phytochemistry*, **59(6):** 579

Somani, L.L. (2004) *Handbook of Biofertilisers.* Agrotech., Udaipur.

Sreekumar R. *et al.,* (2002) *Am. J. Physiol. Endocrinol. Metab.,* **282 :** E1055.

Srivasthava, G.N. *et al.,* (2000) *Indian Traditional veterinary Medicinal Plants* CIMAP, India

Stamopoulos, D.C. (1991) *Jour. Stor. Prod. Res.,* **27:** 199

Steinmuller, C and J. Roesler (1999) *Int. J. Immunopharm.,* **13:** 931

Stipanovic, R.D. *et al.,* (1978) *Phytochemistry,* **17(1):** 151

Stoll, G (2000) *Natural Protection in the Tropics,* Margraf Verlag, Weikersheim

Strunz, G. M. and H. Finlay (1994) *Tetrahedron,* **50(38):** 1113

Stuppner, H. and H. Wagner (1989) *Planta Med.,* **55:** 467, 559

Sukhdev and O. Koul (1997) *Insecticides of Natural Origin,* Harwood Academic Publishers, Australia

Supratman, U. *et al.,* (2001) *Phytochemistry,* **58(2):** 311

Takayanagi, T, Ischikawa, T. and J. Kitajima (2003) *Phytochemistry,* **63(4):** 479

Takeya, K. *et al.,* (1993) *Phytochemistry,* **33(3)**: 613

Tanaguchi M *et al.,* (1996) *Phytochemistry,* **42 :** 843

Taraphdar, A. K. Roy M and P. K. Bhattacharya (2001) *Curr. Sci.,* **80:** 1110

Taura, D.W., Mukhtar, M.D. and O.A. Adoum (2004) *Ife Jour. Sci.,* **6(2):** 115.

Thaker N.A., Patel C.C. and H.R. Patel (1988) *Madras Agri. Jour.,* **75:** 297

Thoison, O. *et al.,* (2004), *Phytochemistry,* **65(14):** 2173

Thomas P.J. (1989) *Chemosystematic and Pharmacognostic Studies on Some Indian Members of the Rubiaceae,* Ph D thesis, M.S.University, Baroda, India

Thomson, R.H. (1971) *Naturally occurring quinones,* A.P. London

Tiwari A.K. (1999) *J. Med. Arom. Plant Sci.,* **21:** 730

Tiwari A.K. (2004) *Curr. Sci.,* **86(8):** 1092

Torres, P *et al.,* (2003) *Phytochemistry,* **64(2):** 463

Tunc, I. and S. Sahinkaya, (1998). *Ent. Exp. et Appl.,* **86:** 183

Tunç, I. *et al.,* (1997) *J. Stor. Prod. Res.,* **33(2):** 181

Umadevi. I and M. Daniel (1993) *Acta Botanica Indica,* **21:**262

Van Driesche, R. G. and T. S. Bellows, Jr. (1996) *Biological Control,* Chapman and Hall, New York. P. 539

Veigas, C. *et al.,* (2003) *J. Nat prod.,* **66(6):** 908

Voragen, A.G.J., Schols, H.A. and H. Gruppen (1993) In *Plant Polymeric Carbohydrates* (Meuser *et al.* Eds). The Royal Society of Chemistry Cambridge. p.3

Wagner H, Wittman, D. and W. Schafer (1975) *Tetrahedron Lett.* 547

Wagner H *et al.*, (1986) *Planta Med.*, **5:** 370

Wagner, H., Bladt, H. and E. M. Zgainski, (1984) *Plant Drug Analysis: A Thin Layer Chromatography Atlas*, Springer-Verlag, Berlin

Walter, T. (1992). *Proceedings of the Symposium RME-Kraft Stoff -Rohstoff.* FICHTE, Technical University, Vienna

Wang, B.H. *et al.*, (1997) *Phytochemistry*, **44(5):** 787

Wang, K. *et al.*, (2003) *J. Nat. Prod.*, **66:** 416

Warthen, J.D. Jr. (1979) *USDA Agri. Rep.* ARM-NE-4

Wasserman M. A. *et al.*, (2003) *Am. J. Cardiol. (Suppl.)*, **91 :** 34

Watanabe, K. *et al.*, (1990) *Pest. Biochem. Physiol.*, **37(3):** 275

Wazir V., Maurya R and R.S. Kapil, (1995) *Phytochemistry*, **38:** 447

Weaver, D.K. *et al.*, (1991). *Jour. of Stored Products Res.*, **27:** 213

Weinzierl, R.A. (1998) In *Biotechnological Control of Insect Pests* (Rechcigl J.E. and N.A. Rechcigl Eds) CRC Press, p. 112

Wenjun, Wu *et al.*, (2001) *Phytochemistry*, **58(8):** 1183

Xu, R., Fazio G.C. and S.P.T. Matsuda (2004) *Phytochemistry*, **65:** 261

Ya, L. *et al.*, (1991) *Phytochemistry*, **30(2):** 719

Yadav P.P. *et al.*, (2004) *Phytochemistry*, **65(4):** 439

Yan W. *et al.*, (1996) *Phytochemistry*, **42:** 1417

Yenesew A *et al.*, (2003) *Pest Management Science*, **59(10):** 1159

Zhang, X *et al.*, (2003) *Phytochemistry*, **64(8):** 1341

Zhu, Y *et al.*, (2005) *Phytochemistry*, **66(9):** 1067

Zidan, Z.H. *et al.*, (1998) *Jour. Egypt Soc. Parasitol.*, **28(1):** 89

□□□

Index

B

E

F

G

H

I

L

N

O

Q

S

U

W

X

Y

Z

□□□